REVIEWS IN MODERN QUANTITATIVE FINANCE

Annual Reviews in Modern Quantitative Finance: Including Current Aspects of Fintech, Risk and Investments

Print ISSN: 3029-2581
Online ISSN: 3029-259X

Series Editor: Andrey Itkin *(New York University, USA)*

This series of books aims to offer a wide view of the current state of modern quantitative finance. Topics of quantitative finance have been significantly extended over the last several years, and nowadays also include fintech, blockchain and distributed ledgers, cryptocurrencies, financial machine learning and artificial intelligence models, etc. This series is conceived to offer reviews of recent/contemporary key results in these areas, as well as original contributions. Other formats acceptable for this series include authored monographs, collected works of noted scholars, surveys of defined subfields of finance, or regular research articles.

Published

Reviews in Modern Quantitative Finance
    edited by Andrey Itkin

Annual Reviews in Modern Quantitative Finance:
Including Current Aspects of Fintech, Risk and Investments

Vol. 1

REVIEWS IN MODERN QUANTITATIVE FINANCE

Editor

Andrey Itkin
New York University, USA

World Scientific

NEW JERSEY · LONDON · SINGAPORE · BEIJING · SHANGHAI · HONG KONG · TAIPEI · CHENNAI · TOKYO

Published by

World Scientific Publishing Co. Pte. Ltd.
5 Toh Tuck Link, Singapore 596224
USA office: 27 Warren Street, Suite 401-402, Hackensack, NJ 07601
UK office: 57 Shelton Street, Covent Garden, London WC2H 9HE

Library of Congress Cataloging-in-Publication Data
Names: Itkin, Andrey, editor.
Title: Reviews in modern quantitative finance / Andrey Itkin, New York University, USA.
Description: Hackensack, NJ : World Scientific, [2024] | Series: Annual reviews in modern quantitative finance: including current aspects of fintech, risk and investments; volume 1 | Includes bibliographical references and index.
Identifiers: LCCN 2023039691 | ISBN 9789811281730 (hardcover) |
    ISBN 9789811281747 (ebook for institutions) | ISBN 9789811281754 (ebook for individuals)
Subjects: LCSH: Finance--Mathematical models. | Finance--Econometric models.
Classification: LCC HG106 .R48 2024 | DDC 332.01/5118--dc23/eng/20230920
LC record available at https://lccn.loc.gov/2023039691

British Library Cataloguing-in-Publication Data
A catalogue record for this book is available from the British Library.

Copyright © 2024 by World Scientific Publishing Co. Pte. Ltd.

All rights reserved. This book, or parts thereof, may not be reproduced in any form or by any means, electronic or mechanical, including photocopying, recording or any information storage and retrieval system now known or to be invented, without written permission from the publisher.

For photocopying of material in this volume, please pay a copying fee through the Copyright Clearance Center, Inc., 222 Rosewood Drive, Danvers, MA 01923, USA. In this case permission to photocopy is not required from the publisher.

For any available supplementary material, please visit
https://www.worldscientific.com/worldscibooks/10.1142/13553#t=suppl

Desk Editors: Nimal Koliyat/Lai Ann

Typeset by Stallion Press
Email: enquiries@stallionpress.com

About the Editor

Andrey Itkin is an Adjunct Professor at NYU, Department of Risk and Financial Engineering and Quantitative Research and Development Lead at Abu Dhabi Investment Authorities. He received his PhD in physics of liquids, gases, and plasma and degree of Doctor of Science in computational physics. During his academic carrier, he published few books and multiple papers on chemical and theoretical physics and astrophysics and later on computational and mathematical finance. Andrey occupied various research and managerial positions in financial industry and also is a member of multiple professional associations in finance and physics.

About the Contributors

Archil Gulisashvili received his PhD degree and Doctor of Science degree from the Tbilisi State University in Tbilisi, Georgia. Currently, he is a professor of mathematics at Ohio University. Prior to joining Ohio University, he has held visiting positions at Boston University, Cornell University, and Howard University. His research interests include financial mathematics, stochastic processes, Schrodinger semigroups, and Feynman–Kac propagators. Gulisashvili has published more than 110 papers in international journals. He is author of a research monograph on stochastic volatility models, co-author of a research monograph on Feynman–Kac propagators, and co-editor of a volume on large deviations and asymptotic methods in finance.

Igor Halperin is an AI researcher and a Group Data Science leader at Fidelity Investments. His research focuses on using methods of reinforcement learning, information theory, and physics for financial problems, such as portfolio optimization, dynamic risk management, and inference of sequential decision-making processes of financial agents. Igor has an extensive industrial and academic experience in statistical and financial modeling, in particular in the areas of option pricing, credit portfolio risk modeling, and portfolio optimization. Prior to joining Fidelity, Igor worked as a Research Professor of Financial Machine Learning at NYU Tandon School of Engineering. Before that, Igor was an Executive Director of Quantitative Research at JPMorgan and a quantitative researcher at Bloomberg LP. Igor has published numerous articles in finance and physics journals and is a frequent speaker at financial conferences. He has co-authored the books *Machine Learning in Finance: From Theory to Practice* (Springer, 2020) and *Credit Risk Frontiers* (Bloomberg LP, 2012). Igor has a Ph.D. in theoretical high energy physics from Tel Aviv University and an M.Sc. in nuclear physics from St. Petersburg State Technical University. In February 2022, Igor was named the Buy-Side Quant of the Year by RISK magazine.

Charles-Albert Lehalle is Global Head, Quantitative Research & Development, at ADIA. He started his career managing AI projects at the Renault research center and moved to the financial industry with the emergence of automated trading in 2005. He was Global Head of Quantitative Research at Crédit Agricole Cheuvreux and Head of Quantitative Research on Market Microstructure at Crédit Agricole Corporate Investment Bank, before moving to Capital Fund Management.

On the academic side, Pr. Lehalle received the 2016 Best Paper Award in Finance from Europlace Institute for Finance (EIF) and has published more than eighty academic papers and book chapters. He co-authored the books *Market Microstructure in Practice* (World Scientific Publisher, 2nd edition 2018), analyzing the main features of modern markets, and *Financial Markets in Practice* (World Scientific Publisher, 2022), explaining how the connected network of intermediaries that makes the financial system is shaping prices formation. Pr. Lehalle studied machine learning for stochastic control during his Ph.D. on nonlinear control and artificial neural networks, and he co-edited with Pr. Agostino Capponi the book *Machine Learning and Data Sciences for Financial Markets A Guide to Contemporary Practices* (Cambridge University Press, 2023). Pr. Lehalle is also a member of the Scientific Directory of the Louis Bachelier Institute, Lecturer at UC Berkeley and Paris 6 Sorbonne Université, and Ecole Polytechnique "Probability and Finance" Master.

Mengda Li is a full-time researcher, or so-called "data scientist/machine learning engineer", and a part-time energy trading speculator. Currently, Li is a Ph.D. student at INRIA Sophia Antipolis, France.

Valerii Salov received his M.S. from the Moscow State University, Department of Chemistry, in 1982 and Ph.D. from the Academy of Sciences of the USSR, Vernadski Institute of Geochemistry and Analytical Chemistry in 1987. He is the author of the articles on analytical, computational, and physical chemistry, the book *Modeling Maximum Trading Profits with C++*, John Wiley and Sons, Inc., Hoboken, New Jersey, 2007, and papers in *Futures Magazine*, ArXiv, Social Science Research Network SSRN-Elsevier.

Anatoly (Alec) Schmidt is an Adjunct Professor at the Finance and Risk Engineering Department of the NYU Tandon School. He also taught at the Financial Engineering Department of the Stevens Institute of Technology and was a visiting professor at Nanyang Technological University and Moscow Financial Academy. Alec holds a Ph.D. in Physics and has worked in the financial industry since 1997, most recently as Lead Research Scientist at Kensho Technologies. Alec published three books, *Quantitative Finance for Physicists: An Introduction* (Elsevier, 2004), *Financial Markets and Trading: Introduction to Market Microstructure and Trading Strategies* (Wiley, 2011), *Modern Equity Investing Strategies* (World Scientific, 2021), and multiple papers on agent-based modeling of financial markets, portfolio management, ESG investing, asset pricing, and trading strategies.

Liuren Wu is the Wollman Distinguished Professor of Finance at Zicklin School of Business, Baruch College, City University of New York. Professor Wu's research interests include option pricing, credit risk and term structure modeling, market microstructure, and general asset pricing. Professor Wu has published over 50 articles, many of them in top finance journals, such as the *Journal of Finance*, the *Journal of Financial Economics*, *Review of Financial Studies*, the *Journal of Financial and Quantitative Analysis*, *Management Science*, and the *Journal of Monetary Economics*. Mr. Wu has worked extensively as consultants in the finance industry, including data vendors, investment banks, and several fixed income, equity, and equity options hedge funds and market making firms. He has developed statistical arbitrage strategies, risk management procedures, optimal trade execution and market making strategies, and quantitative models for pricing fixed income and equity derivative securities.

Xu Zhang is an independent researcher and former(ly) student of Finance and Risk Engineering at NYU Tandon School, New York.

© 2024 World Scientific Publishing Company
https://doi.org/10.1142/9789811281747_fmatter

Contents

About the Editor		v
About the Contributors		vii
Introduction		xiii
Chapter 1.	Multivariate Stochastic Volatility Models and Large Deviation Principles *Archil Gulisashvili*	1
Chapter 2.	Phases of MANES: Multi-Asset Non-Equilibrium Skew Model of a Strongly Nonlinear Market with Phase Transitions *Igor Halperin*	97
Chapter 3.	Mathematics of Embeddings: Spillover of Polarities over Financial Texts *Mengda Li and Charles-Albert Lehalle*	151
Chapter 4.	Optimal ESG Portfolios: Which ESG Ratings to Use? *Anatoly Schmidt and Xu Zhang*	189

Chapter 5.	Centrality of the Supply Chain Network *Liuren Wu*	209
Chapter 6.	Are E-mini S&P 500 Futures Prices Random? *Valerii Salov*	229
Index		367

© 2024 World Scientific Publishing Company
https://doi.org/10.1142/9789811281747_fmatter

Introduction

Dear reader,

We present to your attention a new series of books which aims to offer a wide view of the current state of modern Quantitative Finance. Along with common trends in the financial industry and economical sciences, the topics covered by quantitative finance have been significantly extended over the last several years. Nowadays, in addition to the classical problems in financial engineering and mathematical finance, they also include Fintech, blockchain and distributed ledgers, cryptocurrencies, financial machine learning and artificial intelligence models, etc. Therefore, this series would be conceived to offer reviews of recent/contemporary key results in these areas or, perhaps, some previous but still relevant results either not published yet at all or published only in a form of preprints. Of course, we are open to original contributions on the subject as well. The editors are also keen to collect valuable submissions in other formats including authored monographs, collected works of noted scholars, surveys of various sub-fields in finance, or regular research articles. All articles should aim to be of interest to the broad readership.

Along this effort we are happy to present the first volume of this series which contains six chapters. These chapters are quite different with respect to what area of finance they look at or make research for — exactly what was expected by this series, to reflect the current rather complicated structure of modern quantitative finance. The topics presented in the volume cover both traditional approaches and models (which usually include sophisticated analytic, modern mathematical and numerical methods, etc.) — Chapters 1 and 2 and modern approaches: those inherent to Machine Learning (Chapters 2 and 3), practical aspects of trading equities (Chapters 5 and 6), and economic scenarios generation (Chapter 4). Of course, this is just a minor portion of topics relative to all possible ones, however the editors are in hope that it is representative.

Also, it is worth to underline that some of the presented results are not novel in a sense that they have been obtained some time ago and/or using the market data as of several years back in time. However, since this research was never published on one hand and is still actual and interesting on the other side, we have decided to include it into this volume.

Chapter 1 — Multivariate Stochastic Volatility Models and Large Deviation Principles

The first chapter of this book is developed by Professor Archil Gulisashvili from Department of Mathematics at Ohio University, USA. Being a well-known expert in the area of financial mathematics and stochastic processes, he also contributed to the theory of Schrodinger semigroups and Feynman–Kac propagators. Gulisashvili has published more than hundred papers in international journals and is the author of a research monograph on stochastic volatility models, a co-author of a research monograph on the Feynman–Kac propagators, and co-editor of a volume on large deviations and asymptotic methods in finance.

In his new paper, he establishes a comprehensive sample path large deviation principle (LDP) for processes naturally associated with multivariate time-inhomogeneous stochastic volatility models, e.g. Gaussian models, non-Gaussian fractional models, mixed models, models with reflection, and models in which the volatility process is a solution to a Volterra-type stochastic integral equation. This idea is further extension of the method originated by Dupuis and Ellis, [1] and further developed in [2–6] for families of functionals of Brownian motion. The LDP principle helps us to obtain asymptotic formulas for the distribution function of the first exit time of a log-price process from an open set, prices of multidimensional binary barrier options, call options, Asian options, and the implied volatility. Being asymptotic, such formulas are not exact but capture the leading order asymptotics of the above quantities as the small noise parameter ε tends to zero. The same sample path LDP is proven to be applicable to asymptotically solving Volterra stochastic integral equations with predictable coefficients which depend on other auxiliary stochastic processes.

At the first glance, this chapter looks to be highly theoretical and the main result in Theorem 73 seems to be totally abstract (in a sense that there is neither a clear financial motivation of the propose models, e.g. the Generalized Fractional Heston Models in section "Asian Options", nor the advantage or contribution to financial problems or other type of problems). Therefore, to address this initial (but wrong) impression in section "A Toy Model", the author discusses a simple (toy) model which is an uncorrelated SABR model, [7] (which is also a special case of the Hull–White model, [8]) and obtains a leading order approximation (the small noise limit) for the implied volatility in the toy model. This kind of asymptotic approximation, undoubtedly, can be very useful for both for researchers and practitioners. The editor believes that this chapter provides extensive and interesting new results and contributes to constructing asymptotic solutions for various stochastic volatility models of financial mathematics.

Chapter 2 — Phases of MANES: Multi-Asset Non-Equilibrium Skew Model of a Strongly Nonlinear Market with Phase Transitions

This chapter is written by Dr. Igor Halperin who currently is an AI researcher and a Group Data Science leader at Fidelity Investments. He began his scientific career as a theoretical physicist (he holds Ph.D. in theoretical high energy physics from Tel Aviv University and an M.Sc. in nuclear physics from St. Petersburg State Technical University) which, in the essential degree, explains the topic of this chapter. Igor has published numerous articles in financial and physics journals, co-authored the books *Machine Learning in Finance: From Theory to Practice* (Springer, 2020) and *Credit Risk Frontiers* (Bloomberg LP, 2012). In February 2022, Igor was named the Buy-Side Quant of the Year by RISK magazine. His current research focuses on using methods of reinforcement learning, information theory, and physics for financial problems, such as portfolio optimization, dynamic risk management, and inference of sequential decision-making processes of financial agents.

For last several years, the research of Dr. Halperin has been focused at market models with non-equilibrium dynamics. To briefly explain what he means by that, let us consider some market index, e.g. the SPX index which is a weighted average of individual constituent (stock) prices which this index consists of. It is general practice to assume that the dynamics of market returns of this index would be similar to the dynamics of a single "representative" stock in the market. However, the main problem with this approach is that the dynamics of the mean return of all stocks (i.e. the market return) can only be simply expressed as the mean of dynamics of individual returns if these dynamics are linear. As Dr. Halperin mentions in introduction, "in a more general case, we may think of a market as an ensemble of individual stocks whose individual dynamics are generally non-linear due to market friction effects", as discussed in detail in his paper. Moreover, "these non-linear dynamics for individual stocks are not independent of each other" and specific market mechanisms producing such co-dependencies are also discussed in the paper. The idea of Dr. Halperin is to view such market dynamics as statistical mechanics of an interacting ensemble of self-interacting nonlinear "particles" representing individual stocks. That is where his background as a physicist comes to play to utilize general ideas of statistical physics (looking at market observables as an average over an ensemble of possible realizations).

As such, this chapter brings a lot of terminology inherent to physics but not widely used in finance. This, undoubtedly creates some problems for readers who don't have such background, however, at the same time allows looking at some classical problems of mathematical finance at a different angle and produce a fresh view on them in terms of interdisciplinary science and mathematics.

A particular focus of this chapter is on proposing an analytically tractable and practically oriented model of nonlinear dynamics of a multi-asset market in the limit of a large number of assets. It is assumed that the asset price dynamics are driven by money flows into the market from external investors and their price impact. The author maps this market into an ensemble of interacting nonlinear oscillators which follow the Langevin dynamics, see section "Langevin Dynamics of Interacting Nonlinear Oscillators" of this paper for a more detailed description of the foundations of this approach and also [9]. Assuming the portfolio under consideration is homogeneous, this approach results into the McKean–Vlasov type of equation which is highly nonlinear and describes the dynamics for the market returns. Due to nonlinearity, various new phenomena can exist in such a system when parameters of the model vary, e.g. phase transitions of the first and second kind, [10]. In other words, some characteristics of the system or their derivatives are not continuous anymore and jump at a certain point of time or space.

Since the Langevin approach describes the behavior of the system in terms of the potential function, Dr. Halperin introduces an empirical but tractable potential, same as was proposed in his earlier works in [11,12] (see also [13] for a non-technical presentation) for a single-stock case. And the new Multi-Asset NES (MANES) model enables an analytically tractable framework for a multi-asset market. Then the equilibrium expected market log-return is obtained as a self-consistent mean field of the McKean–Vlasov equation. In this chapter, it is obtained in closed form in terms of parameters that are inferred from market prices of S&P 500 index options. It turns out that such a model is able to accurately fit the market data for either a benign or distressed market environments, while using only a single volatility parameter.

While some criticism can be issued about a lack of comprehensive empirical evidence for the model as well as its potential applications (having in mind that the one-dimensional model has been already introduced in [12]), new coverage of a multi-dimensional case seems to be important. Also, it is unclear whether the choice of the NES potential is supported with other (natural or financial) arguments rather than just tractability. It would also be interesting to better understand the emergence of the phase structure described in this chapter. Could one interpret the double-well NES potential as a source of the phase structure in the first place (i.e. it feels natural that a double potential for each of y_i lends an effective double-well free energy shape for the average return m) or is the main result the extension to the multi-stock case itself? On a related note, it would be interesting to better understand the advantage of using the MANES model vs other models mentioned in the conclusion, such as local, stochastic, or rough volatility models, using some benchmarks. Right now it seems there is a lack of formal empirical validation for the model, just tangential evidence in reproducing a metastable states characteristic of distressed markets. But that rises another important question on whether there exists a principled way in which the model can be tested out of sample. Since the calibrated parameters clearly change as a function of time, it is unclear what the independent prediction to be made after the calibration is. Indeed, the dynamics of the model parameters might be important for the conclusions about the phase structure and the nature of the transitions alluded to section "Fitting Model Parameters Using Option Data".

Despite those and some other points which can be put questionable or discussed in more detail, this chapter itself is a very interesting step toward (a) elaborating a new, inherent to econophysics, approach to understanding the behavior of the equity market and (b) establishing an interesting connection between some phenomena in statistical physics and in equity markets in finance. While bringing specific physics terms and notions, like phase transitions, could be confusing or even misleading unless explained in detail in the original financial terms and justified by comprehensive comparison with the market, Dr. Halperin's findings and results with no doubts deserve to be published in our collection, thus encouraging researchers and practitioners to a wide discussion and possible further development.

Chapter 3 — Mathematics of Embeddings: Spillover of Polarities over Financial Texts

While the previous chapters could be conventionally related to classical mathematical finance, we continue this chapter by presenting the research contributing to another modern field of finance which is machine learning (ML) and, in particular, NLP (natural language processing) — an ML technology that gives computers the ability to interpret, manipulate, and comprehend human language. Putting this very generally, ML systems help people understand massive volumes of data and uncover important patterns within them. This information is then used to enhance business processes, make informed decisions, and assist with prediction tasks. Financial service companies use it to offer better pricing, mitigate risks caused by human error, automate repetitive tasks, and understand customer behavior.

Chapter 3 is provided by Professor Charles-Albert Lehalle — a world class expert in ML and similar areas and also an industry practitioner with an extensive experience. Prof. Lehalle currently serves as the Global Head at Quantitative Research & Development department at Abu Dhabi Investment Authorities, UAE. He started his career managing AI projects at the Renault research center and moved to the financial industry with the emergence of automated trading in 2005, later occupying various senior positions at several banks and hedge funds. On the academic side, Prof. Lehalle received the 2016 Best Paper Award in Finance from Europlace Institute for Finance (EIF) and has published more than eighty academic papers and book chapters: He co-authored the books *Market Microstructure in Practice* (World Scientific Publisher, 2nd edition 2018), analyzing the main features of modern markets, and *Financial Markets in Practice* (World Scientific Publisher, 2022), explaining how the connected network of intermediaries that makes the financial system is shaping prices formation. He also co-edited with Prof. Agostino Capponi the book *Machine Learning and Data Sciences for Financial Markets: A Guide to Contemporary Practices* (Cambridge University Press, 2023).

For readers who are not closely involved into this area of the finance theory and mathematics, it has to be emphasized that text-based numerical scores and features such as sentiment, readability, positivity, negativity, riskiness, and litigiousness play an important role in many financial research and practical applications (see [14,15] for a comprehensive review). As mentioned in [14], the existing practice is to derive word-based features from the asymptotically the loss function of such a learning algorithm is a cross-entropy between the representation of the model and the distribution of the Reference Model. He also defines "frequentists synonyms" — words that have the same context in the considered corpus and tests these concepts on synthetic corpora generated using controlled Markovian models. As the result, he observe that if their identifiability is poor, the cosine similarity between embeddings makes sense, even when it is low: Frequentist synonyms are closer to word from their class than to words of another group of synonyms. As a practical application, these findings are applied to the analysis of financial news. In particular, it is found that embeddings learned on the full body of the news are more reflecting the polarities of the considered lexicon, and they are better reflecting financial polarities than embeddings trained on Wikipedia.

In the end, let me mention that due to high technical quality of this chapter and various justifications made by the author regarding usability of the provided research (e.g. the word2vec can be treated as kind of an "asymptotic solution" for the problem), it can be an interesting and helpful reading for various scientists and practitioners working in this area.

Chapter 4 — Optimal ESG Portfolios: Which ESG Ratings to Use?

Next topic important for modern quantitative finance and represented in this book by this chapter is Economic Scenario Generation (ESG). It is contributed by Dr. Anatoly (Alec) Schmidt who is an Adjunct Professor at the Finance and Risk Engineering Department of the NYU Tandon School (join work with his student Xu Zhang). Alec holds a Ph.D. in Physics and has worked in the financial industry almost 25 years, most recently as Lead Research Scientist at Kensho Technologies. Alec published three books, *Quantitative Finance for Physicists: An Introduction* (Elsevier, 2004), *Financial Markets and Trading: Introduction to Market Microstructure and Trading Strategies* (Wiley, 2011), and *Modern Equity Investing Strategies* (World Scientific, 2021) and multiple papers on agent-based modeling of financial markets, portfolio management, ESG investing, asset pricing, and trading strategies.

As per [16], an ESG is a computer-based model of an economic environment that is used to produce simulations of the joint behavior of financial market values and economic variables. Two common applications are driving the increased utilization of ESGs:

- Market-consistent (risk-neutral) valuation work for pricing complex financial derivatives and insurance contracts with embedded options. These applications are mostly concerned with mathematical relationships input text by applying finance-specific dictionaries such as the Loughran–McDonald (LM) word lists. Word scoring-based numerical features are widely used in regression analysis of regulatory filings, news articles, tweets, etc. It is a testament to the success of this approach that a vast literature in accounting, economics, and finance has exploited these variables, amounting to hundreds of papers in the last two decades. Various papers make the point that lexicons are effective in representing various financial concepts, and the numerical scores of documents are related to financial outcomes and corporate performance, justifying the original ideas in [15].

Several fields benefit from the use of word lists in NLP. For instance, NLP helps in credit analysis of firms using SEC filings and news and also in asset management. Corporate performance is related to textual features such as sentiment, readability, tone, size, risk, and uncertainty. Again, for the detailed analysis, see [14].

Going back to the paper of Prof. Lehalle, one can reveal that it performs a mathematical analysis of the skip-gram word2vec model. This analysis sheds light on how the decision to use such a model makes implicit assumptions on the structure of the language. Besides, under Markovian assumptions discussed in this chapter, he provides a clear theoretical understanding of the formation of embeddings and in particular the way it captures what we call frequentist synonyms. These assumptions allow us to conduct an explicit analysis of the loss function commonly used by these NLP techniques that asymptotically reaches a cross-entropy between the language model and the underlying true generative model.

Perhaps, one could argue that word2vec embeddings are very different from those obtained by using modern architectures, and word2vec is an outdated (or even deprecated) model since 2017. Nevertheless, as the author warns: "For readers focused on deeper language models, it is of importance to understand that embeddings of word2vec can be considered as the first term of a Taylor expansion of the embeddings of higher dimensional models". He also presents a more detailed clarification of this statement to convince the reader "that all the analysis in this paper is locally valid for more complex and deeper models, that are mathematically intractable if considered globally. The conclusions of this paper can hence be considered as "locally valid" for more complex models. The notion of "frequentist synonym" that is usefully developed in this paper has to be read as "local frequentist synonym" for more complex models".

As the main result, the author introduces the concept of Reference Model (that is the uncompressed version of an embedding model) and shows that

within and among financial instruments and less concerned with forward-looking expectations of economic variables.
- Risk management work for calculating business risk, regulatory capital and rating agency requirements. These applications apply real-world models that are concerned with forward-looking potential paths of economic variables and their potential influence on capital and solvency.

An ESG is concerned with simulating future interest rate paths, including yield curves. Financial markets operate within the context of the growth and volatility of economic markets. An ESG typically builds off core default-free interest rate modeling then considers implications of corporate bond yields and returns that include default, transition behavior, and stochastic spreads. Finally, other components, including equity markets, foreign exchange considerations, and economic components, such as inflation and GDP, may be considered. Also, an ESG model creates correlations through direct relationships with other simulated variables.

In his manuscript, Dr. Schmidt considers a problem of a considerable practical interest, namely, optimization of ESG portfolios. This is a timely chapter as the theme of ESG investment progressively becomes more and more important for the asset management industry. The author considers ESG-adjusted portfolio performance metrics and then explores sensitivities of their optimal portfolios to the choice of a particular ESG rating provider. Since at the moment we observe lack of homogeneity in methodologies used by different providers of ESG ratings, and their ESG scores are often different by a wide margin, the quality of the discussed metrics sensitivities play a vital role.

The author's idea with constructing an optimal ESG portfolio (OESGP) is to expand the classical mean variance theory by slightly modifying the objective function subject to minimization. In particular, following [17,18] he adjusts it by adding an extra term which is a product of the portfolio ESG value (PESGV) and the ESG strength parameter γ (which is an investor's choice). PESGV, in turn, is assumed to be the sum of weighted ESG ratings of the portfolio constituents that are offered by several providers. As an example, the author picks the Dow Jones Index and investigates sensitivity of OESGP to the ESG ratings provided by MSCI, S&P Global, and Sustainalytics.

The main result of this chapter claims strong sensitivity of the ESG-optimal portfolios considered in this book to the choice of ESG ratings. Based on these findings, Dr. Schmidt recommends using in-house ESG metrics that are more aligned with the investment philosophy of asset managers. Also, it is found that with γ growing, the OESGP diversity and Sharpe ratio may monotonically decrease. However, the ESG-tilted Sharpe ratio has one or two maximums. The first maximum exists at moderate values of γ and yields a moderately diversified OESGP, which can serve as a criterion for optimal ESG portfolios. The second maximum at large γ corresponds to highly concentrated OESGPs. It appears if portfolio has one or two securities with a lucky combination of high returns and high ESG ratings.

Chapter 5 — Centrality of the Supply Chain Network

This research is provided by Professor Liuren Wu who is the Wollman Distinguished Professor of Finance at Zicklin School of Business, Baruch College, City University of New York. Professor Wu's research interests include option pricing, credit risk and term structure modeling, market microstructure, and general asset pricing. He has published over 50 articles, many of them in top finance journals. He also served as consultant in the finance industry, including data vendors, investment banks, and several fixed income, equity, and equity options hedge funds and market making firms developing statistical arbitrage strategies, risk management procedures, optimal trade execution and market making strategies, and quantitative models for pricing fixed income and equity derivative securities.

In this chapter, the author investigates some measures of a supply chain network and how they can be used for building equity trading strategies based on statistical arbitrage, and/or for portfolio construction. In theory, a supply chain strategy is a process that manages networks and services between suppliers and companies without causing friction. They include decision-making and analytical processes that define roadmaps for products, services, and market interactions. It must involve end-to-end supply chain processes like sourcing goods, delivery, and logistics. Implementing the best supply chain management strategy is key to improving operational efficiency, enabling profits, and reducing overall costs. Every organization needs supply chain management strategies to fulfill demand, build a great network, drive customer loyalty, improve responsiveness, and more.

To build a statistical arbitrage-based strategy, one needs to understand structural relations and connections of financial securities across different companies. As the author mentions, when analyzing the behavior of a particular company, financial analysts and investors pay close attention to the company's major suppliers and customers, from which they strive to infer the potential risks and opportunities for the company. For instance, the existing literature (see the survey in this chapter) argues that stocks that are in economically related supplier and customer industries cross-predict each other's returns, and also the return predictability across supplier–customer industries can be even stronger, e.g. for the corporate bond market.

This chapter strives to define the relative importance, or centrality, of a supplier in the whole supply chain network and understand how the most central suppliers interact with the aggregate economy. For doing so, Professor Wu proposes to examine the supplier–customer relation not in terms of pair-wise connections but from the perspective of an economic supply chain network, with each relation serving as a directed node of the network. This chapter examines the potentials of a list of centrality measures in capturing the major determinants of the supply chain network. Of these measures, he highlights the supplier–customer centrality pair defined based on Kleinberg algorithm [19] to be particularly promising. The supplier centrality of a company is defined as the sum of the customer centrality of all its customers, while the customer centrality of a company is defined as the sum of the supplier centrality of all its suppliers. The two types of centralities can then be solved as the leading eigenvectors of the products of the supplier–customer network matrix.

The main result of this chapter consists in proposing a complete methodology which describes in detail how to extract alpha from a supply chain network.[1] The author presents four measures of supply chain centrality: degree centrality, eigenvector centrality, Kleinberg authority supplier centrality, and Kleinberg hub customer centrality. Then, the detailed comparative analysis of these four measures is presented. It demonstrates that supplier-central companies tend to be more volatile, and their stock performance tends to precede the movements of the aggregate market. The author also illustrates his methodology with the results of numerical experiments supported by a detailed explanation of each step. The results presented in this chapter are of significant interest for both researchers and practitioners.

It seems that the proposed methodology could be further generalized to any other (new) centrality measure. But even in the current state, the results obtained in this chapter can be directly utilized as a source for generating new alpha strategies, constructing predictive signals, clustering the trading universe, etc. Despite in practical examples the analysis is done based on

---

[1]For those readers who are not aware of this terminology, Alpha (α) is a term used in investing to describe an investment strategy's ability to beat the market or its edge. Alpha is thus also often referred to as an "excess return" or the "abnormal rate of return" in relation to a benchmark, when adjusted for risk.

market data back to 2015, still the conclusions made seem to be actual nowadays.

Chapter 6 — Are E-mini S&P 500 Futures Prices Random?

This complex investigation was done by Dr. Valerii Salov who holds a Ph.D. in chemistry from the USSR Academy of Sciences Vernadski Institute of Geochemistry and Analytical Chemistry. He published various papers on analytical, computational, and physical chemistry and also the book *Modeling Maximum Trading Profits with C++* (John Wiley and Sons, Inc., Hoboken, New Jersey, 2007).

In his ample paper, Dr. Salov considers another problem also relevant to equity trading. He investigates chains of the CME Group Time and Sales E-mini S&P 500 futures tick prices and their a-b-c-d-increments. To describe these ticks in a probabilistic manner, he constructs a discrete probability distribution of the price increments based on the Hurwitz zeta function and Dirichlet series. He then discusses randomness of the ticks using the notions of typicality, chaoticness, and stochasticness first introduced by Kolmogorov and Uspenskii [20], further developed by their students, who define randomness in terms of the theory of algorithms.

Dr. Salov's paper contains a lot of interesting reading for traders, mathematicians, programmers, and others interesting in mathematics and the probability theory. His style of presentation, perhaps, is more inherent to physics (so no theorems or proofs of existence and uniqueness) and, in some sense, remind me Martin Gardner's or George Polya's books. So it is kind of encyclopedic description of the problem but as much simple as possible at the same time. This chapter begins with the description of the Futures Trade Tick Chains introducing these types of objects, business definitions, and processes illustrated by a lot of graphs and tables. In the following sections, the author considers in detail a sample statistics of values and increments followed by the description of empirical distributions of b-increments. In section "Theoretical Distributions for b-Increments", he constructs theoretical distributions of b-increments including symmetric, asymmetric, and distributions with additional flexibility. He also draws attention to various "mysterious" power laws revealing various connections between different areas of science, with conclusions made like "chemical kinetics is applied to the Corn futures", and "Zipf's law in linguistics and stock price fluctuations so different at the first glance, are other examples. Such a universalism requires a scalable explanation and the attempt [21] deserves full attention".

As follows from the title, the author's goal is to answer a semi-mathematical (or semi-philosophical) question: Are E-mini S&P 500 Futures Prices Random? One of his final conclusions reads: *While a futures chain is a unique individual object, it is still viewed as a chain of b-increments. Evaluating their empirical distribution, we decompose the chain on parts. What we get statistically describes our unique chain. If we assume that probabilities associate with the found empirical distribution, then our chain is random by definition. There are plenty of other distributions, which the chain does not obey, and then it is recognized non-random with respect to them. One may say that empirical distribution is not a model but the data itself. If we apply it to a next chain, making from data a model, then we can say, if the new chain is the same or not with respect to this empirical distribution. Irrespective on these manipulations with distributions, conceptually, Kolmogorov complexity seems continue working for the chain.*

And later: *No matter how complex and unpredictable the economic and social factors that determine the state of free markets are, their quintessence — time chains of price increments and algebraic sums of increments, i.e. the prices themselves, along with the limited capital, lead to the existence of a trading strategy, generating a maximum profit. This strategy divides the chains on the alternating buying and selling optimal trading elements. Distributions of price increments in these elements are biased comparing with their total sum. This bias is an essential market property offering a profit. Both, the strategy and elements, are objective market properties that reveal the frequency and magnitude of market offerings. The high frequency and magnitude observed until today, time and time again, are prerequisites for the existence of speculative markets tomorrow. Are ES-mini S&P futures prices random?*

Dr. Salov partly answers his own question and also proposes several methods to compare different time intervals and monitor the changing and/or accumulating properties, such as the maximum profit in terms well understood by traders and market analysts. I am very confident that reading his paper would be a pleasure not just for the experts in the field but also for various enthusiast of mathematics and the probability theory.

At the end.

For the reader's convenience, I regrouped all the contributions into their fields of coverage. From this perspective, we have the following:

No.	Topic	Chapters
1	Stochastic/fractional volatility models and their asymptotic solutions	1
2	Equity trading, optimal portfolios and related problems	2, 5, 6
3	Machine learning and NLP	2, 3
4	Economic scenario generation	4

Compare this with "The Future of Quantitative Finance: 5 Trends to Keep an Eye On" — an interesting paper of Radoslav Haralampiev, [22], in which he examines the trends that will shape the future of Quantitative Finance. Radoslav emphasizes five of the most significant trends currently shaping the field of quantitative finance:

1. Machine Learning and Artificial Intelligence;
2. High-Frequency Trading;
3. Cryptocurrencies and Blockchain;
4. ESG Investing;
5. Social Media and Alternative Data.

Hence, surprisingly or not, this volume in a considerable degree reflects these trends (as it should be based on the series title).

In the end, let me wish the readers to have pleasant time reading this book and having in mind a well-known phrase of Vera Nazarian that "Whenever you read a good book, somewhere in the world a door opens to allow in more light".

Andrey Itkin,
Editor-in-Chief,
Tandon School of Engineering, New York University, USA

References

[1] P. Dupuis, and R. S. Ellis. (1997). *A Weak Convergence Approach to the Theory of Large Deviations*. John Wiley & Sons, Inc., New York, NY.

[2] M. Boué, and P. Dupuis. (1998). A variational representation of certain functionals of Brownian motion. *Annals of Probability* 26, 1641–1659.

[3] A. Budhiraja, and P. Dupuis. (2001). A variational representation for positive functionals of infinite-dimensional Brownian motion. *Probability and Mathematical Statistics* 20, 39–61.

[4] A. Budhiraja, and P. Dupuis. (2019). *Analysis and Approximation of Rare Events: Representations and Weak Convergence Methods*. Springer Science+Business Media, LLC (part of Springer Nature), Springer, New York, NY.

[5] A. Budhiraja, P. Dupuis, and V. Maroulas. (2008). Large deviations for infinite-dimensional stochastic dynamical systems. *Annals of Probability* 36, 1390–1420.

[6] A. Budhiraja, P. Dupuis, and V. Maroulas. (2011). Variational representations for continuous time processes. *Annales de l'Institut Henri Poincaré, Probabilités et Statistiques* 47, 725–747.

[7] P. Hagan, D. Kumar, L. Lesniewski, and D. E. Woodward. (2002). Managing smile risk. *Wilmott Magazine* 84–108 (September 2002).

[8] J. Hull, and A. White. (1987). The procong of options on assets with stochastic volatilities. *Journal of Finance* 42, 281–300.

[9] W. T. Coffey, Y. Kalmykov, and J. T. Wandron. (2012). *The Langevin Equation: With Applications to Stochastic Problems in Physics, Chemistry and Electrical Engineering* (3rd edn.). World Scientific Contemporary Chemical Physics, vol. 14. World Scientific, Singapore.

[10] C. Van den Broeck, J. M. R. Parrondo, R. Toral, and R. Kawai. (1997). Nonequilibrium phase transitions induced by multiplicative noise. *Physical Review E* 55, 4084.

[11] I. Halperin, and M. F. Dixon. (2020). Quantum equilibrium-disequilibrium: Asset price dynamics, symmetry breaking, and defaults as dissipative instantons. *Physica A* 537, 122187. https://doi.org/10.1016/j.physa.2019.122187.

[12] I. Halperin. (2022). Non-equilibrium skewness, market crises, and option pricing: Non-linear Langevin model of markets with supersymmetry. *Physica A Statistical Mechanics and its Applications* 594(3), 127065. DOI: 10.1016/j.physa.2022.127065

[13] I. Halperin. (2020). The inverted world of classical quantitative finance: A non-equilibrium and non-perturbative finance perspective. https://arxiv.org/abs/2008.03623.

[14] S. R. Das *et al.* (2022). FinLex: An effective use of word embeddings for financial lexicon generation. *The Journal of Finance and Data Science* 8, 1–11.

[15] T. Loughran, and B. McDonald. (2011). When is a liability not a liability? Textual analysis, dictionaries, and 10-Ks. *The Journal of Finance* 66(1), 35–65.

[16] H. Pedersen *et al.* (2016). *Economic Scenario Generators. A Practical Guide. Society of Actuaries.* Available at https://www.soa.org/globalassets/assets/Files/Research/Projects/research-2016-economic-scenario-generators.pdf.

[17] L. H Pedersen, S. Fitzgibbons, and L. Pomorski. (2021). Responsible investing: The ESG efficient frontier. *Journal of Financial Economics* 142, 572–597.

[18] A. B. Schmidt. (2020). Optimal ESG portfolios: An example for the Dow Jones index. *Journal of Sustainable Finance and Investment.* https://doi.org/10.1080/20430795.2020.1783180.

[19] M. Kleinberg. (1999). Authoritative sources in a hyperlinked environment. *Journal of the ACM* 46(5), 604–632.

[20] N. Kolmogorov, and V. A. Uspenskii. (1987). Algorithms and randomness. *Teoriya Veroyatnostei i ee Primeneniya* 32(3), 425–455 (in Russian) (translated to English by Bernard Seckler. (1987). *Theory of Probability and its Applications* 32(3), 389–412).

[21] Y. Manin. (2014). Zipf's law and Levin's probability distributions. *Functional Analysis and Its Applications* 48(2), 51–66 (in Russian); English translation: (2014). Zipf's Law and L. Levin probability distributions. *Functional Analysis and Its Applications* 48(2), 116–127.

[22] R. Haralampiev. (2022). The future of quantitative finance: 5 trends to keep an eye on. *Quant Factory.* Available at https://medium.com/quant-factory/the-future-of-quantitative-finance-5-trends-to-keep-an-eye-on-c17fd65be664.

© 2024 World Scientific Publishing Company
https://doi.org/10.1142/9789811281747_0001

Chapter 1

Multivariate Stochastic Volatility Models and Large Deviation Principles[*]

Archil Gulisashvili

*Department of Mathematics, Ohio University,
Athens OH 45701, USA
gulisash@ohio.edu*

Abstract

We establish a comprehensive sample path large deviation principle (LDP) for log-price processes associated with multivariate time-inhomogeneous stochastic volatility models. Examples of models for which the new LDP holds include Gaussian models, non-Gaussian fractional models, mixed models, models with reflection, and models in which the volatility process is a solution to a Volterra-type stochastic integral equation. The sample path and small-noise LDPs for log-price processes are used to obtain large deviation-style asymptotic formulas for the distribution function of the first exit time of a log-price process from an open set, multidimensional binary barrier options, call options, Asian options, and the implied volatility. Such formulas capture leading order asymptotics of the above-mentioned important quantities arising in the theory of stochastic volatility models. We also prove a sample path LDP for solutions to Volterra-type stochastic integral equations with predictable coefficients depending on auxiliary stochastic processes.

Keywords: Large deviation principles, stochastic volatility models, Volterra-type equations, small-noise scaling, first exit times, binary barrier options, call options, Asian options, the implied volatility.

---

[*]Dedicated to the memory of Peter Carr.

Introduction

A classical Black–Scholes–Merton model of option pricing assumes that the volatility of a financial asset is constant. Stochastic volatility models provided corrections by taking into account random features of the volatility. In modern stochastic volatility models, the volatility is modeled by a stochastic process.

One of the main objectives in this paper is to introduce and study general multivariate time-inhomogeneous stochastic volatility models. Such a model is described by the following multidimensional stochastic differential equation:

$$dS_t = S_t \circ [b(t, \widehat{B}_t)dt + \sigma(t, \widehat{B}_t)(\bar{C}dW_t + CdB_t)], \quad 0 \leq t \leq T,$$
$$S_0 = s_0 \in \mathbb{R}^m, \qquad (1)$$

where the initial condition $s_0 = (s_0^{(1)}, \ldots, s_0^{(m)})$ is such that $s_i > 0$ for all $1 \leq i \leq m$. The meanings of the symbols appearing in the previous equation are explained in section "Multivariate Stochastic Volatility Models". The model in (1) incorporates various features of numerous known stochastic volatility models (see the discussion in section "Multivariate Stochastic Volatility Models" and the survey in section "Unification of Sample Path Large Deviation Principles for Stochastic Volatility Models"). The interested reader can find in [45] detailed information about classical stochastic volatility models (the Hull–White, the Stein–Stein, and the Heston model). Multivariate models are discussed in [2], see also Chapter 11 in the monograph by Bergomi [7]. This chapter is titled Multi-Asset Stochastic Volatility. Interesting examples of time-inhomogeneous models are the rough Bergomi model introduced in [5] and the super rough Bergomi model (see [6,41]).

Stochastic volatility models are widely used in finance (see, e.g. [7,31,32, 45,47,54,59,60]). The stochastic model in (1) is characterized by the drift map b, the volatility map σ, and the volatility process $\widehat{B}$. Under the conditions formulated in the following section, equation (1) is uniquely solvable, and the solution $S_t = (S_t^{(1)}, \ldots, S_t^{(m)})$, $t \in [0, T]$, is a continuous stochastic process with strictly positive components. The functions $S^{(i)}$, with $1 \leq i \leq m$, can be interpreted as price processes of correlated risky assets in a portfolio or an index. The log-price process associated with the stochastic volatility model introduced in (1) is defined by $X = (X^{(1)}, \ldots, X^{(m)})$, where $X^{(i)} = \log S^{(i)}$ for $1 \leq i \leq m$. The initial condition for the log-price process is denoted by x_0. It is clear that $x_0 = (\log s_0^{(1)}, \ldots, s_0^{(m)})$.

A major part of this paper is devoted to asymptotic analysis of stochastic volatility models. We use sample path and small-noise large deviation principles to perform such an analysis. A sample path large deviation principle (LDP) for a stochastic process characterizes logarithmic asymptotics of the probability that the path of a scaled version of the process belongs to a given set of paths. The theory of sample path large deviations goes back to the celebrated work of Varadhan [77] and Freidlin and Wentzell [33]. For more information about large deviations, see [22,23,26,28,75,78,79]. Our main goal in this paper is to obtain a universal sample path LDP for log-price processes in multivariate stochastic volatility models that unifies the results established in the above-mentioned publications and also provides new results. Such a comprehensive LDP is formulated in section "Sample Path Large Deviation Principles for Log-Price Processes" (see Theorem 24).

Various sample path LDPs are known for log-price processes in stochastic volatility models (see, e.g. [15,16,30,34,38,41–44,51]). I would like to bring the attention of the interested reader to the book [35] titled *Large Deviations and Asymptotic Methods in Finance* and also to the paper of Pham [67] devoted to applications of the large deviation theory in mathematical finance. It is also worth mentioning that there is a rich literature devoted to the applications of large deviation principles to the study of the asymptotic behavior of various quantities arising in finance (see, e.g. [3,19,30,35,38,41–44,49,51,67,70]).

The unification of various large deviation principles is achieved in Theorem 24 due to the wide variety of admissible Volterra-type volatility processes used in this paper. Examples of stochastic models, for which the LDP obtained in Theorem 24 holds, include multivariate Gaussian models, multivariate non-Gaussian fractional models, mixed models, multivariate models with reflection, and models in which the volatility process is a solution to a certain Volterra-type stochastic integral equation. The restrictions imposed on the drift map and the volatility map in Theorem 24 are rather mild (see Assumption A in section "Sample Path Large Deviation Principles for Log-Price Processes"). We also obtain large deviation-style asymptotic formulas for the distribution function of the first exit time of the log-price process from an open set and similar asymptotic formulas for multidimensional binary barrier options.

Most of the volatility processes used in this paper are solutions to certain Volterra-type integral equations (see equation (8)). The coefficients in this equation are predictable maps depending on auxiliary stochastic processes.

We impose special restrictions on equation (8) (see Assumptions (C1)–(C7) introduced and discussed in section "Volatility Processes"). These restrictions are based on Conditions (H1)–(H6) used in the paper of Chiarini and Fischer [17]. However, we had to adapt Conditions (H1)–(H6) to our setting since the stochastic processes employed in [17] are not of Volterra-type. Moreover, we impose an additional restriction (Assumption (C6)) that is not needed in the case of non-Volterra stochastic differential equations studied in [17]. Under Assumptions (C1)–(C7), we prove a sample path LDP for the unique solution to the Volterra-type stochastic integral equation (8) (see Theorem 38). The LDP in Theorem 38 generalizes various known LDPs for Volterra-type processes. Theorems 24 and 38 are the main results obtained in this paper.

The paper of Chiarini and Fischer was an important source of ideas in our work on LDPs for Volterra-type volatility processes. The methods employed in [17] are based on variational representations of functionals of Brownian motion and the weak convergence approach to small-noise LDP problems. We use the same techniques in the proofs of our results concerning LDPs for volatility processes (see Theorems 38 and 41). The weak convergence method was developed by Dupuis and Ellis (see [26]). Various sample path LDPs for families of functionals of Brownian motion were obtained in [10–14].

We next give an overview of the contents of this paper. In section "Multivariate Stochastic Volatility Models", we introduce multivariate stochastic volatility models and their scaled versions. Section "Volatility Processes" deals with volatility processes, scaled volatility processes, controlled counterparts of Volterra-type stochastic integral equations, and special restrictions that are imposed on the volatility models. In section "Sample Path Large Deviation Principles for Log-Price Processes", a sample path LDP for log-price processes is formulated and explained (see Theorem 24), while section "Small-Noise LDPs for Log-Price Processes" analyzes small-noise LDPs for log-price processes. In section "Large Deviation Principles for Volatility Processes", we establish a sample path LDP for solutions to Volterra-type stochastic integral equations (see Theorem 38). This theorem uses the canonical set-up. We do not know whether Theorem 38 holds on any set-up (see Definition 1). It is worth mentioning that for certain less general Volterra-type stochastic integral equations, the LDP in Theorem 38 is valid on any set-up. Exceptional examples here are volatility processes in multivariate Gaussian stochastic volatility models (see Theorem 49), volatility processes in multivariate non-Gaussian fractional models (see Theorem 50), and also Volterra-type stochastic processes used in the paper of Nualart and

Rovira [66] (see also the paper of Rovira and Sanz-Solé [72] devoted to large deviations for stochastic Volterra equations in the plane).

Section "Unification of Sample Path Large Deviation Principles for Stochastic Volatility Models" of this paper is devoted to examples of stochastic volatility models for which the sample path LDP in Theorem 24 holds. It also provides examples of models for which Assumptions (C1)–(C7) formulated in section "Volatility Processes" are satisfied. Section "Unification of Sample Path Large Deviation Principles for Stochastic Volatility Models" is split into several subsections. In sections "Gaussian Stochastic Volatility Models" and "Non-Gaussian Fractional Stochastic Volatility Models", we give a brief overview of one-factor Gaussian models and one-factor non-Gaussian fractional models, respectively. For more information about such models, see [38,41,43]. In section "Unification: Mixed Models", we merge multivariate Gaussian models and multivariate non-Gaussian fractional models and show that Theorem 24 holds for such mixtures. Section "LDPs for Multivariate Gaussian Models and Multivariate Non-Gaussian Fractional Models" is devoted to LDPs for log-price processes in multivariate Gaussian stochastic volatility models on a general set-up under mild restrictions on the drift map, the volatility map, and the volatility process. Similar results are obtained in section "LDPs for Multivariate Gaussian Models and Multivariate Non-Gaussian Fractional Models" for log-price processes in multivariate non-Gaussian fractional models. Note that Theorem 49 is a generalization of Theorem 4.2 in [41]. The former theorem provides an LDP for the log-price process in a multivariate Gaussian stochastic volatility model, while the latter one deals with the one-dimensional case. In section "Unification: Models with Reflection", we prove an LDP for the log-process in a multivariate stochastic volatility model with reflection. The final two subsections of section "Unification of Sample Path Large Deviation Principles for Stochastic Volatility Models" (sections "Unification: Volterra Type SDEs" and "Unification: More Volterra-Type SDEs") deal with volatility processes which are solutions to Volterra-type stochastic integral equations. In section "Unification: Volterra Type SDEs", we discuss the LDP obtained in Zhang [83], while section "Unification: More Volterra-Type SDEs" is devoted to the LDP established in Nualart and Rovira [66].

We would also like to highlight the paper of Jacquier and Pannier [52] in which the authors prove sample path LDPs for solutions to less general Volterra-type stochastic integral equations than equation (8) (see Theorems 3.8 and 3.25 in [52]). Another paper that is worth mentioning is the paper of Catalini and Pacchiarotti [16] in which a sample path LDP for multivariate time-homogeneous Gaussian models is established under stronger restrictions than those employed in our Theorem 49. Interesting results were obtained in the paper of Bayer *et al.* [4]. It was established in the latter paper that a small-noise LDP holds for scaled log processes in certain one-factor Volterra-type stochastic volatility models (see (55),(56), and Corollary 5.5 in [4]). The authors of [4] used Hairer's regularity structures in their work. Corollary 5.5 in [4], under the restriction that the stochastic volatility model is defined on the canonical set-up, follows from the results obtained in this paper (see Theorems 28 and 30 and Remark 31). More details are provided in Remark 58 in section "Unification: Volterra Type SDEs".

Section "Applications" is devoted to applications of the methods developed in this paper to mathematical finance. Here, we perform a small-noise asymptotic analysis of various important objects of study in the theory of stochastic volatility models, for instance, the distribution function of the first exit time of the log-price process from an open set in $\mathbb{R}^m$ (section "First Exit Times"), barrier options (section "Binary Barrier Options"), call options (section "Call Options"), the implied volatility (section "Implied Volatility in the Small-Noise Regime"), and Asian options (section "Asian Options"). We obtain large deviation-style asymptotic formulas for the objects mentioned in the previous sentence. These formulas capture the leading order asymptotic behavior of stochastic volatility models of our interest as the small-noise parameter ε tends to zero. They also provide asymptotic approximations to the above-mentioned objects.

Assumption B (see (114)) plays an important role in this paper. In section "Assumption B, revisited", examples of models for which Assumption B holds are provided. They include multivariate Gaussian models, multivariate generalized fractional Heston models, and mixed models.

In section "A Toy Model" of section "Applications", we study a simple model (a toy model) using the methods developed in this paper. A special uncorrelated SABR model plays the role of the toy model. We obtain various estimates for the rate function in the toy model (see Lemma 79, Corollary 80, and Theorem 81) and also for the small-noise limit of the implied volatility in the toy model (see Theorem 82).

In section "Proof of Theorem 38", we prove Theorem 38, while in section "Proof of Theorem 24", we include the proof of Theorem 24.

Multivariate Stochastic Volatility Models

Let $\mathbb{R}^m$ be m-dimensional Euclidean space equipped with the norm $||\cdot||_m$. For a real $(m \times m)$-matrix M, its Frobenius norm is denoted by $||M||_{m \times m}$ and the symbol M' stands for the transpose of M. We next provide more details about equation (1). This equation is defined on a probability space $(\Omega, \mathcal{F}, \mathbb{P})$ carrying two independent m-dimensional standard Brownian motions W and B with respect to the measure $\mathbb{P}$, and the symbol $\circ$ in (1) stands for the Hadamard (component-wise) product of vectors. The components of the initial condition s_0 of the process S are strictly positive. The matrix $C = (c_{ij})$ in (1) is a real $(m \times m)$-matrix such that $||C||_{m \times m} < 1$. It is clear that the matrix $\text{Id}_m - C'C$ is symmetric and positive definite, and we denote the unique symmetric and positive definite square root of the matrix $\text{Id}_m - C'C$ by $\bar{C}$. The elements of the matrix $\bar{C}$ be denoted by $\bar{c}_{ij}$. Under the previous conditions, the matrix $\bar{C}$ is invertible. By $\{\mathcal{F}_t\}_{0 \leq t \leq T}$ is denoted the augmentation of the filtration generated by the processes W and B (see, e.g. Definition 7.2 in Chapter 2 of [56]). We also use the augmentation of the filtration generated by the process B, and denote it by $\{\mathcal{F}_t^B\}_{0 \leq t \leq T}$. The symbol b in (1) stands for a continuous map defined on $[0, T] \times \mathbb{R}^d$ with values in $\mathbb{R}^m$. We call b the drift map. By σ is denoted a continuous map of $[0, T] \times \mathbb{R}^d$ into the space of $(m \times m)$ real matrices. This map is called the volatility map. The process $\widehat{B} = (\widehat{B}^{(1)}, \ldots, \widehat{B}^{(d)})$ appearing in (1) is a continuous d-dimensional stochastic process defined in terms of Brownian motion B and adapted to the filtration $\{\mathcal{F}_t^B\}_{0 \leq t \leq T}$. The process $\widehat{B}$ is called the volatility process (see Definition 8). We have already mentioned in the introduction that the model in (1) can be interpreted as a time-inhomogeneous stochastic volatility model describing the time behavior of price processes of correlated risky assets. The matrix-valued process $\sigma(t, \widehat{B}_t)$, with $t \in [0, T]$, characterizes the joint volatility of these assets.

The following definition introduces general set-ups (we adopt the terminology used in [71]).

Definition 1. The system $(\Omega, W, B, \mathcal{F}_T, \{\mathcal{F}_t\}_{0 \leq t \leq T}, \mathbb{P})$ is called a set-up associated with the model in (1), while the system $(\Omega, B, \mathcal{F}_T^B, \{\mathcal{F}_t^B\}_{0 \leq t \leq T}, \mathbb{P})$ is called a set-up associated with the volatility process in (1).

In terms of the components $S_t^{(i)}$ of the process S, with $1 \leq i \leq m$, equation (1) can be rewritten as the following system of stochastic differential equations:

$$dS_t^{(i)} = S_t^{(i)}[b_i(t, \widehat{B}_t)dt + \sum_{k,j=1}^{m} \bar{c}_{jk}\sigma_{ij}(t, \widehat{B}_t)dW_t^{(k)} + \sum_{k,j=1}^{m} c_{jk}\sigma_{ij}(t, \widehat{B}_t)dB_t^{(k)}],$$
$$1 \leq i \leq m. \tag{2}$$

Recall that we denoted by X the log-price process associated with the model in (1). We next characterize the dynamics of the process X. For every i with $1 \leq i \leq m$, equation (2) is a linear stochastic differential equation driven by the process

$$G_t^{(i)} = \int_0^t b_i(s, \widehat{B}_s)ds + \sum_{k,j=1}^{m} \bar{c}_{j,k} \int_0^t \sigma_{ij}(s, \widehat{B}_s)dW_s^{(k)}$$
$$+ \sum_{k,j=1}^{m} c_{j,k} \int_0^t \sigma_{ij}(s, \widehat{B}_s)dB_s^{(k)}.$$

This process is a continuous semimartingale since

$$\int_0^T ||b(s, \widehat{B}_s)||_m dt + \int_0^T ||\sigma(s, \widehat{B}_s)||_{m\times m}^2 ds < \infty$$

$\mathbb{P}$-a.s. Next, using the Doléans–Dade formula, we see that

$$S_t^{(i)} = s_0^{(i)} \exp\left\{ \int_0^t b_i(s, \widehat{B}_s)ds - \frac{1}{2}\int_0^t \sum_{j=1}^{m} \sigma_{ij}(s, \widehat{B}_s)^2 ds \right.$$
$$+ \sum_{k,j=1}^{m} \bar{c}_{j,k} \int_0^t \sigma_{ij}(s, \widehat{B}_s)dW_s^{(k)}$$
$$\left. + \sum_{k,j=1}^{m} c_{j,k} \int_0^t \sigma_{ij}(s, \widehat{B}_s)dB_s^{(k)} \right\}. \tag{3}$$

The formula in (3) can be rewritten as follows:

$$S_t^{(i)} = s_0^{(i)} \exp\left\{ \int_0^t b_i(s, \widehat{B}_s)ds - \frac{1}{2}\int_0^t \sum_{j=1}^{m} \sigma_{ij}(s, \widehat{B}_s)^2 ds \right.$$
$$\left. + \left[\int_0^t \sigma(s, \widehat{B}_s)(\bar{C}dW_s + CdB_s)\right]_i \right\}.$$

Recall that for any $m \times m$-matrix D, we denoted by D' the transpose of D. We also denote by $\text{diag}(D)$ the vector whose components are the diagonal elements of D. It is clear that the component a_i of the vector $\operatorname{diag}(\sigma(s,\widehat{B}_s)\sigma(s,\widehat{B}_s)')$, with $1 \leq i \leq m$, is given by $a_i = \sum_{j=1}^{m} \sigma_{ij}(s,\widehat{B}_s)^2$. It follows that the log-price process X is given by

$$X_t = x_0 + \int_0^t b(s,\widehat{B}_s)ds - \frac{1}{2}\int_0^t \operatorname{diag}(\sigma(s,\widehat{B}_s)\sigma(s,\widehat{B}_s)')ds$$

$$+ \int_0^t \sigma(s,\widehat{B}_s)(\bar{C}dW_s + CdB_s), \quad 0 \leq t \leq T. \tag{4}$$

Remark 2. The process obtained from the log-price process X by removing one of the drift terms, more precisely, the term $-\frac{1}{2}\int_0^t \operatorname{diag}(\sigma(s,\widehat{B}_s) \sigma(s,\widehat{B}_s)')ds$, is denoted by $\widehat{X}$. We have

$$\widehat{X}_t = x_0 + \int_0^t b(s,\widehat{B}_s)ds + \int_0^t \sigma(s,\widehat{B}_s)(\bar{C}dW_s + CdB_s). \tag{5}$$

We call the process in (5) the modified log-price process.

Remark 3. In the case where $m = 1$, we use the correlation parameter $\rho \in (-1,1)$ and set $\bar{\rho} = \sqrt{1-\rho^2}$. Then, the equation describing the evolution of the process S is as follows:

$$dS_t = S_t[b(t,\widehat{B}_t)dt + \sigma(t,\widehat{B}_t)(\bar{\rho}dW_t + \rho dB_t)], \quad S_0 = s_0 > 0.$$

Moreover, the log-price process is given by

$$X_t = x_0 + \int_0^t b(s,\widehat{B}_s)ds - \frac{1}{2}\int_0^t \sigma(s,\widehat{B}_s)^2 ds + \int_0^t \sigma(s,\widehat{B}_s)(\bar{\rho}dW_s + \rho dB_s),$$

where $x_0 = \log s_0$.

A modulus of continuity is a non-negative non-decreasing function ω on $[0,\infty)$ such that $\omega(s) \to 0$ as $s \to 0$. Let $x = (t_1, v_1)$ and $y = (t_2, v_2)$ be elements of the space $[0,T] \times \mathbb{R}^d$ equipped with the Euclidean distance $\nu_d(x,y) = \sqrt{(t_1-t_2)^2 + \|v_1-v_2\|_d^2}$. Denote by $\overline{B_d(r)}$ the closed ball centered at $(0,0)$ of radius $r > 0$ in the metric space defined above, and let ω be a modulus of continuity on $[0,\infty)$.

Definition 4. A map $\lambda : [0,T] \times \mathbb{R}^d \mapsto \mathbb{R}^1$ is called locally ω-continuous if for every $r > 0$ there exists $L(r) > 0$ such that for all $x, y \in \overline{B_d(r)}$ the following inequality holds:

$$|\lambda(x) - \lambda(y)| \leq L(r)\omega(\nu_d(x,y)).$$

We next explain what restrictions on the drift map b and the volatility map σ are imposed in this paper. These restrictions are rather mild.

Assumption A: The components of the drift map b and the elements of the volatility map σ are locally ω-continuous on the space $[0,T] \times \mathbb{R}^d$ for some modulus of continuity ω. In addition, the elements of the volatility map σ are not identically zero on $[0,T] \times \mathbb{R}^d$.

Let $\varepsilon \in (0,1]$ be the scaling parameter. The scaled version of the log-price process X is defined by

$$X_t^{(\varepsilon)} = x_0 + \int_0^t b(s, \widehat{B}_s^{(\varepsilon)}) ds - \frac{1}{2}\varepsilon \int_0^t \operatorname{diag}(\sigma(s, \widehat{B}_s^{(\varepsilon)}) \sigma(s, \widehat{B}_s^{(\varepsilon)})') ds$$
$$+ \sqrt{\varepsilon} \int_0^t \sigma(s, \widehat{B}_s^{(\varepsilon)})(\bar{C} dW_s + C dB_s), \tag{6}$$

where $X_0^{(\varepsilon)} = x_0$ for all $s \in (0,1]$. The scaled volatility process $\widehat{B}^{(\varepsilon)}$ appearing in (6) is introduced in the following section (see Definition 10). We also use the process

$$\widehat{X}_t^{(\varepsilon)} = x_0 + \int_0^t b(s, \widehat{B}_s^{(\varepsilon)}) ds + \sqrt{\varepsilon} \int_0^t \sigma(s, \widehat{B}_s^{(\varepsilon)})(\bar{C} dW_s + C dB_s) \tag{7}$$

that is a scaled version of the modified log-price process defined in (5).

Volatility Processes

Our main aim in this section is to introduce the volatility process $\widehat{B}$ that is used in (1). We need several definitions. For a positive integer $p \geq 1$, the symbol $\mathcal{W}^p$ stands for the space of continuous $\mathbb{R}^p$-valued maps on $[0,T]$ equipped with the following norm: $||f|| = \max_{t \in [0,T]} ||f(t)||_p$, $f \in \mathcal{W}^p$. Let B_s, with $s \in [0,T]$, be the coordinate process on $\mathcal{W}^p$. Define a filtration on the space $\mathcal{W}^p$ by $\mathcal{B}_t^p = \sigma(B_s : 0 \leq s \leq t)$, $t \in [0,T]$. The augmentation $\{\widetilde{\mathcal{B}}_t^p\}$ of the filtration $\{\mathcal{B}_t^p\}$ is called the canonical filtration on $\mathcal{W}^p$. Let $\mathbb{P}$ be the Wiener measure on $\widetilde{\mathcal{B}}_T^p$.

Definition 5. The ordered system $(\mathcal{W}^p, B, \widetilde{\mathcal{B}}_T^p, \{\widetilde{\mathcal{B}}_t^p\}, \mathbb{P})$ is called the canonical set-up on $\mathcal{W}^p$.

The set-up introduced in Definition 5 is a special case of a general set-up associated with the volatility process in (1) (see Definition 1). The coordinate process $s \mapsto B_s$ plays the role of p-dimensional standard Brownian motion with respect to the measure $\mathbb{P}$.

Remark 6. One of the reasons why the canonical set-up is employed in this paper is the following known fact. Let Z be an $\{\widetilde{\mathcal{B}}_t^m\}$-adapted continuous stochastic process on $\mathcal{W}^m$ with state space $\mathbb{R}^d$. Then, there exists a process

$\widetilde{Z}$ adapted to the filtration $\{\mathcal{B}_t^m\}$ and indistinguishable from Z. Moreover, there is a functional $j : \mathcal{W}^m \mapsto \mathcal{W}^d$ such that $X = j(B)$ $\mathbb{P}$-a.s. on $\mathcal{W}^m$, and for every $t \in [0, T]$, the functional j is $\widetilde{\mathcal{B}}_t^m/\mathcal{B}_t^d$-measurable. The functional j is generated by the process X. In addition, the canonical probability space plays an important role in the proof of the equality in (188).

We next define the canonical set-up on the space $\Omega = \Omega_1 \times \Omega_2 = \mathcal{W}^m \times \mathcal{W}^m$. Denote the coordinate processes on Ω_1 and Ω_2 by W and B, respectively, and consider the filtration on Ω generated by the process $t \mapsto (W_t, B_t)$, $t \in [0, T]$. Denote by $\{\mathcal{F}_t\}$ the augmentation of this filtration with respect to the measure $\mathbb{P} = \mathbb{P}_1 \times \mathbb{P}_2$, where $\mathbb{P}_1$ and $\mathbb{P}_2$ are the Wiener measures on Ω_1 and Ω_2, respectively. By $\{\mathcal{F}_t^B\}$ is denoted the augmentation of the filtration generated by process $t \mapsto B_t$, $t \in [0, T]$. The processes W and B are independent m-dimensional Brownian motions defined on the space Ω. The canonical set-up on the space $\Omega = \mathcal{W}^m \times \mathcal{W}^m$ is the system $(\Omega, W, B, \mathcal{F}_T, \{\mathcal{F}_t\}, \{\mathcal{F}_t^B\}, \mathbb{P})$.

Let Y be a stochastic process satisfying the following Volterra-type stochastic integral equation on $\mathcal{W}^m$ equipped with the canonical set-up:

$$Y_t = y + \int_0^t a(t, s, V^{(1)}, Y)ds + \int_0^t c(t, s, V^{(2)}, Y)dB_s. \tag{8}$$

In (8), a is a map from the space $[0,T]^2 \times \mathcal{W}^{k_1} \times \mathcal{W}^d$ into the space $\mathbb{R}^d$, while c is a map from the space $[0,T]^2 \times \mathcal{W}^{k_2} \times \mathcal{W}^d$ into the space of $(d \times m)$-matrices. For a matrix M belonging to the latter space, the symbol $||M||_{d \times m}$ stands for its Frobenius norm. Assumption (C1) formulated in the following introduces more restrictions on the maps a and c. The vector $y \in \mathbb{R}^d$ in (8) is the fixed initial condition for the process Y. The processes $V^{(i)}$, with $i = 1, 2$, appearing in (8) are fixed auxiliary continuous stochastic processes on $\mathcal{W}^m$ with state spaces $\mathbb{R}^{k_1}$ and $\mathbb{R}^{k_2}$, respectively. They satisfy the following stochastic differential equations:

$$V_s^{(i)} = V_0^{(i)} + \int_0^s \bar{b}_i(r, V^{(i)})dr + \int_0^s \bar{\sigma}_i(r, V^{(i)})dB_r, \quad i = 1, 2. \tag{9}$$

In (9), $V_0^{(i)} \in \mathbb{R}^{k_i}$ are initial conditions, $\bar{b}_i$ are maps of $[0,T] \times \mathcal{W}^{k_i}$ into $\mathbb{R}^{k_i}$, while $\bar{\sigma}_i$ are maps of $[0,T] \times \mathcal{W}^{k_i}$ into the space of $k_i \times m$-matrices. We assume that equation (9) satisfy Conditions (H1)–(H6) introduced in Chiarini and Fischer [17].

Remark 7. Examples of equation (9), for which Conditions (H1)–(H6) hold true, are provided in Sections 3 and 4 of [17]. For instance, the validity of these conditions is established in [17] for equations with locally Lipschitz coefficients satisfying the sublinear growth condition (see Definitions A1 and A2 in Section 3 of [17], see also (11.1) on p. 128 in [71]). It is also shown in [17] that Conditions (H1)–(H6) hold true for one-dimensional diffusion equations with Hölder dispersion coefficient.

The following definition introduces volatility processes that are used throughout this paper.

Definition 8. The volatility process $\widehat{B}$ appearing in (1) is as follows: $\widehat{B} = GY$, where Y satisfies equation (8), while G is a continuous map from $\mathcal{W}^d$ into itself that is $\widetilde{\mathcal{B}}_t^d/\mathcal{B}_t^d$-measurable for every $t \in [0,T]$.

An example of a map G satisfying the condition in Definition 8 for $d = 1$ is the Skorokhod map (see section "Unification: Models with Reflection"). The measurability condition for G is included in Definition 8 in order for the volatility process $\widehat{B}$ to be adapted to the filtration $\{\widetilde{\mathcal{B}}_t^d\}$.

Let $\varepsilon \in (0,1]$ be a small-noise parameter. A scaled version of equation (8) has the following form:

$$Y_t^{(\varepsilon)} = y + \int_0^t a(t,s,V^{1,\varepsilon},Y^{(\varepsilon)})ds + \sqrt{\varepsilon}\int_0^t c(t,s,V^{2,\varepsilon},Y^{(\varepsilon)})dB_s. \qquad (10)$$

In (10), the process $V^{i,\varepsilon}$, with $i = 1,2$, is a scaled version of the process $V^{(i)}$. It satisfies the equation

$$V_s^{i,\varepsilon} = V_0^{(i)} + \int_0^s \bar{b}_i(r,V^{i,\varepsilon})dr + \sqrt{\varepsilon}\int_0^s \bar{\sigma}_i(r,V^{i,\varepsilon})dB_r. \qquad (11)$$

Remark 9. In [17], more general equations than the equation appearing in (11) are considered. The coefficient maps $\bar{b}$ and $\bar{\sigma}$ in those equations may depend on the scaling parameter ε. We do not study such equations in this paper.

Definition 10. The scaled volatility process $\widehat{B}^{(\varepsilon)}$ is given by $\widehat{B}^{(\varepsilon)} = GY^{(\varepsilon)}$, where G is introduced in Definition 8, while $Y^{(\varepsilon)}$ is the solution to (10).

Controlled counterparts of equations (8)–(11) is also considered. Let $\mathcal{M}^2[0,T]$ be the space of all $\mathbb{R}^m$-valued square-integrable $\{\mathcal{F}_t^B\}$-predictable processes. The controls are chosen from the space $\mathcal{M}^2[0,T]$. Deterministic controls are employed as well. They are functions belonging to the space $L^2([0,T],\mathbb{R}^m)$.

Definition 11. Let $N > 0$. By $\mathcal{M}_N^2[0,T]$ is denoted the class of controls $v \in \mathcal{M}^2[0,T]$ such that

$$\int_0^T \|v_s\|_m^2 ds \leq N \quad \mathbb{P}-\text{a.s.} \qquad (12)$$

Suppose $v \in \mathcal{M}^2[0,T]$. Then, controlled counterparts of equations (8) and (9) are as follows:

$$Y_t^{(v)} = y + \int_0^t a(t,s,V^{1,v},Y^{(v)})ds + \int_0^t c(t,s,V^{2,v},Y^{(v)})v_s ds$$
$$+ \int_0^t c(t,s,V^{2,v},Y^{(v)})dB_s \qquad (13)$$

and

$$V_s^{i,v} = V_0^{(i)} + \int_0^s \bar{b}_i(r,V^{i,v})dr + \int_0^s \bar{\sigma}_i(r,V^{i,v})v_r dr + \int_0^s \bar{\sigma}_i(r,V^{i,v})dB_r,$$
$$i = 1,2.$$

For $v \in \mathcal{M}^2[0,T]$ and $\varepsilon \in (0,1]$, controlled counterparts of equations (10) and (11) satisfy

$$Y_t^{\varepsilon,v} = y + \int_0^t a(t,s,V^{1,\varepsilon,v},Y^{\varepsilon,v})ds + \int_0^t c(t,s,V^{2,\varepsilon,v},Y^{\varepsilon,v})v_s ds$$
$$+ \sqrt{\varepsilon} \int_0^t c(t,s,V^{2,\varepsilon,v},Y^{\varepsilon,v})dB_s \qquad (14)$$

and

$$V_s^{i,\varepsilon,v} = V_0^{(i)} + \int_0^s \bar{b}_i(r,V^{i,\varepsilon,v})dr + \int_0^s \bar{\sigma}_i(r,V^{i,\varepsilon,v})v_r dr$$
$$+ \sqrt{\varepsilon} \int_0^s \bar{\sigma}_i(r,V^{i,\varepsilon,v})dB_r, \quad i = 1,2. \qquad (15)$$

Remark 12. In [17], less general equations than those in (10) and (14) were studied. These equations are as follows:

$$Y_t^{(\varepsilon)} = y + \int_0^t a(s,Y^{(\varepsilon)})ds + \sqrt{\varepsilon} \int_0^t c(s,Y^{(\varepsilon)})dB_s \qquad (16)$$

and

$$Y_t^{\varepsilon,v} = y + \int_0^t a(s,Y^{\varepsilon,v})ds + \int_0^t c(s,Y^{\varepsilon,v})v_s ds + \sqrt{\varepsilon} \int_0^t c(s,Y^{\varepsilon,v})dB_s. \qquad (17)$$

Equations (16) and (17) are not of Volterra-type.

What happens if $\varepsilon = 0$ is explained next. Equations (11) take the following form: $V_s^{i,0} = V_0^{(i)} + \int_0^s \bar{b}_i(r,V^{i,0})dr$, $i = 1,2$. These equations can be solved pathwise, and for every i, all the solutions are the same by the uniqueness. Let us denote the solution by $v^{i,0}$. It follows that $V_s^{i,0} = v^{i,0}(s)$ for $i = 1,2$ and all $s \in [0,T]$.

Suppose $\varepsilon = 0$. Then, equations (14) and (15) can be rewritten as follows:
$$V_s^{i,0,v} = V_0^{(i)} + \int_0^s \bar{b}_i(r, V^{i,0,v}) dr + \int_0^s \bar{\sigma}_i(r, V^{i,0,v}) v_r dr, \quad i = 1, 2,$$
and
$$Y_t^{0,v} = y + \int_0^t a(t, s, V^{1,0,v}, Y^{0,v}) ds + \int_0^t c(t, s, V^{2,0,v}, Y^{0,v}) v_s ds, \qquad (18)$$
respectively. In a special case of a deterministic control f, we have
$$V_s^{i,0,f} = V_0^{(i)} + \int_0^s \bar{b}_i(r, V_r^{i,0,f}) dr + \int_0^s \bar{\sigma}_i(r, V_r^{i,0,f}) f(r) dr. \qquad (19)$$

Remark 13. Under the restrictions imposed on $\bar{b}_i$ and $\bar{\sigma}_i$ in [17], the functional equations
$$\psi_i(s) = V_0^{(i)} + \int_0^s \bar{b}_i(r, \psi_i) dr + \int_0^s \bar{\sigma}_i(r, \psi_i) f(r) dr, \quad i = 1, 2,$$
are uniquely solvable, the solutions $\psi_{i,f}$ belong to the spaces $\mathcal{W}^{k_i}$, and if $f_n \mapsto f$ weakly in $L^2([0,T], \mathbb{R}^m)$, then $\psi_{i,f_n} \mapsto \psi_{i,f}$ in $\mathcal{W}^{k_i}$ for $i = 1, 2$. Therefore, the solution to equation (19) is deterministic, and $V_s^{i,0,f} = \psi_{i,f}(s)$ for all $s \in [0, T]$ (see [17]).

Suppose $v \in \mathcal{M}_N^2[0, T]$ for some $N > 0$. Then, it follows from Girsanov's theorem that for all $0 < \varepsilon \leq 1$, the process
$$B_t^{\varepsilon,v} = B_t + \frac{1}{\sqrt{\varepsilon}} \int_0^t v_s ds, \quad t \in [0, T], \qquad (20)$$
is an m-dimensional Brownian motion on $\mathcal{W}^m$ with respect to a measure $\mathbb{P}^{\varepsilon,v}$ on $\mathcal{F}_T^m$ that is equivalent to the measure $\mathbb{P}$. The process $B^{\varepsilon,v}$ is adapted to the filtration $\{\mathcal{F}_t^B\}$. In a special case where $\varepsilon = 1$, the following notation is used:
$$B_t^{(v)} = B_t + \int_0^t v_s ds, \quad t \in [0, T]. \qquad (21)$$

We next explain what restrictions are imposed on the model for the volatility described by equation (8).

Assumption (C1). For all $(\eta_1, \varphi) \in \mathcal{W}^{k_1} \times \mathcal{W}^d$, the map $(t, s) \mapsto a(t, s, \eta_1, \varphi)$ is Borel measurable, with values in the space $\mathbb{R}^d$, while for all $(\eta_2, \varphi) \in \mathcal{W}^{k_2} \times \mathcal{W}^d$, the map $(t, s) \mapsto c(t, s, \eta_2, \varphi)$ is Borel measurable, with values in the space of $d \times m$-matrices. Moreover, a and c are of Volterra-type in the first two variables. In addition, for every $t \in [0, T]$, $(s, \eta_1, \varphi) \mapsto a(t, s, \eta_1, \varphi)$ and $(s, \eta_2, \varphi) \mapsto c(t, s, \eta_2, \varphi)$ are predictable path functionals mapping the space $[0,t] \times \mathcal{W}^{k_1} \times \mathcal{W}^d$ into the space $\mathbb{R}^d$ and the space $[0,t] \times \mathcal{W}^{k_2} \times \mathcal{W}^d$ into the space of $d \times m$ matrices, respectively. The definition of a predictable path functional can be found in [71] (see Definition (8.3) and Remark (8.4) on p. 122). The requirement above is similar to Convention (8.7) on p. 123 in [71].

Assumption (C2). (a) Let $\eta_1 \in \mathcal{W}^{k_1}$, $\eta_2 \in \mathcal{W}^{k_2}$, and $\varphi \in \mathcal{W}^d$. Then, the following inequalities hold for all $t \in [0,T]$:

$$\int_0^t ||a(t,s,\eta_1,\varphi)||_d \, ds < \infty \quad \text{and} \quad \int_0^T ||c(t,s,\eta_2,\varphi)||_{d \times m}^2 \, ds < \infty. \quad (22)$$

(b) For all fixed $\eta_1 \in \mathcal{W}^{k_1}$ and $\varphi \in \mathcal{W}^d$, the function $t \mapsto \int_0^t a(t,s,\eta_1,\varphi) \, ds$ is a continuous $\mathbb{R}^d$-valued function on $[0,T]$. In addition, for every fixed $t \in [0,T]$, the function $(\eta_1, \varphi) \mapsto \int_0^t a(t,s,\eta_1,\varphi) \, ds$ is continuous on the space $\mathcal{W}^{k_1} \times \mathcal{W}^d$.

(c) Let $\eta_{2,n} \to \eta_2$ in $\mathcal{W}^{k_2}$ and $\varphi_n \to \varphi$ in $\mathcal{W}^d$ as $n \to \infty$. Then, for every $t \in [0,T]$,

$$\int_0^t ||c(t,s,\eta_{2,n},\varphi_n) - c(t,s,\eta_2,\varphi)||_{d \times m}^2 \, ds \to 0 \quad \text{as} \quad n \to \infty.$$

Assumption (C3). (a) For all $0 < \varepsilon \leq 1$, there exists a strong solution to equation (10). (b) Let $v \in M_N^2[0,T]$ for some $N > 0$. Then, any two strong solutions to equation (13) are $\mathbb{P}$-indistinguishable.

Remark 14. The definition of the strong solution used in this paper includes the continuity of the solution.

Remark 15. Assumption (C3)(b) is weaker than the pathwise uniqueness condition employed in [17].

Assumption (C4). For every function $f \in L^2([0,T], \mathbb{R}^m)$ and the functions $\psi_{1,f}$ and $\psi_{2,f}$ defined in Remark 13, the equation

$$\eta(t) = y + \int_0^t a(t,s,\psi_{1,f},\eta) \, ds + \int_0^t c(t,s,\psi_{2,f},\eta) f(s) \, ds, \quad (23)$$

is uniquely solvable in $\mathcal{W}^d$.

Remark 16. It is shown in the following that equation (23) is always solvable. The details can be found in Remark 100. Therefore, only the uniqueness condition must be included in Assumption (C4).

Definition 17. The map $\Gamma_y : L^2([0,T], \mathbb{R}^m) \mapsto \mathcal{W}^d$ is defined by $\Gamma_y f = \eta_f$, where η_f is the unique solution to equation (23).

Assumption (C5). Set $D_N = \{f \in L^2([0,T], \mathbb{R}^m) : \int_0^T \|f(t)\|_m^2 dt \leq N\}$. Then, the restriction of the map Γ_y to D_N is a continuous map from D_N equipped with the weak topology into the space $\mathcal{W}^d$.

Let $v \in M_N^2[0,T]$ for some $N > 0$. Then, there exists a map $g^{(2)} : \mathcal{W}^m \mapsto \mathcal{W}^{k_2}$ satisfying the following conditions: (i) $g^{(2)}(B) = V^{(2)}$, (ii) $g^{(2)}(B^{(v)}) = V^{2,v}$ $\mathbb{P}$-a.s., and (iii) for every $t \in [0,T]$, $g^{(2)}$ is $\widetilde{\mathcal{B}}_t^m / \mathcal{B}_t^{k_2}$-measurable (see Lemma A.1 in [17], see also Theorem 10.4 on p. 126 in [71]).

Remark 18. Suppose Assumption (C3) holds. Then, for every $\varepsilon \in (0,1]$ there exists a map $h^\varepsilon : \mathcal{W}^m \mapsto \mathcal{W}^d$ such that the solution $Y^{(\varepsilon)}$ to equation (13) satisfies $Y^{(\varepsilon)} = h^{(\varepsilon)}(B)$, and the map h^ε is $\widetilde{\mathcal{B}}_t^m / \mathcal{B}_t^d$-measurable for all $t \in [0,T]$ (see Remark 6). It is clear that $Y^{(1)} = Y$. We denote the map $h^{(1)}$ by h. It follows that $Y = h(B)$ $\mathbb{P}$-a.s.

Assumption (C6). The process $t \mapsto \int_0^t c(t, s, g^{(2)}(B^{(v)}), h(B^{(v)})) dB_s^{(v)}$, where $t \in [0,T]$, is continuous.

Assumption (C6) looks rather complicated. A special case, where Assumption (C6) is satisfied, is when the map c does not depend on the variable t. Indeed, in such a case, the correctness of Assumption (C6) follows from the condition for the map c in Assumption (C2)(a) and the continuity properties of stochastic integrals. More examples of the validity of Assumption (C6) are provided in section "Unification of Sample Path Large Deviation Principles for Stochastic Volatility Models".

Assumption (C7). Suppose $0 < \varepsilon_n < 1$, with $n \geq 1$, is a sequence of numbers such that $\varepsilon_n \to 0$ as $n \to \infty$. Let $v^{(n)}$, $n \geq 1$, be a sequence of controls satisfying the condition $v^{(n)} \in \mathcal{M}_N^2[0,T]$ for some $N > 0$ and all $n \geq 1$ (see Definition 11). Then, the family of $\mathcal{W}^d$-valued random variables $Y^{\varepsilon_n, v^{(n)}}$, with $n \geq 1$, is tight in $\mathcal{W}^d$. Moreover, for every $t \in [0,T]$, the following inequality is satisfied:

$$\sup_{n \geq 1} \int_0^t \mathbb{E}\left[\|c(t, s, V^{2,\varepsilon_n, v^{(n)}}, Y^{\varepsilon_n, v^{(n)}})\|_{d \times m}^2\right] ds < \infty. \tag{24}$$

We next formulate several remarks related to Assumptions (C1)–(C7).

Remark 19. Assumptions (C1) and (C2)(a) guarantee that equation (8) exists.

Remark 20. It is easy to see that the first part of Assumption (C2)(b) holds if the following condition is satisfied: Let $t, t' \in [0, T]$, and suppose $\eta_1 \in \mathcal{W}^{k_1}$, $\varphi \in \mathcal{W}^d$, and $t' \to t$. Then,

$$\int_0^T ||a(t', s, \eta_1, \varphi) - a(t, s, \eta_1, \varphi)||_d ds \to 0. \tag{25}$$

Similarly, the second part of Assumption (C2)(b) can be derived from the following condition:

$$\int_0^T ||a(t, s, \eta_{1,n}, \varphi_n) - a(t, s, \eta_1, \varphi)||_d ds \to 0$$

as $n \to 0$ provided that $\eta_{1,n} \to \eta$ in $\mathcal{W}^{k_1}$ and $\varphi_n \to \varphi$ in $\mathcal{W}^d$ as $n \to \infty$.

Remark 21. It follows from Assumption (C2)(a) that for every $t \in [0, T]$, the following condition holds $\mathbb{P}$-a.s.: $\int_0^t ||a(t, s, V^{1,\varepsilon}, Y^{(\varepsilon)})||_d ds + \int_0^t ||c(t, s, V^{2,\varepsilon}, Y^{(\varepsilon)})||_{d\times m}^2 ds < \infty$.

It was shown in [17] that if pathwise uniqueness and existence in the strong sense hold for equation (16) and the control v satisfies (12), then there exists a unique strong solution to equation (17), and this solution can be represented as a measurable functional of the process $B^{\varepsilon,v}$ (see Lemma 1 in Appendix A in [17]). We establish a similar strong solvability result for the Volterra equation (14) (see Lemma 96 and Remark 98).

Remark 22. Using Remark 13 and (18), we see that

$$Y_t^{0,f} = y + \int_0^t a(t, s, \psi_{1,f}, Y^{0,f}) ds + \int_0^t c(t, s, \psi_{2,f}, Y^{0,f}) f(s) ds. \tag{26}$$

It follows from Assumption (C4) that equation (26) can be reduced to equation (23). Therefore, $Y_t^{0,f} = \Gamma_y f(t)$ for all $t \in [0, T]$.

Sample Path Large Deviation Principles for Log-Price Processes

In this section, we formulate and discuss one of the main results of this paper (Theorem 24). The proof of this theorem is given in section "Proof of Theorem 24".

Let us denote by $\mathbb{C}_0^m$ the subspace of the space $\mathcal{W}^m$ consisting of the functions f such that $f(0) = \vec{0}$. Throughout this paper, the symbol $(\mathbb{H}_0^1)^m$ stands for the m-dimensional Cameron–Martin space. A function f from the space $\mathbb{C}_0^m$ belongs to the space $(\mathbb{H}_0^1)^m$ if it has absolutely continuous components and the derivatives of the components on $(0, T)$ are square-integrable with respect to the Lebesgue measure. For $f \in (\mathbb{H}_0^1)^m$, we set $\dot{f} = (\dot{f}_1, \ldots, \dot{f}_m)$ where the symbol $\dot{f}_k$, with $1 \leq k \leq m$, stands for the derivative of the component f_k with respect to the variable t.

In the following definition, we introduce special maps $f \mapsto \mathcal{A}f$ and $f \mapsto \widehat{f}$.

Definition 23. (i) For every function $f \in L^2([0, T]; \mathbb{R}^m)$, the function $\mathcal{A}f \in \mathcal{W}^d$ is defined by $\mathcal{A}f = G(\Gamma_y f)$ where G and Γ_y are introduced in Definitions 8 and 17, respectively. (ii) For every function $f \in (\mathbb{H}_0^1)^m$, the function $\widehat{f} \in \mathcal{W}^d$ is defined by $\widehat{f} = \mathcal{A}\dot{f}$.

The rate function $\widetilde{Q}_T$ governing the large deviation principle for the log-price process depends on the measurable map $\Phi : \mathbb{C}_0^m \times \mathbb{C}_0^m \times \mathcal{W}^d \mapsto \mathbb{C}_0^m$ given by

$$\Phi(l, f, h)(t) = \int_0^t b(s, \widehat{f}(s)) ds + \int_0^t \sigma(s, \widehat{f}(s)) \bar{C} \dot{l}(s) ds$$
$$+ \int_0^t \sigma(s, \widehat{f}(s)) C \dot{f}(s) ds \qquad (27)$$

for all $l, f \in (\mathbb{H}_0^1)^m$, $h = \widehat{f} \in \mathcal{W}^d$, and $0 \leq t \leq T$. For all the remaining triples (l, f, h), we set $\Phi(l, f, h)(t) = 0$ for $t \in [0, T]$. Let $g \in \mathbb{C}_0^m$, and define the function $\widetilde{Q}_T$ by

$$\widetilde{Q}_T(g) = \inf_{l, f \in (\mathbb{H}_0^1)^m} \left[\frac{1}{2} \int_0^T \|\dot{l}(s)\|_m^2 ds + \frac{1}{2} \int_0^T \|\dot{f}(s)\|_m^2 ds : \Phi(l, f, \right.$$
$$\left. \widehat{f}(t)) = g(t), t \in [0, T] \right], \qquad (28)$$

if the equation $\Phi(l, f, \widehat{f}(t)) = g(t)$ is solvable for l and f. If there is no solution, then we set $\widetilde{Q}_T(g) = \infty$. It follows that if the previous equation is solvable, then $g \in (\mathbb{H}_0^1)^m$ and

$$\dot{g}(t) = b(t, \widehat{f}(t)) + \sigma(t, \widehat{f}(t)) \bar{C} \dot{l}(t) + \sigma(t, \widehat{f}(t)) C \dot{f}(t). \qquad (29)$$

The following assertion provides a sample path LDP for log-price processes in general stochastic volatility models.

Theorem 24. *Suppose Assumption A and Assumptions (C1)–(C7) hold true, and the model in (1) is defined on the canonical set-up. Then, the process $\varepsilon \mapsto X^{(\varepsilon)} - x_0$ with state space $\mathcal{W}^m$ satisfies the sample path large deviation principle with speed ε^{-1} and good rate function $\widetilde{Q}_T$ defined in (28).*

The validity of the large deviation principle means that for every Borel measurable subset $\mathcal{A}$ of $\mathcal{W}^m$, the following estimates hold:

$$-\inf_{g \in \mathcal{A}^\circ} \widetilde{Q}_T(g) \leq \liminf_{\varepsilon \downarrow 0} \varepsilon \log \mathbb{P}\left(X^{(\varepsilon)} - x_0 \in \mathcal{A}\right)$$

$$\leq \limsup_{\varepsilon \downarrow 0} \varepsilon \log \mathbb{P}\left(X^{(\varepsilon)} - x_0 \in \mathcal{A}\right) \leq -\inf_{g \in \bar{\mathcal{A}}} \widetilde{Q}_T(g).$$

The symbols $\mathcal{A}^\circ$ and $\bar{\mathcal{A}}$ in the previous estimates stand for the interior and the closure of the set $\mathcal{A}$, respectively.

Remark 25. In Theorem 24, the canonical set-up on the space $\mathcal{W}^m \times \mathcal{W}^m$ is employed. However, if the LDP in Theorem 41 is valid on the set-up associated with the volatility process in (1), then Theorem 24 holds on the set-up associated with equation (1). This follows from the proof of Theorem 24 given in section "Proof of Theorem 24". Examples illustrating the statement formulated above are volatility processes in multivariate Gaussian models (see Theorem 49 in section "LDPs for Multivariate Gaussian Models and Multivariate Non-Gaussian Fractional Models"), volatility processes in multivariate non-Gaussian fractional models (see Theorem 50 in section "LDPs for Multivariate Gaussian Models and Multivariate Non-Gaussian Fractional Models"), and the processes solving Volterra-type stochastic integral equations employed in the paper [66] of Nualart and Rovira (see Theorem 1 in [66], see also section "Unification: More Volterra-Type SDEs" in this paper).

Suppose that for every $(t, u) \in [0, T] \times \mathbb{R}^d$, the matrix $\sigma(t, u)$ is invertible. Then, the expression on the right-hand side of (28) can be simplified. By taking into account (29), we obtain the following equality: $\dot{l}(t) = \bar{C}^{-1}\sigma(t, \widehat{f}(t))^{-1}[\dot{g}(t) - b(t, \widehat{f}(t)) - \sigma(t, \widehat{f}(t))C\dot{f}(t)]$, $t \in [0, T]$. Hence for all $g \in (\mathbb{H}_0^1)^m$,

$$\widetilde{Q}_T(g) = \frac{1}{2} \inf_{f \in (\mathbb{H}_0^1)^m} \int_0^T (\|\bar{C}^{-1}\sigma(s, \widehat{f}(s))^{-1}[\dot{g}(s) - b(s, \widehat{f}(s))$$

$$- \sigma(s, \widehat{f}(s))C\dot{f}(s)]\|_m^2 + \|\dot{f}(s)\|_m^2)ds, \tag{30}$$

and $\widetilde{Q}_T(g) = \infty$ otherwise.

Remark 26. Let $g \in (\mathbb{H}_0^1)^m$. Then, for every fixed function $f \in (\mathbb{H}_0^1)^m$, the integral on the right-hand side of (30) is finite. Indeed, it suffices to prove that

$$\sup_{t \in [0,T]} \|\sigma(s, \widehat{f}(s))^{-1}\|_{m \times m} < \infty. \tag{31}$$

The matrix-valued function $s \mapsto \sigma(s, \widehat{f}(s))$, with $s \in [0, T]$, is continuous. Therefore, the function $s \mapsto |\det(\sigma(s, \widehat{f}(s)))|$ is continuous on $[0, T]$ and bounded away from zero. Next, using the uniform boundedness of the adjugate matrices associated with the matrices $\sigma(s, \widehat{f}(s))$, $s \in [0,T]$, we see that the inequality in (31) holds.

The next statement concerns the continuity of the rate function $\widetilde{Q}_T$.

Lemma 27. *Suppose that for every $(t,u) \in [0,T] \times \mathbb{R}^d$, the matrix $\sigma(t,u)$ is invertible. Then, the function $\widetilde{Q}_T$ defined in (30) is continuous in the topology of the space $(\mathbb{H}_0^1)^m$.*

Proof. The lower semicontinuity of the function $\widetilde{Q}_T$ on the space $(\mathbb{H}_0^1)^m$ can be established using the following two facts: (1) Since $\widetilde{Q}_T$ is a good rate function on the space $\mathbb{C}_0^m$, it is lower semicontinuous on this space. (2) The space $(\mathbb{H}_0^1)^m$ is continuously embedded into the space $\mathbb{C}_0^m$.

We next show that the function $\widetilde{Q}_T$ is upper semicontinuous on the space $(\mathbb{H}_0^1)^m$. By taking into account the representation in (28) and the fact that the greatest lower bound of any family of upper semicontinuous functions is an upper semicontinuous function, we see that it suffices to prove that for any $f \in (\mathbb{H}_0^1)^m$, the function

$$g \mapsto \frac{1}{2} \int_0^T (\|\bar{C}^{-1} \sigma(s, \widehat{f}(s))^{-1} [\dot{g}(s) - b(s, \widehat{f}(s)) - \sigma(s, \widehat{f}(s)) C \dot{f}(s)]\|_m^2$$
$$+ \|\dot{f}(s)\|_m^2) ds \tag{32}$$

is continuous on the space $(\mathbb{H}_0^1)^m$. The previous statement follows from formula (31).

The proof of Lemma 27 is thus completed.

Let $m = 1$, and suppose the volatility function σ satisfies the condition $\sigma(t,u) \neq 0$ for all $(t,u) \in [0,T] \times \mathbb{R}^d$. Then, for every $g \in \mathbb{H}_0^1$, the formula in (30) can be rewritten as follows:

$$\widetilde{Q}_T(g) = \frac{1}{2} \inf_{f \in \mathbb{H}_0^1} \int_0^T \left[\frac{(\dot{g}(s) - b(s, \widehat{f}(s)) - \rho \sigma(s, \widehat{f}(s)) \dot{f}(s))^2}{(1-\rho^2) \sigma(s, \widehat{f}(s))^2} + \dot{f}(s)^2 \right] ds. \tag{33}$$

We also have

$$\widetilde{Q}_T(g) = \infty \quad \text{if} \quad g \in \mathbb{C} \setminus \mathbb{H}_0^1. \tag{34}$$

Small-Noise LDPs for Log-Price Processes

Our goal in this section is to derive from Theorem 24 a small-noise LDP for the process $\varepsilon \mapsto X_T^{(\varepsilon)} - x_0$. In financial mathematics, such LDPs are used in small-noise asymptotic analysis of various path-independent options.

Consider a map $V : \mathbb{C}_0^m \mapsto \mathbb{R}^m$ defined by $V(\varphi) = \varphi(T)$. It is clear that this map is continuous. It follows from Theorem 24 and the contraction principle that the following assertion holds.

Theorem 28. *Under the restrictions in Theorem 24, the process $\varepsilon \mapsto X_T^{(\varepsilon)} - x_0$ with state space $\mathbb{R}^m$ satisfies the small-noise large deviation principle with speed ε^{-1} and good rate function $\widehat{I}_T(x)$, $x \in \mathbb{R}^m$ given by*

$$\widehat{I}_T(x) = \inf_{\{g \in (\mathbb{H}_0^1)^m : g(T) = x\}} \widetilde{Q}_T(g) \tag{35}$$

where the rate function $\widetilde{Q}_T$ is defined in (28).

The expression on the right-hand side of (35) is rather complicated. Our next goal is to show that if $m = 1$, then the formula in (35) can be simplified (see Theorem 30). We do not know whether a similar simplification is possible for $m > 1$. Here our knowledge is fragmentary. More information is given in the following.

Let $m = 1$. For all $y \in \mathbb{R}$ and $f \in \mathbb{H}_0^1$, set

$$\Psi(y, f, \widehat{f}) = \int_0^T b(s, \widehat{f}(s))ds + \rho \int_0^T \sigma(s, \widehat{f}(s))\dot{f}(s)ds$$

$$+ \bar{\rho}\left\{\int_0^T \sigma(s, \widehat{f}(s))^2 ds\right\}^{\frac{1}{2}} y.$$

Define the function $\widetilde{I}_T$ on $\mathbb{R}$ by

$$\widetilde{I}_T(x) = \frac{1}{2} \inf_{y \in \mathbb{R}, f \in \mathbb{H}_0^1} \left\{y^2 + \int_0^T \dot{f}(t)^2 dt : \Psi(y, f, \widehat{f}) = x\right\}, \tag{36}$$

if the equation $\Psi(y, f, \widehat{f}) = x$ is solvable for y and f, and $\widetilde{I}_T(x) = \infty$ otherwise.

Remark 29. The equation $\Psi(y, f, \widehat{f}) = x$ is as follows:

$$\int_0^T b(s, \widehat{f}(s))ds + \rho \int_0^T \sigma(s, \widehat{f}(s))\dot{f}(s)ds + \bar{\rho}\left\{\int_0^T \sigma(s, \widehat{f}(s))^2 ds\right\}^{\frac{1}{2}} y = x. \tag{37}$$

Theorem 30. *Suppose $m = 1$, and the conditions in Theorem 24 hold. Then, for all $x \in \mathbb{R}$, $\widetilde{I}_T(x) = \widehat{I}_T(x)$.*

Proof. We first prove that $\widetilde{I}_T(x) \leq \widehat{I}_T(x)$ for $x \in \mathbb{R}$. Fix $x \in \mathbb{R}$ and $g \in \mathbb{H}_0^1$ with $g(T) = x$. If the equation

$$\Phi(l, f, \widehat{f})(t) = g(t), \quad t \in [0, T], \tag{38}$$

is not solvable for l and f, then $\widetilde{Q}_T(g) = \infty$. If equation (38) is solvable and f is such that $\int_0^T \sigma(s, \widehat{f}(s))^2 ds = 0$, then (27) implies the equality $\int_0^T b(s, \widehat{f}(s)) ds = x$. Hence the equation $\Psi(y, f, \widehat{f}) = x$, where f is the function mentioned above, holds with any $y \in \mathbb{R}$. Next, suppose equation (38) is solvable, and the function f is such that $\int_0^T \sigma(s, \widehat{f}(s))^2 ds > 0$. It follows that the equation $\Psi(y, f, \widehat{f}) = x$, with the same function f, is uniquely solvable for y, and the unique solution satisfies

$$\int_0^T \sigma(s, \widehat{f}(s)) \dot{l}(s) ds = \left\{ \int_0^T \sigma(s, \widehat{f}(s))^2 ds \right\}^{\frac{1}{2}} y.$$

Moreover, $y^2 \leq \int_0^T \dot{l}(s)^2 ds$. Finally, we see that the reasoning above shows that the inequality $\widetilde{I}_T(x) \leq \widehat{I}_T(x)$ holds for all $x \in \mathbb{R}$.

We next prove that the opposite inequality holds as well. Let $x \in \mathbb{R}$, and suppose the equation

$$\Psi(y, f, \widehat{f}) = x \tag{39}$$

is not solvable for y and f. Then, we have $\widetilde{I}_T(x) = \infty$, and hence $\widehat{I}_T(x) \leq \widetilde{I}_T(x)$. Next, suppose equation (39) is solvable for y and f, and the function f satisfies the condition $\int_0^T \sigma(s, \widehat{f}(s))^2 ds = 0$. Then, $x = \int_0^T b(s, \widehat{f}(s)) ds$. Set $g(t) = \int_0^t b(s, \widehat{f}(s)) ds$. It is not hard to see that $g \in \mathbb{H}_0^1$, $g(T) = x$, and (38) holds for g, with the same function f and any function $l \in \mathbb{H}_0^1$. Let us next assume that equation (39) is solvable for y and f, and the function f satisfies $\int_0^T \sigma(s, \widehat{f}(s))^2 ds > 0$. Then, there exists $l \in \mathbb{H}_0^1$ such that

$$\dot{l}(s) = \left\{ \int_0^T \sigma(u, \widehat{f}(u))^2 du \right\}^{-\frac{1}{2}} \sigma(s, \widehat{f}(s)) y$$

for all $s \in [0, T]$. It follows that $\int_0^T \dot{l}(s)^2 ds = y^2$. Set

$$g(t) = \int_0^t b(s, \widehat{f}(s)) ds + \rho \int_0^t \sigma(s, \widehat{f}(s)) \dot{f}(s) ds + \bar{\rho} \int_0^t \sigma(s, \widehat{f}(s)) \dot{l}(s) ds,$$

with the same functions l and f as above. Then, $g \in \mathbb{H}_0^1$. It is not hard to see that

$$g(T) = \int_0^T b(s, \widehat{f}(s)) ds + \rho \int_0^T \sigma(s, \widehat{f}(s)) \dot{f}(s) ds + \bar{\rho} \left\{ \int_0^T \sigma(s, \widehat{f}(s))^2 ds \right\}^{\frac{1}{2}} y$$
$$= \Psi(y, f, \widehat{f}) = x.$$

In addition, the functions l and f solve equation (38). Finally, summarizing what was said above, we see that $\widehat{I}_T(x) \leq \widetilde{I}_T(x)$ for all $x \in \mathbb{R}$.

This completes the proof of Theorem 30.

Remark 31. The expression for the rate function $\widetilde{I}_T$ on the right-hand side of (36) can be given a simpler form. Note that for every $f \in \mathbb{H}_0^1$, the function $s \mapsto \sigma(s, \widehat{f}(s))$ is continuous. Define
$$Q_1 = \{f \in \mathbb{H}_0^1 : \sigma(s, \widehat{f}(s)) \neq 0 \text{ for at least one } s \in [0, T]\}$$
and $Q_2 = \{f \in \mathbb{H}_0^1 : \sigma(s, \widehat{f}(s)) = 0 \text{ for all } s \in [0, T]\}$. It is clear that $\mathbb{H}_0^1 = Q_1 \cup Q_2$. Set $Q_3(x) = \{f \in Q_2 : x = \int_0^T b(s, \widehat{f}(s)) ds\}$. It is not hard to see that (36) implies the following equality: $\widetilde{I}_T(x) = \min\{A_1(x), A_2(x)\}$ where
$$A_1(x) = \frac{1}{2} \inf_{f \in Q_1} \left[\frac{(x - \int_0^T b(s, \widehat{f}(s)) ds - \rho \int_0^T \sigma(s, \widehat{f}(s)) \dot{f}(s) ds)^2}{\bar{\rho}^2 \int_0^T \sigma(s, \widehat{f}(s))^2 ds} + \int_0^T \dot{f}(t)^2 dt \right]$$
and
$$A_2(x) = \begin{cases} \frac{1}{2} \inf_{\{f \in Q_3(x)\}} \int_0^T \dot{f}(t)^2 dt & \text{if } Q_3(x) \neq \emptyset \\ \infty & \text{if } Q_3(x) = \emptyset. \end{cases}$$

In a special case, where $\sigma(s, z) \neq 0$ for all $(s, z) \in [0, T] \times \mathbb{R}^d$, we have $Q_2 = \emptyset$, and hence
$$\widetilde{I}_T(x) = \frac{1}{2} \inf_{f \in \mathbb{H}_0^1} \left[\frac{(x - \int_0^T b(s, \widehat{f}(s)) ds - \rho \int_0^T \sigma(s, \widehat{f}(s)) \dot{f}(s) ds)^2}{\bar{\rho}^2 \int_0^T \sigma(s, \widehat{f}(s))^2 ds} + \int_0^T \dot{f}(t)^2 dt \right] \quad (40)$$
for all $x \in \mathbb{R}$.

The next statement follows from Theorems 28, 30, and (40).

Theorem 32. *Suppose Assumption A and Assumptions (C1)–(C7) hold true, and the model in (1) is defined on the canonical set-up. Suppose also that $\sigma(s, z) \neq 0$ for all $(s, z) \in [0, T] \times \mathbb{R}^d$. Then, the process $\varepsilon \mapsto X_T^{(\varepsilon)} - x_0$ with state space $\mathbb{R}^1$ satisfies the small-noise large deviation principle with speed ε^{-1} and good rate function $\widehat{I}_T(x)$, $x \in \mathbb{R}^m$ given by the formula in (40).*

The following lemma concerns the continuity of the rate function in Theorem 32.

Lemma 33. *Suppose $\sigma(s,z) \neq 0$ for all $(s,z) \in [0,T] \times \mathbb{R}^d$. Then, the function $\widetilde{I}_T$ is continuous on $\mathbb{R}$.*

Proof. The function $\widetilde{I}_T$ is lower semicontinuous since it is a good rate function. Moreover, it is upper semicontinuous being equal to the greatest upper bound of a family of continuous functions on $\mathbb{R}$.

This completes the proof of Lemma 33.

Next, we turn our attention to the case where $m > 1$. As we have already mentioned, our knowledge in this case is incomplete. First of all, it is not clear how to choose the map Ψ. One of the acceptable candidates is as follows:

$$\Psi(y, f, \widehat{f}) = \int_0^T b(s, \widehat{f}(s))ds + \int_0^T \sigma(s, \widehat{f}(s)) C \dot{f}(s) ds$$
$$+ \left\{ \int_0^T \|\sigma(s, \widehat{f}(s))\bar{C}\|_*^2 ds \right\}^{\frac{1}{2}} y, \quad y \in \mathbb{R}^m, \quad f \in (\mathbb{H}_0^1)^m, \tag{41}$$

where $\|\cdot\|_*$ is the operator norm on the space of $m \times m$-matrices. The usefulness of the operator norm employed in (41) is clear in the following (see the proof of Lemma 35). Let us also define a map $\widetilde{I}_T : \mathbb{R}^m \mapsto \mathbb{R}$ by

$$\widetilde{I}_T(x) = \frac{1}{2} \inf_{y \in \mathbb{R}^m, f \in (\mathbb{H}_0^1)^m} \left\{ \|y\|_m^2 + \int_0^T \|\dot{f}(t)\|_m^2 dt : \Psi(y, f, \widehat{f}) = x \right\}, \tag{42}$$

if the equation $\Psi(y, f, \widehat{f}) = x$ is solvable for y and f, and by $\widetilde{I}_T(x) = \infty$ otherwise.

Lemma 34. *For every $x \in \mathbb{R}^m$, the inequality $\widetilde{I}_T(x) \leq \widehat{I}_T(x)$ holds for the functions defined in (35) and (42), respectively.*

Proof. The proof of Lemma 34 is similar to that of the first part of Theorem 30. We include this proof for the sake of completeness. Fix $x \in \mathbb{R}^m$ and $g \in (\mathbb{H}_0^1)^m$, with $g(T) = x$. If the equation

$$\Phi(l, f, \widehat{f})(t) = g(t), \quad t \in [0,T], \tag{43}$$

is not solvable for l and f, then $\widetilde{Q}_T(g) = \infty$. If equation (43) is solvable, and the function f is such that $\int_0^T \|\sigma(s, \widehat{f}(s))\|_*^2 ds = 0$, then for every $s \in [0,T]$, $\sigma(s, \widehat{f}(s)) = 0$. It follows from (27) that $\int_0^T b(s, \widehat{f}(s))ds = x$. Hence the equation $\Psi(y, f, \widehat{f}) = x$, where f is the function mentioned above, holds with any $y \in \mathbb{R}^m$. Next, suppose equation (43) is solvable and $\int_0^T \|\sigma(s, \widehat{f}(s))\|_*^2 ds > 0$. Then, we have $\int_0^T \|\sigma(s, \widehat{f}(s))\bar{C}\|_*^2 ds > 0$. Hence the equation $\Psi(y, f, \widehat{f}) = x$, with the same function f as above, is uniquely solvable for y, and the unique solution satisfies

$$\int_0^T \sigma(s, \widehat{f}(s))\bar{C}\dot{l}(s)ds = \left\{\int_0^T \|\sigma(s, \widehat{f}(s))\bar{C}\|_*^2 ds\right\}^{\frac{1}{2}} y. \qquad (44)$$

Now, it is not hard to prove using Hölder's inequality that the equality in (44) implies the estimate $\|y\|_m^2 \leq \int_0^T \|\dot{l}(s)\|_m^2 ds$.

Finally, we see that the reasoning above shows that the inequality $\widetilde{I}_T(x) \leq \widehat{I}_T(x)$ holds for all $x \in \mathbb{R}$.

This completes the proof of Lemma 34.

Our next goal is to provide examples of volatility maps σ such that $\widetilde{I}_T(x) = \widehat{I}_T(x)$ for all $x \in \mathbb{R}^m$.

Lemma 35. *Let $m > 1$, and suppose that for all $t \in [0,T]$ and $z \in \mathbb{R}^d$, the volatility map is given by $\sigma(t,z) = \xi(t,z)O(t,z)\bar{C}^{-1}$, where $O(t,z)$ are orthogonal $m \times m$-matrices and ξ is a real function on $[0,T] \times \mathbb{R}^d$. Let us also assume that the function ξ and the map O are continuous on $[0,T] \times \mathbb{R}^m$. Then, $\widetilde{I}_T(x) = \widehat{I}_T(x)$ for all $x \in \mathbb{R}^m$.*

Proof. It suffices to show that $\widehat{I}_T(x) \leq \widetilde{I}_T(x)$ for $x \in \mathbb{R}^m$ (see Lemma 34). Under the conditions in Lemma 35, we have

$$\Psi(y, f, \widehat{f}) = \int_0^T b(s, \widehat{f}(s))ds + \int_0^T \sigma(s, \widehat{f}(s))C\dot{f}(s)ds$$
$$+ \left\{\int_0^T \xi(s, \widehat{f}(s))^2 ds\right\}^{\frac{1}{2}} y. \qquad (45)$$

The equality in (45) can be established by using the fact that for any orthogonal matrix O, $\|O\|_* = 1$.

Let $x \in \mathbb{R}^m$, and suppose the equation $\Psi(y, f, \widehat{f}) = x$ is not solvable for y and f. Then $\widetilde{I}_T(x) = \infty$, and hence $\widehat{I}_T(x) \leq \widetilde{I}_T(x)$. Now, suppose the equation $\Psi(y, f, \widehat{f}) = x$ is solvable for y and f, and the condition $\int_0^T \xi(s, \widehat{f}(s))^2 ds = 0$ is satisfied. Then, we have $x = \int_0^T b(s, \widehat{f}(s))ds$. Set $g(t) = \int_0^t b(s, \widehat{f}(s))ds$. It is not hard to see that $g \in (\mathbb{H}_0^1)^m$, $g(T) = x$, and (43) holds for g, with the same function f and any function $l \in (\mathbb{H}_0^1)^m$. Let us next assume that the equation $\Psi(y, f, \widehat{f}) = x$ is solvable for y and f, and the condition $\int_0^T \xi(s, \widehat{f}(s))^2 ds > 0$ is satisfied. Then, there exists $l \in (\mathbb{H}_0^1)^m$

such that

$$\dot{l}(s) = \left\{ \int_0^T \xi(u, \widehat{f}(u))^2 du \right\}^{-\frac{1}{2}} \xi(s, \widehat{f}(s))O(s, \widehat{f}(s))'y \quad (46)$$

for all $s \in [0, T]$. It follows that

$$\int_0^T \sigma(s, \widehat{f}(s))\bar{C}\dot{l}(s)ds = \int_0^T \xi(s, \widehat{f}(s))O(s, \widehat{f}(s))\dot{l}(s)ds$$

$$= \left\{ \int_0^T \xi(s, \widehat{f}(s))^2 ds \right\}^{\frac{1}{2}} y. \quad (47)$$

Moreover, using (46) and the fact that for every orthogonal matrix O and $y \in \mathbb{R}^m$, the equality $\|Oy\|_m^2 = \|y\|_m^2$ holds, we obtain the equality $\int_0^T \|\dot{l}(s)\|_m^2 ds = \|y\|_m^2$. Set

$$g(t) = \int_0^t b(s, \widehat{f}(s))ds + \int_0^t \sigma(s, \widehat{f}(s))C\dot{f}(s)ds + \int_0^t \sigma(s, \widehat{f}(s))\bar{C}\dot{l}(s)ds$$

where the functions l and f are as above. Then we have $g \in (\mathbb{H}_0^1)^m$. Next, using (47) we obtain

$$g(T) = \int_0^T b(s, \widehat{f}(s))ds + \int_0^T \sigma(s, \widehat{f}(s))C\dot{f}(s)ds + \left\{ \int_0^T \xi(s, \widehat{f}(s))^2 ds \right\}^{\frac{1}{2}} y$$

$$= \Psi(y, f, \widehat{f}) = x.$$

In addition, the functions l and f solve the equation $\Phi(l, f, \widehat{f})(t) = g(t)$, $t \in [0, T]$. Finally, by taking into account the reasoning above, we see that $\widehat{I}_T(x) \leq \widetilde{I}_T(x)$ for all $x \in \mathbb{R}$.

This completes the proof of Lemma 35.

Remark 36. Suppose the conditions in Lemma 35 hold, and the function ξ satisfies $\xi(t, z) > 0$ for all $(t, z) \in [0, T] \times \mathbb{R}^d$. Then, we have

$$\widetilde{I}_T(x)$$
$$= \frac{1}{2} \inf_{f \in (\mathbb{H}_0^1)^m} \left[\frac{\|x - \int_0^T b(s, \widehat{f}(s))ds - \int_0^T \xi(s, \widehat{f}(s))O(s, \widehat{f}(s))\bar{C}^{-1}C\dot{f}(s)ds\|_m^2}{\int_0^T \xi(s, \widehat{f}(s))^2 ds} \right.$$
$$\left. + \int_0^T \|\dot{f}(t)\|_m^2 dt \right].$$

The previous formula can be established by taking into account (36), (45) and reasoning as in Remark 31.

Remark 37. Suppose the conditions in Remark 36 hold for the stochastic model considered in Lemma 35. Suppose also that the model is uncorrelated, that is, the condition $C = 0$ holds. Then, the process S satisfies the following equation:

$$dS_t = S_t \circ [b(t, \widehat{B}_t)dt + \xi(t, \widehat{B}_t)O(t, \widehat{B}_t)dW_t], \quad 0 \le t \le T, \quad S_0 = s_0 \in \mathbb{R}^m. \tag{48}$$

In addition, the rate function $\widetilde{I}_T$ in Remark 36 is given by

$$\widetilde{I}_T(x) = \frac{1}{2} \inf_{f \in (\mathbb{H}_0^1)^m} \left[\frac{||x - \int_0^T b(s, \widehat{f}(s))ds||_m^2}{\int_0^T \xi(s, \widehat{f}(s))^2 ds} + \int_0^T ||\dot{f}(t)||_m^2 dt \right]. \tag{49}$$

Note that although the model in (48) depends on the family O of orthogonal matrices used in Lemma 35, the rate function $\widetilde{I}_T$ given by the expression in (49) is independent of that family.

We next provide a special example of a model, for which the formula in Remark 36 holds true. Similar more complicated models can also be constructed. Let $d = 1$, $m = 2$, and choose $C = \begin{bmatrix} \frac{1}{2} & 0 \\ 0 & \frac{1}{2} \end{bmatrix}$. Then $\bar{C} = \begin{bmatrix} \frac{1}{2}\sqrt{3} & 0 \\ 0 & \frac{1}{2}\sqrt{3} \end{bmatrix}$ and $\bar{C}^{-1} = \begin{bmatrix} \frac{2\sqrt{3}}{3} & 0 \\ 0 & \frac{2\sqrt{3}}{3} \end{bmatrix}$. Let b be a drift map satisfying Assumption A, and suppose $\xi(t, z)$ is a real strictly positive ω-continuous function on $[0, T] \times \mathbb{R}$. Consider the following family of orthogonal 2×2-matrices: $O(z) = \begin{bmatrix} \cos z & -\sin z \\ \sin z & \cos z \end{bmatrix}$, where $z \in \mathbb{R}$, and set

$$\sigma(t, z) = \xi(t, z)O(z)\bar{C}^{-1}, \quad (t, z) \in [0, T] \times \mathbb{R}.$$

Then, the stochastic volatility model in (1) takes the following form:

$$dS_t = S_t \circ \left(b(t, \widehat{B}_t)ds + \xi(t, \widehat{B}_t) \begin{bmatrix} \cos \widehat{B}_t & -\sin \widehat{B}_t \\ \sin \widehat{B}_t & \cos \widehat{B}_t \end{bmatrix} \left(dW_t + \frac{\sqrt{3}}{3} dB_t \right) \right),$$

where $0 \leq t \leq T$ and $S_0 = s_0 \in \mathbb{R}^2$. Here we use the equality $\bar{C}^{-1}C = \begin{bmatrix} \frac{\sqrt{3}}{3} & 0 \\ 0 & \frac{\sqrt{3}}{3} \end{bmatrix}$. It is easy to see that the rate function in Remark 37 is given by

$$\widetilde{I}_T(x) = \frac{1}{2} \inf_{f \in (\mathbb{H}_0^1)^m}$$

$$\times \left\{ \frac{\|x - \int_0^T b(s, \widehat{f}(s))ds - \frac{\sqrt{3}}{3} \int_0^T \xi(s, \widehat{f}(s)) \begin{bmatrix} \cos \widehat{f}(s) & -\sin \widehat{f}(s) \\ \sin \widehat{f}(s) & \cos \widehat{f}(s) \end{bmatrix} \dot{f}(s)ds\|_m^2}{\int_0^T \xi(s, \widehat{f}(s))^2 ds} \right.$$

$$\left. + \int_0^T \|\dot{f}(t)\|_m^2 dt \right\}.$$

Large Deviation Principles for Volatility Processes

The main result of this section is Theorem 38 that provides a LDP for the solution to equation (10). Theorem 38 uses the canonical set-up.

Theorem 38. *Suppose Assumptions (C1)–(C7) hold, and let $Y^{(\varepsilon)}$ with $Y_0^{(\varepsilon)} = y$ be the process solving equation (10). Then, the process $Y^{(\varepsilon)}$ satisfies a sample path large deviation principle with speed ε^{-1} and good rate function defined on $\mathcal{W}^d$ by*

$$I_y(\varphi) = \inf_{\{f \in L^2([0,T], \mathbb{R}^m) : \Gamma_y(f) = \varphi\}} \frac{1}{2} \int_0^T \|f(t)\|_m^2 dt \qquad (50)$$

if $\{f \in L^2([0,T], \mathbb{R}^m) : \Gamma_y(f) = \varphi\} \neq \emptyset$, and $I_y(\varphi) = \infty$ otherwise.

Remark 39. The goodness of the rate function I_y can be shown as follows. Consider sublevel sets of I_y given by $L_c = \{\varphi \in \mathcal{W}^d : I_y(\varphi) \leq c\}$ for $c > 0$. Then, we have $L_c = \cap_{\varepsilon > 0} \Gamma_y(D_{2c+\varepsilon})$. Every set $D_{2c+\varepsilon}$ is compact in the weak topology of the space $L^2([0,T], \mathbb{R}^m)$. It follows from Assumption (C5) that the set $\Gamma_y(D_{2c+\varepsilon})$ is compact in the space $\mathcal{W}^d$. Therefore, the set L_c is compact in $\mathcal{W}^d$ since this set can be represented as the intersection of compacts sets.

Corollary 40. *Suppose Assumptions (C1)–(C7) hold, and let $\widehat{B}^\varepsilon$ be the volatility process (see Definition 8). Then, the process $\widehat{B}^\varepsilon$ satisfies a sample path LDP with speed ε^{-1} and good rate function given for $\varphi \in \mathcal{W}^d$ by*

$$J_y(\varphi) = \inf_{\{f \in L^2([0,T], \mathbb{R}^m) : \mathcal{A}f = \varphi\}} \frac{1}{2} \int_0^T \|f(t)\|_m^2 dt \qquad (51)$$

if $\{f \in L^2([0,T], \mathbb{R}^m) : \mathcal{A}f = \varphi\} \neq \emptyset$, and $J_y(\varphi) = \infty$ otherwise. In (51), $\mathcal{A}$ is the map introduced in Definition 23.

Corollary 40 follows from Theorem 38 and the contraction principle.

The following assertion can be derived from Theorem 38. It concerns a sample path large deviation principle for the process

$$\varepsilon \mapsto (\sqrt{\varepsilon}W, \sqrt{\varepsilon}B, \widehat{B}^\varepsilon), \quad \varepsilon \in (0, 1], \tag{52}$$

where W and B are independent m-dimensional Brownian motions appearing in (4), while $\widehat{B}^\varepsilon$ is the scaled volatility process (see Definition 8). The state space of the process in (52) is $\mathcal{W}^m \times \mathcal{W}^m \times \mathcal{W}^d$.

Theorem 41. *Under the restrictions in Theorem 38, the process in (52) satisfies a sample path large deviation principle with speed ε^{-1} and good rate function defined on $\mathcal{W}^m \times \mathcal{W}^m \times \mathcal{W}^d$ by*

$$\widetilde{I}_y(\varphi_1, \varphi_2, \varphi_3) = \frac{1}{2}\int_0^T \|\dot{\varphi}_1(t)\|_m^2 dt + \frac{1}{2}\int_0^T \|\dot{\varphi}_2(t)\|_m^2 dt \tag{53}$$

in the case where $\varphi_1, \varphi_2 \in (H_0^1)^m$ and $\varphi_3 = \widehat{\varphi_2}$, and by $\widetilde{I}_y(\varphi_1, \varphi_2, \varphi_3) = \infty$ otherwise.

Remark 42. Theorem 41 is an important ingredient in the proof of the LDP in Theorem 24 given in the following section.

Proof of Theorem 41. To derive Theorem 41 from Theorem 38, we use the following $(d+2m)$-dimensional system:

$$\begin{cases} G_t^\varepsilon = \sqrt{\varepsilon}W_t \\ Z_t^\varepsilon = \sqrt{\varepsilon}B_t \\ Y_t^\varepsilon = y + \int_0^t a(t,s,V^{1,\varepsilon},Y^\varepsilon)ds + \sqrt{\varepsilon}\int_0^t c(t,s,V^{2,\varepsilon},Y^\varepsilon)dB_s. \end{cases} \tag{54}$$

The third equation (54) is equation (10). In addition, W and B are independent m-dimensional standard Brownian motions appearing in (4). We can rewrite the model in (54) so that Theorem 38 can be applied to it. The new representation depends on $2m$-dimensional standard Brownian motion (W, B). We also use $2m$-dimensional deterministic controls $F = (f_1, f_2) \in L^2([0,T]; \mathbb{R}^m) \times L^2([0,T]; \mathbb{R}^m)$. It is not hard to see that the system in (23) becomes the following $(2m+d)$-dimensional system:

$$\begin{cases} \eta_1(t) = \int_0^t f_1(u)du \\ \eta_2(t) = \int_0^t f_2(u)du \\ \eta_3(t) = y + \int_0^t a(t,s,\psi_{1,f_2},\eta_3)ds + \int_0^t c(t,s,\psi_{2,f_2},\eta_3)f_2(s)ds. \end{cases}$$

The corresponding map $\widetilde{\Gamma}_y$ is given on $\mathcal{W}^m \times \mathcal{W}^m$ by

$$\widetilde{\Gamma}_y(F)(t) = \left(\int_0^t f_1(u)du, \int_0^t f_2(u)du, \Gamma_y(f_2)(t) \right), \quad t \in [0,T]$$

(see Definition 17).

Now let $\varphi_1, \varphi_2 \in (H_0^1)^m$ and $\varphi_3 \in \mathcal{W}^d$. Then, the equation $\widetilde{\Gamma}_y(F)(t) = (\varphi_1, \varphi_2, \varphi_3)$ has a unique solution given by $f_1 = \dot{\varphi}_1$ and $f_2 = \dot{\varphi}_2$. It follows that $\varphi_3 = \Gamma_y(\dot{\varphi}_2)$. Therefore, Theorem 38 shows that the process $\varepsilon \mapsto (\sqrt{\varepsilon}W, \sqrt{\varepsilon}B, Y^\varepsilon)$ satisfies the large deviation principle with speed ε^{-1} and good rate function $\widetilde{I}_y$ defined in (53).

Finally, we finish the proof of Theorem 41 by using the previous reasoning, Definition 23, and the contraction principle.

Our next goal is to discuss volatility models for which Theorems 38 and 41 hold on any set-up. Consider the following non-Volterra stochastic differential equation:

$$V_t^{(\varepsilon)} = y + \int_0^t a(s, V^{(\varepsilon)})ds + \sqrt{\varepsilon}\int_0^t c(s, V^{(\varepsilon)})dB_s, \tag{55}$$

where a is a map from the space $[0,T] \times \mathcal{W}^d$ into the space $\mathbb{R}^d$, while c is a map from the space $[0,T] \times \mathcal{W}^d$ into the space of $(d \times m)$-matrices. Let us assume that the maps a and c are locally Lipschitz and satisfy the sublinear growth condition (see the definitions in [17], A1 and A2 in Section 3, or in [71], (12.2) and (12.3) on p. 132). A sample path LDP for the process $\varepsilon \mapsto V_\cdot^{(\varepsilon)}$ was established on the canonical set-up in [17], Theorem 3.1.

Theorem 43. *Under the restrictions formulated above, the LDP in [17], Theorem 3.1 holds on any set-up* $(\Omega, \mathcal{B}, \mathcal{F}_T^B, \{\mathcal{F}_t^B\}_{0 \leq t \leq T}, \mathbb{P})$ *(see Definition 1).*

Corollary 44. *Suppose the conditions formulated above are satisfied for equation (55) defined on a general set-up* $(\Omega, \mathcal{B}, \mathcal{F}_T^B, \{\mathcal{F}_t^B\}_{0 \leq t \leq T}, \mathbb{P})$. *Then, the LDP in Theorem 38 holds for the process* $\varepsilon \mapsto V^{(\varepsilon)}$ *solving equation (55), while the LDP in Theorem 41 holds for the process* $\varepsilon \mapsto (\sqrt{\varepsilon}W, \sqrt{\varepsilon}B, V^{(\varepsilon)})$.

Proof of Theorem 43. We have already mentioned that Theorem 43 holds on the canonical set-up [17]. It suffices to show that for every $\varepsilon \in (0,T]$ and

$y \in \mathbb{R}^d$, the distribution of the random variable $V_{\cdot}^{(\varepsilon)}$, with values in the space $\mathcal{W}^d$, does not depend on the set-up. By taking into account Theorem 12.1 on p. 132 in [71], we establish that equation (55) is pathwise exact (see Definition 9.4 on p. 124 in [71]). Next, we can use Theorem 10.4 in [71] to prove that for any $\varepsilon \in (0,1]$ and any fixed initial condition $y \in \mathbb{R}^d$, there exists a measurable functional $F_y^{(\varepsilon)} : \mathcal{W}^m \mapsto \mathcal{W}^d$ such that $F_y^{(\varepsilon)}(B) = Y^\varepsilon$. The functional $F_y^{(\varepsilon)}$ does not depend on the set-up. Finally, using the previous equality and the fact that the distribution of Brownian motion B with respect to the measure $\mathbb{P}$ is the same for all set-ups, we see that the distribution of Y^ε in $\mathcal{W}^d$ does not depend on the set-up. It follows that since the LDPs in Theorems 38 and 41 hold for the processes $\varepsilon \mapsto V_{\cdot}^{(\varepsilon)}$ and $\varepsilon \mapsto (\sqrt{\varepsilon}W_{\cdot}, \sqrt{\varepsilon}B_{\cdot}, V_{\cdot}^{(\varepsilon)})$ defined on the canonical set-up, they also hold on any set-up.

This completes the proof of Theorem 43.

Remark 45. The fact that the distribution of the solution does not depend on the set-up was established in [76] for less general equations than that in (55) and under stronger restrictions on the coefficient maps a and c (see the equation in formula (1.3) and Corollary 5.1.3 in [76]).

Unification of Sample Path Large Deviation Principles for Stochastic Volatility Models

Our aim in this section is to provide examples of log-price processes and volatility processes to which the LDPs in Theorems 24 and 41 can be applied. This section is divided into several subsections. It is organized as follows. In sections "Gaussian Stochastic Volatility Models" and "Non-Gaussian Fractional Stochastic Volatility Models", we overview one-factor Gaussian models studied in [41] and one-factor non-Gaussian fractional models introduced in [38]. These subsections are auxiliary. They provide examples of Volterra-type kernels and processes used in section "Unification: Mixed Models" devoted to mixtures of multivariate Gaussian models and multivariate non-Gaussian fractional models. Theorem 47 obtained in section "Unification: Mixed Models" is one of the main results in this paper. It follows from Theorem 47 that the LDPs in Theorems 24 and 38 can be applied to log-price processes and volatility processes in mixed models (see Remark 48) defined on the canonical set-up. In section "LDPs for Multivariate Gaussian Models and Multivariate Non-Gaussian Fractional Models", we show that Theorem 24 holds for the log-price process associated with a multivariate

Gaussian model defined on any set-up (see Theorem 49). This theorem is more general than the LDP for multivariate Gaussian models obtained in [16]. Section "Unification: Models with Reflection" concerns stochastic volatility models with reflection. Finally, in sections "Unification: Volterra Type SDEs" and "Unification: More Volterra-Type SDEs", we discuss the LDPs obtained in [83] and [66].

Gaussian Stochastic Volatility Models: In this subsection, we discuss Gaussian stochastic volatility models studied in [41]. Let K be a real function on $[0, T]^2$. We call the function K an admissible Hilbert–Schmidt kernel if the following conditions hold: (a) K is Borel measurable on $[0, T]^2$, (b) K is Lebesgue square-integrable over $[0, T]^2$, (c) for every $t \in (0, T]$, the slice function $s \mapsto K(t, s)$, with $s \in [0, T]$, belongs to the space $L^2[0, T]$, and (d) for every $t \in (0, T]$, the slice function is not almost everywhere zero. If an admissible kernel K satisfies the condition $K(t, s) = 0$ for all $s > t$, then K is called an admissible Volterra kernel. Any such kernel K generates a Hilbert–Schmidt operator

$$\mathcal{K}(f)(t) = \int_0^t K(t, s) f(s) ds, \quad f \in L^2[0, T], \quad t \in [0, T], \tag{56}$$

and a Volterra Gaussian process

$$\widehat{B}_t = \int_0^t K(t, s) dB_s, \quad t \in [0, T]. \tag{57}$$

It is clear that the process in (57) is adapted to the filtration $\{\mathcal{F}_t^B\}_{0 \leq t \leq T}$. This process is used as the volatility process in a Gaussian model. The scaled volatility process is defined as follows: $\widehat{B}_t^{(\varepsilon)} = \sqrt{\varepsilon} \widehat{B}_t$ for $t \in [0, T]$.

In this paper, only continuous volatility processes are used. We next formulate Fernique's condition guaranteeing that the process in (57) is a continuous Gaussian process. Let X_t, $t \in [0, T]$, be a square integrable stochastic process on $(\Omega, \mathcal{F}, \mathbb{P})$. The canonical pseudo-metric δ associated with this process is defined by the formula $\delta^2(t, s) = \mathbb{E}[(X_t - X_s)^2]$ for $(t, s) \in [0, T]^2$. Suppose η is a modulus of continuity on $[0, T]$ such that $\delta(t, s) \leq \eta(|t - s|)$ for $t, s \in [0, T]$). Suppose also that for some $b > 1$, the following inequality holds:

$$\int_b^\infty \eta\left(u^{-1}\right) (\log u)^{-\frac{1}{2}} \frac{du}{u} < \infty. \tag{58}$$

It was announced by Fernique [29] that a Gaussian process X satisfying the previous condition is a continuous stochastic process. The first proof was published by Dudley [25] (see also [63] and the references therein). By the Itô isometry, the following equality holds for the process $\widehat{B}$:

$\delta^2(t,s) = \int_0^T (K(t,u) - K(s,u))^2 du$, $t, s \in [0, T]$. The L^2-modulus of continuity of the kernel K is defined on $[0, T]$ by

$$M_K(\tau) = \sup_{t,s \in [0,T]: |t-s| \leq \tau} \int_0^T (K(t,u) - K(s,u))^2 du$$

for all $\tau \in [0, T]$.

Assumption F. The kernel K in (57) is an admissible Volterra kernel such that the estimate $M_K(\tau) \leq \eta^2(\tau)$, $\tau \in [0, T]$, holds for some modulus of continuity η satisfying Fernique's condition.

Under Assumption F, the process $\widehat{B}$ defined by (57) is a continuous Gaussian process. Moreover, the operator $\mathcal{K}$ introduced in (56) is compact from the space $L^2[0, T]$ into the space $\mathbb{C}[0, T]$. Important examples of Volterra Gaussian processes are classical fractional processes, e.g. fractional Brownian motion or the Riemann–Liouville fractional Brownian motion. For $0 < H < 1$, fractional Brownian motion B^H is a centered Gaussian process with the covariance function given by $C_H(t,s) = \frac{1}{2}(t^{2H} + s^{2H} - |t-s|^{2H})$, $t, s \geq 0$. The process B^H was first implicitly considered by Kolmogorov in [57] and was studied by Mandelbrot and van Ness in [62]. The constant H is called the Hurst parameter. Fractional Brownian motion is a Volterra type process. This was established by Molchan and Golosov (see [65] and also [21]). The corresponding Volterra kernel K_H is as follows: for $\frac{1}{2} < H < 1$,

$$K_H(t,s) = \sqrt{\frac{H(2H-1)}{\int_0^1 (1-x)^{1-2H} x^{H-\frac{3}{2}} dx}} s^{\frac{1}{2}-H} \int_s^t (u-s)^{H-\frac{3}{2}} u^{H-\frac{1}{2}} du,$$
$$\times 0 < s < t,$$

while for $0 < H < \frac{1}{2}$,

$$K_H(t,s) = \sqrt{\frac{2H}{(1-2H) \int_0^1 (1-x)^{-2H} x^{H-\frac{1}{2}} dx}} \left[\left(\frac{t}{s}\right)^{H-\frac{1}{2}} (t-s)^{H-\frac{1}{2}} \right.$$
$$\left. - \left(H - \frac{1}{2}\right) s^{\frac{1}{2}-H} \int_s^t u^{H-\frac{3}{2}} (u-s)^{H-\frac{1}{2}} du \right], \quad 0 < s < t.$$

An equivalent representation of the kernel K_H is the following:

$$K_H(t,s) = C_H(t-s)^{H-\frac{1}{2}} \left(\frac{s}{t}\right)^{\frac{1}{2}-H} {}_2F_1\left(\frac{1}{2}-H, 1, H+\frac{1}{2}, \frac{t-s}{t}\right),$$

where ${}_2F_1$ is the Gauss hypergeometric function (see, e.g. (3.11) in [53]).

The Riemann–Liouville fractional Brownian motion is defined by the following formula:

$$R_t^H = \Gamma(H + 1/2)^{-1} \int_0^t (t-s)^{H-\frac{1}{2}} dB_s, \quad t \geq 0, \tag{59}$$

where $0 < H < 1$, and the symbol Γ stands for the gamma function. This stochastic process was introduced by Lévy [58]. More information about the process R^H can be found in [61,68]. Fractional Brownian motion and the Riemann–Liouville fractional Brownian motion are continuous stochastic processes. In the case, where $0 < H < \frac{1}{2}$, they are called rough processes since their paths are more rough than those of standard Brownian motion. In [41], we introduced a new class of Gaussian stochastic volatility models (the class of super rough models). Super rough models were also considered in [6]. In a super rough model, the modulus of continuity associated with the volatility process $\widehat{B}$ grows near zero faster than any power function. Interesting examples of super rough processes can be obtained using Gaussian processes defined by Mocioalca and Viens [64]. It was established in [64] that if $\eta \in \mathbb{C}^2(0,T)$ is a modulus of continuity on $[0,T]$ such that the function $x \mapsto (\eta^2)'(x)$ is positive and non-increasing on $(0,T)$, then the process $\widehat{B}_t^{(\eta)} = \int_0^t \tau(t-s) dB_s$, $t \in [0,T]$, with $\tau(x) = \sqrt{(\eta^2)'(x)}$, is a Gaussian process satisfying the following conditions: (i) $c_1 \eta(|t-s|) \leq \delta(t,s) \leq c_2 \eta(|t-s|)$ for some $c_1, c_2 > 0$, (ii) $X_0 = 0$, and (iii) the process X is adapted to the filtration $\{\mathcal{F}_t^B\}_{0 \leq t \leq T}$.

A typical example of a modulus of continuity that grows near zero faster than any power is the logarithmic modulus of continuity given by

$$\eta_\beta(x) = \left(\log \frac{1}{x}\right)^{-\frac{\beta}{2}}, \quad 0 \leq x < 1, \quad \beta > 0.$$

It is clear that in this case, the function τ_β is determined from the equality

$$\tau_\beta^2(x) = \beta x^{-1} \left(\log \frac{1}{x}\right)^{-\beta-1}, \quad 0 \leq x < 1.$$

In [64], the Volterra Gaussian process with the kernel τ_β was called logarithmic Brownian motion. If $\beta > 1$, then Fernique's condition is satisfied and the process is continuous. In [41], we called a Gaussian stochastic volatility model, in which the logarithmic Brownian motion with $\beta > 1$ is the volatility process, a logarithmic Gaussian stochastic volatility model. The logarithmic model is super rough (more details can be found in [41]). An interesting example of a super rough Gaussian model is the model where the volatility is described by the Wick exponential of a constant multiple of the logarithmic Brownian motion (see [41]). The previous model is similar in structure to the rough Bergomi model introduced in [5], and it may be called the super rough Bergomi model. More details can be found in [41] and [6]. A celebrated Stein and Stein model (see [73] and also the discussion in [42]) was one of the first examples of a Gaussian model. For a super rough version of the Stein and Stein model, see [41].

Non-Gaussian Fractional Stochastic Volatility Models. This class of one-factor stochastic volatility models was studied in the paper of Gerhold et al. [38]. The volatility process in such a model is given by $\hat{B}_t = \int_0^t K(t,s) U(V_s) ds$, where $U : \mathbb{R} \mapsto [0, \infty)$ is a continuous non-negative function and K is an admissible kernel with the modulus of continuity in L^2 satisfying the Hölder condition. The process V in the formula above is the unique solution to the stochastic differential equation

$$dV_t = \bar{b}(V_t)dt + \bar{\sigma}(V_t)dB_t, \quad t \in [0,T],$$

where $V_0 = v_0 > 0$. It is assumed that the following conditions hold (see Section 4.2 in [17] and also [38]):

(i) The function $\bar{b} : \mathbb{R} \mapsto \mathbb{R}$ is locally Lipschitz on $\mathbb{R}$, satisfies the sublinear growth condition, and $\bar{b}(0) > 0$.
(ii) The function $\bar{\sigma} : \mathbb{R} \mapsto [0,\infty)$ is locally Lipschitz continuous on $\mathbb{R} - \{0\}$, satisfies the sublinear growth condition, $\bar{\sigma}(0) = 0$, and $\bar{\sigma}(x) \neq 0$ for all $x \neq 0$.
(iii) The function $\bar{\sigma}$ satisfies the Yamada–Watanabe condition.

A well-known example of the process V is the CIR process for which $\bar{b}(x) = a_1 - a_2 x$ and $\bar{\sigma}(x) = a_3\sqrt{x}$. It is assumed in [38] that the model in (4) does not have the drift term ($b = 0$), while the volatility function σ is time-homogeneous. The drift-less fractional Heston models studied in [1,18,40] are special cases of the models described above. In fractional Heston models, the volatility is the fractional integral operator applied to the CIR process. A different generalization of the Heston model (a rough Heston model) is due to El Euch and Rosenbaum (see [27]). In the rough Heston model, the fractional integral operator is applied to the CIR equation, and not to the CIR process. We do not know whether the LDP in Theorem 24 holds the log-price process in the rough Heston model.

For a small-noise parameter ε, we define the scaled version $V^{(\varepsilon)}$ of the process V as the solution to the following equation: $dV_t^{(\varepsilon)} = \bar{b}(V_t^{(\varepsilon)})dt + \sqrt{\varepsilon}\bar{\sigma}(V_t^{(\varepsilon)})dB_t$, with the initial condition given by $V_0^{(\varepsilon)} = v_0 > 0$. The scaled volatility process in the non-Gaussian fractional model is given by

$\widehat{B}_t^{(\varepsilon)} = \int_0^t K(t,s)U(V_s^{(\varepsilon)})ds$, while the scaled log-price process is as follows:

$$X_t^{(\varepsilon)} = -\frac{1}{2}\varepsilon \int_0^t \sigma(\widehat{B}_t^{(\varepsilon)})^2 dt + \sqrt{\varepsilon}\int_0^t \sigma(\widehat{B}_t^{(\varepsilon)})(\bar{\rho}dW_t + \rho dB_t).$$

In addition, the map $f \mapsto \widehat{f}$ is defined by $\widehat{f}(t) = \int_0^t K(t,s)U(\varphi_f(s))ds$, where $t \in [0,T]$, $f \in \mathbb{H}_0^1$, and φ_f is the unique solution to the ODE $\dot{v} = \bar{b}(v) + \bar{\sigma}(v)\dot{f}$, $f \in \mathbb{H}_0^1[0,T]$.

Unification: Mixed Models. In this section, we introduce a new class of volatility models. A model belonging to this class may be called a mixture of a multivariate Gaussian stochastic volatility model and a multivariate non-Gaussian fractional model.

Let K_i, with $0 \leq i \leq d$, and $\{K_{ij}\}$, with $1 \leq i \leq d$ and $1 \leq j \leq m$, be families of admissible Volterra-type Hilbert–Schmidt kernels such that Assumption F holds for them. Define an $(d \times k)$-matrix by $\mathcal{K} = (K_{ij})$. Suppose that V is an auxiliary k-dimensional continuous process defined on the space $\mathcal{W}^m$ equipped with the canonical set-up. Suppose also that Conditions (H1)–(H6) in [17] are satisfied for the equation defining the process V (see (9) and Remark 7). Let U be a continuous map from $\mathbb{R}^k$ into $\mathbb{R}^d$, and consider the following stochastic model for the volatility process: $Y_t = (Y_t^{(1)}, \ldots, Y_t^{(d)})$ where $t \in [0,T]$ and

$$Y_t^{(i)} = x_i + \int_0^t K_i(t,s)U_i(V_s)ds + \sum_{j=1}^m \int_0^t K_{ij}(t,s)dB_s^{(j)}, \quad 1 \leq i \leq d. \quad (60)$$

The volatility model introduced in (60) is a special example of the models described in (8). Indeed, we can assume that the map a in (8) does not depend on the fourth variable and its components are given by $a_i(t,s,u,v) = K_i(t,s)U_i(u)$, where $1 \leq i \leq d$, $t,s \in [0,T]$, and $u \in \mathbb{R}^k$. We can also assume that the matrix function c does not depend on the third and fourth variables, and its elements are defined by $c_{ij}(t,s,u,v) = K_{ij}(t,s)$, where $t,s \in [0,T]$, $1 \leq i \leq d$, and $1 \leq j \leq m$. The scaled version of the process in (60) is given by

$$Y_t^{i,\varepsilon} = x_i + \int_0^t K_i(t,s)U_i(V_s^{(\varepsilon)})ds + \sqrt{\varepsilon}\sum_{j=1}^m \int_0^t K_{ij}(t,s)dB_s^{(j)},$$

$$1 \leq i \leq d. \quad (61)$$

The model for the volatility process described in (60) is more general than the models considered in [38,41]. Unlike the latter models, the model in (60) is multidimensional, the restrictions on V and U are weaker than those in [38], and the volatility process in (60) is a mixture of volatility processes in the above-mentioned models.

Remark 46. By assuming that $U = 0$ in (61), we obtain the scaled volatility process in a multivariate Gaussian stochastic volatility model. This process is defined by

$$\widetilde{Y}_t^{i,\varepsilon} = x_i + \sqrt{\varepsilon} \sum_{j=1}^{m} \int_0^t K_{ij}(t,s) dB_s^{(j)}, \quad 1 \le i \le d. \tag{62}$$

On the other hand, the scaled volatility process in a multivariate non-Gaussian fractional model can be obtained from (60) by setting $K_{ij} = 0$ for all $1 \le i \le d$ and $1 \le j \le m$. This process is given by

$$\widehat{Y}_t^{i,\varepsilon} = x_i + \int_0^t K_i(t,s) U_i(V_s^{(\varepsilon)}) ds, \quad 1 \le i \le d. \tag{63}$$

Theorem 47. *Assumptions (C1)–(C7) hold true for the mixed volatility model introduced in (60).*

Proof. It is not hard to see that Assumptions (C1) and (C2) are satisfied for the model in (60). Assumption (C3) is satisfied as well. Note that for every $\varepsilon \in (0,1]$, we have an equality in (61), and not an equation. The process $Y^{(\varepsilon)}$ defined in (61) is continuous. The previous statement follows from the continuity of the function U and the process $V^{(\varepsilon)}$ and from Assumption F. Next, using (61) and the fact that pathwise uniqueness and existence in the strong sense hold for the equation defining the process V (see Condition (H3) in [17]), we derive Assumption (C3).

The validity of Assumption (C4) can be checked as follows. Let $f \in L^2([0,T], \mathbb{R}^m)$. Then, equation (23) becomes the following equality:

$$\eta_f^{(i)}(t) = x_i + \int_0^t K_i(t,s) U_i(\psi_f(s)) ds + \sum_{j=1}^{m} \int_0^t K_{ij}(t,s) f_j(s) ds,$$

$$1 \le i \le d, \tag{64}$$

where ψ_f solves the equation $\psi(s) = v_0 + \int_0^s \bar{b}(r,\psi) dr + \int_0^s \bar{\sigma}(r,\psi) f(r) dr$. The latter equation is uniquely solvable and the solution ψ_f is in the space $\mathcal{W}^k$. The previous statement follows from the results obtained in [17]. The equality in (64) implies the validity of Assumption (C4). We also have $\Gamma_x f = \eta_f$. Here η_f is given by (64).

We next turn our attention to Assumption (C5). Suppose that $f_n \in L^2([0,T], \mathbb{R}^m)$, with $n \ge 1$, is a sequence of control functions such that

$f_n \mapsto f$ weakly in $L^2([0,T], \mathbb{R}^m)$. We have from (64) that

$$\eta_{f_n}^{(i)}(t) = x_i + \int_0^t K_i(t,s) U_i(\psi_{f_n}(s)) ds + \sum_{j=1}^m \int_0^t K_{ij}(t,s) f_j^{(n)}(s) ds,$$

$$1 \leq i \leq d. \tag{65}$$

It is shown next that for every $1 \leq i \leq d$, $\eta_{f_n}^{(i)} \mapsto \eta_f^{(i)}$ in $\mathcal{W}^1$ as $n \to \infty$. First, observe that since the kernels appearing in (65) are admissible, and, moreover, the Hilbert–Schmidt operators appearing in (64) and (65) are compact maps from $L^2([0,T], \mathbb{R}^1)$ into $\mathcal{W}^1$, the last term on the right-hand side of (65) tends to the last term on the right-hand-side of (64) as $n \to \infty$. We can also prove that the same conclusion holds for the second terms on the right-hand sides of (65) and (64) by using the fact that $\psi_{f_n} \to \psi_f$ in $\mathcal{W}^k$ (this follows from the results obtained in [17]). Hence, for all $1 \leq i \leq d$, $U_i(\psi_{f_n}) \mapsto U_i(\psi_f)$ in $\mathcal{W}^1$ as $n \to \infty$. Therefore, Assumption (C5) is satisfied.

The validity of Assumption (C6) follows from the following: (a) Every kernel K_{ij} is admissible and satisfies Fernique's condition and (b) the process $s \mapsto B_s^v$ is standard Brownian motion with respect to the measure $\mathbb{P}^v$.

Finally, we prove that Assumption (C7) holds. Let $\varepsilon_n \to 0$ as $n \to \infty$, and let $v^{(n)} \in \mathcal{M}^2[0,T]$, with $n \geq 1$, be such that for some $N > 0$,

$$\sup_{n \geq 1} \int_0^T \|v_s^{(n)}\|_m^2 ds \leq N \tag{66}$$

$\mathbb{P}$-a.s. Our first goal is to prove that the sequence $n \mapsto Y^{n,v^n}$ is tight in $\mathcal{W}^d$. It suffices to show that the sequences of components $n \mapsto Y^{i,n,v^n}$, with $1 \leq i \leq d$, are tight in $\mathcal{W}^1$. For every $1 \leq i \leq d$, we have

$$Y_t^{i,n,v^n} = x_i + \int_0^t K_i(t,s) U_i(V_s^{\varepsilon_n, v^n}) ds + \sum_{j=1}^m \int_0^t K_{ij}(t,s) v_s^{j,n} ds$$

$$+ \sqrt{\varepsilon_n} \sum_{j=1}^m \int_0^t K_{ij}(t,s) dB_s^{(j)}. \tag{67}$$

In the rest of the proof, we use the fact that for a finite number of tight sequences $A_n^{(k)}$, with $1 \leq k \leq m$ and $n \geq 1$, of random elements in a normed space, the sum $\sum_{k=1}^m A_n^{(k)}$ is a tight family of random elements.

For a fixed index $1 \leq i \leq d$, consider the sequence Y^{i,n,v^n}, $n \geq 1$, of random elements in $\mathcal{W}^1$. Our goal is to prove that this sequence is tight. By taking into account what was said above, we see that it suffices to show that the following sequences of random elements in $\mathcal{W}^1$ are tight:

$J_{1,n}^{(i)}(t) = \int_0^t K_i(t,s) U_i(V_s^{\varepsilon_n, v^n}) ds$, $J_{2,n}^{(i)}(t) = \sum_{j=1}^m \int_0^t K_{ij}(t,s) v_s^{j,n} ds$, and $J_{3,n}^{(i)}(t) = \sqrt{\varepsilon_n} \sum_{j=1}^m \int_0^t K_{ij}(t,s) dB_s^{(j)}$.

Let us begin with the sequence $n \mapsto J_{1,n}^{(i)}$. Using the fact that the sequence of random elements $n \mapsto V^{\varepsilon_n, v^n}$ is tight in $\mathcal{W}^k$ (Condition (H6) in [17]), and U_i is a continuous map from $\mathbb{R}^k$ into $\mathbb{R}$, we see that the sequence $n \mapsto U_i(V^{\varepsilon_n, v^n})$ is tight in $\mathcal{W}^1$. The latter statement follows from Prokhorov's theorem and Corollary 3 on p. 9 in [9]. Therefore, for every $\varepsilon > 0$ there exists a compact set $C_1^{(i)}$ in $\mathcal{W}^1$ such that $\mathbb{P}(U_i(V^{\varepsilon_n, v^n}) \in C_1^{(i)}) \geq 1 - \varepsilon$ for all $n \geq 1$. Denote by $C_2^{(i)}$ the image of $C_1^{(i)}$ by the Hilbert–Schmidt operator with the kernel K_i. Then, $C_2^{(i)}$ is a compact subset of $\mathcal{W}^1$, by the compactness of the above-mentioned operator. Therefore, $\mathbb{P}(J_{1,n}^{(i)}(\cdot) \in C_2^{(i)}) \geq \mathbb{P}(U_i(V^{\varepsilon_n, v^n}) \in C_1^{(i)}) \geq 1 - \varepsilon$. The previous estimates show that the sequence $n \mapsto J_{1,n}^{(i)}$ is tight in $\mathcal{W}^1$.

Let us next consider the sequence $n \mapsto J_{2,n}^{(i)}$. Since we assumed that (66) holds, there exists a set $\widetilde{\Omega} \subset \Omega$ of full measure such that for every $1 \leq j \leq m$, the family $v_{\cdot}^{j,n}(\omega)$, with $n \geq 1$ and $\omega \in \widetilde{\Omega}$, is uniformly bounded in $L^2[0,T]$. Therefore, for every $1 \leq i \leq d$, the image of this family with respect to the Hilbert–Schmidt operator with the kernel K_{ij} is a compact subset of $\mathcal{W}^1$. Now, it easily follows that the sequence of random elements $n \mapsto J_{2,n}^{(i)}$ is tight in $\mathcal{W}^1$.

Finally, we turn our attention to the sequence $n \mapsto J_{3,n}^{(i)}$. The sum appearing in the definition of $J_{3,n}^{(i)}$ is a continuous stochastic process on $\Omega = \mathcal{W}^m$ with state space $\mathcal{W}^1$. Therefore, the sequence of random elements $n \mapsto J_{3,n}^{(i)}(\cdot)$ on Ω with values in $\mathcal{W}^1$ converges in $\mathcal{W}^1$ to the identically zero random element. Let g be a bounded continuous real function on $\mathcal{W}^1$. Then $\mathbb{E}[g(J_{3,n}^{(i)}(\cdot))] = 0$ by the bounded convergence theorem, and hence, the sequence of random elements $n \mapsto J_{3,n}^{(i)}$ is weakly convergent. It follows from Prokhorov's theorem that this sequence is tight in $\mathcal{W}^1$.

Summarizing what was said above, we conclude that for every $1 \leq i \leq d$, the sequence of random elements $n \mapsto Y^{i,n,v^n}$ defined in (67) is tight in $\mathcal{W}^1$. This establishes the first part of Assumption (C7). It remains to show that the second part of Assumption (C7) holds. This means that we must prove the estimate in (24). In our special case, this estimate reduces to the following:

$$\sup_{t \in [0,T]} \sum_{i=1}^d \sum_{j=1}^m \int_0^t K_{ij}(t,s)^2 ds < \infty.$$

To obtain the previous inequality, it suffices to show that

$$\sup_{t\in[0,T]} \int_0^t K_{ij}(t,s)^2 ds < \infty \qquad (68)$$

for all $1 \leq i \leq d$ and $1 \leq j \leq m$. The inequality in (68) follows from the fact that the function $t \mapsto \int_0^t K_{ij}(t,s)^2 ds$, with $t \in [0,T]$, $1 \leq i \leq d$, and $1 \leq j \leq m$, is the variance function of the continuous Gaussian process $t \mapsto \int_0^t K_{ij}(t,s) dB_s^j$, and the variance function of such a process is continuous. Therefore, Assumption (C7) is satisfied.

This completes the proof of Theorem 47.

Remark 48. Since Assumptions (C1)–(C7) hold for the model in (60), Theorem 38 can be applied to the scaled volatility process defined by $\widehat{B}^{(\varepsilon)} = Y^{(\varepsilon)}$ (see (61)), while Theorem 41 holds for the process $(\sqrt{\varepsilon}W, \sqrt{\varepsilon}B, \widehat{B}^{(\varepsilon)})$, provided that canonical set-up is used. Moreover, the LDP in Theorem 24 is valid for the stochastic volatility model in (1), with the process $\widehat{B} = Y$ as the volatility process, if the model is defined on the space $\Omega = \mathcal{W}^m \times \mathcal{W}^m$ equipped with the $2m$-dimensional canonical set-up. As corollaries, we obtain sample path LDPs for multivariate Gaussian models and multivariate non-Gaussian fractional models on the canonical set-up. It is shown in the following subsection that for multivariate Gaussian models and for certain multivariate non-Gaussian fractional models, Theorems 24, 38 and 41 hold on any set-up.

LDPs for Multivariate Gaussian Models and Multivariate Non-Gaussian Fractional Models: Recall that the scaled volatility process in a multivariate Gaussian stochastic volatility model is given by $\widehat{B}^{(\varepsilon)} = \widetilde{Y}^{(\varepsilon)}$ where the latter process is defined in (62). In this subsection, we prove the following assertion.

Theorem 49. *The LDP in Theorem 24 holds for a multivariate Gaussian model defined on any set-up.*

Proof. For one-factor Gaussian models, Theorem 49 was established in [41], Theorem 4.2. We only sketch the proof of Theorem 49 for multivariate models and leave filling in the necessary details to the interested reader. The first step in the proof of Theorem 49 is to establish that the LDP in Theorem 41 holds on any set-up. Then, we can derive Theorem 49 by observing that the proof of Theorem 24 given in section "Proof of Theorem 24" uses only the set-up utilized in Theorem 41.

Let us take $U = 0$ in (60). Then, the expression in (60) represents the volatility vector $\widehat{B}$ in the multivariate Gaussian model. Consider the random vector $\mathcal{X} = (W, B, \widehat{B})$ on Ω with values in the space $\Lambda = \mathcal{W}^m \times \mathcal{W}^m \times \mathcal{W}^d$. Reasoning as in the proof of Theorem 6.8 in [41], we can show that $\mathcal{X}$ is a centered Gaussian random vector. Actually, $\mathcal{X}$ is a Gaussian random vector in a smaller space $\widetilde{\mathcal{G}}$. This is explained next (see also the proof on p. 63 of [41]). Let $H = L^2([0,T], \mathbb{R}^m) \times L^2([0,T], \mathbb{R}^m)$. Then H is a separable Hilbert space equipped with the norm

$$||(h_1, h_2)||_H = \sqrt{||h_1||^2_{L^2([0,T],\mathbb{R}^m)} + ||h_2||^2_{L^2([0,T],\mathbb{R}^m)}}.$$

Let $j : H \mapsto \Lambda$ be the map defined by $j(h_0, h_1) = (g_0, g_1, g_2)$ where for every $t \in [0,T]$, $g_0(t) = \int_0^t h_0(s)ds$, $g_1(t) = \int_0^t h_1(s)ds$, and $g_2(t) = \int_0^t K(t,s)h_2(s)ds$. The map j is an injection. Set $\widetilde{\mathcal{G}} = \overline{j(H)}$ where the closure is taken in the space Λ. Then, $\widetilde{\mathcal{G}}$ is a separable Banach space. It was established in [41] for one-dimensional Gaussian models that $\mathcal{X}$ is a Gaussian random vector in the space $\widetilde{\mathcal{G}}$ (see Theorem 6.11 in [41]). It is not hard to see that the same result holds in the multivariate case. We can also find the covariance operator $\widehat{K}$ by imitating the proof of Theorem 6.14 in [41]. Let ζ be the distribution of the random vector $\mathcal{X}$ on the measurable space $(\widetilde{\mathcal{G}}, \mathcal{B}(\widetilde{\mathcal{G}}))$. Then, ζ is a Gaussian measure. Using the same ideas as in the proof of Theorem 6.19 in [41], we can prove that the quadruple $(\widetilde{\mathcal{G}}, H, j, \zeta)$ is an abstract Wiener space. Finally, applying the known LDP for abstract Wiener spaces (see the references before Theorem 6.19 in [41]), we establish the validity of the LDP in Theorem 41 on any set-up.

This completes the proof of Theorem 49.

We next turn our attention to multivariate non-Gaussian fractional models. The volatility process in such a model is given by $\widehat{B}^{(\varepsilon)} = \widehat{Y}^{(\varepsilon)}$ where the latter process is defined in (62). Recall that the process $V^{(\varepsilon)}$ appearing in (62) is as follows:

$$V_t^{(\varepsilon)} = y + \int_0^t a(s, V^{(\varepsilon)})ds + \sqrt{\varepsilon} \int_0^t c(s, V^{(\varepsilon)})dB_s, \qquad (69)$$

where a is a map from the space $[0,T] \times \mathcal{W}^d$ into the space $\mathbb{R}^d$, while c is a map from the space $[0,T] \times \mathcal{W}^d$ into the space of $(d \times m)$-matrices. It is assumed in the rest of this subsection that the maps a and c are locally Lipschitz and satisfy the sublinear growth condition.

Theorem 50. *Under the conditions formulated above, the LDP in Theorem 24 holds for a multivariate Gaussian model defined on any set-up.*

Proof. It was shown in Corollary 44 that if the maps a and c are locally Lipschitz and satisfy the sublinear growth condition, then the LDP in Theorem 41 holds for the process $\varepsilon \mapsto (\sqrt{\varepsilon}W, \sqrt{\varepsilon}B, V^{(\varepsilon)}_\cdot)$ defined on any set-up. Therefore, the LDP in Theorem 24 holds true on any set-up (see Remark 25).

The proof of Theorem 50 is thus completed.

Unification: Models with Reflection. Large deviation principles for log-price processes and volatility processes in one-factor stochastic volatility models with reflection were obtained in [42]. We next establish similar LDPs for multivariate models with reflection.

Let O be an open set in $\mathbb{R}^d$, with the boundary ∂O and the closure $\overline{O}$. Suppose also that a vector field K of reflecting directions is given on ∂O. The main restrictions that we impose on the model with reflection are as follows: (i) The unique solvability of Skorokhod's problem for (O, K, f), where $f \in \mathcal{W}^d$ and $f(0) \in \overline{O}$. (ii) The continuity of the Skorokhod map Γ in the space $\mathcal{W}^d$. All the necessary definitions can be found in Chapter 2 of [69]. In a multivariate stochastic volatility model with reflection, the scaled volatility process is a reflecting diffusion in $\overline{O}$ given by $\widehat{B}^{(\varepsilon)}_t = (\Gamma U^{(\varepsilon)})(t)$, $t \in [0, T]$ where the process $t \mapsto U^\varepsilon_t$ is the unique continuous solution to the following d-dimensional stochastic differential equation:

$$dU^{(\varepsilon)}_t = \hat{a}(t, (\Gamma U^{(\varepsilon)})(t))dt + \sqrt{\varepsilon}\hat{c}(t, (\Gamma U^{(\varepsilon)})(t))dB_t, \quad U^\varepsilon_0 = y \in \overline{O} \qquad (70)$$

(see [69], (2.5) and (2.6) in Section 2). In (70), $\hat{a}$ is a map from $[0,T] \times \mathbb{R}^d$ into $\mathbb{R}^d$, while $\hat{c}$ maps $[0,T] \times \mathbb{R}^d$ into the space of $d \times m$-matrices. We assume that the maps $a(t, \varphi) = \hat{a}(t, (\Gamma \varphi)(t)$ and $c(t, \varphi) = \hat{a}(t, (\Gamma \varphi)(t)$, where $t \in [0, T]$ and $\varphi \in \mathcal{W}^d$, are locally Lipschitz continuous in the second variable, uniformly in time, and satisfy the sublinear growth condition in the second variable, uniformly in time (see Definitions (A1) and (A2) in Section 3 of [17]). Then, Conditions (H1)–(H6) in [17] are satisfied (see Section 3 of [17]), and hence Theorems 24 and 38 hold true for the log-price process and the volatility process in a stochastic volatility model with reflection, under the restrictions mentioned above. More information about multidimensional reflecting diffusions and also examples of uniquely solvable Skorokhod's problems can be found in Chapters 2 and 3 of [69]. Note that Skorokhod's problem for the half-line is uniquely solvable, and the Skorokhod map $\Gamma : \mathcal{W}^1 \mapsto \mathcal{W}^1$ is defined by $(\Gamma f)(t) = f(t) - \min_{s \in [0,t]}(f(s) \wedge 0)$, $t \in [0, T]$ (see [69] for more details). The Skorokhod map Γ is continuous from the space $\mathcal{W}^1$ into itself. It is also $\widetilde{\mathcal{F}}_t^1/\mathcal{F}_t^1$-measurable for every $t \in [0,T]$. The previous statements follow from Lemma 1.1.1 in [69].

A special example of a one-factor stochastic volatility model with reflection is one of the three versions of the Stein and Stein model (see [42]). The volatility process in this model is the instantaneously reflecting Ornstein–Uhlenbeck process.

Remark 51. In a model with reflection, the map G, appearing in Definitions 8 and 10, is the Skorokhod map Γ. In the other models considered in this paper, the map G is the identity map.

Unification: Volterra Type SDEs. Consider the following multidimensional Volterra-type stochastic differential equation:

$$Y_t = y + \int_0^t a(t,s,Y_s)ds + \int_0^t c(t,s,Y_s)dB_s. \tag{71}$$

Equation (71) is a special case of equation (8). We also use a scaled version of equation (71), that is, the equation

$$Y_t^\varepsilon = y + \int_0^t a(t,s,Y_s^\varepsilon)ds + \sqrt{\varepsilon}\int_0^t c(t,s,Y_s^\varepsilon)dB_s, \tag{72}$$

and a scaled controlled version given by

$$Y_t^{\varepsilon,v} = y + \int_0^t a(t,s,Y_s^{\varepsilon,v})ds + \int_0^t c(t,s,Y_s^{\varepsilon,v})v_s ds + \sqrt{\varepsilon}\int_0^t c(t,s,Y_s^{\varepsilon,v})dB_s. \tag{73}$$

In the papers of Wang [82] and Zhang [83], the following conditions were formulated:

(H1) For some $p > 2$, there exists $C_T > 0$ such that for all $x, y \in \mathbb{R}^d$ and $s, t \in [0,T]$,

$$\|a(t,s,x) - a(t,s,y)\|_d \leq C_T K_1(t,s)\rho^{\frac{1}{p}}(\|x-y\|_d^p), \tag{74}$$

$$\|c(t,s,x) - c(t,s,y)\|_{d\times m}^2 \leq C_T K_2(t,s)\rho^{\frac{2}{p}}(\|x-y\|_d^p), \tag{75}$$

and

$$\int_0^t (\|a(t,s,0)\|_d + \|c(t,s,0)\|_{d\times m}^2)ds \leq C_T, \tag{76}$$

where K_i, with $i = 1, 2$, are two positive functions on $[0,T]^2$ for which

$$\int_0^t \left[K_1(t,s)^{\frac{p}{p-1}} + K_2(t,s)^{\frac{p}{p-2}}\right]ds \leq C_T, \quad t \in [0,T]. \tag{77}$$

In addition, $\rho : \mathbb{R}^+ \mapsto \mathbb{R}^+$ is a concave function satisfying

$$\int_{0+} \rho(u)^{-1} du = \infty. \tag{78}$$

(H2) For all $t, t', s \in [0, T]$ and $x \in \mathbb{R}^d$,

$$||a(t, s, x) - a(t', s, x)||_d \leq F_1(t', t, s)(1 + ||x||_d), \tag{79}$$

$$||c(t, s, x) - c(t', s, x)||^2_{d \times m} \leq F_2(t', t, s)(1 + ||x||^2_d), \tag{80}$$

and for some $C > 0$ and $\theta > 1$,

$$\int_0^t (||a(t, s, 0)||^\theta_d + ||c(t, s, 0)||^{2\theta}_{d \times m}) ds < C. \tag{81}$$

The functions F_i, $i = 1, 2$, in (79) and (80) are two positive functions on $[0, T]^3$ satisfying the condition

$$\int_0^{t \wedge t'} (F_1(t', t, s) + F_2(t', t, s)) ds \leq C|t - t'|^\gamma \tag{82}$$

for some $\gamma > 0$.

Remark 52. It was shown in [82] that if Condition (H1) holds, then there exists a unique progressively measurable solution Y to equation (71). Moreover, it was established in [82] that if Conditions (H1) and (H2) hold, then the unique solution Y to equation (71) has a δ-Hölder continuous version for any $\delta \in (0, \frac{1}{p} \wedge \frac{\theta-1}{2\theta} \wedge \frac{\gamma}{2})$.

In [83], Zhang obtained a sample path LDP for the unique solution

$$\varepsilon \mapsto Y^{(\varepsilon)}_\cdot(\cdot), \quad \varepsilon \in (0, 1], \tag{83}$$

to equation (72), under Conditions (H1) and (H2) and two extra conditions (H3) and (H4) (see Theorem 1.2 in [83]). Note that the initial condition $y \in \mathbb{R}^d$ plays the role of a variable in the process defined by (83). The state space of the process in (83) is the space of continuous maps from $[0, T] \times \mathbb{R}^d$ into $\mathbb{R}^d$. Using Theorem 1.2 established in [83] and the contraction principle, we can obtain a sample path LDP for the process

$$\varepsilon \mapsto Y^{(\varepsilon)}_\cdot, \quad \varepsilon \in (0, 1], \tag{84}$$

with the initial condition $y \in \mathbb{R}^d$ that is fixed. The state space of the process in (84) is the space $\mathcal{W}^d$.

It is shown in the following that Conditions (H3) and (H4) are not needed in the LDP for the process in (84).

Remark 53. In the remaining part of this paper, we employ a weaker condition than Condition (H2). In the new condition, the restriction $t, t', s \in [0, T]$ in (79) and (80) is replaced by the restriction $0 \leq s \leq t, t' \leq T$. We denote the new condition by $(\widehat{H}2)$. By analyzing the main results obtained in the paper [82] of Wang, one can see that these results hold true with Condition (H2) replaced by Condition $(\widehat{H}2)$.

We next show that Conditions (H1) and $(\widehat{H}2)$ imply Assumptions (C1)–(C7).

Lemma 54. *Suppose Conditions (H1) and $(\widehat{H}2)$ hold true for the maps a and c appearing in (71). Then, Assumptions (C1)–(C7) introduced in section "Volatility Processes" are satisfied.*

The following corollary follows from Lemma 54.

Corollary 55. *Under Conditions (H1) and $(\widehat{H}2)$, the LDPs in Theorems 38 and 41 hold for the process in (84).*

Proof of Lemma 54. Some of the ideas utilized in the proof of Lemma 54 are borrowed from [82,83]. Suppose Conditions (H1) and $(\widehat{H}2)$ are satisfied. Then, it is clear that Assumption (C1) holds. For equation (71), the inequalities in (C2)(a) are as follows:

$$\int_0^t ||a(t, s, \varphi(s))||_d ds < \infty \quad \text{and} \quad \int_0^t ||c(t, s, \varphi(s))||_{d \times m}^2 ds < \infty$$

for all $t \in [0, T]$ and $\varphi \in \mathcal{W}^d$. These inequalities can be obtained by integrating the estimates

$$||a(t, s, \varphi(s))||_d \leq ||a(t, s, 0)||_d + CK_1(t, s)(1 + ||\varphi(s)||_d)$$

and

$$||c(t, s, \varphi(s))||_{d \times m}^2 \leq 2||c(t, s, 0)||_{d \times m}^2 + CK_2(t, s)(1 + ||\varphi(s)||_d^2)$$

over the interval $[0, T]$, and then using (76) and (77). The previous estimates follow from the inequalities established in [82] (see the bottom part of p. 1064 in [82]).

We next prove the first statement in Assumption (C2)(b). Suppose $t' \leq t$. The case where $t < t'$ is similar. We have

$$\| \int_0^t a(t,s,\varphi(s))ds - \int_0^{t'} a(t',s,\varphi(s))ds \|_d$$

$$\leq \int_{t'}^t \|a(t,s,\varphi(s)) - a(t,s,0)\|_d ds + \int_{t'}^t \|a(t,s,0)\|_d ds$$

$$+ \int_0^{t'} \|a(t,s,\varphi(s)) - a(t',s,\varphi(s))\|_d ds. \tag{85}$$

Suppose $t' \to t$. Then, the three terms on the right-hand side of (85) tend to zero. For the first term, this follows from (74), for the second one, we can use (81) and Hölder's inequality, while for the third one, one can use the estimates in (79) and (82), with $0 \leq s \leq t' \leq t \leq T$. Note that in the previous reasoning we used Condition $(\widehat{H2})$. This completes the proof of the first statement in Assumption (C2)(b).

The second condition in (C2)(b) and the condition in (C2)(c) follow from (74), (75), (77), and the concavity of the function ρ. This shows that Assumption (C2) holds true.

It was established in [82] that under the restrictions in (H1) and (H2) equation (71) is strongly uniquely solvable (see Theorems 1.1 and 1.3 in [82]). We have already mentioned that it is possible to replace (H2) by $(\widehat{H2})$ in the previous statement (see Remark 53). Similarly, for every $\varepsilon \in (0,1]$, equation (72) is strongly uniquely solvable. This proves the validity of Assumption (C3)(a). The unique solvability of equation (73) for any $v \in M_N^2[0,T]$ was mentioned on p. 2240 in [83]. This fact can be established by using Picard's iterative method. It follows that the statement in Assumption (C3)(b) holds true.

We next show that Assumption (C4) is satisfied. Let $\eta_1, \eta_2 \in \mathcal{W}^d$, $f \in L^2([0,T], \mathbb{R}^m)$, and suppose that for all $t \in [0,T]$,

$$\eta_1(t) = y + \int_0^t a(t,s,\eta_1(s))ds + \int_0^t c(t,s,\eta_1(s))f(s)ds$$

and

$$\eta_2(t) = y + \int_0^t a(t,s,\eta_2(s))ds + \int_0^t c(t,s,\eta_2(s))f(s)ds.$$

We must prove that the equality $\eta_1(t) = \eta_2(t)$ holds for all $t \in [0,T]$ (see Remark 100). We have

$$\|\eta_1(t) - \eta_2(t)\|_d^p \leq \int_0^t \|a(t,s,\eta_1(s)) - a(t,s,\eta_2(s))\|_d ds$$

$$+ \int_0^t ||c(t,s,\eta_1(s)) - c(t,s,\eta_2(s))||_{d\times m}||f(s)||_m ds$$

$$\leq C_T \int_0^t K_1(t,s)\rho^{\frac{1}{p}}(||\eta_1(t) - \eta_2(t)||_d^p) ds$$

$$+ \sqrt{C_T} \int_0^t K_2(t,s)^{\frac{1}{2}}\rho^{\frac{1}{p}}(||\eta_1(t) - \eta_2(t)||_d^p)||f(s)||_m ds. \quad (86)$$

It is not hard to see using Hölder's inequality that (86) implies the following:

$$||\eta_1(t) - \eta_2(t)||_d^p \leq C_1 \left\{ \int_0^t K_1(t,s)^{\frac{p}{p-1}} ds \right\}^{p-1} \int_0^t \rho(||\eta_1(s) - \eta_2(s)||_d^p)$$

$$+ C_1 \left\{ \int_0^t K_2(t,s)\rho^{\frac{2}{p}}(||\eta_1(s) - \eta_2(s)||_d^p) ds \right\}^{\frac{p}{2}} \left\{ \int_0^t ||f(s)||_m^2 ds \right\}^{\frac{p}{2}}$$

$$\leq C_2 \int_0^t \rho(||\eta_1(s) - \eta_2(s)||_d^p) ds + C_2 \left\{ \int_0^t K_2(t,s)^{\frac{p}{p-2}} ds \right\}^{\frac{p-2}{2}}$$

$$\times \int_0^t \rho(||\eta_1(s) - \eta_2(s)||_d^p) ds \leq C_3 \int_0^t \rho(||\eta_1(s) - \eta_2(s)||_d^p) ds.$$

Finally, applying Bihari's inequality (see [8], see also Lemma 2.1 in [82]) we obtain

$$||\eta_1(t) - \eta_2(t)||_d^p = 0 \quad \text{for all } t \in [0,T].$$

This completes the proof of (C4).

Our next goal is to prove that Assumption (C5) holds. Suppose $f_n \to f$ weakly in $L^2([0,T], \mathbb{R}^m)$. Set $\eta_n = \Gamma_y f_n$ and $\eta = \Gamma_y f$. Then, we have $\sup_{n\geq 1} ||f_n||_2 < \infty$. Moreover,

$$\eta_n(t) - \eta(t) = \int_0^t [a(t,s,\eta_n(s)) - a(t,s,\eta(s))] ds + \int_0^t [c(t,s,\eta_n(s))$$

$$- c(t,s,\eta(s))] f_n(s) ds + \int_0^t c(t,s,\eta(s))[f_n(s) - f(s)] ds.$$

Using the estimates obtained in the proof of (C4), we see that

$$||\eta_n(t) - \eta(t)||_d^p \leq C \int_0^t \rho(||\eta_n(s) - \eta(s)||_d^p) ds + A_n \quad (87)$$

where $C > 0$ is a constant independent of t and n, and

$$A_n = C \sup_{t \in [0,T]} || \int_0^t c(t,s,\eta(s))[f_n(s) - f(s)] ds ||_d^p, \quad n \geq 1. \quad (88)$$

We next prove that $\sup_{n\geq 1} A_n < \infty$. Indeed, the following estimates hold:

$$A_n \leq C_1 \left(\sup_{t \in [0,T]} \int_0^t \|c(t,s,\eta(s))\|_{d\times m}^2 ds \right)^{\frac{p}{2}}$$

$$\leq C_2 \left(\sup_{t \in [0,T]} \int_0^t \|c(t,s,\eta(s)) - c(t,s,0)\|_{d\times m}^2 ds \right)^{\frac{p}{2}}$$

$$+ C_2 \left(\sup_{t \in [0,T]} \int_0^t \|c(t,s,0)\|_{d\times m}^2 ds \right)^{\frac{p}{2}} \leq C_3$$

$$+ C_3 \left(\sup_{t \in [0,T]} \int_0^t K_2(t,s) \rho^{\frac{2}{p}}(\|\eta(s)\|_d^p) ds \right)^{\frac{p}{2}}$$

$$\leq C_3 + C_4 \int_0^T \rho(\|\eta(s)\|_d^p) ds \leq C_5 + C_5 \int_0^T \|\eta(s)\|_d^p ds.$$

In the previous estimates, we used (75), (76), (77), the concavity of the function ρ, and the fact that $\|f\|_2 + \sup_{n\geq 1} \|f_n\|_2 < \infty$. It follows that $\sup_{n\geq 1} A_n < \infty$.

Using (87) and Bihari's inequality, we obtain $\|\eta_n(t) - \eta(t)\|_d^p \leq G^{-1}(G(A_n) + Ct)$ for all $n \geq 1$ and $t \in [0,T]$, where C is the constant in (87), and $G(x) = \int_{x_0}^x \frac{dz}{\rho(z)}$ for $x \geq 0$ and $x_0 > 0$. Hence

$$\sup_{t \in [0,T]} \|\eta_n(t) - \eta(t)\|_d^p \leq G^{-1}(G(A_n) + CT), \quad n \geq 1. \tag{89}$$

Let us consider the following family $\mathcal{A}$ of maps: $s \mapsto c(t,s,\eta(s))\mathbb{1}_{\{s \leq t\}}$, $t \in [0,T]$. The set $\mathcal{A}$ is a pre-compact subset of the space $L^2([0,T], \mathbb{R}^{d\times m})$. Indeed, we can establish the previous statement by constructing an ε-net for the set $\mathcal{A}$ for every $\varepsilon > 0$. This can be done by using estimates for the map c similar to those in (85) and reasoning as in the proof of the validity of the first statement in Assumption (C2)(b). We take into account the conditions in (75), (77), (80), (81), and (82) in the proof of the existence of the ε-net. Note that the estimates in (80) and (82) are employed in the previous proof only in the case where $0 \leq s \leq t, t' \leq T$. This means that it suffices to use Condition $(\widehat{H}2)$ instead of Condition (H2) in this part of the proof of Lemma 54. It follows from the previous reasoning and the weak convergence of f_n to f that $A_n \to 0$ where A_n is defined in (88). Next, using (78) we see that $G(A_n) \to -\infty$ as $n \to \infty$. Therefore, $G(A_n) + CT \to -\infty$, and hence

$$G^{-1}(G(A_n) + CT) \to 0, \quad \text{as} \quad n \to \infty. \tag{90}$$

Finally, using (89) and (90) we see that Assumption (C5) holds true.

To prove that Assumption (C6) is satisfied, we first observe that under Condition (H1), the following inequality holds for the solution Y to equation (71):
$$\sup_{t\in[0,T]} \mathbb{E}\left[||Y_t||_d^q\right] < \infty$$
for all $q \geq 2$. This was shown in [82], Lemma 2.2. We also have $Y = h(B)$, where h is a measurable map from $\mathcal{W}^m$ into $\mathcal{W}^d$ (see Remark 18). Therefore, the following inequality holds true:
$$\sup_{t\in[0,T]} \mathbb{E}\left[||h(B)(t)||_d^q\right] < \infty.$$

Let v be a control satisfying (12). Recall that the control v defines a new Brownian motion B^v with respect to the measure $\mathbb{P}^v$ that is equivalent to the measure $\mathbb{P}$ (see (21)). Our next goal is to show that we can replace B by B^v and $\mathbb{P}$ by $\mathbb{P}^v$ in the previous inequality. Indeed, it is not hard to see that for every $t \in [0,T]$, the CDF of the random variable $||h(B)(t)||_d$ with respect to the measure $\mathbb{P}$ is equal to that of the random variable $||h(B^v)(t)||_d$ with respect to the measure $\mathbb{P}^v$. It follows from the previous remark that
$$\sup_{t\in[0,T]} \mathbb{E}_{\mathbb{P}^v}\left[||h(B^v)(t)||_d^q\right] < \infty \qquad (91)$$
for all $q \geq 2$. It was established in the proof of Theorem 1.3 on p. 1067 of [82] that if (H1) and (H2) hold (one can use $(\widehat{H}2)$ instead of (H2)), then $t \mapsto \int_0^t c(t,s,Y_s)dB_s$ is a continuous stochastic process. Hence the process $t \mapsto \int_0^t c(t,s,h(B)(s))dB_s$ is continuous. Actually, the same proof works if we replace the Brownian motion B and the measure $\mathbb{P}$ in the previous stochastic integral by the Brownian motion B^v and the measure $\mathbb{P}^v$. Here we use the inequality in (91). This establishes the validity of Assumption (C6).

Our final goal is to prove that Assumption (C7) holds. Since Assumptions (C1)–(C6) hold true, equation (73) is strongly uniquely solvable for every $\varepsilon \in (0,1]$ and every control v satisfying the condition in (12) (see Lemma 96 and Remark 98). We next show that if $\varepsilon_n \to 0$, and a sequence of controls v^n is such that
$$\sup_{n\geq 1} \int_0^T ||v_s^n||_m^2 ds \leq N \qquad (92)$$
$\mathbb{P}$-a.s., then the following inequality holds:
$$\sup_{n\geq 1} \int_0^t \mathbb{E}\left[||c(t,s,Y_s^{\varepsilon_n,v^n})||_{d\times m}^2\right] ds < \infty, \quad t \in [0,T]. \qquad (93)$$

In the next estimates, the constants $\tau > 0$ may change from line to line. It follows from (73), (92), and (76) that for every $q > p$ and $t \in [0,T]$, we have

$$\mathbb{E}[||Y_t^{\varepsilon_n,v^n}||_d^q] \leq \tau + \tau \left\{\int_0^t ||a(t,s,0)||_d ds\right\}^q$$

$$+ \tau \mathbb{E}\left[\left\{\int_0^t ||a(t,s,Y_t^{\varepsilon_n,v^n}) - a(t,s,0)||_d ds\right\}^q\right]$$

$$+ \tau \mathbb{E}\left[\left\{\int_0^t ||c(t,s,Y_t^{\varepsilon_n,v^n}) - c(t,s,0)||_{d\times m}^2 ds\right\}^{\frac{q}{2}}\right]$$

$$+ \tau \mathbb{E}\left[\left\{\int_0^t ||c(t,s,0)||_{d\times m}^2 ds\right\}^{\frac{q}{2}}\right]$$

$$+ \tau \mathbb{E}\left[\left\{||\int_0^t c(t,s,Y_s^{\varepsilon,v})dB_s||_d\right\}^q\right]$$

$$\leq \tau + \tau \mathbb{E}\left[\left\{\int_0^t ||a(t,s,Y_t^{\varepsilon_n,v^n}) - a(t,s,0)||_d ds\right\}^q\right]$$

$$+ \tau \mathbb{E}\left[\left\{\int_0^t ||c(t,s,Y_t^{\varepsilon_n,v^n}) - c(t,s,0)||_{d\times m}^2 ds\right\}^{\frac{q}{2}}\right]$$

$$+ \tau \mathbb{E}\left[\left\{||\int_0^t c(t,s,Y_s^{\varepsilon_n,v^n})dB_s||_d\right\}^q\right]. \qquad (94)$$

In order to estimate the last term in (94), we use the Burkholder–Davis–Gundy inequality. This gives

$$\mathbb{E}\left[\left\{||\int_0^t c(t,s,Y_s^{\varepsilon,v^n})dB_s||_d\right\}^q\right] \leq \tau \mathbb{E}\left[\left\{\int_0^t ||c(t,s,Y_t^{\varepsilon_n,v^n})||_{d\times m}^2 ds\right\}^{\frac{q}{2}}\right]$$

$$\leq \tau \mathbb{E}\left[\left\{\int_0^t ||c(t,s,Y_t^{\varepsilon_n,v^n}) - c(t,s,0)||_{d\times m}^2 ds\right\}^{\frac{q}{2}}\right]$$

$$+ \tau \mathbb{E}\left[\left\{\int_0^t ||c(t,s,0)||_{d\times m}^2 ds\right\}^{\frac{q}{2}}\right]. \qquad (95)$$

By taking into account (94), (95), Hölder's inequality, (H1), and reasoning as in the proof of Lemma 2.2 in [82], we obtain $\mathbb{E}[||Y_t^{\varepsilon_n,v^n}||_d^q] \leq \tau + \tau \int_0^t \mathbb{E}[||Y_s^{\varepsilon_n,v^n}||_d^q] ds$. Now, using Grönwall's inequality, we establish the

following estimate:
$$\sup_{n\geq 1} \sup_{t\in[0,T]} \mathbb{E}[||Y_t^{\varepsilon_n,v^n}||_d^q] < \infty, \quad q \geq 2. \tag{96}$$

Our next goal is to prove (93). Using (H1), Hölder's inequality, and the inequality $\rho^{\frac{1}{p}}(u^p) \leq C(1+u)$ (see [82], p. 1064), we obtain

$$\int_0^t \mathbb{E}\left[||c(t,s,Y_s^{\varepsilon_n,v^n})||_{d\times m}^2\right] ds \leq 2\int_0^t \mathbb{E}\left[||c(t,s,Y_s^{\varepsilon_n,v^n}) - c(t,s,0)||_{d\times m}^2\right] ds$$

$$+ 2\int_0^t \mathbb{E}\left[||c(t,s,0)||_{d\times m}^2\right] ds$$

$$\leq \tau + \tau \left\{ \int_0^t (\mathbb{E}[||Y_s^{\varepsilon_n,v^n}||_d^2])^{\frac{p}{2}} ds \right\}^{\frac{2}{p}}$$

$$\leq \tau + \tau \left\{ \int_0^t \mathbb{E}[||Y_s^{\varepsilon_n,v^n}||_d^p] ds \right\}^{\frac{2}{p}}. \tag{97}$$

Now, it is clear that (96) and (97) imply (93). It remains to prove that the family Y^{ε_n,v^n} is tight. We first prove that for all $t',t \in [0,T]$ there exist constants $\alpha > 0$, $\beta > 0$, and $\gamma > 0$ such that

$$\sup_{n\geq 1} \mathbb{E}[||Y_{t'}^{\varepsilon_n,v^n} - Y_t^{\varepsilon_n,v^n}||_d^\alpha] \leq \beta |t' - t|^{1+\beta}.$$

We have

$$Y_{t'}^{\varepsilon_n,v^n} - Y_t^{\varepsilon_n,v^n} = \left[\int_0^{t'} a(t',s,Y_s^{\varepsilon_n,v^n})ds - \int_0^t a(t,s,Y_s^{\varepsilon_n,v^n})ds\right]$$

$$+ \left[\int_0^{t'} c(t',s,Y_s^{\varepsilon_n,v^n})v_s^n ds - \int_0^t c(t,s,Y_s^{\varepsilon_n,v^n})v_s^n ds\right]$$

$$+ \sqrt{\varepsilon_n}\left[\int_0^{t'} c(t',s,Y_s^{\varepsilon_n,v^n})dB_s - \int_0^t c(t,s,Y_s^{\varepsilon_n,v^n})dB_s\right]$$

$$= J_1^{(n)}(t',t) + J_2^{(n)}(t',t) + J_3^{(n)}(t',t).$$

To estimate $J_3^{(n)}(t',t)$, we use (96) and reason as in the proof of Theorem 1.3 in [82]. This gives the following:

$$\sup_{n\geq 1} \mathbb{E}[||J_3^{(n)}(t',t)||_d^q] \leq C|t'-t|^\delta \tag{98}$$

for $q > \max(p, \frac{2}{\gamma})$ and $\delta = \min\{\frac{1}{p}, \frac{\theta-1}{2\theta}, \frac{\gamma}{2}\}$. The constant C in (98) does not depend on n. The last statement follows from (96). The inequality similar to that in (98) can also be established for $J_1^{(n)}(t',t)$ and $J_2^{(n)}(t',t)$. Therefore, the conditions in Kolmogorov's tightness criterion are satisfied for the sequence Y^{ε_n, v^n}, $n \geq 1$. It follows that Assumption (C7) holds true.

The proof of Lemma 54 is thus completed.

Remark 56. The following example of a model satisfying Conditions (H1) and (H2) was provided in [83], p. 2242. Consider a stochastic volatility model introduced in (1), with the volatility process that is the unique solution to equation (71), under the following restrictions:

(i) The coefficient maps in (71) are given by
$$a(t, s, x) = K_H(t, s) U_1(s, x), \quad c(t, s, x) = K_H(t, s) U_2(s, x), \qquad (99)$$

where $0 \leq s \leq t \leq 1$ and $x \in \mathbb{R}^d$.

(ii) In (99), K_H is the kernel in the Molchan–Golosov representation of fractional Brownian motion with Hurst parameter $H \in (0, 1)$ (see section "Gaussian Stochastic Volatility Models").

(iii) U_1 and U_2 are maps from $[0, t] \times \mathbb{R}^d$ into $\mathbb{R}^d$ and from $[0, t] \times \mathbb{R}^d$ into $\mathbb{R}^d \times \mathbb{R}^m$, respectively. They satisfy the following conditions: There exist $\eta > 0$ and $C > 0$ such that

$$||U_1(s, x) - U_1(s', x)||_d + ||U_2(s, x) - U_2(s', x)||_{d \times m}$$
$$\leq C|s - s'|^\eta (1 + ||x||_d) \qquad (100)$$

and

$$||U_1(s, x) - U_1(s, y)||_d + ||U_2(s, x) - U_2(s, y)||_{d \times m} \leq C||x - y||_d$$
$$(101)$$

for all $x, y \in \mathbb{R}^d$ and $s, s' \in [0, 1]$ (see p. 2242 in [83]).

Since the model described above satisfies Conditions (H1) and (H2), Theorem 24 holds for the log-price process in the model.

Remark 57. Equation (71), with the coefficient functions a and c similar to those in (99), was studied in [20] in the case where $d = m = 1$. It was established in [20] that there exists a unique continuous strong solution to this equation if U_1 and U_2 satisfy (101), and the kernel K is such that the operator $f \mapsto \int_0^t K(t, s) f(s) ds$ has certain smoothing properties on $L^1[0, T]$ and $L^2[0, T]$.

Remark 58. In [4], the following one-factor drift-less stochastic volatility model was considered (we use the notation adopted in this paper):

$$dS_t = S_t \sigma(Y_t)(\bar{\rho} dW_t + \rho dB_t),$$

$$Y_t = y + \int_0^T K_H(t,s) U_1(Y_s) ds + \int_0^t K_H(t,s) U_2(Y_s) dB_s,$$

where the function f is smooth, while the functions U_1 and U_2 are from the space $\mathbb{C}_b^{(3)}$ (see [4], p. 813). Here the symbol $\mathbb{C}_b^{(3)}$ stands for the space of three times continuously differentiable functions on $\mathbb{R}$, with bounded derivatives. It follows that the functions U_1 and U_2 satisfy the estimates in (100) and (101). Therefore, if the canonical set-up is used, then the small-noise LDP for the scaled log-price process obtained in Corollary 5.5 in [4] is a special case of small-noise LDPs established in section "Small-Noise LDPs for Log-Price Processes" of this paper. One can use Lemma 54, Theorems 28, 30, and Remark 31 to justify the previous statement.

Unification: More Volterra-Type SDEs. In [66], Nualart and Rovira formulated the following restrictions on the coefficients in (71):

(H_1) The map a is measurable from $\{0 \leq s \leq t \leq T\} \times \mathbb{R}^d$ to $\mathbb{R}^d$, while the map c is measurable from $\{0 \leq s \leq t \leq T\} \times \mathbb{R}^d$ to $\mathbb{R}^{d \times m}$.

(H_2) The maps a and c are Lipschitz in x uniformly in the other variables, that is,

$$||c(t,s,x) - c(t,s,y)||_{d \times m} + ||a(t,s,x) - a(t,s,y)||_d \leq K||x-y||_d$$

for some constant $K > 0$, all $x, y \in \mathbb{R}^d$, and all $0 \leq s \leq t \leq T$.

(H_3) The maps a and c are α-Hölder continuous in t on $[s,T]$ uniformly in the other variables. This means that there exists a constant $K > 0$ such that

$$||c(t,s,x) - c(r,s,x)||_{d \times m} + ||a(t,s,x) - a(r,s,x)||_d \leq K|t-r|^\alpha$$

for all $x \in \mathbb{R}^d$ and $s \leq t, r \leq T$ where $0 < \alpha \leq 1$.

(H_4) There exists a constant $K > 0$ such that

$$||c(t,s,x) - c(r,s,x) - c(t,s,y) + c(r,s,y)||_{d \times m} \leq K|t-r|^\gamma ||x-y||_d$$

for all x, y $\in \mathbb{R}^d$ and $T \geq t, r \geq s$ where $0 < \gamma \leq 1$.

(H_5) $a(t,s,x_0)$ and $c_j(t,s,x_0)$ are bounded.

A sample path LDP was established in [66] for the unique solution to equation (72) under Conditions (H_1)–(H_5) (see Theorem 1 in [66]).

We next show that only Conditions (H$_1$)–(H$_3$) and (H$_5$) are needed in order Theorem 1 in [66] to be true.

Lemma 59. *Conditions* (H$_1$)–(H$_3$) *and* (H$_5$) *used in [66] imply Conditions* (H1) *and* ($\widehat{H}$2) *formulated in section "Unification: Volterra Type SDEs".*

Proof. Let us assume that Conditions (H$_1$)–(H$_3$) and (H$_5$) hold. Choose $\rho(u) = u$, any $p > 2$, and $K_1(t,s) = K_2(t,s) = \mathbb{1}_{\{0 \leq s \leq t \leq T\}}$. Then, there exists a constant $C > 0$ depending on the constant K in Condition (H$_2$) and such that (74) and (75) hold. The equality in (78) clearly holds, while (77) also holds with some constant $C > 0$. The validity of (76) follows from the fact that Conditions (H$_2$) and (H$_5$) imply the boundedness of $a(t,s,0)$ and $c_j(t,s,0)$. This completes the proof of Condition (H1) used in [82].

Next, observe that for
$$F_1(t',t,s) = K|t-t'|^\alpha \mathbb{1}_{\{0 \leq s \leq t \leq T\}} \mathbb{1}_{\{0 \leq s \leq t' \leq T\}}$$
and
$$F_2(t',t,s) = K'|t-t'|^{2\alpha} \mathbb{1}_{\{0 \leq s \leq t \leq T\}} \mathbb{1}_{\{0 \leq s \leq t' \leq T\}},$$
the estimates in (79) and (80) hold true for $s \leq t, t' \leq T$, with $K' > 0$ depending on the constant K appearing in Condition (H$_3$). The previous statement follows from Condition (H$_3$). In addition, it is easy to see that the estimates in (81) and (82) are valid. Therefore, Condition ($\widehat{H}$2) in Remark 53 is satisfied.

The proof of Lemma 59 is thus completed.

Remark 60. It follows from Lemmas 54 and 59 and Theorem 38 that the sample path LDP formulated in Theorem 1 in [66] holds for the solution to equation (71) under Conditions (H$_1$)–(H$_3$) and (H$_5$) provided that the canonical set-up is used. Hence, under the previous restrictions, Condition (H$_4$) can be removed from Theorem 1 in [66]. We do not know whether the same conclusion can be reached if the model in [66] is defined on a general set-up.

Applications

First Exit Times

In this subsection, we obtain a large deviation-style formula for the distribution function of the first exit time of the log-price process from an open set in $\mathbb{R}^m$. Such formulas go back to the known results on first exit time due to Freidlin and Wentzell (see [33,80,81]). Suppose $X_t^{(\varepsilon)}$, with $t \in [0,T]$, is the scaled log-price process (see (6)) starting at $x_0 \in \mathbb{R}^m$. It is assumed in the rest of this section that the conditions in Theorem 24 are satisfied, and, moreover, for all $(t, u) \in [0, T] \times \mathbb{R}^d$, the matrix $\sigma(t, u)$ is invertible. Then, formula (30) holds for the rate function $\widetilde{Q}_T$ in Theorem 24. Let O be a proper open subset of $\mathbb{R}^m$ such that $x_0 \in O$.

Definition 61.

(i) For every $\varepsilon \in (0, 1]$, the first exit time of the scaled log-price process from the set O is defined by $\tau^{(\varepsilon)} = \inf\{s \in (0, T] : X_s^{(\varepsilon)} \notin O\}$ if the previous set is not empty and by $\tau^{(\varepsilon)} = \infty$ otherwise.

(ii) For every $\varepsilon \in (0, 1]$, the first exit time probability function is defined by $v_\varepsilon(t) = \mathbb{P}(\tau^{(\varepsilon)} \leq t)$, $t \in (0, T]$.

In the book of Freidlin and Wentzell [33], the following restriction on an open set $O \subset \mathbb{R}^m$ was used: There exist interior points of the complement of O arbitrarily close to every point of the boundary of O (see [33], Example 3.5). The previous condition can be formulated as follows:
$$\partial O = \partial(\text{ext}(O)), \qquad (102)$$
where ext(O) is the set of interior points of the complement of O, and, for a set $D \subset \mathbb{R}^m$, the symbol ∂D stands for the boundary of D.

Let us fix $t \in (0, T]$, and put $\mathcal{A}_t = \{f \in \mathbb{C}_0^m : f(s) \notin O - x_0 \text{ for some } s \in (0, t]\}$. Then, $\mathcal{A}_t$ is a closed subset of the space $\mathbb{C}_0^m$. Its interior $\mathcal{A}_t^\circ$ consists of the maps $f \in \mathcal{A}_t$ for which there exists $s < t$ with $f(s) \notin \text{cl}(O) - x_0$. Here the symbol cl($O$) stands for the closure of the set O in the space $\mathbb{R}^m$. The boundary bd($\mathcal{A}_t$) of the set $\mathcal{A}_t$ in the space $\mathbb{C}_0^m$ consists of the maps $f \in \mathbb{C}_0^m$ which hit the set $\partial O - x_0$ before t, or at $s = t$, but never exit the set cl(O) $- x_0$ before t.

A Borel set $A \subset \mathbb{C}_0^m$ is called a set of continuity for the rate function $\widetilde{Q}_T$ if the following equality holds:
$$\inf_{g \in A^\circ} \widetilde{Q}_T(g) = \inf_{g \in \bar{A}} \widetilde{Q}_T(g).$$
It follows from the LDP in Theorem 24 that for any set of continuity A,
$$\varepsilon \log \mathbb{P}\left(X^{(\varepsilon)} - x_0 \in A\right) = -\inf_{g \in A} \widetilde{Q}_T(g) + o(1) \text{ as } \varepsilon \to 0. \qquad (103)$$

The following theorem provides a sample path large deviation-style formula for the first exit time probability function.

Theorem 62. *Suppose an open set $O \subset \mathbb{R}^m$ is such that the condition in (102) holds. Then, the set $\mathcal{A}_t$ is a set of continuity for the rate function $\widetilde{Q}_T$, and hence*
$$\varepsilon \log \mathbb{P}(\tau^{(\varepsilon)} \leq t) = -\inf_{g \in \mathcal{A}_t} \widetilde{Q}_T(g) + o(1) \text{ as } \varepsilon \to 0. \qquad (104)$$

Proof. In the proof of Theorem 62, we borrow some ideas from the proof of Theorem 2.16 in [41] and also take into account Example 3.5 in [33]. Our first goal is to provide a sufficient condition for a Borel set $A \subset \mathbb{C}_0^m$ to be a set of continuity for the rate function $\widetilde{Q}_T$. The following statement can be obtained using the continuity of the function $\widetilde{Q}_T$ on the space $(\mathbb{H}_0^1)^m$ (see Lemma 27).

Lemma 63. *Suppose a Borel set $A \subset \mathbb{C}_0^m$ is such that for every $h \in \mathrm{bd}\,(A) \cap (\mathbb{H}_0^1)^m$, there exists a sequence $h_n \in A^\circ \cap (\mathbb{H}_0^1)^m$ for which $\lim_{n \to \infty} h_n = h$ in the space $(\mathbb{H}_0^1)^m$. Then, the set A is a set of continuity for the rate function $\widetilde{Q}_T$.*

Let us continue the proof of Theorem 62. It is not hard to see that
$$\{\tau^{(\varepsilon)} \leq t\} = \{X^{(\varepsilon)} - x_0 \in \mathcal{A}_t\}. \tag{105}$$
We next prove that the set $\mathcal{A}_t$ is a set of continuity for $\widetilde{Q}_T$. Let $f \in \mathrm{bd}\,(\mathcal{A}_t) \cap (\mathbb{H}_0^1)^m$, and let $t_0 \in (0, t]$ be such that $f(t_0) \in \partial O - x_0$ and $f(u) \in \mathrm{cl}\,(O) - x_0$ for all $u \in (0, t]$. Using (102), we see that for every $n \geq 1$ there is a point $x_n \in \mathrm{ext}(O) - x_0$ such that $\|x_n - f(t_0)\|_m \leq \frac{1}{n}$ for all $n \geq 1$. Define a sequence of functions on $[0, T]$ by the following formula: $f_n(t) = f(t) + \frac{t}{t_0}(x_n - f(t_0))$, $n \geq 1$. It is easy to see that for every $n \geq 1$, we have $f_n \in (\mathbb{H}_0^1)^m$. Moreover, $f_n \to f$ in $(\mathbb{H}_0^1)^m$ as $n \to \infty$. Since $f_n(t_0) = x_n \in \mathrm{ext}(O) - x_0$, the map f_n exits the set $\mathrm{cl}\,(O) - x_0$ before t. This is clear for $t_0 < t$, while for $t_0 = t$, we can use the continuity of f. It follows that $f_n \in \mathcal{A}_t^\circ \cap (\mathbb{H}_0^1)^m$ for all $n \geq 1$. Next, using Lemma 63, we see that $\mathcal{A}_t$ is a set of continuity for the rate function $\widetilde{Q}_T$. Finally, the equality in (104) follows from (103).

Binary Barrier Options

Our goal in this subsection is to obtain a large deviation-style formula in the small-noise regime for multidimensional binary barrier options. Suppose that the model in (1) describes the dynamics of price processes associated with a portfolio of correlated assets. Let $\varepsilon \in (0, 1]$, and consider the scaled m-dimensional asset price process $t \mapsto S_t^{(\varepsilon)}$ and the scaled log-price process $t \mapsto X_t^{(\varepsilon)}$. The latter process is given by the expression in (6).

We study the small-noise asymptotic behavior of binary up-and-in barrier options. Similar results can be obtained for up-and-out, down-and-in, and down-and-out options. We refer the reader to [41] where one-dimensional Gaussian models are considered.

Denote by $\mathbb{R}_+^m$ the subset of $\mathbb{R}^m$ consisting of all the vectors $s = (s_1, \ldots, s_m) \in \mathbb{R}^m$ such that $s_i > 0$ for all $1 \leq i \leq m$, and let $O \subset \mathbb{R}_+^m$ be an open set satisfying the condition in (102). The boundary ∂O of the set O plays the role of the barrier. Throughout this section we assume that the model in (1) satisfies the restrictions imposed in Theorem 24.

Let us suppose that for every $\varepsilon \in (0, 1]$ the initial condition s_0 for the process $t \mapsto S_t^{(\varepsilon)}$ is such that $s_0 \in O$.

Definition 64. Let O be an open set in $\mathbb{R}_+^m$ satisfying the condition in (102). In a small-noise setting, a binary up-and-in barrier option pays a fixed amount of cash, say one dollar, if the m-dimensional asset price process $S^{(\varepsilon)}$ hits the barrier ∂O at some time during the life of the option.

For every $\varepsilon \in (0, 1]$, the payoff of a binary up-and-in barrier option at the maturity is $\{S_t^{(\varepsilon)} \in \partial O \text{ for some } t \in [0, T]\}$. Therefore, the price $B(\varepsilon)$ of the option at $t = 0$ is given by

$$B(\varepsilon) = e^{-rT} \mathbb{P}(S_t^{(\varepsilon)} \in \partial O \text{ for some } t \in [0, T]) \qquad (106)$$

where $r > 0$ is the interest rate. It is not hard to see using (106) that

$$B(\varepsilon) = e^{-rT} \mathbb{P}(S_t^{(\varepsilon)} \notin O \text{ for some } t \in [0, T]). \qquad (107)$$

It is clear that a binary up-and-in barrier option contract depends on r, s_0, the maturity T of the option, and the barrier ∂O. We study the asymptotic behavior of the price of the barrier option as $\varepsilon \to 0$. Barrier options are path-dependent options.

Our next goal is to rewrite the expression on the right-hand side of (107) in terms of the log-price process. The resulting equality is as follows:

$$B(\varepsilon) = e^{-rT} \mathbb{P}(X_t^{(\varepsilon)} - x_0 \notin \widetilde{O} - x_0 \text{ for some } t \in [0, T]) \qquad (108)$$

where $\widetilde{O}$ is the open subset of $\mathbb{R}^m$ defined by

$$\widetilde{O} = \{x = (x_1, \ldots, x_m) \in \mathbb{R}^m : (e^{x_1}, \ldots, e^{x_m}) \in O\}.$$

It is easy to see that the set $\widetilde{O} - x_0$ satisfies the condition in (102).

The price of a binary up-and-in barrier option is related to the exit time probability function $\tau_{\widetilde{O}}$ of the log-price process $X^{(\varepsilon)}$ from the set $\widetilde{O}$. Indeed, it was shown in the proof of Theorem 62 that

$$\{\tau_{\widetilde{O}}^{(\varepsilon)} \leq T\} = \{X^{(\varepsilon)} - x_0 \in \mathcal{A}_T\}, \tag{109}$$

where

$$\mathcal{A}_T = \{f \in \mathbb{C}_0^m : f(s) \notin \widetilde{O} - x_0 \text{ for some } s \in [0, T]\} \tag{110}$$

(see (105)).

The next assertion provides a large deviation-style formula in the small-noise regime for the price $B(\varepsilon)$ of the binary up-and-in barrier option given by (107).

Theorem 65. *The following asymptotic formula holds:*

$$\varepsilon \log B(\varepsilon) = -\inf_{g \in \mathcal{A}_T} \widetilde{Q}_T(g) + o(1) \text{ as } \varepsilon \to 0,$$

where $\widetilde{Q}_T$ is the rate function in Theorem 24, and the set $\mathcal{A}_T$ is defined by (110).

Theorem 65 can be easily derived from (108), (109), (110), and Theorem 62.

Call Options

In this subsection, we consider the model in (4) with $m = 1$ and $d \geq 1$. It is assumed that $b(t, u) = r$ where $r \geq 0$ is the interest rate. This means that we turn our attention to d-factor stochastic volatility models of financial mathematics. More precisely, the models that we study in this subsection are the following:

$$\frac{dS_t}{S_t} = rdt + \sigma(t, \widehat{B}_t)(\bar{\rho}dW_t + \rho dB_t), \quad t \in [0, T]. \tag{111}$$

In (111), $\rho \in (-1, 1)$ is the correlation parameter, $\bar{\rho} = \sqrt{1 - \rho^2}$, and $\widehat{B}$ is the d-dimensional volatility process introduced in Definition 8. The scaled version of the model in (111) is as follows:

$$\frac{dS_t^{(\varepsilon)}}{S_t^{(\varepsilon)}} = rdt + \sqrt{\varepsilon}\sigma(t, \widehat{B}_t^{(\varepsilon)})(\bar{\rho}dW_t + \rho dB_t), \quad t \in [0, T], \tag{112}$$

where $\varepsilon \in (0, 1]$ and $\widehat{B}^{(\varepsilon)}$ is the scaled volatility process (see Definition 10).

The price of the call option in the small-noise regime is the following function of the strike $K > 0$, the maturity $T > 0$, and the small-noise parameter $\varepsilon \in (0, 1]$:

$$C(\varepsilon, T, K) = \mathbb{E}[(S_T^{(\varepsilon)} - K)^+]. \tag{113}$$

We assume that K and T are fixed and study the asymptotic behavior of the call price as $\varepsilon \to 0$. The following assumption is used in the sequel:

Assumption B: For every $\alpha > 0$ there exists $\varepsilon_0 \in (0, 1]$ depending only on α and such that for all $0 < \varepsilon < \varepsilon_0$, the following estimate holds true:

$$\mathbb{E}\left[\exp\left\{\alpha \int_0^T \sigma(s, \widehat{B}_s^{(\varepsilon)})^2 ds\right\}\right] \leq M, \tag{114}$$

where $M > 0$ is a constant depending only on α.

Lemma 66. *Suppose Assumption B holds. Then, for every $C > 0$ there exists $\varepsilon_1 \in (0, 1]$ depending on C and such that for every $\widehat{C}$, with $0 < \widehat{C} \leq C$, the stochastic exponential*

$$\mathcal{E}(t, \varepsilon, \widehat{C}) = \exp\left\{-\frac{1}{2}\widehat{C}^2 \int_0^t \sigma(s, \widehat{B}_s^{(\varepsilon)})^2 ds + \widehat{C} \int_0^t \sigma(s, \widehat{B}_s^{(\varepsilon)})(\bar{\rho} dW_t + \rho dB_t)\right\},$$

$$t \in [0, T]$$

is an $\{\mathcal{F}_t\}$-martingale for all $\varepsilon \leq \varepsilon_1$.

Proof. Set $\alpha = \frac{1}{2}C^2$ and $U(t, \varepsilon, C) = C\sigma(t, \widehat{B}_t^{(\varepsilon)})$, $t \in [0, T]$. Then, Assumption B implies that there exists $\varepsilon_1 \in (0, 1]$ such that Novikov's condition is satisfied for the process $t \mapsto U(t, \varepsilon, C)$, with $0 < \varepsilon \leq \varepsilon_1$. It follows that the stochastic exponential in Lemma 66 is an $\{\mathcal{F}_t\}$-martingale for all $\varepsilon \leq \varepsilon_1$. The same proof works for any $\widehat{C} < C$ since if the inequality in (114) holds for some $\alpha > 0$ and $\varepsilon < \varepsilon_0$, it also holds for any $0 < \hat{\alpha} < \alpha$ and $\varepsilon < \varepsilon_0$.

The proof of Lemma 66 is thus completed.

Using the Doléans–Dade formula, we see that for every $\varepsilon \in (0, 1]$, the discounted scaled asset price process is given by

$$e^{-rt} S_t^{(\varepsilon)} = s_0 \exp\left\{-\frac{\varepsilon}{2} \int_0^t \sigma(s, \widehat{B}_s^{(\varepsilon)})^2 ds + \sqrt{\varepsilon} \int_0^t \sigma(s, \widehat{B}_s^{(\varepsilon)}) \right.$$

$$\left. \times (\bar{\rho} dW_t + \rho dB_t)\right\}, \tag{115}$$

where $t \in [0, T]$.

Remark 67. Applying Lemma 66, with $C = \sqrt{\varepsilon}$, we see that if Assumption B holds, then for small enough values of ε, the discounted asset price process in (115) is a martingale. It follows that the measure $\mathbb{P}$ in the model defined in (112), with small enough values of ε, is risk-neutral.

Remark 68. It is not hard to see, using the formula in (40), that for the model in (111), with

$$\sigma(s,z) \neq 0 \quad \text{for all} \quad (s,z) \in [0,T] \times \mathbb{R}^d, \tag{116}$$

the rate function $\widetilde{I}_T$ can be represented as follows:

$$\widetilde{I}_T(x) = \frac{1}{2} \inf_{f \in \mathbb{H}_0^1} \left[\frac{(x - r - \rho \int_0^T \sigma(s, \widehat{f}(s)) \dot{f}(s) ds)^2}{\bar{\rho}^2 \int_0^T \sigma(s, \widehat{f}(s))^2 ds} + \int_0^T \dot{f}(t)^2 dt \right]. \tag{117}$$

Moreover, the function in (117) is continuous on $\mathbb{R}$ (see Lemma 33).

Remark 69. Suppose we do not impose the restriction in (116) on the model in (111). Then, the formula in (117) holds for all $x \in \mathbb{R}$ such that the set Q_2 is empty (see Remark 31). If $Q_2 \neq \emptyset$, then (37) implies that the set $Q_3(x)$ can be non-empty only if $x = r$. Therefore, the formula in (117) holds for all $x > r$. In addition, the function $\widetilde{I}_T$ is continuous on the set $x > r$. The previous statement can be established exactly as Lemma 33.

The next assertion provides an LDP-style formula for the price of the scaled call option.

Theorem 70. *Suppose Assumption A and Assumptions (C1)–(C7) hold true for the model in (111) defined on the canonical set-up. Further suppose that the condition in (116) holds and Assumption B is satisfied. Then, the following asymptotic formula is valid:*

$$\varepsilon \log C(\varepsilon, T, K) = - \inf_{x \geq k} \widetilde{I}_T(x) + o(1) \tag{118}$$

as $\varepsilon \to 0$ where k is the log-moneyness defined by $k = \log \frac{K}{s_0}$ and $\widetilde{I}_T$ is the good rate function given by (40).

Proof. We first establish a lower large deviation estimate for the scaled call price. It is not hard to see that for every $\delta > 0$,

$$C(\varepsilon, T, K) \geq \delta \mathbb{P}(S_T^{(\varepsilon)} > K + \delta) = \delta \mathbb{P}\left(X_T^{(\varepsilon)} - x_0 > \log \frac{K + \delta}{s_0} \right).$$

It follows from the LDP in Theorem 32 that

$$\liminf_{\varepsilon \to 0} \varepsilon \log C(\varepsilon, T, K) \geq - \inf_{x > \log \frac{K+\delta}{s_0}} \widetilde{I}_T(x). \tag{119}$$

Since (119) holds for all $\delta > 0$ and the function $\widetilde{I}_T$ is continuous on $\mathbb{R}$ (see Remark 68), we derive the following estimate from (119):

$$\liminf_{\varepsilon \to 0} \varepsilon \log C(\varepsilon, T, K) \geq - \inf_{x > k} \widetilde{I}_T(x). \qquad (120)$$

The formula in (120) is a lower LDP-style estimate for the scaled call price.

We next obtain a similar upper estimate. We borrow some ideas from the proof on page 1131 in [44]. Let $p > 1$ and $q = \frac{p}{p-1}$. Using Hölder's inequality, we get

$$C(\varepsilon, T, K) \leq \left\{ \mathbb{E}\left[(S_T^{(\varepsilon)})^p \right] \right\}^{\frac{1}{p}} \mathbb{P}\left(S_T^{(\varepsilon)} \geq K \right)^{\frac{1}{q}}.$$

Therefore,

$$\limsup_{\varepsilon \to 0} \varepsilon \log C(\varepsilon, T, K) \leq \frac{1}{p} \limsup_{\varepsilon \to 0} \varepsilon \log \mathbb{E}\left[(S_T^{(\varepsilon)})^p \right]$$
$$+ \frac{1}{q} \limsup_{\varepsilon \to 0} \varepsilon \log \mathbb{P}\left(S_T^{(\varepsilon)} \geq K \right). \qquad (121)$$

The second term on the right-hand side of (121) can be estimated using the LDP in Theorem 32 as in the proof of (120). This gives

$$\limsup_{\varepsilon \to 0} \varepsilon \log \mathbb{P}\left(S_T^{(\varepsilon)} \geq K \right) \leq - \inf_{x \geq k} \widetilde{I}_T(x). \qquad (122)$$

Note that, by the continuity of the rate function, we have

$$\inf_{x > k} \widetilde{I}_T(x) = \inf_{x \geq k} \widetilde{I}_T(x). \qquad (123)$$

Our next goal is to estimate the first term on the right-hand side of (121). Using (115), we see that for every $p > 1$, all $t \in [0, T]$, and all $\varepsilon \in (0, 1]$,

$$\mathbb{E}\left[(S_t^{(\varepsilon)})^p \right] \leq s_0^p e^{prt} \mathbb{E}\left[\mathcal{E}(t, \varepsilon, 2p) \right]^{\frac{1}{2}} \mathbb{E}\left[\exp\left\{ (2p^2 - p) \int_0^t \sigma(s, \widehat{B}_s^{(\varepsilon)})^2 ds \right\} \right]^{\frac{1}{2}}.$$

It follows from Assumption B and Lemma 66 that there exist $\varepsilon_p \in (0, 1]$ and $M_p > 0$ depending on p and such that

$$\mathbb{E}\left[(S_t^{(\varepsilon)})^p \right] \leq s_0^p e^{pr} M_p \qquad (124)$$

for all $t \in [0, T]$ and $0 < \varepsilon \leq \varepsilon_p$. It is easy to see that (124) implies the following equality:

$$\limsup_{\varepsilon \to 0} \varepsilon \log \mathbb{E}\left[(S_T^{(\varepsilon)})^p \right] = 0 \qquad (125)$$

for every $p > 1$. Now, using (121), (122), and (125), and assuming that $q \to 1$, we get the upper estimate

$$\limsup_{\varepsilon \to 0} \varepsilon \log C(\varepsilon, T, K) \leq - \inf_{x \geq k} \widetilde{I}_T(x). \qquad (126)$$

Finally, using (120), (123), and (126), we obtain formula (118).

This completes the proof of Theorem 70.

The restrictions in Theorem 70 include the condition in (116). The class of stochastic volatility models for which this condition is satisfied includes stochastic volatility models, with exponential volatility function, e.g. the Scott model (see [74]), the rough Bergomi model (see [5]), and the super rough Bergomi model (see [6,41]). On the other hand, the condition in (116) is not satisfied for certain classical stochastic volatility models, for instance, the Stein and Stein model (see [73]) and the Heston model (see [48]). In the former model, the volatility function is $\sigma(x) = x$, for all $x \in \mathbb{R}$, while in the latter one, the volatility function is given by $\sigma(x) = \sqrt{x}$, for all $x \geq 0$.

The next assertion explains what happens if we remove the condition in (116) from Theorem 70.

Theorem 71. *Suppose Assumption A and Assumptions (C1)–(C7) hold true for the model in (111) defined on the canonical set-up. Further suppose that Assumption B is satisfied. Then, the formula*

$$\varepsilon \log C(\varepsilon, T, K) = - \inf_{x \geq k} \widetilde{I}_T(x) + o(1)$$

as $\varepsilon \to 0$ holds for all $K > s_0 e^r$.

Proof. Theorem 71 can be established using the same methods as in the proof of Theorem 70 and by taking into account Remark 69. This gives the formula in (118) for all $k > r$ that is equivalent to $K > s_0 e^r$.

The proof of Theorem 71 is thus completed.

Implied Volatility in the Small-Noise Regime

Our main objective in this subsection is to use fundamental results of Gao and Lee (see [36]) which provide relations between the asymptotic behavior of the logarithm of the call price with respect to various parametrizations and the asymptotic behavior of the implied volatility.

Let us consider the scaled stochastic volatility model described in (112). We assume that $s_0 = 1$ and $r = 0$. The previous normalization is employed in [36]. It follows that $k = \log K$. Recall that the small-noise call option price $C(\varepsilon, T, K)$ is defined by (113). Set

$$\widetilde{C}(\varepsilon, T, k) = C(\varepsilon, T, e^k) = \mathbb{E}[(S_T^{(\varepsilon)} - e^k)^+], \qquad (127)$$

with $k \in \mathbb{R}$.

It follows from Theorem 71 that for every $k > 0$,

$$\log \frac{1}{\widetilde{C}(\varepsilon, T, k)} = \frac{1}{\varepsilon} \inf_{x \geq k} \widetilde{I}_T(x) + o\left(\frac{1}{\varepsilon}\right) \qquad (128)$$

as $\varepsilon \to 0$. Recall that in this subsection, we assume that $r = 0$.

In [36], the following formula is used for the call price C_{BS} in the Black–Scholes model (see (3.1) in [36]):

$$C_{BS}(v, k) = \frac{1}{\sqrt{2\pi}} \int_{-\infty}^{d_1} e^{-\frac{y^2}{2}} dy - \frac{e^k}{\sqrt{2\pi}} \int_{-\infty}^{d_2} e^{-\frac{y^2}{2}} dy, \qquad (129)$$

where

$$d_1 = \frac{-k + \frac{1}{2} v^2}{v} \quad \text{and} \quad d_2 = \frac{-k - \frac{1}{2} v^2}{v}.$$

The formula in (129) represents the call price in the Black–Scholes model as a function of the log-strike $k \in \mathbb{R}$ and the dimensionless implied volatility $v > 0$ (see formula (3.1) in [36]).

Let $\sigma > 0$ be the volatility parameter in the classical Black–Scholes model. It is clear that if we replace v by $\sigma\sqrt{T}$ in (129), the formula in (129) becomes the classical Black–Scholes formula for the call price.

A generally accepted small-noise parametrization of the implied volatility v is as follows: $v(\varepsilon) = \sqrt{\varepsilon}\sigma$. Our next goal is to take into account the scaling in the Black–Scholes formula. Set

$$\widehat{C}_{BS}(\varepsilon, T, k, \sigma) = \frac{1}{\sqrt{2\pi}} \int_{-\infty}^{d_1(\varepsilon, k, \sigma)} e^{-\frac{y^2}{2}} dy - \frac{e^k}{\sqrt{2\pi}} \int_{-\infty}^{d_2(\varepsilon, k, \sigma)} e^{-\frac{y^2}{2}} dy \qquad (130)$$

where

$$d_1(\varepsilon, T, k, \sigma) = \frac{-k + \frac{1}{2}\varepsilon T \sigma^2}{\sqrt{\varepsilon T}\sigma} \quad \text{and} \quad d_2(\varepsilon, T, k, \sigma) = \frac{-k - \frac{1}{2}\varepsilon T \sigma^2}{\sqrt{\varepsilon T}\sigma}.$$

Definition 72. Let $\widetilde{C}$ be the scaled call price function defined in (127). Given $k \in \mathbb{R}$, $T > 0$, and $\varepsilon \in (0, 1]$, the implied volatility in the small-noise setting associated with the function $\widetilde{C}$ is the value of the volatility parameter σ in (130) for which $\widetilde{C}(\varepsilon, T, k) = \widehat{C}_{BS}(\varepsilon, T, k, \sigma)$. The implied volatility is denoted by $V(\varepsilon, T, k)$.

The formulas above show that we consider the case of fixed k and T and the dimensionless implied volatility v parametrized as follows:

$$v(\varepsilon) = \sqrt{\varepsilon T} V(\varepsilon, T, k), \quad \varepsilon \in (0, 1]. \tag{131}$$

The following statement provides an asymptotic formula for the implied volatility V as $\varepsilon \to 0$.

Theorem 73. *Suppose Assumption A and Assumptions (C1)–(C7) hold true for the model in (111) defined on the canonical set-up. Further suppose that Assumption B is satisfied. Then, the following formula holds for every $k > 0$ and $T > 0$:*

$$\lim_{\varepsilon \to 0} V(\varepsilon, T, k) = \frac{k}{\sqrt{2T \inf_{x \geq k} \widetilde{I}_T(x)}}. \tag{132}$$

Proof. It follows from (128), (130), and Definition 72 that

$$\log \frac{1}{C_{BS}(v(\varepsilon), k)} = \frac{1}{\varepsilon} \inf_{x \geq k} \widetilde{I}_T(x) + o\left(\frac{1}{\varepsilon}\right), \quad \varepsilon \to 0, \tag{133}$$

where $v(\varepsilon)$ is given by (131). It is not hard to see that the conditions in Corollary 7.2 in [36] hold. It follows from (133) and formulas (7.6) and (7.8) in [36] that as $\varepsilon \to 0$,

$$v(\varepsilon) = \frac{\sqrt{\varepsilon} k}{\sqrt{2 \inf_{x \geq k} \widetilde{I}_T(x)}} + o(\sqrt{\varepsilon}).$$

Next, using (131), we see that the formula in (132) can be obtained from the previous equality.

Corollary 74. *Suppose the conditions in Theorem 73 hold. Suppose also that the restriction in (116) is satisfied. If $r = 0$, and the model is uncorrelated ($\rho = 0$), then for all $k > 0$ and $T > 0$,*

$$\lim_{\varepsilon \to 0} V(\varepsilon, T, k) = \frac{k}{\sqrt{2T \widetilde{I}_T(k)}}. \tag{134}$$

Proof. It follows from the conditions in Corollary 74 that the rate function in (117) is increasing on $(0, \infty)$. Since the rate function is continuous (see Lemma 33), we have

$$\inf_{x \geq k} \widetilde{I}_T(x) = \widetilde{I}_T(k). \tag{135}$$

Now, it is clear that (134) follows from (132) and (135).

A Toy Model

In this subsection, we consider a simple model (a toy model) and discuss the applicability of Theorem 73 to the toy model. We choose a special uncorrelated SABR model as the toy model.

The SABR model was introduced in [46]. The toy model analyzed in this subsection is the following special case of the SABR model:

$$dS_t = X_t S_t dW_t,$$
$$dX_t = X_t dB_t, \tag{136}$$

where $0 \leq t \leq T$, $S_0 = 1$, $X_0 = 1$, and W and B are independent standard Brownian motions. The toy model in (136) is the SABR model with $\nu = 1$, $\beta = 1$, and $\rho = 0$ (see the notation in [46]).

The toy model described in (136) is one of the Gaussian models used in this paper. Indeed, it follows from (136) that

$$dS_t = \sigma(t, B_t) S_t dW_t, \quad 0 \leq t \leq T, \tag{137}$$

where

$$\sigma(t, u) = \exp\left\{-\frac{1}{2}t + u\right\}, \quad (t, u) \in [0, T] \times \mathbb{R}^1. \tag{138}$$

The Gaussian model in (137) is drift-less and uncorrelated ($\rho = 0$). Moreover, the volatility process $\widehat{B}$ in the toy model is the standard Brownian motion B, and the volatility function σ is given by (138). The scaled volatility process is as follows: $\widehat{B}_t^{(\varepsilon)} = \sqrt{\varepsilon} B_t$, with $0 \leq t \leq T$ and $\varepsilon \in (0, 1]$.

Remark 75. The SABR model in (136) is a special case of the Hull–White model studied in [50] (see also [45]).

Our next goal is to apply Theorem 73 to the toy model. It is easy to see that Assumption A is satisfied. The reader can find the formulation of Assumption A after Definition 4. It is also not hard to check the validity of the estimate in (114). Therefore, Assumption B is satisfied. Moreover, Assumptions (C1)–(C7) formulated in section "Volatility Processes" are also satisfied. The previous statement follows from the fact that the toy model is Gaussian and from Theorem 47.

We next describe the mapping $f \mapsto \widehat{f}$ associated with the toy model. By taking into account the equalities $\widehat{B}_t = B_t = \int_0^t dB_s$, we see that for the toy model, equation (23) has the following form: $\eta_f = \int_0^t f(s) ds$. The previous statement can be derived from the fact that in equation (8), $c(t, s) = 1$ for $s \leq t$, $y = 0$, and $a = 0$. Now, using Definition 23, we see that $\widehat{f} = f$, for all $f \in \mathbb{H}_0^1$.

Since in the toy model, $\sigma(t, u) \neq 0$, for all $(t, u) \in [0, T] \times \mathbb{R}$, the rate function $\widetilde{I}_T$ appearing in Theorem 73 has the following form:

$$\widetilde{I}_T(x) = \frac{1}{2} \inf_{f \in \mathbb{H}_0^1} \left[\frac{x^2}{\int_0^T \exp\{-t + 2f(t)\} dt} + \int_0^T \dot{f}(t)^2 dt \right], \quad x \in \mathbb{R}. \quad (139)$$

The previous formula can be obtained from (40). It follows from Corollary 74 that the formula in (134) holds for the toy model.

Remark 76. It would be interesting to find a simple explicit representation for the rate function $\widetilde{I}_T(k)$, $k > 0$. However, we do not know how to obtain such a representation. In the remaining part of this subsection, we establish estimates from above and below for the rate function in the toy model.

We first obtain an estimate from the following for the rate function. It is not difficult to prove that for any $f \in \mathbb{H}_0^1$ the following inequality holds:

$$T \int_0^T \dot{f}(t)^2 dt \geq \left(\max_{t \in [0,T]} |f(t)| \right)^2.$$

Therefore, (139) implies that for every $k > 0$ we have

$$\widetilde{I}_T(k) \geq \frac{1}{2} \min_{a \geq 0} \left[\frac{k^2}{(1 - e^{-T}) e^{2a}} + \frac{1}{T} a^2 \right].$$

The minimization problem in the formula above can be easily solved. The resulting estimate is as follows:

$$\widetilde{I}_T(k) \geq \frac{1}{2} \left[\frac{k^2}{(1 - e^{-T}) e^{2a(k)}} + \frac{1}{T} a(k)^2 \right], \quad (140)$$

where $a(k) > 0$ is such that

$$a(k) e^{2a(k)} = \frac{T k^2}{1 - e^{-T}}. \quad (141)$$

It is easy to prove that the equalities in (140) and (141) imply the following estimate:

$$\widetilde{I}_T(k) \geq \frac{1}{2T} [a(k) + a(k)^2]. \quad (142)$$

Define a function on the set $[0, \infty)$ by $h(u) = u e^u$. The function h is strictly increasing and continuous. It follows from (141) that

$$a(k) = \frac{1}{2} h^{-1} \left(\frac{2 T k^2}{1 - e^{-T}} \right). \quad (143)$$

In (143), the symbol h^{-1} stands for the inverse function of the function h.

Remark 77. Using the Lagrange inversion formula, we can represent the function $h^{-1}(y)$ by its Taylor series at $y = 0$. This gives

$$h^{-1}(y) = \sum_{n=1}^{\infty} (-1)^{n-1} \frac{n^{n-1}}{n!} y^n. \qquad (144)$$

It follows from the ratio test that the radius of convergence R of the alternating power series appearing in (144) is given by $R = \frac{1}{e}$. Moreover, it is not hard to prove that for $0 < y < \frac{1}{e}$, the absolute values of the terms of the series in (144) decrease. The proof is based on the fact that the sequence $(1 + \frac{1}{n})^n$ increases and its limit as $n \to \infty$ is equal to e. It follows from what was said above that

$$h^{-1}(y) \geq y - y^2, \qquad (145)$$

provided that $0 < y < \frac{1}{e}$.

Remark 78. By taking into account the previous reasoning, especially the estimates in (142), (143), and (145), we obtain an estimate from the following for the rate function $\widetilde{I}_T$ in the toy model.

Lemma 79. *Suppose $T > 0$ and $0 < k < \sqrt{\frac{1-e^{-T}}{2Te}}$. Then, the following estimate holds true:*

$$\widetilde{I}_T(k) \geq \frac{k^2(e-1)}{2e(1-e^{-T})}. \qquad (146)$$

Proof. The condition in Lemma (79) is equivalent to the following: $\frac{2Tk^2}{1-e^{-T}} < \frac{1}{e}$. Now, (143) and (145) imply that

$$a(k) \geq \frac{1}{2}\left(\frac{2Tk^2}{1-e^{-T}} - \frac{4T^2k^4}{(1-e^{-T})^2}\right) \geq \frac{Tk^2}{1-e^{-T}}\left(1 - \frac{1}{e}\right) = \frac{Tk^2(e-1)}{e(1-e^{-T})}.$$

Finally, (142) implies the estimate in (146).

This completes the proof of Lemma 79.

Our next goal is to obtain an estimate from above for the rate function in the toy model.

Corollary 80. *Let $T > 0$ and $k > 0$. Then,*

$$\widetilde{I}_T(k) \leq \frac{k^2}{1-e^{-T}}. \qquad (147)$$

Proof. The estimate in (147) can be derived from (139) by plugging the function f that is identically equal to zero into the expression on the right-hand side of (139).

Theorem 81. *For $T > 0$ and $0 < k < \sqrt{\frac{1-e^{-T}}{2Te}}$, the following two-sided estimates hold for the rate function in the toy model:*

$$\frac{k^2(e-1)}{2e(1-e^{-T})} \leq \widetilde{I}_T(k) \leq \frac{k^2}{1-e^{-T}}. \tag{148}$$

Theorem 81 can be derived from (146) and (147).

The next assertion provides two-sided estimates by constant functions for the small-noise limit of the implied volatility in the toy model.

Theorem 82. *For all $T > 0$ and $0 < k < \sqrt{\frac{1-e^{-T}}{2Te}}$, the following inequalities are valid:*

$$\frac{\sqrt{1-e^{-T}}}{\sqrt{2T}} \leq \lim_{\varepsilon \to 0} V(\varepsilon, T, k) \leq \frac{\sqrt{e(1-e^{-T})}}{\sqrt{T(e-1)}}. \tag{149}$$

Proof. The estimates in (149) follow from Theorem 81 and the equality in (134).

This completes our analysis of the toy model.

Asian Options

In this subsection, we study a small-noise asymptotic behavior of the price of an Asian option in the stochastic volatility model defined by (111). A small-noise Asian option is a path-dependent option with the payoff $(\frac{1}{T}\int_0^T S_t^{(\varepsilon)} dt - K)^+$. Here $K > 0$ is the strike, $T > 0$ is the maturity of the option, $\varepsilon \in (0, 1]$ is the scaling parameter, and it is assumed that K and T are fixed. The price of the Asian option in the small-noise setting is given by

$$A(\varepsilon, K, T) = e^{-rT} \mathbb{E}\left[\left(\frac{1}{T}\int_0^T S_t^{(\varepsilon)} dt - K\right)^+\right].$$

For every $a > 0$, define a subset of $\mathbb{C}$ by

$$G(a, T) = \left\{f \in \mathbb{C} : \frac{1}{T}\int_0^T \exp\{f(t)\} dt > a\right\}.$$

It is easy to see that the set $G(a, T)$ is open in $\mathbb{C}[0, T]$, and its closure in $\mathbb{C}[0, T]$ is given by

$$\overline{G(a, T)} = \left\{f \in \mathbb{C} : \frac{1}{T}\int_0^T \exp\{f(t)\} dt \geq a\right\}.$$

It is easy to see that for the model in (111), the representation of the rate function $\widetilde{Q}_T$ given in (33) and (34) is as follows:

$$\widetilde{Q}_T(g) = \frac{1}{2} \inf_{f \in \mathbb{H}_0^1} \int_0^T \left[\frac{(\dot{g}(s) - r - \rho\sigma(s, \widehat{f}(s))\dot{f}(s))^2}{(1-\rho^2)\sigma(s, \widehat{f}(s))^2} + \dot{f}(s)^2 \right] ds \quad (150)$$

for all $g \in \mathbb{H}_0^1$, and $\widetilde{Q}_T(g) = \infty$ for all $g \in \mathbb{C}\setminus\mathbb{H}_0^1$.

We next characterize small-noise asymptotics of the price of the Asian option.

Theorem 83. *Suppose Assumption A and Assumptions (C1)–(C7) hold true for the model in (111) defined on the canonical set-up. Further suppose the condition in (116) holds and Assumption B is satisfied. Then, the following asymptotic formula is valid:*

$$\varepsilon \log A(\varepsilon, K, T) = - \inf_{f \in G(\mathcal{K}, T)} \widetilde{Q}_T(f) + o(1) \quad (151)$$

as $\varepsilon \to 0$ where the good rate function $\widetilde{Q}_T$ is defined in (33) and (34) and $\mathcal{K}$ is the moneyness given by $\mathcal{K} = \frac{K}{s_0}$.

Proof. We first obtain a lower large deviation estimate for $A(\varepsilon, K, T)$. Let $\delta > 0$. It is easy to see that

$$A(\varepsilon, K, T) \geq \delta e^{-rT} \mathbb{P}\left(\frac{1}{T}\int_0^T S_t^{(\varepsilon)} dt > K + \delta\right)$$

$$= \delta e^{-rT} \mathbb{P}\left(\frac{1}{T}\int_0^T \exp\{X_t^{(\varepsilon)} - x_0\} dt > (K+\delta)s_0^{-1}\right)$$

$$= \delta e^{-rT} \mathbb{P}\left(X^{(\varepsilon)} - x_0 \in G((K+\delta)s_0^{-1}, T)\right).$$

Next, using the LDP in Theorem 24, we obtain

$$\liminf_{\varepsilon \to 0} \varepsilon \log A(\varepsilon, K, T) \geq - \inf_{\delta > 0} \inf_{f \in G((K+\delta)s_0^{-1}, T)} \widetilde{Q}_T(f). \quad (152)$$

Let $f \in G(Ks_0^{-1}, T)$. Then, $f \in G((K+\delta)s_0^{-1}, T)$ for some $\delta > 0$. Here δ depends on f. Next, using (152), we see that

$$\liminf_{\varepsilon \to 0} \varepsilon \log A(\varepsilon, K, T) \geq - \inf_{f \in G(Ks_0^{-1}, T)} \widetilde{Q}_T(f).$$

Recall that the moneyness is defined by $\mathcal{K} = \frac{K}{s_0}$. Then, we have

$$\liminf_{\varepsilon \to 0} \varepsilon \log A(\varepsilon, K, T) \geq - \inf_{f \in G(\mathcal{K}, T)} \widetilde{Q}_T(f). \tag{153}$$

The formula in (153) provides a lower LDP-style estimate for the price of the Asian option.

We next obtain the corresponding upper estimate. Let $p > 1$ and $q = \frac{p}{p-1}$. It is not hard to see that

$$A(\varepsilon, K, T) \leq e^{-rT} T^{-\frac{1}{p}} \left\{ \int_0^T \mathbb{E}\left[|S_t^{(\varepsilon)}|^p\right] dt \right\}^{\frac{1}{p}} \mathbb{P}\left(\frac{1}{T} \int_0^T S_t^{(\varepsilon)} \geq K\right)^{\frac{1}{q}}$$

$$= e^{-rT} T^{-\frac{1}{p}} \left\{ \int_0^T \mathbb{E}\left[|S_t^{(\varepsilon)}|^p\right] dt \right\}^{\frac{1}{p}} \mathbb{P}(X^{(\varepsilon)} - x_0 \in \overline{G(\mathcal{K}, T)})^{\frac{1}{q}}.$$

Therefore,

$$\limsup_{\varepsilon \to 0} \varepsilon \log A(\varepsilon, K, T) \leq \frac{1}{p} \limsup_{\varepsilon \to 0} \varepsilon \log \int_0^T \mathbb{E}\left[|S_t^{(\varepsilon)}|^p\right] dt$$
$$+ \frac{1}{q} \limsup_{\varepsilon \to 0} \varepsilon \log \mathbb{P}(X^{(\varepsilon)} - x_0 \in \overline{G(\mathcal{K}, T)}). \tag{154}$$

Using the LDP in Theorem 24 and the estimate in (154), we obtain

$$\limsup_{\varepsilon \to 0} \varepsilon \log A(\varepsilon, K, T) \leq \frac{1}{p} \limsup_{\varepsilon \to 0} \varepsilon \log \int_0^T \mathbb{E}\left[|S_t^{(\varepsilon)}|^p\right] dt$$
$$- \frac{1}{q} \inf_{f \in \overline{G(\mathcal{K}, T)}} \widetilde{Q}_T(f). \tag{155}$$

Our next goal is to estimate the first term on the right-hand side of (155). Using (124), we get

$$\int_0^T \mathbb{E}\left[|S_t^{(\varepsilon)}|^p\right] dt \leq s_0^p e^{prT} T M_p.$$

It follows that

$$\limsup_{\varepsilon \to 0} \varepsilon \log \int_0^T \mathbb{E}\left[|S_t^{(\varepsilon)}|^p\right] dt = 0. \tag{156}$$

Now, (155) and (156) imply

$$\limsup_{\varepsilon \to 0} \varepsilon \log A(\varepsilon, K, T) \leq - \inf_{f \in \overline{G(\mathcal{K}, T)}} \widetilde{Q}_T(f). \tag{157}$$

Our next goal is to prove that

$$\inf_{f \in \overline{G(\mathcal{K},T)}} \widetilde{Q}_T(f) = \inf_{f \in G(\mathcal{K},T)} \widetilde{Q}_T(f). \tag{158}$$

It follows from (34) that (158) can be rewritten as follows:

$$\inf_{f \in \overline{G(\mathcal{K},T)} \cap \mathbb{H}_0^1} \widetilde{Q}_T(f) = \inf_{f \in G(\mathcal{K},T) \cap \mathbb{H}_0^1} \widetilde{Q}_T(f). \tag{159}$$

Let us recall that since the volatility function σ in the model in (4) satisfies the condition in (116), the rate function defined in (33) is continuous on the space $\mathbb{H}_0^1$ (see Lemma 27). It is not hard to prove that the set $G(\mathcal{K},T) \cap \mathbb{H}_0^1$ is dense in the set $\overline{G(\mathcal{K},T)} \cap \mathbb{H}_0^1$. It follows that the equalities in (159) and (158) hold. Finally, (153), (157), and (158) imply (151).

The proof of Theorem 83 is thus completed.

Our next goal is to describe what happens if we remove the condition in (116) from the formulation of Theorem 83. In such an environment, the formula in (150) may not hold for all $f \in \mathbb{H}_0^1$. However, the set of functions for which (150) does not hold consists of at most one function. Indeed, suppose the set B_2 defined in Remark 31 is not empty. Let $f \in B_2$. Then, the equation $\Phi(l, f, \widehat{f}(t)) = g(t)$ in (28) becomes $rt = g(t)$. Here we use (27). It follows that the representation in (150) holds for all functions $g \in \mathbb{H}_0^1$, except, maybe, the function $\tilde{g}(t) = rT$, $t \in [0,T]$. Moreover, by reasoning as in the proof of Lemma 27, we can establish that the rate function in formula (150) is continuous on the set $\mathbb{H}_0^1 \setminus \{\tilde{g}\}$.

Theorem 84. *Suppose Assumption A and Assumptions (C1)–(C7) hold true for the model in (111) defined on the canonical set-up. Further suppose that Assumption B is satisfied. Then, for all $K > \frac{s_0}{rT}(e^{rT} - 1)$, the following asymptotic formula holds:*

$$\varepsilon \log A(\varepsilon, K, T) = - \inf_{f \in \overline{G(\mathcal{K},T)}} \widetilde{Q}_T(f) + o(1) \tag{160}$$

as $\varepsilon \to 0$ where the good rate function $\widetilde{Q}_T$ is defined in (33) and (34) and $\mathcal{K}$ is the moneyness given by $\mathcal{K} = \frac{K}{s_0}$.

Remark 85. The restriction on the strike K in Theorem 84 is needed in order to take into account the existence of the exceptional function $\tilde{g}(t) = rT$. This is explained in the proof in the following. If $r = 0$, then this restriction becomes $K > s_0$.

Proof of Theorem 84. Using the discussion before the formulation of Theorem 84, we see that Theorem 84 can be established exactly as

Theorem 83. The only difference is that the strike K should be restricted so that the exceptional function $\tilde{g}$ does not belong to the set $\overline{G(\mathcal{K},T)}$. The previous condition means that $\frac{1}{T}\int_0^T e^{rt}dt < \frac{K}{s_0}$. The previous inequality is equivalent to $K > \frac{s_0}{rT}(e^{rT}-1)$.

This completes the proof of Theorem 84.

Assumption B, Revisited

In this subsection, we provide examples of stochastic volatility models for which the asymptotic formulas for call options, the implied volatility, and Asian options obtained in sections "Call Options" to "Asian Options" hold true. Recall that the results obtained in those subsections use Assumption B together with Assumption A and Assumptions (C1)–(C7). Our main goal in this subsection is to study scaled volatility processes $\widehat{B}^{(\varepsilon)}$ for which Assumption B holds. However, our knowledge here is far from being complete.

In the following definition, we introduce a sublinear growth condition for the volatility function in (111).

Definition 86. It is said that the volatility function $\sigma : [0,T] \times \mathbb{R}^d \mapsto \mathbb{R}$ in (111) satisfies a sublinear growth condition uniformly with respect to $t \in [0,T]$ provided that there exists a constant $C > 0$ such that $\sigma(t,x)^2 \leq C(1+||x||_d^2)$ for all $(t,x) \in [0,T] \times \mathbb{R}^d$.

The sublinear growth condition in Definition 86 allows us to work with a simpler inequality than that in Assumption B.

Remark 87. Suppose the sublinear growth condition holds for the volatility function σ. Suppose also that the following condition is satisfied: For every $\beta > 0$, there exists $\varepsilon_1 \in (0,1]$ depending only on β and such that for all $0 < \varepsilon < \varepsilon_1$, the following estimate holds true:

$$\mathbb{E}\left[\exp\left\{\beta \int_0^T ||\widehat{B}_s^{(\varepsilon)}||_d^2 ds\right\}\right] \leq L, \tag{161}$$

where $L > 0$ is a constant independent of ε. Then, Assumption B holds. The previous statement can be easily established.

Remark 88. It is not hard to see that if the condition in (161) holds for a finite family of scaled volatility processes, then the same condition holds for the sum of those processes. In addition, if the condition in (161) is satisfied for every component of the process $\widehat{B}^{(\varepsilon)}$, then it is satisfied for the process $\widehat{B}^{(\varepsilon)}$.

Example 1. Multivariate Gaussian Models.

The scaled volatility process in a multivariate Gaussian stochastic volatility model is given by $\widehat{B}^{(\varepsilon)} = (\widehat{B}^{1,(\varepsilon)}, \ldots, \widehat{B}^{d,(\varepsilon)})$ where

$$\widehat{B}_t^{i,(\varepsilon)} = x_i + \sqrt{\varepsilon} \sum_{j=1}^m \int_0^t K_{ij}(t,s) dB_s^{(j)}, \quad t \in [0,T], \quad 1 \leq i \leq d.$$

In the previous formula, $\{K_{ij}\}$, with $1 \leq i \leq d$ and $1 \leq j \leq m$, is a family of admissible Volterra-type Hilbert–Schmidt kernels for which Fernique's condition is satisfied (see (58)).

Lemma 89. *The inequality in (161) holds for a multivariate Gaussian model defined on any set-up.*

Proof. It follows from Remark 88 that it suffices to prove the following: For every $\beta > 0$, there exists $\varepsilon_1 \in (0,1]$ depending only on β and such that for all $0 < \varepsilon < \varepsilon_1$, $1 \leq i \leq d$, and $1 \leq j \leq m$,

$$\mathbb{E}\left[\exp\left\{\beta\varepsilon \int_0^T |\int_0^s K_{ij}(s,u) dB_u^{(j)}|^2 ds\right\}\right] \leq L$$

for some $L > 0$ independent of ε. The previous inequality was established in Lemma 34 in [44]. Note that the Gaussian Volterra processes used in [44] satisfy a stronger condition than Assumption F used in this paper (see section "Gaussian Stochastic Volatility Models" where Assumption F is introduced). However, the proof of Lemma 34 in [44] does not change if we assume that the Gaussian Volterra process satisfies Assumption F.

The proof of Lemma 89 is thus completed.

Example 2. Generalized Fractional Heston Models.

The Heston model was introduced in [48]. The asset price process S and the variance process V in the Heston model satisfy the following system of stochastic differential equations:

$$dS_t = S_t \left[r dt + \sqrt{V_t}\left(\sqrt{1-\rho^2} dW_t + \rho dB_t\right)\right],$$
$$dV_t = \kappa(\theta - V_t) dt + \eta\sqrt{V_t} dB_t. \tag{162}$$

In (162), W and B are independent Brownian motions, $\kappa > 0$, $\theta > 0$, and $\eta > 0$ are positive parameters, $r \geq 0$ is the interest rate, and $\rho \in (-1,1)$ is the correlation coefficient. The initial conditions for the processes S and V are denoted by s_0 and v_0, respectively, and it is assumed that $s_0 > 0$ and

$v_0 > 0$. The variance process V in the Heston model is the Cox–Ingersoll–Ross process (CIR-process). More information about the CIR-process can be found in [45]. A scaled version of the Heston model is as follows:

$$dS_t^{(\varepsilon)} = S_t^{(\varepsilon)} \left[rdt + \sqrt{\varepsilon}\sqrt{V_t^{(\varepsilon)}} \left(\sqrt{1-\rho^2}dW_t + \rho dB_t \right) \right],$$

$$dV_t^{(\varepsilon)} = \kappa \left(\theta - V_t^{(\varepsilon)} \right) dt + \sqrt{\varepsilon}\eta\sqrt{V_t^{(\varepsilon)}}dB_t. \tag{163}$$

It is assumed in (163) that $S_0^{(\varepsilon)} = s_0$ and $V_0^{(\varepsilon)} = v_0$ for all $\varepsilon \in (0,1]$.

Remark 90. Note that for every $\varepsilon \in (0,1]$, the scaled variance process $V^{(\varepsilon)}$ is a CIR process, with the same parameters κ and θ as the process V and with the parameter η replaced by $\sqrt{\varepsilon}\eta$.

Our next goal is to introduce generalized fractional Heston models. We consider such models defined on the canonical set-up. Let K be a non-negative admissible Volterra-type kernel satisfying Assumption F (see section "Gaussian Stochastic Volatility Models"). An important example of such a kernel is the kernel of the Riemann–Liouville fractional Brownian motion. Recall that this kernel is given by the following formula: $K_H(t,s) = \Gamma(H+1/2)^{-1}(t-s)^{H-\frac{1}{2}}\mathbb{1}_{s<t}$ (see (59)). Another interesting example is the kernel $\widetilde{K}(t,s) = |K(t,s)|$ where K is an admissible Volterra-type Hilbert–Schmidt kernel satisfying Assumption F. It is not hard to prove that the kernel $\widetilde{K}$ is also an admissible Volterra-type Hilbert–Schmidt kernel satisfying Assumption F. A special example of such a kernel is given by $\widetilde{K}_H(t,s) = |K_H(t,s)|$ where K_H is the Volterra-type kernel of fractional Brownian motion (see section "Gaussian Stochastic Volatility Models").

Let V be a CIR-process with parameters κ, θ, and η defined in (162), and let $V^{(\varepsilon)}$ be a scaled version of this process defined in (163). Consider the following stochastic model for the volatility process:

$$\widehat{B}_t = x + \int_0^t K(t,s)V_s ds, \quad x \geq 0. \tag{164}$$

The scaled version of the process in (164) is given by

$$\widehat{B}_t^{(\varepsilon)} = x + \int_0^t K(t,s)V_s^{(\varepsilon)} ds, \quad x \geq 0, \quad \varepsilon \in (0,1]. \tag{165}$$

Definition 91. The stochastic model given by

$$dS_t = S_t \left[rdt + \sqrt{V_t} \left(\sqrt{1-\rho^2}dW_t + \rho dB_t \right) \right],$$

is called a generalized fractional Heston model. The scaled version of the model in (166) is as follows:

$$dV_t = \kappa(\theta - V_t)dt + \eta\sqrt{V_t}dB_t,$$
$$\widehat{B}_t = x + \int_0^t K(t,s)V_s ds \qquad (166)$$

$$dS_t^{(\varepsilon)} = S_t^{(\varepsilon)}\left[rdt + \sqrt{\widehat{B}_t^{(\varepsilon)}}\left(\sqrt{1-\rho^2}dW_t + \rho dB_t\right)\right],$$
$$dV_t^{(\varepsilon)} = \kappa(\theta - V_t^{(\varepsilon)})dt + \sqrt{\varepsilon}\eta\sqrt{V_t^{(\varepsilon)}}dB_t,$$
$$\widehat{B}_t^{(\varepsilon)} = x + \int_0^t K(t,s)V_s^{(\varepsilon)}ds. \qquad (167)$$

Remark 92. If $K = K_H$, where K_H is the kernel of the Riemann–Liouville fractional Brownian motion, then the model in (166) is the fractional Heston model studied in [40].

Using Theorem 47 and Remark 48, we see that Assumptions (C1)–(C7) hold true for the volatility model introduced in (164). Therefore, Theorems 38 and 41 hold for the scaled volatility process $\widehat{B}^{(\varepsilon)}$ appearing in (165). It follows that the scaled log-price process in the generalized fractional Heston model satisfies the LDP in Theorem 24, provided that the model is defined on the canonical set-up.

Lemma 93. *Assumption B holds for the generalized fractional Heston model.*

Proof. For the sake of simplicity, we assume that $T = 1$. The statement in Lemma (93) means the following: For every $\alpha > 0$ there exists $\varepsilon_0 \in (0,1]$ depending only on α and such that for all $0 < \varepsilon < \varepsilon_0$, the following estimate holds true:

$$\mathbb{E}\left[\exp\left\{\alpha\int_0^1 ds \int_0^s K(s,u)V_u^{(\varepsilon)}du\right\}\right] \leq M, \qquad (168)$$

where $M > 0$ is a constant depending only on α.

The scaled CIR-process $V^{(\varepsilon)}$ appearing in (168) is given by $V_t^{(\varepsilon)} = \varepsilon V_t$, $t \in (0,1]$, $\varepsilon \in (0,1]$ (see the last line on page 2 in [39]). Let us denote the expression on the left-hand side of (168) by $H(\varepsilon, \alpha)$. It is not hard to show that $H(\varepsilon, \alpha) \leq \mathbb{E}\left[\exp\left\{\alpha C\varepsilon \max_{u\in[0,1]} V_u\right\}\right]$, where $C = \sup_{s\in[0,1]} \int_0^s K(s,u)^2 du\}^{\frac{1}{2}}$. We have $C < \infty$ (see the proof of (68)).

Next, analyzing the reasoning above, we see that in order to complete the proof of Lemma 93, it suffices to show that there exists $\delta > 0$ such that

$$\mathbb{E}\left[\exp\left\{\delta \max_{u\in[0,1]} V_u\right\}\right] < \infty. \tag{169}$$

We establish the inequality in (169) using Proposition 2.1 in [24]. The authors of [24] use a scaled process $X^{(\varepsilon)}$ satisfying $dX_t^{(\varepsilon)} = b(X_t^{(\varepsilon)})dt + 2\varepsilon\sqrt{|X_t^{(\varepsilon)}|}dB_t$, $X_0^{(\varepsilon)} = a \geq 0$ for all $\varepsilon \in (0,1]$. Set $b(u) = \kappa(\theta - u)$, $\varepsilon = \frac{\eta}{2}$, and $a = v_0$. Then, we have

$$X_t^{(\frac{\eta}{2})} = V_t, \quad 0 \leq t \leq 1. \tag{170}$$

It was established in [24], Proposition 2.1 that there exists $\lambda > 0$ such that for every $\varepsilon > 0$,

$$\mathbb{E}\left[\exp\left\{\lambda \varepsilon^{-2} \max_{t\in[0,1]} X_t^{(\varepsilon)}\right\}\right] \leq \exp\left\{\varepsilon^{-2} + 1\right\}. \tag{171}$$

Next, using (170) and (171), with $\varepsilon = \frac{\eta}{2}$, we obtain

$$\mathbb{E}\left[\exp\left\{\frac{4\lambda}{\eta^2} \max_{t\in[0,1]} V_t\right\}\right] \leq \exp\left\{\frac{4}{\eta^2} + 1\right\}.$$

Therefore, the estimate in (169) holds, with $\delta = \frac{4\lambda}{\eta^2}$.
The proof of Lemma 93 is thus completed.

Proof of Theorem 38

The proof of Theorem 38 is similar in structure to that of Theorem 2.1 in [17]. However, there are certain differences between the proofs since the models that we use in this paper are more general than those studied in [17].

The next assertion (Theorem 94) states that Laplace's principle holds for the process $Y^{(\varepsilon)}$. It is known that if the rate function is good, then the LDP in Theorem 38 and Laplace's principle in Theorem 94 are equivalent. Since the rate function I_y is good (see Remark 39), Theorem 38 can be derived from Theorem 94.

Theorem 94. *Suppose the conditions in Theorem 38 hold. Then, for all bounded and continuous functions* $F : \mathcal{W}^d \mapsto \mathbb{R}$,

$$\lim_{\varepsilon \to 0} -\varepsilon \log \mathbb{E}\left[\exp\left\{-\frac{1}{\varepsilon} F(Y^{(\varepsilon)})\right\}\right] = \inf_{\phi \in \mathcal{W}^d} [I_y(\phi) + F(\phi)]$$

where the rate function I_y *is defined by* (50).

Proof. Lower bound in Laplace's principle. We have to show that

$$\lim_{\varepsilon \to 0} -\varepsilon \log \mathbb{E}\left[\exp\left\{-\frac{1}{\varepsilon}F(Y^{(\varepsilon)})\right\}\right] \geq \inf_{\phi \in \mathcal{W}^d}[I_y(\phi) + F(\phi)]. \tag{172}$$

It suffices to prove that for any sequence $\varepsilon_n \in (0, 1]$, $n \geq 1$, such that $\varepsilon_n \to 0$ as $n \to \infty$, there exists a subsequence along which the inequality in (172) holds. Let $Y^{(\varepsilon_n)}$, with $n \geq 1$, be the strong solution to equation (10) with $\varepsilon = \varepsilon_n$ (see Assumption (C3)). Then, there exists a map $h^{(n)} : \mathcal{W}^m \mapsto \mathcal{W}^d$ such that it is $\widetilde{\mathcal{B}}_t^m/\mathcal{B}_t^d$-measurable for all $t \in [0, T]$, and, moreover,

$$Y^{(\varepsilon_n)} = h^{(n)}(B) \quad \mathbb{P}-\text{a.s.} \tag{173}$$

(see Remark 6).

We use a variational representation of functionals of Brownian motion (see, e.g. Theorem 3.6 in the paper of Budhiraja and Dupuis [11]).

Theorem 95. *Let f be a bounded Borel measurable real function on $\mathcal{W}^m$. Then,*

$$-\log \mathbb{E}[\exp\{-f(B)\}] = \inf_{v \in \mathcal{M}^2[0,T]} \mathbb{E}\left[\frac{1}{2}\int_0^T \|v_s\|_m^2 ds + f\left(B + \int_0^\cdot v_s ds\right)\right].$$

For the function $F : \mathcal{W}^d \mapsto \mathbb{R}$ appearing in the formulation of Theorem 94, the representation formula in Theorem 95 implies the following:

$$-\varepsilon_n \log \mathbb{E}\left[\exp\left\{-\frac{1}{\varepsilon_n}F(Y^{(\varepsilon_n)})\right\}\right]$$

$$= -\varepsilon_n \log \mathbb{E}\left[\exp\left\{-\frac{1}{\varepsilon_n}F \circ h^{(n)}(B)\right\}\right]$$

$$= \varepsilon_n \inf_{v \in \mathcal{M}^2[0,T]} \mathbb{E}\left[\frac{1}{2}\int_0^T \|v_s\|_m^2 ds + \frac{1}{\varepsilon_n}F \circ h^{(n)}\left(B + \int_0^\cdot v_s ds\right)\right]$$

$$= \inf_{v \in \mathcal{M}^2[0,T]} \mathbb{E}\left[\frac{1}{2}\int_0^T \|v_s\|_m^2 ds + F \circ h^{(n)}\left(B + \frac{1}{\sqrt{\varepsilon_n}}\int_0^\cdot v_s ds\right)\right]. \tag{174}$$

The last equality in (174) is obtained by passing from v to $\frac{v}{\sqrt{\varepsilon_n}}$.

Fix $\delta > 0$. It can be shown exactly as in [17] that there exists $N > 0$ such that for every $n \geq 1$, a control $v^{(n)}$ can be found satisfying

$$v^{(n)} \in M_N^2[0, T] \tag{175}$$

and

$$-\varepsilon_n \log \mathbb{E}\left[\exp\left\{-\frac{1}{\varepsilon_n} F(Y^{(\varepsilon_n)})\right\}\right]$$
$$\geq \mathbb{E}\left[\frac{1}{2}\int_0^T \|v_s^{(n)}\|_m^2 ds F \circ h^{(n)}\left(B + \frac{1}{\sqrt{\varepsilon_n}}\int_0^{\cdot} v_s^{(n)} ds\right)\right] - \delta. \quad (176)$$

The controls $v^{(n)}$ in (176) depend on δ. Let N and v_n be such as in (175) and (176). For every $n \geq 1$, consider the process B^{ε_n, v^n} defined by (20). For the sake of shortness, we use the symbol $B^{(n)}$ instead of B^{ε_n, v^n}. The process $B^{(n)}$ is an m-dimensional Brownian motion on $\mathcal{W}^m$ with respect to a measure $\mathbb{P}^{(n)}$ on $\mathcal{B}_T^m$ that is equivalent to the measure $\mathbb{P}$. This process is given by $B_s^{(n)} = B_s + \frac{1}{\sqrt{\varepsilon_n}}\int_0^s v_u^{(n)} du$, $s \in [0,T]$. The process $B^{(n)}$ is adapted to the filtration $\{\mathcal{B}_t^m\}$.

Fix $n \geq 1$, and consider the following scaled controlled stochastic integral equation:

$$Y_t^{\varepsilon_n, v^{(n)}} = y + \int_0^t a\left(t, s, V^{1,\varepsilon_n, v^{(n)}}, Y^{\varepsilon_n, v^{(n)}}\right) ds$$
$$+ \int_0^t c\left(t, s, V^{2,\varepsilon_n, v^{(n)}}, Y^{\varepsilon_n, v^{(n)}}\right) v_s^{(n)}\right) ds$$
$$+ \sqrt{\varepsilon_n} \int_0^t c\left(t, s, V^{2,\varepsilon_n, v^{(n)}}, Y^{\varepsilon_n, v^{(n)}}\right) dB_s \quad (177)$$

where the processes $V^{i,\varepsilon_n, v^{(n)}}$, with $i = 1, 2$, satisfy

$$V_s^{i,\varepsilon_n, v^{(n)}} = V_0^{(i)} + \int_0^s \bar{b}_i\left(r, V^{i,\varepsilon_n, v^{(n)}}\right) dr + \int_0^s \bar{\sigma}_i\left(r, V^{i,\varepsilon_n, v^{(n)}}\right) v_r^{(n)} dr$$
$$+ \sqrt{\varepsilon_n} \int_0^s \bar{\sigma}_i\left(r, V^{i,\varepsilon_n, v^{(n)}}\right) dB_r. \quad (178)$$

We next show that for every $n \geq 1$, equation (177) possesses a strong solution. This solution is unique, by Assumption (C3)(b). We also provide a representation formula for the unique solution.

Lemma 96. *Let $v \in \mathcal{M}^2[0,T]$ be such that (12) holds, and suppose Assumptions (C1)–(C3) and (C6) are satisfied. Then, equation (13) has a strong solution $Y^{(v)}$ with $Y_0^{(v)} = y$. Moreover, the following formula holds $\mathbb{P}$-a.s.: $Y^{(v)} = h\left(B^{(v)}\right)$, where h and $B^{(v)}$ are defined in Remark 18 and (21), respectively.*

Remark 97. Lemma 96 is similar to Lemma A.1 in [17].

Remark 98. The existence of the strong solution $Y^{\varepsilon_n,v^{(n)}}$ to equation (177) and the representation formula

$$Y^{\varepsilon_n,v^{(n)}} = h^{(n)}(B^{\varepsilon_n,v^{(n)}}), \quad n \geq 1, \tag{179}$$

where $h^{(n)}$ and $B^{\varepsilon,v}$ are defined in (173) and (20), respectively, can be established as follows. Fix $n \geq 1$, and replace the control v by $\frac{v^{(n)}}{\sqrt{\varepsilon_n}}$, the maps $\bar{\sigma}_i$ and c by $\sqrt{\varepsilon_n}\bar{\sigma}_i$ and $\sqrt{\varepsilon_n}c$, respectively, and also replace h by $h^{(n)}$. After making such replacements, we can apply Lemma 96 to establish the formula in (179).

Proof of Lemma 96. Using (8) and (9), we obtain

$$h(B) = y + \int_0^\cdot a(\cdot, s, V^{(1)}, h(B)) ds + \int_0^\cdot c(\cdot, s, V^{(2)}, h(B)) dB_s. \tag{180}$$

In (180), $V^{(i)}$ are the unique solutions to the equations

$$V_s^{(i)} = V_0^{(i)} + \int_0^s \bar{b}_i(r, V^{(i)}) dr + \int_0^s \bar{\sigma}_i(r, V^{(i)}) dB_r, \quad i = 1, 2. \tag{181}$$

Since the conditions in Theorem 10.4 in [71] are satisfied for equation (181), there exist maps $g^{(i)} : \mathcal{W}^m \mapsto \mathcal{W}^{k_i}$, with $i = 1, 2$, such that (i) the map $g^{(i)}$ is $\widetilde{\mathcal{B}}_t^m/\mathcal{B}_t^{k_i}$-measurable for every $t \in [0,T]$, (ii) $V^{(i)} = g^{(i)}(B)$ for $i = 1, 2$, and (iii) for every $i = 1, 2$,

$$g^{(i)}(B^{(v)}) = V_0^{(i)} + \int_0^s \bar{b}_i(r, g^{(i)}(B^{(v)})) dr + \int_0^s \bar{\sigma}_i(r, g^{(i)}(B^{(v)})) dB_s^{(v)}$$

$$= V_0^{(i)} + \int_0^s \bar{b}_i(r, g(i)(B^{(v)})) dr + \int_0^s \bar{\sigma}_i(r, g^{(i)}(B^{(v)})) v_s ds$$

$$+ \int_0^s \bar{\sigma}_i(r, g^{(i)}(B^{(v)})) dB_s. \tag{182}$$

Next, using (182), we see that for every $i = 1, 2$, the equality $V^{i,v} = g^{(i)}(B^{(v)})$ holds. Therefore, equation (180) can be rewritten as follows:

$$h(B) = y + \int_0^\cdot a(\cdot, s, g^{(1)}(B), h(B)) ds + \int_0^\cdot c(\cdot, s, g^{(2)}(B), h(B)) dB_s. \tag{183}$$

It is not hard to see that if we could replace B by $B^{(v)}$ in (183), then Lemma 96 is established. Here we take into account Assumption (C3)(b). There is no problem in replacing B by $B^{(v)}$ in the expression on the left-hand side of (183) and in the first integral on the right-hand side of (183).

We next explain how to deal with the stochastic integral appearing in (183). By the second inequality in (22),

$$\int_0^t c_{lj}(t,s,g^{(2)}(B),h(B))^2 ds < \infty \quad \mathbb{P}-\text{a.s.} \tag{184}$$

for all $t \in [0,T]$, $1 \leq l \leq d$, and $1 \leq j \leq m$. It is clear that the paths of the process $h(B)$ are continuous. Therefore, (183) and the first restriction on the function a in Assumption (C2)(b) imply that the paths of the process

$$t \mapsto \int_0^t c(t,s,g^{(2)}(B),h(B))dB_s \tag{185}$$

are continuous as well.

Using Lemma 10.1 in [71], we can show that for every rational number $r \in [0,T]$ there exists a functional $u_r : \mathcal{W}^m \mapsto \mathcal{W}^d$ satisfying the following conditions:

(i) For every r, the functional u_r is $\widetilde{\mathcal{B}}_t^m/\mathcal{B}_t^d$-measurable for all $t \in [0,T]$.
(ii) $\mathbb{P}$-a.s. on the space $\mathcal{W}^m$, the equalities

$$u_r(B). = \int_0^\cdot c(r,s,g^{(2)}(B),h(B))\mathbb{1}_{\{s \leq r\}}dB_s$$

and

$$u_r(B^{(v)}). = \int_0^\cdot c(r,s,g^{(2)}(B^{(v)}),h(B^{(v)}))\mathbb{1}_{\{s \leq r\}}dB_s^{(v)}$$

hold for all r. The existence of the previous stochastic integrals follows from the second inequality in (22).

Using (ii), we see that $\mathbb{P}$-a.s. on $\mathcal{W}^m$, $u_r(B)(r) = \int_0^r c(r,s,g^{(2)}(B), h(B))dB_s$ and, in addition, $u_r(B^{(v)})(r) = \int_0^r c(r,s,g^{(2)}(B^{(v)}),h(B^{(v)}))dB_s^{(v)}$ for all $r \in [0,T]$. It follows from the continuity of the process in (185) and Assumption (C6) that the following statement holds true $\mathbb{P}$-a.s. on $\mathcal{W}^m$: For every $t \in [0,T]$ and every sequence r_i of rational numbers in $[0,T]$ such that $r_i \to t$,

$$\lim_{r_i \to t} u_{r_i}(B)(r_i) = \int_0^t c(t,s,g^{(2)}(B),h(B))dB_s^{(v)}$$

and

$$\lim_{r_i \to t} u_{r_i}(B^{(v)})(r_i) = \int_0^t c(t,s,g^{(2)}(B^{(v)}),h(B^{(v)}))dB_s^{(v)}. \tag{186}$$

Therefore, (183) implies that

$$h(B) = y + \int_0^\cdot a(\cdot,s,g^{(1)}(B),h(B))ds + u(B) \quad \mathbb{P}-\text{a.s.,} \tag{187}$$

where $u(B) = \lim_{r_i \to t} u_{r_i}(B)(r_i)$. It follows from (187) that $u : \mathcal{W}^m \mapsto \mathcal{W}^d$ is an $\widetilde{\mathcal{B}}_t^m / \mathcal{B}_t^d$-measurable functional for every $t \in [0,T]$. Since B is the coordinate process on $\mathcal{W}^m$, the mapping $s \mapsto B_s$ is the identity mapping on $\mathcal{W}^m$. Therefore, $u(\eta)(t) = \lim_{r_i \to t} u_{r_i}(\eta)(r_i)$ for all $t \in [0,T]$ $\mathbb{P}$-a.s. It follows that $u(B^{(v)}) = \lim_{r_i \to t} u_{r_i}(B^{(v)})(r_i)$ $\mathbb{P}$-a.s. The equality in (187) is a statement about some measurable functional of B and the measure $\mathbb{P}$. Since the distribution of B with respect to $\mathbb{P}$ is the same as the distribution of $B^{(v)}$ with respect to $\mathbb{P}^v$, we can replace B by $B^{(v)}$ and $\mathbb{P}$ by $\mathbb{P}^{(v)}$ in (187). Next, using (186) we see that

$$h(B^{(v)}) = y + \int_0^{\cdot} a(\cdot, s, g^{(1)}(B^{(v)}), h(B^{(v)})) ds$$
$$+ \int_0^t c(t, s, g^{(2)}(B^{(v)}), h(B^{(v)})) dB_s^{(v)}. \quad (188)$$

Finally, by comparing (13) and (188) and using Assumption (C3)(b), we see that the following equality holds: $Y^{(v)} = h\left(B^{(v)}\right)$.

This completes the proof of Lemma 96.

We next return to the proof of Theorem 94. Using formula 179 we can rewrite the estimate in (176) as follows:

$$-\varepsilon_n \log \mathbb{E}\left[\exp\left\{-\frac{1}{\varepsilon_n} F(Y^{(\varepsilon_n)})\right\}\right]$$
$$\geq \mathbb{E}\left[\frac{1}{2} \int_0^T \|v_s^{(n)}\|_m^2 ds + F\left(Y^{\varepsilon_n, v^{(n)}}\right)\right] - \delta. \quad (189)$$

Next, reasoning as in [17] we can show that the sequence $(Y^{\varepsilon_n, v^{(n)}}, V^{1,\varepsilon_n, v^{(n)}}, V^{2,\varepsilon_n, v^{(n)}}, v^{(n)})$, $n \geq 1$, is tight as a family of random variables with values in $\mathcal{W}^d \times \mathcal{W}^{k_1} \times \mathcal{W}^{k_2} \times D_N$ for some $N > 0$ where D_N is defined in Assumption (C5). The tightness of $\{Y^{\varepsilon_n, v^{(n)}}\}$ follows from Assumption (C7), while the tightness of $\{V^{1,\varepsilon_n, v^{(n)}}\}$ and $\{V^{2,\varepsilon_n, v^{(n)}}\}$ follows from (H6) in [17]. Hence, possibly taking a subsequence, we see that the sequence of random variables $(Y^{\varepsilon_n, v^{(n)}}, V^{1,\varepsilon_n, v^{(n)}}, V^{2,\varepsilon_n, v^{(n)}}, v^{(n)})$ converges in distribution as $n \to \infty$ to a $(\mathcal{W}^d \times \mathcal{W}^{k_1} \times \mathcal{W}^{k_2} \times D_N)$-valued random variable $(\widehat{Y}, \widehat{V}^{(1)}, \widehat{V}^{(2)}, v)$ that is defined on some probability space $(\widehat{\Omega}, \widehat{\mathcal{F}}, \widehat{\mathbb{P}})$.

Lemma 99. *The processes $\widehat{V}^{(1)}$, $\widehat{V}^{(2)}$, and $\widehat{Y}$ satisfy the following system of equations:*

$$\widehat{V}_s^{(i)} = V_0^{(i)} + \int_0^s \bar{b}_i(r, \widehat{V}^{(i)}) dr + \int_0^s \bar{\sigma}_i(r, \widehat{V}^{(i)}) v_r dr, \quad i = 1, 2,$$
$$\widehat{Y}_t = y + \int_0^t a(t, s, \widehat{V}^{(1)}, \widehat{Y}) ds + \int_0^t c(t, s, \widehat{V}^{(2)}, \widehat{Y}) v_s ds \quad (190)$$

$\widehat{\mathbb{P}}$-a.s. on $\widehat{\Omega}$.

Proof. In the proof of Lemma 99, we borrow some ideas from the proof on p. 1132 in [17]. However, there are significant differences between those proofs because of the difficulties caused by the extra processes $V^{(1)}$ and $V^{(2)}$.

For $\widehat{V}^{(i)}$, with $i = 1, 2$, the validity of the statement in Lemma 99 was established in [17] (see the proof of formula (12) in [17]). We next prove the same statement for $\widehat{Y}$. Let $t \in [0, T]$, and consider the map $\Psi_t : \mathcal{W}^d \times \mathcal{W}^{k_1} \times \mathcal{W}^{k_2} \times D_N \mapsto \mathbb{R}$ defined by

$$\Psi_t(\varphi, \tau_1, \tau_2, f) = \|\varphi(t) - y - \int_0^t a(t, s, \tau_1, \varphi) ds$$
$$- \int_0^t c(t, s, \tau_2, \varphi) f(s) ds\|_d \wedge 1.$$

It is clear that this map is bounded. Our next goal is to show that it is continuous. Let $\varphi_n \to \varphi$ in $\mathcal{W}^d$, $\tau_1^{(n)} \to \tau_1$ in $\mathcal{W}^{k_1}$, $\tau_2^{(n)} \to \tau_2$ in $\mathcal{W}^{k_2}$. Suppose also that $f_n \in D_N$, $f \in D_N$, and $f_n \to f$ in the weak topology of $L^2([0, T], \mathbb{R}^m)$ (see Assumption (C5) for the definition of D_N). We have

$$|\Psi_t(\varphi_n, \tau_1^{(n)}, \tau_2^{(n)}, f_n) - \Psi_t(\varphi, \tau_1, \tau_2, f)|$$
$$\leq \|\varphi_n(t) - \varphi(t)\|_d$$
$$+ \|\int_0^t a(t, s, \tau_1^{(n)}, \varphi_n) ds - \int_0^t a(t, s, \tau_1, \varphi) ds\|_d$$
$$+ \int_0^t \|c(t, s, \tau_2^{(n)}, \varphi_n) - c(t, s, \tau_2, \varphi)\|_{d \times m} \|f_n(s)\|_m ds$$
$$+ \|\int_0^t c(t, s, \tau_2, \varphi)(f_n(s) - f(s)) ds\|_d. \tag{191}$$

The first term on the right-hand side of (191) tends to zero as $n \to \infty$ because $\varphi_n \to \varphi$ in $\mathcal{W}^d$. To prove the same for the second term, we use Assumption (C2)(b) for the vector function a. The third term can be handled by using Hölder's inequality, the boundedness of the family f_n in the space $L^2([0, T], \mathbb{R}^m)$, and Assumption (C2)(c) for the matrix function c. Finally, the fourth term on the right-hand side of (155) tends to zero as $n \to \infty$ by the L^2-condition for the matrix function c in Assumption (C2)(a) and because $f_n \to f$ weakly in $L^2([0, T], \mathbb{R}^m)$. This establishes the continuity of the map Ψ_t. Next, we use the continuous mapping theorem for the weak convergence to show that

$$\lim_{n \to \infty} \mathbb{E}[\Psi_t(Y^{\varepsilon_n, v^n}, V^{1, \varepsilon_n, v^n}, V^{2, \varepsilon_n, v^n}, v^n)] = \mathbb{E}_{\widehat{\mathbb{P}}}[\Psi_t(\widehat{Y}, \widehat{V}^1, \widehat{V}^2, v)]. \tag{192}$$

It follows from the definition of Ψ_t that

$$\mathbb{E}[\Psi_t(Y^{\varepsilon_n,v^n}, V^{1,\varepsilon_n,v^n}, V^{2,\varepsilon_n,v^n}, v^n)]$$
$$\leq \mathbb{E}\left[||Y_t^{\varepsilon_n,v^n} - y - \int_0^t a(t,s,V^{1,\varepsilon_n,v^n}, Y^{\varepsilon_n,v^n})ds\right.$$
$$\left. - \int_0^t c(t,s,V^{2,\varepsilon_n,v^n}, Y^{\varepsilon_n,v^n})v_s^n ds||_d\right].$$

Now, using (13) we obtain

$$\mathbb{E}[\Psi_t(Y^{\varepsilon_n,v^n}, V^{1,\varepsilon_n,v^n}, V^{2,\varepsilon_n,v^n}, v^n)]$$
$$\leq \sqrt{\varepsilon_n}\mathbb{E}\left[||\int_0^t c(t,s,V^{2,\varepsilon_n,v^n}, Y^{\varepsilon_n,v^n})dB_s||_d\right]$$
$$\leq \sqrt{\varepsilon_n}\sqrt{\int_0^t \mathbb{E}[||c(t,s,V^{2,\varepsilon_n,v^n}, Y^{\varepsilon_n,v^n})||_d^2]ds}. \qquad (193)$$

For every $t \in [0,T]$, the last expression in (193) tends to zero as $n \to \infty$ by (175) and Assumption (C7). Therefore, (192) implies that for every $t \in [0,T]$, $\Psi_t(\widehat{Y}, \widehat{V}, v) = 0$ $\widehat{\mathbb{P}}$-a.s., and hence for every $t \in [0,T]$, $\widehat{Y}$ satisfies the second equation (190) $\widehat{\mathbb{P}}$-a.s. Since $\widehat{Y}$ maps $\widehat{\Omega}$ into $\mathcal{W}^d$, equation (190) holds for all $t \in [0,T]$ $\widehat{\mathbb{P}}$-a.s.

This completes the proof of Lemma 99.

We are finally ready to finish the proof of the lower bound in Laplace's principle in Theorem 94. We follow the proof at the bottom of p. 1133 in [17]. Using Fatou's lemma for the convergence in distribution, we obtain

$$\liminf_{n\to\infty} \mathbb{E}\left[\int_0^T ||v_s^n||_m^2 ds\right] \geq \mathbb{E}_{\widehat{\mathbb{P}}}\left[\int_0^T ||v_s||_m^2 ds\right]. \qquad (194)$$

It follows from Lemma 99 and (189) that

$$\liminf_{n\to\infty} -\varepsilon_n \log \mathbb{E}\left[\exp\left\{-\frac{1}{\varepsilon_n}F(Y^{\varepsilon_n})\right\}\right]$$
$$\geq \liminf_{n\to\infty} \mathbb{E}\left[\frac{1}{2}\int_0^T ||v_s^n||_m^2 ds + (Y^{\varepsilon_n,v^n})\right] - \delta. \qquad (195)$$

Since F is continuous and bounded, we can use (195), the continuous mapping theorem, and (194) to get

$$\liminf_{n\to\infty} -\varepsilon_n \log \mathbb{E}\left[\exp\left\{-\frac{1}{\varepsilon_n}F(Y^{\varepsilon_n})\right\}\right] \geq \mathbb{E}_{\widehat{\mathbb{P}}}\left[\frac{1}{2}\int_0^T ||v_s||_m^2 ds + F(\widehat{Y})\right] - \delta.$$
$$(196)$$

Let us recall that under the restrictions imposed on the functions $\bar{b}_i$ and $\bar{\sigma}_i$ in [17], the functional equations

$$\psi_i(s) = V_0^i + \int_0^s \bar{b}_i(r, \psi_i) dr + \int_0^s \bar{\sigma}_i(r, \psi_i) f(r) dr, \quad i = 1, 2, \tag{197}$$

are uniquely solvable. Moreover, for every $i = 1, 2$, the solution $\psi_{i,f}$ belongs to the space $\mathcal{W}^{k_i}$, and if $f_n \to f$ weakly in $L^2([0, T], \mathbb{R}^m)$, then $\psi_{i,f_n} \to \psi_{i,f}$ in $\mathcal{W}^{k_i}$. Therefore, the solutions to equation (19) are deterministic, and they coincide with the functions $\psi_{i,f}$ (see Remark 13).

Let v be the control process appearing in Lemma 99. Then, $\widehat{\mathbb{P}}$-almost all paths of v belong to the space $L^2([0, T], \mathbb{R}^m)$. For every such path f, (197) and the first equation (190) show that $\widehat{V}^i = \psi_{i,v}$ $\widehat{\mathbb{P}}$-a.s. Moreover, the previous equality and the second equation (190) imply that $\widehat{Y} = \Gamma_y(v)$ $\widehat{\mathbb{P}}$-a.s. Now, we can estimate the expectation on the right-hand side of (196) by the essential greatest lower bound of the integrand and use the equality $\widehat{Y} = \Gamma_y(v)$. This gives

$$\liminf_{n \to \infty} -\varepsilon_n \log \mathbb{E}\left[\exp\left\{-\frac{1}{\varepsilon_n} F(Y^{\varepsilon_n})\right\}\right]$$
$$\geq \inf_{\{(\varphi, f) \in \mathcal{W}^d \times L^2 : \varphi = \Gamma_y(f)\}} \left\{\frac{1}{2} \int_0^T \|f(s)\|_m^2 ds + F(\varphi)\right\} - \delta$$
$$\geq \inf_{\varphi \in \mathcal{W}^d} \{I_y(\varphi) + F(\varphi)\} - \delta,$$

where I_y is the rate function defined in (50).

This completes the proof of the lower bound in Laplace's principle in Theorem 94, since the previous estimate holds for an arbitrary $\delta > 0$.

Remark 100. It can be shown that equation (23) is solvable for any deterministic control $f \in L^2([0, T], \mathbb{R}^m)$ by using the same ideas as above. Therefore, only the uniqueness condition should be included in Assumption (C4). A similar remark can be found on p. 1133 in [17].

Upper bound in Laplace's principle. We have to show that

$$\limsup_{\varepsilon \to 0} -\varepsilon \log \mathbb{E}\left[\exp\left\{-\frac{1}{\varepsilon} F(Y^\varepsilon)\right\}\right] \leq \inf_{\varphi \in \mathcal{W}^d} [I_y(\varphi) + F(\varphi)] \tag{198}$$

for all bounded and continuous functions $F : \mathcal{W}^d \mapsto \mathbb{R}$. Exactly as in the proof on p. 1134 in [17], we assume that the greatest lower bound in (198) is finite and prove that for every fixed $\delta > 0$ there exists $\varphi \in \mathcal{W}^d$ for which

$$I_y(\varphi) + F(\varphi) \leq \inf_{\zeta \in \mathcal{W}^d}(I_y(\zeta) + F(\zeta)) + \frac{\delta}{2} < \infty. \tag{199}$$

We can also choose, for the function φ defined above, a control $f \in L^2([0,T], \mathbb{R}^m)$ such that

$$\frac{1}{2}\int_0^T \|f(s)\|_m^2 ds \leq I_y(\varphi) + \frac{\delta}{2}$$

and $\varphi = \Gamma_y(f)$. It follows from (199) that

$$I_y(\Gamma_y(f)) + F(\Gamma_y(f)) \leq \inf_{\zeta \in \mathcal{W}^d}(I_y(\zeta) + F(\zeta)) + \frac{\delta}{2} < \infty. \tag{200}$$

Let us suppose that $\varepsilon_n \in (0,1]$ and $\varepsilon_n \to 0$. For every $n \geq 1$, denote by $Y^{\varepsilon_n,f}$ the unique strong solution to equation (10). The existence of the solution follows from Lemma 96. Next, reasoning exactly as in the proof of the lower bound, we establish that the family $(Y^{\varepsilon_n,f}, V^{1,\varepsilon_n,f}, V^{2,\varepsilon_n,f}, f)$, with $n \geq 1$, is tight in $\mathcal{W}^d \times \mathcal{W}^{k_1} \times \mathcal{W}^{k_2} \times D_N$ for some number $N > 0$ where D_N is defined in Assumption (C5). Hence, possibly taking a subsequence, $(Y^{\varepsilon_n,f}, V^{1,\varepsilon_n,f}, V^{2,\varepsilon_n,f}, f)$ converges in distribution to a $\mathcal{W}^d \times \mathcal{W}^{k_1} \times \mathcal{W}^{k_2} \times D_N$-valued random variable $(\widehat{Y}, \widehat{V}^1, \widehat{V}^2, f)$ that is defined on some probability space $(\widehat{\Omega}, \widehat{\mathcal{F}}, \widehat{\mathbb{P}})$. We can also prove that $\widehat{\mathbb{P}}$-a.s. on $\widehat{\Omega}$ we have

$$\widehat{V}_s^i = V_0^i + \int_0^s \bar{b}_i(r, \widehat{V}^i)dr + \int_0^s \bar{\sigma}_i(r, \widehat{V}^i)f(r)dr$$

and

$$\widehat{Y}_t = y + \int_0^t a(t, s, \widehat{V}^1, \widehat{Y})ds + \int_0^t c(t, s, \widehat{V}^2, \widehat{Y})f(s)ds \tag{201}$$

for all $t \in [0, T]$. The solutions to the equations for $\widehat{V}^i$ in (201), with $i = 1, 2$, are deterministic, and we have $\widehat{V}^i = \psi_{i,f}$ $\widehat{\mathbb{P}}$-a.s. on $\widehat{\Omega}$. It follows that the second equation (201) can be rewritten as follows:

$$\widehat{Y}_t = y + \int_0^t a(t, s, \psi_{1,f}, \widehat{Y})ds + \int_0^t c(t, s, \psi_{2,f}, \widehat{Y})f(s)ds.$$

Therefore, $\widehat{Y} = \Gamma_y(f)$ $\widehat{\mathbb{P}}$-a.s. on $\widehat{\Omega}$. Next, using (174) and Remark 98 and reasoning as at the end of the proof of the upper bound in Laplace's principle in [17] (see p. 1134 in [17]), we see that

$$\limsup_{\varepsilon \to 0} -\varepsilon \log \mathbb{E}\left[\exp\left\{-\frac{1}{\varepsilon}F(Y^\varepsilon)\right\}\right] \leq I_y(\Gamma_y(f)) + \frac{\delta}{2} + \lim_{n \to \infty}\mathbb{E}[F(Y^{\varepsilon_n,f})]. \tag{202}$$

Recall that the sequence $Y^{\varepsilon_n,f}$ converges in distribution to $\Gamma_y(f)$. Now, using the continuity theorem for the convergence in distribution and (200), we obtain from (202) that the following estimate holds:

$$\limsup_{\varepsilon \to 0} -\varepsilon \log \mathbb{E}\left[\exp\left\{-\frac{1}{\varepsilon}F(Y^\varepsilon)\right\}\right] \leq I_y(\Gamma_y(f)) + \frac{\delta}{2} + F(\Gamma_y(f)])$$
$$\leq \inf_{\zeta \in \mathcal{W}^d}(I_y(\zeta) + F(\zeta)) + \delta.$$

The upper estimate in Laplace's principle easily follows from the previous inequalities.

The proof of Theorems 38 and 94 is thus completed.

Proof of Theorem 24

We use the extended contraction principle in the proof of Theorem 24 (see Theorem 4.2.23 in [22] or Lemma 2.1.4 in [23], see also [37]). The main idea is to consider a family of discrete approximations to the functional Φ given by (27). Let $n \geq 2$, and define the functional $\Phi_n : \mathbb{C}_0^m \times \mathbb{C}_0^m \times \mathcal{W}^d \mapsto \mathbb{C}_0^m$ as follows. For $(l, f, h) \in \mathbb{C}_0^m \times \mathbb{C}_0^m \times \mathcal{W}^d$ and $\frac{jT}{n} < t \leq \frac{(j+1)T}{n}$, with $1 \leq j \leq n-1$, set

$$\Phi_n(l,f,h)(t) = \frac{T}{n}\sum_{k=0}^{j-1} b\left(\frac{kT}{n}, h\left(\frac{kT}{n}\right)\right) + \left(t - \frac{jT}{n}\right) b\left(\frac{jT}{n}, h\left(\frac{jT}{n}\right)\right)$$
$$+ \sum_{k=0}^{j-1} \sigma\left(\frac{kT}{n}, h\left(\frac{kT}{n}\right)\right) \bar{C}\left[l\left(\frac{(k+1)T}{n}\right) - l\left(\frac{kT}{n}\right)\right]$$
$$+ \sigma\left(\frac{jT}{n}, h\left(\frac{jT}{n}\right)\right) \bar{C}\left[l(t) - l\left(\frac{jT}{n}\right)\right]$$
$$- \sum_{k=0}^{j-1} \sigma\left(\frac{kT}{n}, h\left(\frac{kT}{n}\right)\right) C\left[f\left(\frac{(k+1)T}{n}\right) - f\left(\frac{kT}{n}\right)\right]$$
$$+ \sigma\left(\frac{jT}{n}, h\left(\frac{jT}{n}\right)\right) C\left[f(t) - f\left(\frac{jT}{n}\right)\right], \quad (203)$$

and for $0 \leq t \leq \frac{T}{n}$, put $\Phi_n(l,f,h)(t) = tb(0, h(0)) + \sigma(0, h(0))\bar{C}l(t) + \sigma(0, h(0))Cf(t)$. It is not hard to see that for every $n \geq 2$, the map Φ_n is continuous.

The rest of the proof consists of three parts. First, we show that the term

$$-\frac{1}{2}\varepsilon \int_0^t \mathrm{diag}(\sigma(s, \widehat{B}_s^{(\varepsilon)})\sigma(s, \widehat{B}_s^{(\varepsilon)})')ds \quad (204)$$

appearing in (6) can be removed, and the resulting process $\widehat{X}^\varepsilon$ given by (7) satisfies the same large deviation principle as the process X^ε. It suffices to establish that the processes $X^\varepsilon - x_0$ and $\widehat{X}^\varepsilon - x_0$ are exponentially equivalent, that is, for every $\delta > 0$,

$$\limsup_{\varepsilon \to 0} \varepsilon \log \mathbb{P}\left(\varepsilon \sup_{t \in [0,T]} \|\int_0^t \operatorname{diag}(\sigma(s, \widehat{B}_s^\varepsilon)\sigma(s, \widehat{B}_s)')ds\|_d \geq \delta\right) = -\infty. \tag{205}$$

The second step in the proof of Theorem 24 is as follows. Recall that the function $\widetilde{I}_y$ is defined on the space $\mathbb{C}_0^m \times \mathbb{C}_0^m \times \mathcal{W}^d$ as follows:

$$\widetilde{I}_y(l, f, h) = \frac{1}{2}\int_0^T \|\dot{l}(t)\|_m^2 dt + \frac{1}{2}\int_0^T \|\dot{f}(t)\|_m^2 dt,$$

when $l, f \in (H_0^1)^m$ and $h = \hat{f}$, and by $\widetilde{I}_y(l, f, h) = \infty$ otherwise (see (53)). It is established that the sequence of functionals Φ_n given by (203) approximates the functional Φ in (27) in the following sense: For every $\alpha > 0$,

$$\lim_{n \to \infty} \sup_{\{(l,f,h):\widetilde{I}_y(l,f,h) \leq \alpha\}} \|\Phi(l, f, h) - \Phi_n(l, f, h)\|_{\mathcal{W}^m} = 0. \tag{206}$$

Finally, we prove that $\Phi_n(\sqrt{\varepsilon}W, \sqrt{\varepsilon}B, \widehat{B}^\varepsilon)$ are exponentially good approximations of $\widehat{X}^\varepsilon$. The latter statement means the following: For all $\delta > 0$,

$$\lim_{n \to \infty} \limsup_{\varepsilon \to 0} \varepsilon \log \mathbb{P}\left(\sup_{t \in [0,T]} \|\widehat{X}_t^\varepsilon - \Phi_n(\sqrt{\varepsilon}W, \sqrt{\varepsilon}B, \widehat{B}^\varepsilon)(t)\|_m \geq \delta\right) = -\infty. \tag{207}$$

It is not hard to see that Theorem 24 can be derived from (205), (206), (207), Theorem 41, and the extended contraction principle.

Proof of (205). The proof is similar to that used in Section 5 of [44]. It follows from the continuity of the functions σ_{ij} that there exists a positive even function η defined on $[0, \infty)$ and satisfying the following conditions: η is strictly increasing and continuous on $[0, \infty)$; $\eta(u) \to \infty$ as $u \to \infty$; $\sigma_{ij}(t, z)^2 \leq \eta(\|z\|_d)$ for all $1 \leq i, j \leq m$, $t \in [0, T]$, and $z \in \mathbb{R}^d$. The inverse function of the function η is denoted by η^{-1}. The latter function is defined on $[\eta(0), \infty)$.

We have
$$\sup_{t\in[0,T]} \|\int_0^t \operatorname{diag}(\sigma(s,\widehat{B}_s^\varepsilon)\sigma(s,\widehat{B}_s)')ds\|_d \le \int_0^T \sum_{i,j=1}^m \sigma_{ij}(s,\widehat{B}_s^\varepsilon)^2 ds$$
$$\le Tm^2\eta\left(\max_{t\in[0,T]}\|\widehat{B}_t^\varepsilon\|_d\right). \quad (208)$$

It follows from (208) that for $\varepsilon < \varepsilon_0$,
$$\mathbb{P}\left(\varepsilon \sup_{t\in[0,T]} \|\int_0^t \operatorname{diag}(\sigma(s,\widehat{B}_s^\varepsilon)\sigma(s,\widehat{B}_s)')ds\|_d \ge \delta\right)$$
$$\le \mathbb{P}\left(\max_{t\in[0,T]}\|\widehat{B}_t^\varepsilon\|_d \ge \eta^{-1}\left(\frac{\delta}{Tm^2\varepsilon}\right)\right). \quad (209)$$

Let us denote by U the expression on the left-hand side of (205). Using (209), we get
$$U \le \limsup_{\varepsilon\to 0} \varepsilon \log \mathbb{P}\left(\max_{t\in[0,T]}\|\widehat{B}_t^\varepsilon\|_d \ge \eta^{-1}\left(\frac{\delta}{Tm^2\varepsilon}\right)\right). \quad (210)$$

Fix $N > 0$. Then, for $\varepsilon < \varepsilon_N$, we have $\eta^{-1}\left(\frac{\delta}{Tm^2\varepsilon}\right) \ge N$. Therefore, (210) implies that
$$U \le \limsup_{\varepsilon\to 0} \varepsilon \log \mathbb{P}\left(\max_{t\in[0,T]}\|\widehat{B}_t^\varepsilon\|_d \ge N\right).$$

Set $A_N = \{\varphi \in \mathcal{W}^d : \max_{t\in[0,T]} \|\varphi(t)\|_d \ge N\}$. The set A_N is a closed subset of $\mathcal{W}^d$. Next, using Corollary 40 we obtain
$$U \le -\inf_{\{\varphi\in A_N\}} J_y(\varphi). \quad (211)$$

It remains to prove that
$$K_N = \inf_{\{\varphi\in A_N\}} J_y(\varphi) \to \infty \quad \text{as} \quad N\to\infty. \quad (212)$$

The sequence K_N, $N \ge 1$, is positive and non-decreasing. In order to prove (212), it suffices to show that the sequence K_N is unbounded. We reason by contradiction. Suppose $K_N \le C_1$ for some $C_1 > 0$ and all $N \ge 1$. Then, there exist sequences $\varphi_N \in A_N$ and $f_N \in L^2([0,T],\mathbb{R}^m)$ such that $\varphi_N = \mathcal{A}f_N$ and $\sup_N \|f_N\|_{L^2([0,T],\mathbb{R}^m)} < C_2$ for some $C_2 > 0$. It follows that there exists a weakly convergent subsequence $f_{N_k} \in L^2([0,T],\mathbb{R}^m)$. Next, using Assumption (C5), the continuity of the map G on the space $\mathcal{W}^d$, and the equality $\varphi_N = \mathcal{A}f_N$, we see that the sequence φ_{N_k} converges in the space $\mathcal{W}^d$. Hence, $\max_k \|\varphi_{N_k}\|_{\mathcal{W}^d} < \infty$. The previous inequality contradicts the assumption that $\varphi_N \in A_N$ for all $N \geq 1$. Therefore, (212) holds true. Finally, (211) implies that $U = -\infty$.

This completes the proof of (205).

Proof of (191). It is not hard to see that

$$\Phi_n(l, f, \widehat{f})(t) = \int_0^t b_n(s, \widehat{f}(s))ds + \int_0^t \sigma_n(s, \widehat{f}(s))\bar{C}\dot{l}(s)ds$$
$$+ \int_0^t \sigma_n(s, \widehat{f}(s))C\dot{f}(s)ds \qquad (213)$$

where the maps b_n and σ_n are given by

$$b_n(s, \widehat{f}(s)) = \sum_{k=0}^{n-1} b\left(\frac{kT}{n}, \widehat{f}\left(\frac{kT}{n}\right)\right) \mathbb{1}_{\frac{kT}{n} < s \leq \frac{(k+1)T}{n}}$$

and

$$\sigma_n(s, \widehat{f}(s)) = \sum_{k=0}^{n-1} \sigma\left(\frac{kT}{n}, \widehat{f}\left(\frac{kT}{n}\right)\right) \mathbb{1}_{\frac{kT}{n} < s \leq \frac{(k+1)T}{n}}.$$

Recall that

$$\Phi(l, f, \widehat{f})(t) = \int_0^t b(s, \widehat{f}(s))ds + \int_0^t \sigma(s, \widehat{f}(s))\bar{C}\dot{l}(s)ds$$
$$+ \int_0^t \sigma(s, \widehat{f}(s))C\dot{f}(s)ds \qquad (214)$$

(see (27)). For every $\alpha > 0$, set

$$D_\alpha = \left\{\tau \in (\mathbb{H}_0^1)^m : \int_0^T \|\dot\tau(t)\|_m^2 dt \leq 2\alpha\right\}.$$

Note that l and f appearing in (191) belong to D_α. Next, we take into account that $f \mapsto \widehat{f}$ is a compact map from any closed ball in $(\mathbb{H}_0^1)^m$ into $\mathcal{W}^d$. The previous statement follows from Assumption (C5) and the definition of the map $f \mapsto \widehat{f}$. Applying the Arzelà–Ascoli theorem, we obtain the following formulas:

$$\sup_{h \in D_\alpha} \sup_{\{t \in [0,T]\}} \|\widehat{f}(t)\|_d < M_\alpha$$

and

$$q_{\alpha,n} = \sup_{f \in D_\alpha} \sup_{\{t,u \in [0,T]: |t-u| \leq \frac{T}{n}\}} \|\widehat{f}(t) - \widehat{f}(u)\|_d \to 0 \qquad (215)$$

as $n \to \infty$. Finally, using Assumption A, (213), (214), and reasoning as in the proofs of Lemmas 6.23 and 6.24 in [41], we obtain

$$\sup_{\{(l,f,\widehat{f}):\widetilde{I}_y(l,f,\widehat{f})\leq\alpha\}} ||\Phi(l,f,\widehat{f})(t) - \Phi_n(l,f,\widehat{f})(t)||_{\mathcal{W}^m}$$

$$\leq \sup_{\{l,f\in D_\alpha\}} ||\Phi(l,f,\widehat{f})(t) - \Phi_n(l,f,\widehat{f})(t)||_{\mathcal{W}^m} \leq \zeta\omega\left(\frac{T}{n} + q_{\alpha,n}\right) \quad (216)$$

where the constant $\zeta > 0$ does not depend on n and ω is the modulus of continuity in Assumption A. Now, (191) follows from (215) and (216).

Proof of (207). The structure of the proof of the formula in (207) is the same as in a similar proof of Lemma 6.25 in [41]. Note that in [41], the case where $d = 1$ is considered, while in this paper, we deal with multivariate models. Moreover, the volatility process in [41] is Gaussian. It is explained in the following how to take into account these differences. We need the following two lemmas.

Lemma 101. *Let $0 < r < r_0$ where r_0 is a small number. Suppose q is the function on $(0, r_0]$ defined in the proof of Lemma 6.26 in [41]. Then, the following equality holds:*

$$\limsup_{r\to 0}\limsup_{\varepsilon\to 0} \varepsilon \log \mathbb{P}\left(\sup_{t\in[0,T]} ||\widehat{B}_t^\varepsilon||_d \geq 2^{-1}q(r)\right) = -\infty. \quad (217)$$

Lemma 102. *For every $\beta > 0$, the following formula is valid:*

$$\limsup_{n\to\infty}\limsup_{\varepsilon\to 0} \varepsilon \log \mathbb{P}\left(\max_{t,s\in[0,T]:|t-s|\leq\frac{T}{n}} ||\widehat{B}_t^\varepsilon - \widehat{B}_s^\varepsilon||_d \geq \beta\right) = -\infty. \quad (218)$$

Remark 103. The function q appearing in the formulation of Lemma 101 is positive, strictly decreasing, and continuous (see the proof of Lemma 6.26 in [41]). Moreover, $\lim_{r\to 0} q(r) = \infty$. For the statements similar to those in Lemmas 101 and 102, see (6.50) and (6.51) in [41].

Proof of Lemma 101. The proof of (217) is similar to that of (205). For every $r \in (0, r_0]$, define a closed subset of $\mathcal{W}^d$ by $E_r = \{\varphi \in \mathcal{W}^d : \sup_{t\in[0,T]} ||\varphi(t)||_d \geq 2^{-1}q(r)\}$ and set $N_r = \limsup_{\varepsilon\to 0} \varepsilon \log \mathbb{P}(\sup_{t\in[0,T]} ||\widehat{B}_t^\varepsilon||_d \geq 2^{-1}q(r))$. The large deviation principle in Corollary 40 implies that $N_r = -\inf_{\varphi\in E_r} J_y(\varphi)$ where J_y is the rate function defined in (51). Set $K_r = -N_r$. Then, K_r is a non-negative non-increasing function on $(0, r_0]$. It remains to prove that $\lim_{r\to 0} K_r = \infty$. We reason by contradiction. Suppose $K_r \leq M$ for all $r \in (0, r_0]$. Then, exactly as in the proof of the formula in (205), we see that there exist a sequence $r_k \to 0$ and a sequence $\varphi_k \in N_{r_k}$ such that the set $\{\varphi_k\}$ is compact in $\mathcal{W}^d$. It follows that the set $\{\varphi_k\}$ is bounded in $\mathcal{W}^d$. This means that we arrived at a contradiction since $\|\varphi_k\|_{\mathcal{W}^d} \geq 2^{-1} q(r_k)$ and $q(r_k) \to \infty$ as $k \to \infty$. It follows that $\lim_{r \to 0} N_r = -\infty$.

This completes the proof of Lemma 101.

Proof of Lemma 102. For every $n \geq 1$, define a closed subset of $\mathcal{W}^d$ by

$$V_n = \left\{ g \in \mathcal{W}^d : \max_{t,s \in [0,T] : |t-s| \leq \frac{T}{n}} \|g(t) - g(s)\|_d \geq \beta \right\}$$

and set

$$U_n = \limsup_{\varepsilon \to 0} \varepsilon \log \mathbb{P}\left(\max_{t,s \in [0,T] : |t-s| \leq \frac{T}{n}} \|\widehat{B}_t^\varepsilon - \widehat{B}_s^\varepsilon\|_d \geq \beta \right).$$

Using the large deviation principle established in Corollary 40, we can prove the following equality: $U_n = -\inf_{\varphi \in V_n} J_y(\varphi)$ where J_y is the rate function defined in (51). Set $K_n = -U_n$. Then, K_n is a non-negative non-decreasing sequence.

Our next goal is to prove that $K_n \to \infty$ as $n \to \infty$. It suffices to show that the sequence K_n is unbounded. To prove the previous statement, we reason by contradiction like in the proof of the formula in (205). Suppose $K_n \leq C$ for all $n \geq 1$. Then, there exists a sequence $\varphi_{n_k} \in V_{n_k}$ that is compact in $\mathcal{W}^d$ (see the proof of (205)). Applying the Arzelà–Ascoli theorem, we obtain

$$\sup_{k \geq 1} \max_{t,s \in [0,T] : |t-s| \leq \frac{T}{m}} \|\varphi_{n_k}(t) - \varphi_{n_k}(s)\|_d \to 0$$

as $m \to \infty$. Therefore, there exists m_0 such that

$$\max_{t,s \in [0,T] : |t-s| \leq \frac{T}{m}} \|\varphi_{n_k}(t) - \varphi_{n_k}(s)\|_d < \frac{\beta}{2} \tag{219}$$

for all $k \geq 1$ and $m \geq m_0$. Let k_0 be such that $n_k > m_0$ for all $k \geq k_0$. Then, (219) implies that

$$\max_{t,s \in [0,T] : |t-s| \leq \frac{T}{n_k}} \|\varphi_{n_k}(t) - \varphi_{n_k}(s)\|_d < \frac{\beta}{2} \tag{220}$$

for all $k \geq k_0$. It is not hard to see that (220) contradicts the condition $\varphi_{n_k} \in V_{n_k}$ for all $k \geq 1$.

This completes the proof of Lemma 102.

Let us return to the proof of (207). Using (6) and (213), we obtain

$$\widehat{X}^\varepsilon_t - \Phi_n(\sqrt{\varepsilon}W, \sqrt{\varepsilon}B, \widehat{B}^\varepsilon)(t) = \int_0^t [b(s, \widehat{B}^\varepsilon_s) - b_n(s, \widehat{B}^\varepsilon_s)]ds + \sqrt{\varepsilon}\int_0^t [\sigma(s, \widehat{B}^\varepsilon_s)$$

$$-\sigma_n(s, \widehat{B}^\varepsilon_s)]\bar{C}dW_s + \sqrt{\varepsilon}\int_0^t [\sigma(s, \widehat{B}^\varepsilon_s)$$

$$-\sigma_n(s, \widehat{B}^\varepsilon_s)]CdB_s, \quad t \in [0, T]. \quad (221)$$

By analyzing the formula in (221), we observe that in order to prove (207), it suffices to show that for all $1 \leq i, j, k \leq m$ and $\kappa > 0$,

$$\lim_{n\to\infty} \limsup_{\varepsilon\to 0} \varepsilon \log \mathbb{P}\left(\sup_{t\in[0,T]} |\int_0^t [b_i(s, \widehat{B}^\varepsilon_s) - b_i^{(n)}(s, \widehat{B}^\varepsilon_s)]ds| \geq \kappa\right) = -\infty,$$

$$(222)$$

$$\lim_{n\to\infty} \limsup_{\varepsilon\to 0} \varepsilon \log \mathbb{P}\left(\sqrt{\varepsilon}\sup_{t\in[0,T]} |\int_0^t [\sigma_{ij}(s, \widehat{B}^\varepsilon_s) - \sigma_{ij}^{(n)}(s, \widehat{B}^\varepsilon_s)]dW_k(s)| \geq \kappa\right)$$

$$= -\infty, \quad (223)$$

and

$$\lim_{n\to\infty} \limsup_{\varepsilon\to 0} \varepsilon \log \mathbb{P}\left(\sqrt{\varepsilon}\sup_{t\in[0,T]} |\int_0^t [\sigma_{ij}(s, \widehat{B}^\varepsilon_s) - \sigma_{ij}^{(n)}(s, \widehat{B}^\varepsilon_s)]dB_k(s)| \geq \kappa\right)$$

$$= -\infty. \quad (224)$$

The proofs of the equalities in (222)–(224) are similar to those of (6.38)–(6.40) in [41]. These proofs are long and involved, but all the necessary details are given in [41]. However, there are also certain differences in the proofs in [41] and in this paper. First of all, the process $\varepsilon \mapsto \sqrt{\varepsilon}\widehat{B}$ in [41] should be replaced by the process $\varepsilon \mapsto \widehat{B}^\varepsilon$. We also use Lemmas 101 and 102 instead of (6.50) and Corollary 6.22 in [41]. The stopping time $\xi^{\varepsilon,m,r}$ defined in (6.42) of [41] is replaced by the following stopping time: For $\varepsilon \in (0, 1]$, $n \geq 2$, and $0 < r < r_0$, we set

$$\xi^{\varepsilon,m,r} = \inf_{s\in[0,T]} \left\{\frac{r}{q(r)}||\widehat{B}^\varepsilon_s||_d + ||\widehat{B}^\varepsilon_s - \widehat{B}^\varepsilon_{\frac{[nsT^{-1}]T}{n}}||_d > r\right\}.$$

Note that everywhere in the proof of (218) we assume that the filtration $\{\mathcal{F}^B_t\}$ is hidden in the background. By taking into account the remarks mentioned above, we can complete the proof of (207) by reasoning as in the proof of Lemma 6.25 in [41]. We do not provide more details here and leave finding a detailed proof as an exercise for the interested reader.

Finally, we can finish the proof of Theorem 24 by using Theorem 38 and the formulas in (205), (206), and (207) and applying the extended contraction principle.

Acknowledgments

I am indebted to Peter Friz and Stefan Gerhold for the valuable remarks. I also thank the anonymous referee for reading this paper and for providing useful comments which significantly contributed to improving this paper.

References

[1] E. Alòs, and Y. Yang. (2017). A fractional Heston models with $H > \frac{1}{2}$. *Stochastics* 89, 384–399.
[2] M. Asai, M. McAleer, and J. Yu. (2006). Multivariate stochastic volatility: A review. *Econometric Reviews* 2/3, 145–175.
[3] P. Baldi, and L. Caramellino. (2011). General Freidlin-Wentzell large deviations and positive diffusions. *Statistics & Probability Letters* 81, 1218–1229.
[4] C. Bayer, P. K. Friz, P. Gassiat, J. Martin, and B. Stemper. (2020). A regularity structure for rough volatility. *Mathematical Finance* 30, 782–832.
[5] C. Bayer, P. K. Friz, and J. Gatheral. (2016). Pricing under rough volatility. *Quantitative Finance* 16, 887–904.
[6] C. Bayer, F. Harang, and P. Pigato. (2021). Log-modulated rough stochastic volatility models. *SIAM Journal on Financial Mathematics* 12, 1257–1284.
[7] L. Bergomi. (2016). *Stochastic Volatility Modeling*. CRC Press, Taylor and Francis Group.
[8] I. Bihari. (1956). A generalization of a lemma of Bellman and its application to uniqueness problem of differential equations. *Acta Mathematica Hungarica* (Academy of Sciences) 7, 71–94.
[9] P. Billingsley. (1971). *Weak Convergence of Measures: Applications in Probability*. Society for Industrial and Applied Mathematics, Philadelphia, Pennsylvania.
[10] M. Boué, and P. Dupuis. (1998). A variational representation of certain functionals of Brownian motion. *Annals of Probability* 26, 1641–1659.
[11] A. Budhiraja, and P. Dupuis. (2001). A variational representation for positive functionals of infinite-dimensional Brownian motion. *Probability and Mathematical Statistics* 20(1), 39–61.
[12] A. Budhiraja, and P. Dupuis. (2019). *Analysis and Approximation of Rare Events: Representations and Weak Convergence Methods*. Springer Science+Business Media, LLC (Part of Springer Nature). New York, NY.
[13] A. Budhiraja, P. Dupuis, and V. Maroulas. (2008). Large deviations for infinite-dimensional stochastic dynamical systems. *Annals of Probability* 36, 1390–1420.
[14] A. Budhiraja, P. Dupuis, and V. Maroulas. (2011). Variational representations for continuous time processes. *Annales de l'Institut Henri Poincaré Probabilités et Statistiques* 47, 725–747.
[15] G. Catalini, and B. Pacchiarotti. (2022). Asymptotics for multifactor Volterra type stochastic volatility models. Available on arXiv:2109.09448.
[16] M. Cellupica, and B. Pacchiarotti. (2021). Pathwise asymptotics for Volterra type stochastic volatility models. *Journal of Theoretical Probability* 34, 682–727.
[17] A. Chiarini, and M. Fischer. (2014). On large deviations for small noise Itô processes. *Advances in Applied Probability* 46, 1126–1147.
[18] F. Comte, L. Coutin, and E. Renault. (2010). Affine fractional stochastic volatility models. *Annals of Finance* 8, 1–42.

[19] G. Conforti, S. De Marco, and J.-D. Deuschel. (2015). On small-noise equations with degenerate limiting system arising from volatility models. In: P. K. Friz, J. Gatheral, A. Gulisashvili, A. Jacquier, and J. Teichmann (eds.) *Large Deviations and Asymptotic Methods in Finance.* Springer International Publishing, Switzerland, pp. 473–505.

[20] L. Coutin, and L. Decreusefond. (2001). Stochastic Volterra equations with singular kernels. In: *Stochastic Analysis and Mathematical Physics.* Progress in Probability, vol. 50. Birkhäuser Boston, Boston, MA, pp. 39–50.

[21] L. Decreusefond, and A. S. Üstünel. (1999). Stochastic analysis of the fractional Brownian motion. *Potential Analysis* 10, 177–214.

[22] A. Dembo, and O. Zeitouni. (2009). *Large Deviations Techniques and Applications.* Springer Science & Business Media. Springer-Verlag, New York, NY.

[23] J.-D. Deuschel, and D. W. Stroock. (1989). *Large Deviations.* Academic Press, Boston.

[24] C. Donati-Martin, A. Rouault, M. Yor, and M. Zani. (2004). Large deviations for squares of Bessel and Ornstein-Uhlenbeck processes. *Probability Theory and Related Fields* 129, 261–289.

[25] R. M. Dudley. (1967). The sizes of compact subsets of Hilbert space and continuity of Gaussian processes. *Journal of Functional Analysis* 1, 290–330.

[26] P. Dupuis, and R. S. Ellis. (1997). *A Weak Convergence Approach to the Theory of Large Deviations.* John Wiley & Sons, Inc., New York, NY.

[27] O. El Euch, and M. Rosenbaum. (2019). The characteristic function of rough Heston models. *Mathematical Finance* 29, 3–38.

[28] J. Feng, and T. G. Kurtz. (2006). *Large Deviations for Stochastic Processes.* American Mathematical Society, Providence, RI.

[29] X. Fernique. (1964). Continuité des processus Gaussiens. *Comptes Rendus de l'Académie des Sciences* 258, 6058–6060.

[30] M. Forde, and H. Zhang. (2017). Asymptotics for rough stochastic volatility models. *SIAM Journal on Financial Mathematics* 8, 114–145.

[31] J.-P. Fouque, G. Papanicolaou, and R. Sircar. (2011). *Derivatives in Financial Markets with Stochastic Volatility.* Cambridge University Press, Cambridge, UK.

[32] J.-P. Fouque, G. Papanicolaou, R. Sircar, and K. Sølna. (2011). *Multiscale Stochastic Volatility for Equity, Interest Rates, and Credit Derivatives.* Cambridge University Press, Cambridge, UK.

[33] M. I. Freidlin, and A. D. Wentzell. (1998). *Random Perturbations of Dynamical Systems.* Springer-Verlag, NewYork.

[34] P. K. Friz, P. Gassiat, and P. Pigato. (2021). Precise asymptotics: Robust stochastic volatility models. *Annals of Applied Probability* 31, 896–940.

[35] P. K. Friz, J. Gatheral, A. Gulisashvili, A. Jacquier, and J. Teichmann (Eds.). (2015). *Large Deviations and Asymptotic Methods in Finance.* Springer Proceedings in Mathematics and Statistics, Vol. 110. Springer International Publishing, Switzerland.

[36] K. Gao, and R. Lee. (2014). Asymptotics of implied volatility to arbitrary order. *Finance and Stochastics* 18, 349–392.

[37] J. Garcia. (2004). An extension of the contraction principle. *Journal of Theoretical Probability* 17, 403–434.

[38] S. Gerhold, C. Gerstenecker, and A. Gulisashvili. (2021). Large deviations for fractional volatility models with non-Gaussian volatility driver. *Stochastic Processes and Their Applications* 142, 580–600.

[39] S. Gerhold, F. Hubalek, and R. B. Paris. (2022). The running maximum of the Cox-Ingersoll-Ross process with some properties of the Kummer function. *Journal of Inequalities and Special Functions* 13, 1–18.

[40] H. Guennoun, A. Jacquier, P. Roome, and F. Shi. (2018). Asymptotic behavior of the fractional Heston model. *SIAM Journal on Financial Mathematics* 9, 1017–1045.

[41] A. Gulisashvili. (2021). Time-inhomogeneous Gaussian stochastic volatility models: Large deviations and super-roughness. *Stochastic Processes and Their Applications* 139, 37–79.

[42] A. Gulisashvili. (2020). Large deviation principles for stochastic volatility models with reflection and three faces of the Stein and Stein model (submitted for publication). Available on arXiv:2006.15431.

[43] A. Gulisashvili. (2020). Gaussian stochastic volatility models: Scaling regimes, large deviations, and moment explosions. *Stochastic Processes and Their Applications* 130, 3648–3686.

[44] A. Gulisashvili. (2018). Large deviation principle for Volterra type fractional stochastic volatility models. *SIAM Journal on Financial Mathematics* 9, 1102–1136.

[45] A. Gulisashvili. (2012). *Analytically Tractable Stochastic Stock Price Models.* Springer-Verlag, Berlin.

[46] P. Hagan, D. Kumar, L. Lesniewski, and D. E. Woodward. (2002). Managing smile risk. Wilmott Magazine (September 2002) 84–108.

[47] P. Henry-Labordère. (2009). *Analysis, Geometry, and Modeling in Finance: Advanced Methods in Option Pricing.* Chapman & Hall/CRC, Boca Raton.

[48] S. L. Heston. (1993). A closed-form solution for options with stochastic volatility, with applications to bond and currency options. *Review of Financial Studies* 6, 327–343.

[49] B. Horvath, A. Jacquier, and C. Lacombe. (2019). Asymptotic behavior of randomized fractional volatility models. *Journal of Applied Probability* 56, 496–523.

[50] J. Hull, and A. White. (1987). The procong of options on assets with stochastic volatilities. *Journal of Finance* 42, 281–300.

[51] A. Jacquier, M. S. Pakkanen, and H. Stone. (2018). Pathwise large deviations for the rough Bergomi model. *Journal of Applied Probability* 55, 1078–1092.

[52] A. Jacquier, and A. Pannier. (2020). Large and moderate deviations for stochastic Volterra systems. Available on arXiv:2004.10571.

[53] C. Jost. (2007). Integral Transformations of Volterra Gaussian Processes. PhD dissertation, University of Helsinki.

[54] C. Kahl. (2008). *Modeling and Simulation of Stochastic Volatility in Finance.* Universal-Publishers, Boca Raton, FL.

[55] G. Kallianpour. (1980). *Stochastic Filtering Theory.* Springer Science+Business Media, New York.

[56] I. Karatzas, and S. E. Shreve. (1991). *Brownian Motion and Stochastic Calculus,* 2nd edn. Springer-Verlag, New York, NY.

[57] A. N. Kolmogorov. (1940). Wienersche Spiralen und einige andere interessante Kurven im Hilbertschen Raum. Doklady Academy of Sciences USSR 26, 115–118.

[58] P. Lévy. (1953). *Random Functions: General Theory with Special Reference to Laplacian Random Functions.* University of California Publications in Statistics. University of California Press, Berkeley, CA.

[59] A. Lewis. (2000). *Option Valuation under Stochastic Volatility: With Mathematica Code.* Finance Press, Newport Beach, CA.

[60] A. Lewis.(2016). *Option Valuation under Stochastic Volatility II: With Mathematica Code.* Finance Press, Newport Beach, CA.

[61] S. C. Lim, and V. M. Sithi. (1995). Asymptotic properties of the fractional Brownian motion of Riemann–Liouville type. *Physics Letters A* 206, 311–317.

[62] B. Mandelbrot, and J. W. van Ness. (1968). Fractional Brownian motions, fractional noises and applications. *SIAM Review* 10, 422–437.

[63] M. B. Marcus, and L. A. Shepp. (1972). Sample behavior of Gaussian processes. *Proceedings of the Sixth Berkeley Symposium on Mathematical Statistics and Probability*, Vol. 2. University of California Press, pp. 423–441.

[64] O. Mocioalca, and F. Viens. (2005). Skorokhod integration and stochastic calculus beyond the fractional Brownian scale. *Journal of Functional Analysis* 222, 385–434.

[65] G. M. Molchan, and Y. I. Golosov. (1969). Gaussian stationary processes with asymptotic power spectrum. *Soviet Mathematics: Doklady* 10, 133–137.

[66] D. Nualart, and C. Rovira. (2000). Large deviations for stochastic Volterra equations. *Bernoulli* 6, 339–355.

[67] H. Pham. (2010). Large deviations in mathematical finance. *Third SMAI European Summer School in Financial Mathematics*, Paris, 23–27 August, 2010. Available on https://www.lpsm.paris/pageperso/pham/GD-finance.pdf, January 5, 2010.

[68] J. Picard. (2011). Representation formulae for the fractional Brownian motion. *Séminaire de Probabilités XLIII*. Springer-Verlag, pp. 3–70.

[69] A. Pilipenko. (2014). *An Introduction to Stochastic Differential Equations with Reflection*. Potsdam University Press, Potsdam, Germany.

[70] S. Robertson. (2010). Sample path large deviations and optimal importance sampling for stochastic volatility models. *Stochastic Processes and Their Applications* 120, 66–83.

[71] L. C. G. Rogers, and D. Williams. (2000). *Diffusions, Markov Processes, and Matringales: Volume 2, Itô Calculus*. Cambridge University Press, Cambridge, UK.

[72] C. Rovira, and M. Sanz-Solé. (2000). Large deviations for stochastic Volterra equations in the plane. *Potential Analysis* 12, 359–383.

[73] E. M. Stein, and J. C. Stein. (1991). Stock price distributions with stochastic volatility: An analytic approach. *The Review of Financial Studies* 4, 727–752.

[74] L. O. Scott. (1987). Option pricing when the variance changes randomly: Theory, estimation and an application. *Journal of Financial and Quantitative Analysis* 22, 419–438.

[75] D. W. Stroock. (1984). *An Introduction to the Theory of Large Deviations*. Springer-Verlag, New York.

[76] D. W. Stroock, and S. R.S. Varadhan. (2006). *Multidimensional Diffusion Processes*. Springer-Verlag, Berlin.

[77] S. R. S. Varadhan. (1966). Asymptotic probabilities and differential equations. *Communications on Pure and Applied Mathematics* 19, 261–286.

[78] S. R. S. Varadhan. (1984). *Large Deviations and Applications*. SIAM, Philadelphia.

[79] S. R. S. Varadhan. (2017). *Large Deviations*. AMS Courant Lecture Notes.

[80] A. D. Ventsel', and M. I. Freidlin. (1970). On small random perturbations of dynamical systems. *Russian Mathematical Surveys* 25, 1–56.

[81] A. D. Ventsel', and M. I. Freidlin. (1972). Some problems concerning stability under small random perturbations. *Theory of Probability and Its Applications* 17, 269–283.

[82] Z. Wang. (2008). Existence and uniqueness of solutions to stochastic Volterra equations with singular kernels and non-Lipschitz coefficients. *Statistics & Probability Letters* 78, 1062–1071.

[83] X. Zhang. (2008). Euler schemes and large deviations for stochastic Volterra equations with singular kernels. *Journal of Differential Equations* 244, 2226–2250.

Chapter 2

Phases of MANES: Multi-Asset Non-Equilibrium Skew Model of a Strongly Nonlinear Market with Phase Transitions

Igor Halperin

Fidelity Investments, Boston, USA
igor.halperin@fmr.com

Abstract

This chapter presents an analytically tractable and practically oriented model of nonlinear dynamics of a multi-asset market in the limit of a large number of assets. The asset price dynamics are driven by money flows into the market from external investors and their price impact. This leads to a model of a market as an ensemble of interacting nonlinear oscillators with the Langevin dynamics. In a homogeneous portfolio approximation, the mean field treatment of the resulting Langevin dynamics produces the McKean–Vlasov equation as a dynamic equation for market returns. Due to the strong nonlinearity of the McKean–Vlasov equation, the resulting dynamics give rise to ergodicity breaking and first- or second-order phase transitions under variations of model parameters. Using a tractable potential of the Non-Equilibrium Skew (NES) model previously suggested by the author for a single-stock case, the new Multi-Asset NES (MANES) model enables an analytically tractable framework for a multi-asset market. The equilibrium expected market log return is obtained as a self-consistent mean field of the McKean–Vlasov equation and derived in closed form in terms of parameters that are inferred from market prices of S&P 500 index options. The model is able to accurately fit the market data for either benign or distressed market environments, while using only a single volatility parameter.

Keywords: Non-equilibrium market dynamics, Langevin dynamics, mean field, statistical mechanics, McKean–Vlasov equation.

Introduction

When financial practitioners and academics talk about the behavior of the market, they normally refer to the price behavior of stock market indexes such as the S&P 500 or the Dow Jones index. For both these market indexes (and other similar ones), returns are defined as weighted averages of returns of their constituent stocks, the differences being in the chosen universe of stocks and the weighting scheme. In particular, the S&P 500 index weighs all stocks by their total market capitalization and is therefore heavily influenced by mega-stocks.

Given that a market index such as the S&P 500 (the SPX index) is a weighted average of individual stock prices, one might be tempted to assume that the dynamics of market returns would be similar to the dynamics of a single "representative" stock in the market. Such interchangeability of modeling returns for the whole market versus modeling returns for single stocks is commonly assumed by both academics and practitioners alike. In particular, in both the theory and practice of derivatives pricing, the same models such as stochastic volatility models are used for either single stocks or market indexes, albeit with different parameters.

However, the dynamics of the mean return of all stocks (i.e. the market return) can only be simply expressed as the mean of dynamics of individual returns if these dynamics are *linear*. In a more general case, we may think of a market as an ensemble of individual stocks whose individual dynamics are generally *nonlinear* due to market friction effects as discussed in more detail in the following. Furthermore, these nonlinear dynamics for individual stocks are not independent of each other (again, specific market mechanisms producing such co-dependencies is presented in the following). Therefore, in general the market dynamics should be viewed as statistical mechanics of an interacting ensemble of self-interacting nonlinear "particles" representing individual stocks. Quantities such as market returns would be computed in such a framework as ensemble averages.

For nonlinear interacting systems, ensemble averages do *not* in general reduce to some sort of expectations within a single particle dynamics. Therefore, a relation between dynamic properties of a markets index such as SPX and properties of a "typical" or "representative" stock is in general unknown. However, in statistical physics, there exists an approach that essentially *constructs* such effective single-particle dynamics starting with an initial multi-particle interacting system. This approach, or rather a family of approaches, is known in physics as *mean field approximation* or MFA for short. Mean field approximations used in statistical physics are known to become exact in the limit when the number N of particles in the systems goes to infinity,

$N \to \infty$, known as the thermodynamic limit. As the S&P 500 index is composed of $N = 500$ stocks, such value of N might be already sufficiently large to make the MFA qualitatively or even quantitatively accurate.

In this chapter, I present a simple and tractable model that builds the dynamics of a market index as the dynamics of the mean log-return of all stocks in the market. In other words, the market index is identified with the mean field in an ensemble of stocks that compose the market index. The way the MFA is constructed and used in this work is similar to how it is used in one of its most canonical applications to the classical Ising model, where the mean field is identified with the mean magnetization of Ising spins, and computed as a solution of a certain self-consistency equation (see, e.g. [14] or [16]).

Similarly, in the model presented in the following, the MFA gives rise to the mean field (i.e. the equilibrium expected market log-return) as a solution of a particular MFA self-consistency equation. Different from the Ising model that deals with binary spins without self-interactions, the model developed in the following deals with continuous real-valued nonlinear (i.e. self-interacting) oscillators as building blocks representing individual stocks. Nonetheless, the mean field approximation applied to the model of this chapter similarly produces a self-consistency equation whose solution gives the predicted equilibrium market log-return. Moreover, many other details of the model developed in this chapter will also bring strong analogies with the Ising model, including the phase structure of the model that admits both first- and second-order phase transitions in different regimes of parameters, and the values of critical exponents.

In this work, the market index portfolio is considered as an ensemble of stocks with individual log-returns y_i ($i = 1, \ldots, N$). They can be viewed as "particles" with "positions" y_i. The key input to modeling ensembles of interacting particles in statistical physics is a potential function $U(y_1, \ldots, y_N)$. The choice of the potential defines all further properties of a model, including both static and dynamic properties. In general, a potential $U(y_1, \ldots, y_N)$ can be decomposed into a sum of self-interaction potentials and interaction potentials. For example, if only pairwise interactions are allowed, the decomposition takes the following form:

$$U(y_1, \ldots, y_N) = \sum_{i=1}^{N} V(y_i) + \frac{g}{2N} \sum_{i,j=1}^{N} V_{\text{int}}(y_i - y_j), \qquad (1)$$

where $V(y_i)$ is a self-interaction potential for stock i, $V_{\text{int}}(y_i - y_j)$ is a pairwise interaction potential, and g is a coupling constant that regulates the strength of interactions in the system.

In this chapter, I motivate the choices for both self-interaction and interaction potentials using the analysis of money flows and their price impact in a multi-asset market. This approach follows the previous work by the author [7,8][1] where nonlinear models of a single stock price dynamics were obtained starting with a similar analysis of money flows and their impact. As was shown in [7], a combined effect of money flows from external investors and their price impact gives rise to a nonlinear (cubic) drift $\mu(x)$ in the diffusion law for the stock price. This translates into stochastic Langevin dynamics where diffusion in the price space is described as a Brownian motion of a particle that is additionally subject to an external nonlinear potential $V(x)$. The latter is defined to satisfy the relation $\mu(x) = -\partial V/\partial x$, therefore a cubic drift $\mu(x)$ translates into a quartic polynomial as a model of the potential $V(x)$.

In [8], a closely related model called the Non-Equilibrium Skew (NES) model was presented for the log-return space. Unlike more traditional approaches in physics where a potential is an input and stationary or non-stationary (transition) probability distributions are outputs, the NES model starts with a parameterized model for a stationary distribution, chosen to be a square of a simple two-component Gaussian mixture. The potential in the NES model is given by a negative of a logarithm of this Gaussian mixture. This provides a flexible and highly analytically tractable self-interaction potential $V(y)$ with five parameters, which can produce either single-well or double-well potentials that have, respectively, either one or two local minima. The outputs of the model are transition probabilities for a pre-asymptotic regime, before settling to an equilibrium steady state (which is in fact an input to the model as mentioned above) in the long run. As was shown in [8], these pre-asymptotic, non-equilibrium corrections to the asymptotic steady-state return distribution impact estimated moments of the return distribution, such as variance, skewness and kurtosis, which explains the name of the NES model.

This chapter presents a multi-asset extension of the NES model, to be referred to as the MANES model.[2] In this framework, the NES potential serves as a self-interaction potential $V(y)$ in equation (1), while the interaction potential $V_{\text{int}}(y_i - y_j)$ is found to be quadratic $V_{\text{int}}(y) = \frac{1}{2}y^2$. The MANES model can be viewed as a new statistical mechanics model with a highly tractable potential that can describe different dynamics depending

---

[1] See also [9] for a non-technical presentation.
[2] While the NES model developed in [8] is a single-stock model, its performance was explored in [8] using options on the S&P 500 indexes rather than single stocks. In this chapter, the single-stock model proposed in [8] will be properly used as a building block of a multi-asset MANES model.

on the model parameters. In particular, the self-interaction potential in the MANES model can be of a double-well form, depending on model parameters. In the latter case, the model behavior is similar to the Desai–Zwanzig model of interacting double-well anharmonic oscillators [2].

As is shown in detail in the following, the approach developed in this chapter offers a number of insights. First, it links the dynamics of market returns with the dynamics of an equivalent single stock that arises in the mean field approximation to the multi-particle dynamics of a market made of N stocks. Second, the MFA approximation applied to a multi-particle interacting system with the potential (1) produces a nonlinear extension of the classical Fokker–Planck equation called the McKean–Vlasov equation [1]. Because the McKean–Vlasov equation is nonlinear, it leads to ergodicity breaking and phase transitions [1]. As the mean field approximation employed in the McKean–Vlasov equation is accurate in the thermodynamical limit $N \to \infty$, this suggests that the dynamics of the market index can be successfully modeled as the mean field dynamics for a large ensemble of particles/stocks.

This approach is able to produce a large set of dynamics scenarios. In addition to a benign market regime of small fluctuations around some stationary or time-varying deterministic level (trend), the model also admits regimes of large fluctuations involving both first- and second-order phase transitions that can be realized in different parameter regimes.

Furthermore, I show how the model can be used in practice by calibrating it to market quotes on options on market indexes such as S&P 500 (SPX) options. Using a homogeneous approximation, the mean field potential arising with the McKean–Vlasov equation describing an ensemble of nonlinear oscillators with the NES self-interaction potentials can be represented as an effective *single-stock* NES potential with parameters that are modified ("renormalized") by interactions. This provides the aforementioned missing link between the dynamics of the market index and the dynamics of a "representative" stock that mimics the index. When parameters of this effective single-stock NES potential are inferred from the market options data, they are used to compute the equilibrium mean field of the MANES model. The latter is interpreted as a model-based, option-implied prediction of the equilibrium log-return of the market and can be used as a signal driving asset allocation decisions. Furthermore, other moments of the index return distribution inferred from market option quotes can also be used as predictive signals that can be used for investment decisions.

This chapter is organized as follows. Section "Nonlinear Stochastic Dynamics of Market Returns" gives the derivation of nonlinear stochastic dynamics of a multi-asset market that is driven by money flows and their impact. These dynamics are then reformulated as multi-particle Langevin and Fokker–Planck dynamics in section "Markets as Interacting Nonlinear Oscillators". Section "NES and MANES: Multi-Asset dynamics with the Non-Equilibrium Skew potential" introduces the NES potential as a tractable approximation to a nonlinear self-interaction potential obtained with the approach of section "Nonlinear Stochastic Dynamics of Market Returns". Section "McKean–Vlasov equation for the mean-field dynamics" provides a derivation of the McKean–Vlasov equation — a nonlinear version of the Fokker–Planck equation for a multi-particle system that arises within a mean field approximation. Section "Phase Transitions of MANES" derives the self-consistency equations resulting from using the NES potential within the McKean–Vlasov equation and then explores the phase structure of the model including both first- and second-order phase transitions in different parameter regimes and computes critical exponents. Section "Fitting Model Parameters Using Option Data" derives a closed-form relation for the equilibrium expected log-return of the market in terms of parameters obtained by calibration to index options and considers examples of calibration to the market data on the SPX options. Section "Summary and Outlook" concludes.

Nonlinear Stochastic Dynamics of Market Returns

Asset price dynamics with money flows and price impact

Let $\mathbf{x}_t$ with components x_{it} be a vector of asset values with $i = 1, \ldots, N$ in an investment universe of N assets at the beginning of the period $[t, t+\Delta t]$, where t is the current time and Δt is a time step size. Let us start with a discrete-time asset price dynamics in the presence of outside investors described by the following equations:

$$\mathbf{x}_{t+\Delta t} = (1 + \mathbf{r}_t \Delta t) \circ (\mathbf{x}_t + \mathbf{a}_t \circ \mathbf{x}_t \Delta t),$$
$$\mathbf{r}_t = r + \mathbf{w}\mathbf{z}_t + \mathbf{f}(\mathbf{a}_t) + \frac{1}{\sqrt{\Delta t}} \varepsilon_t, \qquad (2)$$

where $\circ$ stands for the element-wise product. Here the first equation in (2) defines the change of asset values in the time step $[t, t+\Delta t]$ as a composition of two changes to their time-t values $\mathbf{x}_t$. First, at the beginning of the interval, investors adjust positions in each stock i by buying (or selling, depending on the sign of a_{it}) the amount $a_{it} x_{it} \Delta t$ of the stock (so that the action vector $\mathbf{a}_t$ is defined as an instantaneous rate of change with the dimension of inverse time). Therefore, immediately after that the asset value is deterministically changed to $\mathbf{x}_t^+ := \mathbf{x}_t + \mathbf{a}_t \circ \mathbf{x}_t \Delta t$. After that, the new portfolio $\mathbf{x}_t^+ = \mathbf{x}_t + \mathbf{a}_t \circ \mathbf{x}_t \Delta t$ grows at rate $\mathbf{r}_{t+\Delta t}$. The latter is given by the second of equation (2) that defines the vector of equity returns as a combination of a risk-free rate r, predictors $\mathbf{z}_t$ with weights $\mathbf{w}$, a multivariate noise $\varepsilon_t \sim \mathcal{N}(\cdot|\mathbf{0}, \mathbf{\Sigma})$, and a vector-valued market impact factor $\mathbf{f}(\mathbf{a}_t)$ with components $f_i(a_{ti})$. A particular form for function $\mathbf{f}(\mathbf{a}_t)$ is presented in the following section, while in this section, I proceed with a general form of this function.[3]

Equations (2) can also be written for increments $\Delta \mathbf{x}_t = \mathbf{x}_{t+\Delta t} - \mathbf{x}_t$:

$$\Delta \mathbf{x}_t = \mathbf{x}_t \circ (r + \mathbf{w}\mathbf{z}_t + \mathbf{f}(\mathbf{a}_t) + \mathbf{a}_t) \Delta t + \mathbf{x}_t \circ \sqrt{\Delta t}\, \varepsilon_t, \qquad (3)$$

where terms $\sim (\Delta t)^2$ are omitted assuming that the time step Δt is small enough to justify a continuous-time limit $\Delta t \to 0$. In the strict limit $\Delta t = dt \to 0$, equation (3) transforms into the following stochastic differential equation (SDE):

$$d\mathbf{x}_t = \mathbf{x}_t \circ (r + \mathbf{w}\mathbf{z}_t + \mathbf{f}(\mathbf{a}_t) + \mathbf{a}_t) \, dt + \mathbf{x}_t \circ \boldsymbol{\sigma} \, d\mathbf{W}_t, \qquad (4)$$

where $\mathbf{W}_t$ is a standard N-dimensional Brownian motion and $\boldsymbol{\sigma}$ is a $N \times N$ volatility matrix such that $\boldsymbol{\sigma}\boldsymbol{\sigma}^T = \mathbf{\Sigma}$.

Next consider a vector $\mathbf{y}_t$ of period-T log-returns y_{it} with $i = 1, \ldots, N$, defined as follows:

$$y_{it} = \log \frac{x_{it}}{x_{i,t-T}}. \qquad (5)$$

Using Itô's lemma, we obtain the SDE for the log-return vector $\mathbf{y}_t$:

$$d\mathbf{y}_t = \left(r + \mathbf{w}\mathbf{z}_t + \mathbf{f}(\mathbf{a}_t) + \mathbf{a}_t - \frac{\dot{\mathbf{x}}_{t-T}}{\mathbf{x}_{t-T}} - \frac{1}{2}\text{Tr}\,\mathbf{\Sigma} \right) dt + \boldsymbol{\sigma}\, d\mathbf{W}_t, \qquad (6)$$

where $\dot{\mathbf{x}}_{t-T}$ stands for the derivative of the time-lagged asset price $\mathbf{x}_{t-T}$ with respect to the calendar time t. Note that without control $\mathbf{a}_t$ and the price impact $\mathbf{f}(\mathbf{a}_t)$, equation (4) becomes a standard multivariate lognormal process with the growth rate given by $r + \mathbf{w}\mathbf{z}_t$, while equation (6) becomes a multivariate normal process with a time-dependent drift due to the term $\dot{\mathbf{x}}_{t-T}/\mathbf{x}_{t-T}$.

On the other hand, equation (2) or its continuous-time limits (4) or (6) describe a *controlled* system where investors buy or sell stocks at rates a_i at each time step. If investors are rational or bounded-rational, their decisions a_{it} would depend on the previous performance of assets. The problem of finding *optimal* control $\mathbf{a}_t$ from the viewpoint of market investors could be further formalized by specifying their utility (reward) functions and then solving corresponding Bellman or Hamilton–Jacobi–Bellman (HJB) equations. Instead of following this route, in this chapter,

---

[3]The impact function $\mathbf{f}(\mathbf{a}_t)$ may also depend on other state variables, however we will neglect such additional dependencies, see the following section for more details.

I employ a more phenomenological way and choose a simple functional form of the optimal investment rate $\mathbf{a}_t$ as a function of state variables based on general arguments.

To this end, our specification of investment rates $\mathbf{a}_t$ should encode some simple stylized facts about retail investors. In particular, they usually buy stocks when prices go up (i.e. log-returns $\mathbf{y}_t$ are positive) and sell when prices go down and log-returns are negative. This can be formalized by specifying a parametric function $\mathbf{a}_t = \mathbf{a}(\mathbf{y}_t)$ where a possible dependence on other state variables can be encoded in parameters of this function. Assuming for simplicity that their actions are perfectly asymmetric with respect to such scenarios, it produces asymmetry relations $\mathbf{a}(-\mathbf{y}_t) = -\mathbf{a}(\mathbf{y}_t)$ that should be imposed on all admissible functional specifications $\mathbf{a}(\mathbf{y}_t)$. In other words, $\mathbf{a}(\mathbf{y}_t)$ should be an odd function.

Based on this, the following specification of the investment rate a_{it} for stock i as a simple function of log-return y_{it} is used in this work:

$$a_{it} = \phi_i y_{it} + \lambda_i y_{it}^3 + \frac{\kappa_i}{N} \sum_{j=1}^{N} y_{jt}, \qquad (7)$$

where $\phi_i, \lambda_i \geq 0$ are parameters capturing, respectively, a linear and nonlinear dependence of investors' allocation decisions on the log-return of asset i, and parameter $\kappa_i \geq 0$ determines how it depends on the average performance of other assets (i.e. performance of the market). As is shown in the following, parameters λ_i control nonlinear effects, while parameters κ_i control interactions in the system. Note that the functional specification (7) can be viewed as a low-order Taylor expansion of a general function, where a constant term and a coefficient in front of y_{it}^2 are set to zero in order to produce an odd function.

The definition of the investor flow rate given by equation (7) both refines the previous similar specifications suggested in [7,8] for a single-stock (1D) case and extends the approach suggested in the previous work to the current case of a multi-asset market with N stocks. Furthermore, the model developed in the following assumes that N is large, which seems to be a good assumption if, for example, we consider the case $N = 500$ by the number of constituents in the S&P 500 index.

Price impact model with dumb money

To complete the model specification, we need to define the model of market impact $\mathbf{f}(\mathbf{a_t})$. First, note that on the grounds of dimensional analysis, as both

$\mathbf{a}_t$ and $\mathbf{f}$ have dimensions of rates, it means that, in addition to the ratios $\mathbf{a}_t$, the market impact can only depend on some dimensionless variables such as, e.g. a ratio of the asset price x_{it} to average trading volume over a lookback time window $[t-T, t]$. In what follows, the market impact $\mathbf{f}(\mathbf{a}_t)$ is modeled as a parameterized function of $\mathbf{a}_t$ with constant parameters, thus neglecting such possible additional dependencies.[4]

We need a functional specification of the price impact function $f_i(a_i)$ that captures the most important effects of impact of market flows on market prices. First, we want to ensure consistency with the presence of momentum in stock prices. As typically a good recent performance of a stock leads to an increased demand, in a short run, this typically further increases returns of this stock. Therefore, at least until cumulative flows are small enough, one should expect a positive co-dependence between market flows and asset returns.

However, such positive co-dependence will only persist until the stock becomes "saturated" or "crowded". "Crowding" in a stock occurs when many market participants simultaneously hold large positions in this stock. During periods of market downturns or high volatility, when many market participants simultaneous unwind or reduce their positions in a crowded stock, it creates a further downward price pressure on the stock, producing diminishing or even negative returns on positions in the stock.

To capture both effects discussed above, we need to produce scenarios where money flows initially produce a positive impact on asset returns but switch to a negative impact once cumulative inflows exceed some threshold value. Such a model would be consistent with the "dumb money" effect of [4] that predicts that an initial flow into a stock should increase expected returns, but a continuous buildup of inflows into the stock (a "crowding") leads to diminishing long-term returns.

To capture such saturation effects, a proper impact function should depend on previous inflows. Let $\bar{a}_{it}^{\tau}$ be the cumulative inflow rate in the stock i over the last τ periods excluding the current period:

$$\bar{a}_{it}^{\tau} := \sum_{t'=1}^{\tau} a_{i,t-t'}. \tag{8}$$

Now consider a simple model that produces an increasing impact for small values of a_{it} until the sum of a_{it} and $\bar{a}_{it}^{\tau}$ does not exceed some fixed value $\hat{a}_i$,

---

[4]If needed or desired, such dependencies can be re-installed by making parameters of a model for $\mathbf{f}(\mathbf{a}_t)$ dependent on these variables.

and a decreasing impact for larger values:

$$f_i(a_i) = -\eta_i \left(|a_{it} + \bar{a}_{it}^\tau - \hat{a}_i| - |\bar{a}_{it}^\tau - \hat{a}_i| \right), \qquad (9)$$

where $\eta_i > 0$ is a parameter. This piece-linear function vanishes at $a_i = 0$ and $a_i = 2\left(\hat{a}_i - \bar{a}_{it}^\tau\right)$ and reaches its maximum equal to $f_{i\star} = \eta_i \left(|\hat{a}_i - \bar{a}_{it}^\tau|\right)$ at $a_i = \hat{a}_i - \bar{a}_{it}^\tau$. For large values of $|a_i|$, this function asymptotically behaves as $f_i(a_i) \sim -\eta_i |a_i|$.

While the functional form (9) produces the desired behavior, it amounts to a non-differentiable function. We can consider a soft relaxation of this function that produces a differentiable approximation:

$$f_\beta(a_i) = -\eta_i \left(H_\beta(a_i - b_i) - H_\beta(-b_i) \right), \quad b_i := \hat{a}_i - \bar{a}_{it}^\tau \qquad (10)$$

where the function $H_\beta(z)$ is defined as follows:

$$H_\beta(z) = \frac{1}{\beta} \log \left(\frac{e^{\beta z} + e^{-\beta z}}{2} \right) = \frac{1}{\beta} \log \cosh(\beta z). \qquad (11)$$

Note that this function provides a soft differentiable relaxation of function $f(z) = |z|$ so that $H_\infty(z) = |z|$. Furthermore, for finite values of β, the function $H_\beta(z)$ is convex with a minimum at $z = 0$ and $H_\beta(0) = 0$. For small values of z, the Taylor expansion of $H_\beta(z)$ produces

$$H_\beta(z) = \frac{\beta}{2} z^2 + \frac{\beta^3}{12} z^4 + O(z^6), \qquad (12)$$

while for large values of $|z|$, we have $H_\beta(z) \to |z|$. Therefore, the impact function (10) produces the following behavior:

$$f_\beta(0) = 0, \quad a_\star := \arg\max_z f_\beta(z) = b_i, \quad f_\beta(a_\star) = \eta_i H_\beta(b_i). \qquad (13)$$

For small values of the argument, we obtain

$$f_\beta(a_i) = -\eta_i \left(-\beta b_i \left(1 + \frac{\beta^2 b_i^2}{3}\right) a_i + \frac{\beta}{2} \left(1 + \beta^2 b_i^2\right) a_i^2 - \frac{\beta^3 b_i}{3} a_i^3 + \frac{\beta^3}{12} a_i^4 + \cdots \right) \qquad (14)$$

and for $|a_i| \gg 1$, we have $f_\beta(a_i) = -\eta_i |a_i|$. Both (9) and its relaxed form (10) describe a concave impact function that vanishes at $a_i = 0$, reaches its peak at $a_i = b_i$, and then decreases and eventually becomes negative for larger values $a_i \geq 2b_i$.[5]

---

[5]A similar profile could be reached with an inverted parabola function such as $f(z) = az(b-z)$ with parameters $a, b > 0$, however I choose (10) over such specification as I want to have an asymptotically linear, rather that quadratic, behavior of $f(z)$.

Note for what follows the asymptotic behavior of the combination $a_{it} + f_i(a_{it})$ that enters the SDE (6):

$$a_i + f_i(a_i)) = \begin{cases} \left(1 + \eta_i \beta b_i \left(1 + \frac{\beta^2 b_i^2}{3}\right)\right) a_i \\ \quad - \frac{\beta \eta_i}{2}\left(1 + \beta^2 b_i^2\right) a_i^2 + \frac{\beta^3 b_i \eta_i}{3} a_i^3 + \cdots, & \text{for } |a_i| \ll 1 \\ a_i - \eta_i |a_i|, & \text{for } |a_i| \gg 1. \end{cases} \quad (15)$$

Using the cubic law (7) for the flow rate, this can also be written as a function of $y_i = y_{it}$:

$$a_i + f_i(a_i)) = \begin{cases} \xi_i y_i + \rho_i y_i^2 + \zeta_i y_i^3 \\ \quad + g_i \left(\frac{1}{N} \sum_{j=1}^N y_j - y_i\right) + \cdots, & \text{for } |y_i| \ll 1 \\ (1 - \eta_i) \lambda_i y_i^3, & \text{for } y_i \to \infty \\ (1 + \eta_i) \lambda_i y_i^3, & \text{for } y_i \to -\infty \end{cases} \quad (16)$$

where I retained only a linear dependence on the "mean field" $\frac{1}{N} \sum_{j=1}^N y_j$, and parameters are defined in terms of previously defined model parameters as follows:

$$\xi_i := (\phi_i + \kappa_i)\left(1 + \eta_i \beta b_i \left(1 + \frac{\beta^2 b_i^2}{3}\right)\right), \quad g_i := \kappa_i \left(1 + \eta_i \beta b_i \left(1 + \frac{\beta^2 b_i^2}{3}\right)\right),$$

$$\rho_i := -\frac{\beta \eta_i}{2}\left(1 + \beta^2 b_i^2\right)\phi_i^2, \quad \zeta_i := \frac{\beta^3 b_i \eta_i \phi_i^3}{3} + \lambda_i \left(1 + \eta_i \beta b_i \left(1 + \frac{\beta^2 b_i^2}{3}\right)\right).$$

$$(17)$$

The functional form of dependence on log-returns y_i found in equation (16) will be used in the following section to introduce a nonlinear potential that will be then used for analysis of the joint dynamics of all stocks in the market.

Markets as Interacting Nonlinear Oscillators

Langevin dynamics of interacting nonlinear oscillators

Focusing on an idealized market portfolio made of N identical stocks, let us consider the case when all model parameters $\xi_i, g_i, \rho_i, \mathbf{w}_{i\cdot}, h_i$, etc. are the same for all stocks, so that we write them as ξ, g, ρ, etc.[6] Furthermore, while parameters defined in equation (17) are generally time-dependent, here I neglect such potential time dependence, and treat them as constants.

---

[6] The homogeneity assumption will be lifted in the following in section "Partition function and mean field without homogeneity".

The stochastic differential equation (6) that describes the continuous-time stochastic dynamics of any individual stock log-return $y_i = y_{it}$ in this setting can be represented as an (overdamped) Langevin equation [11]. In the Langevin approach, the diffusion drift is obtained as a negative gradient of a potential function $U(\mathbf{y}_t)$:

$$dy_i = -\frac{\partial U(\mathbf{y}_t)}{\partial y_i} dt + h dW_{it}, \qquad (18)$$

where h is a volatility parameter,[7] W_{it} is a standard Brownian motion, and $U(\mathbf{y}_t)$ is a potential that is decomposed into a sum of self-interaction and pair interaction terms:

$$U(\mathbf{y}_t) = \sum_{i=1}^{N} V(y_i) + \frac{g}{2N} \sum_{i,j=1}^{N} V_{\text{int}}(y_i - y_j). \qquad (19)$$

Here $V(y_i)$ stands for a single-stock self-interaction potential, g is an interaction constant, and $V_{\text{int}}(y_i - y_j)$ is an interaction potential. The Langevin equation (18) describes diffusion of N identical particles placed in a common potential $V(y)$ and in addition interacting via a pairwise interaction potential V_{int}. Using equation (19), we can also write equation (18) as follows:

$$dy_i = -\left(\frac{\partial V(y_i)}{\partial y_i} + \frac{g}{N} \sum_{j=1}^{N} \frac{\partial V_{\text{int}}(y_i - y_j)}{\partial y_i}\right) dt + h dW_{it}. \qquad (20)$$

In our setting, the choices for potentials V, V_{int} are informed by equations (6) and (16) adapted to a homogeneous portfolio setting (with the coupling constant g obtained similarly from equation (17)). In particular, interactions are induced by the third term in the control law (7). The interaction potential $V_{\text{int}}(y_i - y_j)$ is therefore given by the Curie–Weiss quadratic potential:

$$V_{\text{int}}(y_i - y_j) = \frac{1}{2}(y_i - y_j)^2. \qquad (21)$$

While the interaction potential V_{int} in equation (21) is quadratic, the self-interaction potential $V(y)$ as suggested by equation (16) should have higher-order non-linearities:

$$V(y) = \begin{cases} -\theta y - \frac{\xi}{2} y^2 - \frac{\rho}{3} y^3 - \frac{\zeta}{4} y^4 + \cdots, & \text{for } |y| \ll 1 \\ -(1-\eta)\frac{\lambda}{4} y^4, & \text{for } y \to \infty \\ -(1+\eta)\frac{\lambda}{4} y^4, & \text{for } y_i \to -\infty, \end{cases} \qquad (22)$$

---

[7] I changed here the notation for the volatility parameters from σ_i to h, as parameters σ_i will be utilized differently in the following.

where θ is an additional parameter given by the sum of all $\mathbf{a}_t$-independent terms in equation (6). Therefore, for small values of log-returns, the self-potential (22) behaves as a quartic potential. It also scales as y^4 asymptotically for $|y| \to \infty$, though with different coefficients from those appearing in its small-y expansion given by the first line in equation (22).

Instead of literally using equation (22) as the specification of a self-interaction potential $V(y)$, in this chapter, I will use a somewhat different potential that, while retaining the important nonlinearity of equation (22), also offers better analytical tractability, as well as fixes some potential issues with equation (22). These issues are related to the asymptotic behavior of the potential $V(y)$ at $y \to \pm\infty$, as is discussed next.

First, note that expression (22) implies that in order to have a confining potential that grows as $y \to \infty$, we need to have parameter $\eta > 1$.[8] On the other hand, even if we take $\eta > 1$, the potential (22) is unbounded from below for large *negative* values $y \to -\infty$. Such an unbounded behavior could describe a corporate default or bankruptcy event when the stock price drops to zero, which would be equivalent to the strict limit $y \to -\infty$. This behavior of the model is obtained as a direct consequence of the specification of the cubic flow rate (7) which produces a strong negative selloff rate for large negative values of log-returns y which, together with the price impact function $f(a_t) \propto -\eta|a_t|$, can drive the stock price all the way to zero or equivalently to the infinite negative log-return $y \to -\infty$.

Note however that for all practical purposes, instead of being identified with price drops to the strict zero level, corporate defaults or bankruptcies may be associated with events when the stock price drops to a very small but a non-zero value (e.g. a few cents given the previous price of \$10, or \$1 given the previous price of \$50, etc.). In terms of log-returns y, such events would correspond to sudden drops to a large negative value. The infinite value of y obtained when the stock price hits the exact zero is therefore solely due to singularity of the transformation $y_t = \log(x_t/x_{t-T})$ at $x_t = 0$. If a small but non-vanishing stock price is used as the default boundary, the range of practically admissible values of y becomes finite.

The last observation implies that we can replace a potential that may become unbounded as $y \to -\infty$ with a *confining* (i.e. growing at $y \to \pm\infty$) potential where corporate defaults would correspond to events of reaching a certain arbitrarily negative but finite threshold $y_\star < 0$. If the probability of

---

[8] While scenarios with an unbounded potential as $y \to \infty$ may describe market bubbles that can only occur for short periods of time, they will not be pursued in this work.

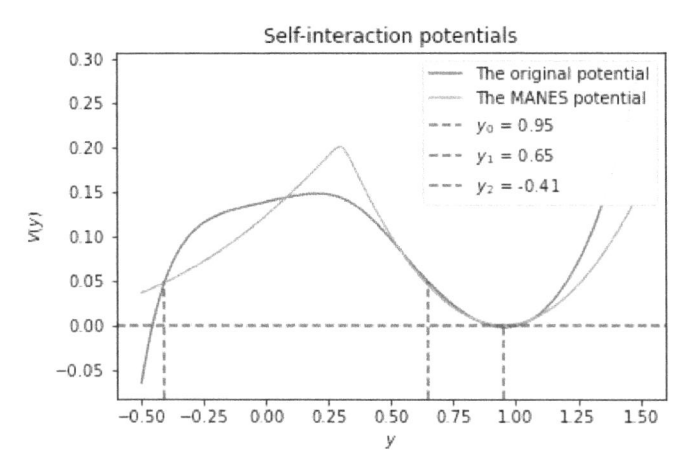

Figure 1. Comparison between the original unbounded potentials $V(y)$ in equation (22) and the confining MANES potentials (23). Both the dynamics of local fluctuations around the minimum y_0 and transition probabilities to move from y_1 to y_2 would be similar for the two potentials if potential barriers are sufficiently tall and have similar heights.

reaching this barrier is small enough to ensure consistency with the market data, further details of model behavior for yet smaller values of y, which could otherwise differentiate between a confining and unbounded potential, would be immaterial for all practical purposes.

This is illustrated in Figure 1 that shows the original potential $V(y)$ according to equation (22) alongside a tractable confining potential referred to as the MANES potential, to be presented in details in the following section. Depending on the parameters, the MANES potential can have one or more minima. In Figure 1, parameters are such that the MANES potential is a double-well potential with a second minimum at a negative value of y, which is not shown in Figure 1. The important point to emphasize here is that as long as barriers for both potentials are sufficiently high and have similar heights, both the dynamics of small fluctuations around the potential minimum at y_0 and probabilities of transitions from y_1 to y_2 will be similar in both models. When these probabilities are small, details of behavior in the left tail between the two potentials (an unbounded original potential vs a confining MANES potential) would have a negligible impact on observable consequences of the model.

Therefore, instead of literally using equation (22) as a specification of the self-interaction potential $V(y)$, this work will assume a confining potential that may be a single-well or a multi-well potential, depending on the model parameters. Confining potentials are easier to work with than non-confining ones, as they give rise to discrete spectra and stationary states in the long run

$t \to \infty$, which would not exist for non-confining potentials. For a multi-well potential with two or more local minima, the model dynamics would be similar to the Desai–Zwanzig model [2] which chooses a symmetric quartic double-well potential as a model of $V(y)$ and its generalized version in [6] that considers more complex multi-well potentials. The following section provides details of the MANES potential that will be used later in this chapter for the analysis of dynamics and phase structure of the model.

NES and MANES: Multi-Asset dynamics with the Non-Equilibrium Skew potential

In this chapter, the dynamics implied by the Langevin equation (20) are explored using a simple and highly tractable nonlinear self-interaction potential $V(y)$ given by the logarithm of a two-component Gaussian Mixture (GM):

$$V(y) = -h^2 \log\left[(1-a)\phi(y|\mu_1 T, \sigma_1^2 T) + a\phi(y|\mu_2 T, \sigma_2^2 T)\right] - h^2 \log C + V_0, \tag{23}$$

where $\phi(y|\mu, \sigma^2)$ is a Gaussian density, C is a constant that will be defined below, and V_0 is another constant chosen such that the minimum value of $V(y)$ is zero. This potential was introduced in [8] within the context of a model for a single stock called the Non-Equilibrium Skew (NES) model. The choice (23) is motivated by both the ease of the analytical treatment and the flexibility of the parametric family described by equation (23) which, depending on parameters, describes either a single-well anharmonic potential or a double-well potential. As market flows and their price impact produce a nonlinear potential, the parametric family of potentials presented by equation (23) make it possible for the model to capture these effects. As this work can be considered a multi-asset (MA) generalization of the NES model, in what follows, I will refer to the form in equation (23) as the MANES potential.

As explained in more detail in [8], the MANES potential (23) can also be represented as

$$V(y) = -h^2 \log \Psi_0(y) + V_0,$$
$$\Psi_0(y) := C\left[(1-a)\phi(y|\mu_1 T, \sigma_1^2 T) + a\phi(y|\mu_2 T, \sigma_2^2 T)\right], \tag{24}$$

where $\Psi_0(y)$ is the ground state wave function (WF) of a quantum mechanical system corresponding to the classical stochastic dynamics with potential

$V(y)$ [8].[9] In what follows, the WF (24) will be occasionally referred to as the MANES WF. The stationary state of the original classical stochastic dynamics is then given by its square $\Psi_0^2(y)$. The constant C introduced in (23) is in fact a normalization constant that can be obtained from the requirement that the ground state WF Ψ_0 is squared-normalized, i.e. $\int dy \Psi_0^2(y) = 1$. Note that while $\Psi_0(y)$ is proportional to a *two*-component Gaussian mixture, its square is proportional to a *three*-component Gaussian mixture:

$$\Psi_0^2(y) = \frac{C^2}{2\sqrt{\pi T}} \left[\frac{(1-a)^2}{\sigma_1} \phi\left(y|\mu_1 T, \frac{\sigma_1^2}{2}T\right) + \frac{a^2}{\sigma_2} \phi\left(y|\mu_2 T, \frac{\sigma_2^2}{2}T\right) \right.$$
$$\left. + \frac{2a(1-a)}{\sqrt{(\sigma_1^2+\sigma_2^2)/2}} e^{-\frac{(\mu_1-\mu_2)^2 T}{2(\sigma_1^2+\sigma_2^2)}} \phi\left(y|\mu_3 T, \frac{\sigma_3^2}{2}T\right) \right] \quad (25)$$

where the additional third Gaussian component has the following mean and variance:

$$\mu_3 := \frac{\mu_1 \sigma_2^2 + \mu_2 \sigma_1^2}{\sigma_1^2 + \sigma_2^2}, \quad \frac{\sigma_3^2}{2} = \frac{\sigma_1^2 \sigma_2^2}{\sigma_1^2 + \sigma_2^2}. \quad (26)$$

The normalization condition thus fixes the value of the constant C as follows:

$$C^2 = \frac{2\sqrt{\pi T}}{\Omega}, \quad \text{where } \Omega = \frac{(1-a)^2}{\sigma_1} + \frac{a^2}{\sigma_2} + \frac{2a(1-a)}{\sqrt{(\sigma_1^2+\sigma_2^2)/2}} e^{-\frac{(\mu_1-\mu_2)^2 T}{2(\sigma_1^2+\sigma_2^2)}}. \quad (27)$$

The three-component Gaussian mixture density $\Psi_0^2(y)$ given by equation (25) can be written more compactly using Gaussian mixture weights:

$$\omega_1 := \frac{(1-a)^2}{\sigma_1 \Omega}, \quad \omega_2 := \frac{a^2}{\sigma_2 \Omega},$$
$$\omega_3 := \frac{2a(1-a)}{\Omega\sqrt{(\sigma_1^2+\sigma_2^2)/2}} e^{-\frac{(\mu_1-\mu_2)^2 T}{2(\sigma_1^2+\sigma_2^2)}}, \quad \sum_{i=1}^{3} \omega_i = 1. \quad (28)$$

This produces the following three-component Gaussian mixture model for the stationary distribution $p_s(y) = \Psi_0^2(y)$:

$$p_s(y) = \sum_{k=1}^{3} \omega_k \phi\left(y|\mu_k T, \frac{\sigma_k^2}{2}T\right). \quad (29)$$

Examples of trial ground state WFs Ψ_0 and resulting potentials $V(y)$ are shown in Figure 2.

---

[9] A Gaussian mixture can approximate a ground state wave function of a particle placed in either a single-well or a double-well potential. Double-well potentials play a special role in statistical physics and quantum mechanics and are often used to model tunneling phenomena, see, e.g. [10] or [16]. In particular, a symmetric double-well is described by a symmetric version of Ψ_0 with $a = 1/2$ and $\mu_1 = -\mu_2$, $\sigma_1 = \sigma_2$. For other choices of model parameters, the GM model for function Ψ_0 can fit a variety of shapes including both a unimodal and bimodal shapes.

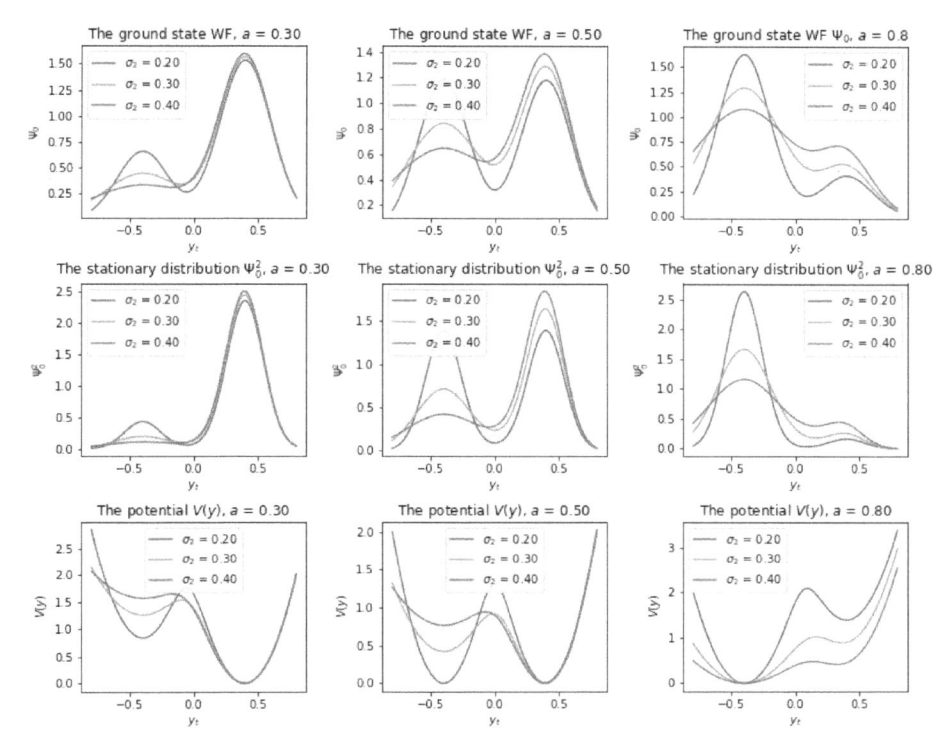

Figure 2. The ground state wave function $\Psi_0(y)$, the stationary distribution $\Psi_0^2(y)$ and the Langevin potential (23) as a function of the log-return y_t, for a few values of the asymmetry parameter a, and with different values of parameters σ_2, with fixed values $\sigma_1 = 0.2, \mu_1 = 0.4, \mu_2 = -0.4, T = 1$. Graphs on the left describe a healthy stock where the right minimum is a global minimum. Graphs on the right correspond to a severely distressed stock, when the left minimum becomes a global minimum. Graphs in the middle column describe intermediate scenarios. In particular, when $a = 0.5$ and $\sigma_1 = \sigma_2$, the resulting potential shown in the line is symmetric. For a multi-asset setting, this potential gives rise to a spontaneous breaking of the $\mathbb{Z}_2$ symmetry $y_t \to -y_t$.

While this expression produces a nonlinear behavior for small positive or negative values of y, its limiting behavior at $y \to \pm\infty$ is rather simple and coincides with a harmonic (quadratic) potential:

$$V(y)|_{y\to-\infty} = h^2 \frac{(y-\mu_2 T)^2}{2\sigma_2^2 T}, \quad V(y)|_{y\to\infty} = h^2 \frac{(y-\mu_1 T)^2}{2\sigma_1^2 T} \quad (\mu_2 < \mu_1). \tag{30}$$

The fact that the limiting behavior of the potential coincides with a harmonic potential as $y \to \pm\infty$ means that in this asymptotic regime the model behavior is described by a harmonic oscillator and thus is fully analytically tractable.

For small values $y \ll 1$, the NES potential (23) can provide a good approximation to a quartic potential that we found for small values $y \ll 1$ in the previous analysis, see equation (22). On the other hand, the asymptotic behavior of the NES potential (23) which coincides with a harmonic potential as $|y| \to \infty$ is different from the asymptotically quartic potential implied by equation (22). However, if the "physics" of the system is determined by a region of small values of y, replacing an asymptotically quartic potential by an asymptotically harmonic potential should produce a negligible impact on observable consequences of the model while significantly simplifying the analysis.

As Gaussian mixtures are known to be universal approximations for an arbitrary non-negative functions given enough components, this implies that an *arbitrary* potential that asymptotically coincides with a harmonic oscillator potential can be represented as a negative logarithm of a Gaussian mixture. We can refer to such class of potentials as Log-Gaussian Mixture (LGM) potentials. In this chapter, I only consider a two-component LGM potential.[10]

While with general parameters $\mu_1 \neq \mu_2, \sigma_1 \neq \sigma_2$ and $a \neq 1/2$ the potential (24) is non-symmetric under reflections $y \to -y$, it becomes symmetric, with $V(-y) = V(y)$, for the special choice $\mu_1 = -\mu_2 = \mu$, $\sigma_1 = \sigma_2 = \sigma$ and $a = 1/2$, see Figure 2. It is useful for what follows to consider a potential which is only slightly asymmetric. This can be done by considering the following specification of model parameters in (24):

$$a = \frac{1}{2} + \varepsilon_a, \quad \mu_{1,2} = \pm\mu + \varepsilon_\mu, \quad \sigma^2_{1,2} = \sigma^2 \pm \varepsilon_\sigma \quad (31)$$

with small parameters $\varepsilon_a, \varepsilon_\mu, \varepsilon_\sigma \ll 1$. Using these parameters in equation (24), we obtain, to the linear order in the asymmetry,

$$V(y) = V^{(s)}(y) - B_0 y \quad (32)$$

where $V^{(s)}(y) = -h^2 \log \Psi_0^{(s)}(y) + V_0$ is a symmetric potential with $V^{(s)}(y) = V^{(s)}(-y)$,

$$\Psi_0^{(s)}(y) := \frac{C}{2}\left[\phi(y|\mu T, \sigma^2 T) + \phi(y|-\mu T, \sigma^2 T)\right] \quad (33)$$

---

[10] A requirement of an asymptotic harmonic oscillator behavior could be seen as a potential limitation for the LGM class of potentials, as many interesting potentials have a different asymptotic behavior. To this point, we can note that an onset of such a quadratic regime can always be pushed further away by a proper rescaling of the coordinate, while for small or moderate values of a new rescaled argument, the dynamics can still be arbitrarily nonlinear and driven by the number of Gaussian components and their parameters.

is a symmetric ground state wave function, and parameter B_0 is a linear function of ε_a, ε_μ, ε_σ:

$$B_0 = \frac{h^2}{\sigma^2 T}\left[\left(1 - \frac{\mu^2}{\sigma^2}\right)\varepsilon_\mu + \frac{\mu}{\sigma^2}\left(1 - \frac{\mu^2}{2\sigma^2}\right)\varepsilon_\sigma\right] - \frac{2\mu h^2}{\sigma^2}\varepsilon_a. \quad (34)$$

Equation (32) shows that a slightly asymmetric potential corresponding to model parameters in equation (31) can be approximated for small values of y by adding a linear term to the potential $V^{(s)}(y)$ obtained in the symmetric limit, where the coefficient B_0 can be directly computed from the original model parameters. As the term $B_0 y$ in equation (32) can be interpreted as the contribution of an external field B_0 to the potential energy of the oscillator y, this implies that the dynamics in a slightly asymmetric potentials $V(y)$ can be approximated by the dynamics in a symmetric potential $V^{(s)}(y)$ with an additional fictitious external field B_0. This observation is used in the following.

The Fokker–Planck equation for MANES

Now, after we specified the particular self-interaction MANES potential $V(y)$ given by equation (23), we proceed with an equivalent probabilistic approach to the dynamics described by the Langevin equation (20). Such a probabilistic method is provided by a corresponding Fokker–Planck equation (FPE). This approach is presented in the following two sections. Note that in both of them the explicit form of the potential $V(y)$ is not used, and thus all equations in this section and section "McKean–Vlasov equation for the mean-field dynamics" are general and valid for an arbitrary confining potential $V(y)$.

The FPE corresponding to the Langevin equation (20) is a linear partial differential equation for the joint probability $P(\mathbf{y}, t) = P(y_1, \ldots, y_N, t)$ of a state $\mathbf{y} = [y_1, \ldots, y_N]$ describing N stocks at time t, given an initial position $\mathbf{y}_0$ at time $t = 0$:

$$\frac{\partial P(\mathbf{y}, t)}{\partial t} = \sum_{i=1}^{N} \frac{\partial}{\partial y_i}\left[\left(\frac{\partial V(y_i)}{\partial y_i} + \frac{g}{N}\sum_{j=1}^{N}\frac{\partial V_{\text{int}}(y_i - y_j)}{\partial y_i}\right)\right.$$
$$\left. \times P(\mathbf{y}, t) + \frac{h^2}{2}\frac{\partial}{\partial y_i}P(\mathbf{y}, t)\right]. \quad (35)$$

For applications, the most interesting probability distributions for a system of identical nonlinear oscillators are one-particle density $P^{(1)}$ and pair-density function $P^{(2)}$ defined as follows:

$$P^{(1)}(y,t) = \int dy_2, \ldots, dy_N P(y, y_2, \ldots, y_N, t),$$

$$P^{(2)}(y, y', t) = \int dy_3, \ldots, dy_N P(y, y', y_3, \ldots, y_N, t). \tag{36}$$

Integrating over $y_2, \ldots, y_N$ in the FPE (35), we obtain

$$\frac{\partial P^{(1)}(y,t)}{\partial t} = \frac{\partial}{\partial y}\left[\frac{\partial V(y)}{\partial y}P^{(1)}(y,t) + \frac{h^2}{2}\frac{\partial}{\partial y}P^{(1)}(y,t) \right. $$
$$\left. + g\int dy' \frac{\partial V_{\text{int}}(y-y')}{\partial y} P^{(2)}(y,y',t)\right]. \tag{37}$$

Note that while a single-particle FPE is a partial differential equation (PDE) for a one-particle density, equation (37) is an integro-differential equation that relates two different densities $P^{(1)}$ and $P^{(2)}$. Due to multi-body interactions whose strength is controlled by the coupling constant g, the densities $P^{(1)}$, and $P^{(2)}$ are coupled in equation (37) which in fact represents the first equation in an infinite hierarchy of the BBGKY type, see, e.g. [14].

McKean–Vlasov equation for the mean-field dynamics

To proceed, I follow the traditional approach in the physics literature (see, e.g. [6,12]) and rely on the mean field approximation (MFA) where the probability density factorizes into a product of single-particle densities:

$$P(y_1, \ldots, y_N, t) = \prod_{i=1}^{N} p(y_i, t). \tag{38}$$

Therefore, with the MFA, dynamics amounts to a system of N independent and identical particles, such that the coordinate of any of them is given by the average $\frac{1}{N}\sum_{i=1}^{N} y_i$ of all original (interacting) particles in the system. This approximation becomes exact in the thermodynamic limit $N \to \infty$.

Note that the mean field $\frac{1}{N}\sum_{i=1}^{N} y_i$ of a homogeneous system of identical nonlinear oscillators can be viewed as a reasonable proxy to the market returns, which are usually proxied by returns of the S&P 500 index. Of course, the real stock market is quite heterogeneous, and furthermore different firms are weighted in the S&P 500 index by their total capitalization. Therefore, identifying the mean field $\frac{1}{N}\sum_{i=1}^{N} y_i$ for a homogeneous system with market returns is *not* expected to provide a good approximation for the price dynamics of any individual stock. However, more importantly, the present mean field approach emphasizes the difference between a single-stock dynamics and the group dynamics of a market made of many stocks while still retaining a link between them and keeping the whole approach practical.

Plugging the MFA ansatz (38) into equation (37), we obtain

$$\frac{\partial p(y,t)}{\partial t} = \frac{\partial}{\partial y}\left[\frac{\partial V(y)}{\partial y}p(y,t) + \frac{h^2}{2}\frac{\partial}{\partial y}p(y,t)\right.$$
$$\left. + gp(y,t)\frac{\partial}{\partial y}\int dy' V_{\text{int}}(y-y')p(y',t)\right]. \quad (39)$$

This nonlinear integro-differential equation holds for an arbitrary interaction potential $V_{\text{int}}(y-y')$. For a particular case of the quadratic Curie–Weiss potential (21), equation (39) produces the following equation:

$$\frac{\partial p(y,t)}{\partial t} = \frac{\partial}{\partial y}\left[\frac{\partial V(y)}{\partial y}p(y,t) + \frac{h^2}{2}\frac{\partial}{\partial y}p(y,t) + g\left(y-\langle y\rangle_p\right)p(y,t)\right],$$
$$\langle y\rangle_p := \int dy\, y p(y,t). \quad (40)$$

This equation is nonlinear due to the dependence of the coefficient $(y - \langle y\rangle_p)$ in the last term on the density $p(y,t)$. The nonlinear Fokker–Planck equation (40) is known in the literature as the McKean–Vlasov equation [1,3,13]. Critically important is the fact the unlike the initial *linear* FPE equation (35) for a finite N-particle system, the the McKean–Vlasov equation that describes the thermodynamic limit $N \to \infty$ is *nonlinear*. As a result, it produces far richer dynamics including phase transitions [1].

Note that if we formally treat $\langle y\rangle_p$ as an independent parameter, the McKean–Vlasov equation (40) can be viewed as a linear FPE with the following "effective" potential:

$$V_{\text{eff}}(y) = V(y) + g\left(\frac{y^2}{2} - my\right), \quad m := \langle y\rangle_p. \quad (41)$$

The stationary density for a given value of $m = \langle y\rangle_p$ is therefore obtained as a Boltzmann distribution with the "inverse temperature" $\beta = 2/h^2$:

$$p(y|m) = \frac{1}{Z(m)}e^{-\frac{2}{h^2}\left[V(y)+g\left(\frac{y^2}{2}-my\right)\right]},$$
$$Z(m) := \int dy\, e^{-\frac{2}{h^2}\left[V(y)+g\left(\frac{y^2}{2}-my\right)\right]}. \quad (42)$$

This solution should satisfy the self-consistency condition for $m = \langle y \rangle_p = \int dy\, y p(y, t)$:

$$m = \frac{1}{Z(m)} \int dy\, y\, e^{-\frac{2}{h^2}\left[V(y)+g\left(\frac{y^2}{2}-my\right)\right]}. \quad (43)$$

Once the value of m is found by solving the self-consistency equation (43), it is substituted back to equation (42) to find the stationary density. The number of equilibrium states in the system is therefore given by the number of solutions to equation (43). The solution of this equation for the MANES potential is presented in section "Phase Transitions of MANES".

Renormalization of the single-stock NES potential by interactions

As was noted above, the McKean–Vlasov equation (40) can be viewed as a single-particle linear FPE with the effective potential (41) that I repeat here for convenience:

$$V_{\text{eff}}(y) = V(y) + g\left(\frac{y^2}{2} - my\right), \quad m := \langle y \rangle_p, \quad (44)$$

provided m is found from self-consistency equations that would be introduced in the following. However, prior to this, we can note that for the specific choice of the NES potential (23), the effective potential $V_{\text{eff}}(y)$ will have exactly the same functional form as the single-stock potential $V(y)$, albeit with different parameters. Using the physics nomenclature, one can say that parameters of the original single-particle system are *renormalized* by interactions which are represented by the second term in equation (44).

To find the new renormalized parameters, we write equation (44) using equation (23) as follows:

$$\begin{aligned}
V_{\text{eff}}(y) &= -h^2 \log\left[(1-a)\phi(y|\mu_1 T, \sigma_1^2 T) + a\phi(y|\mu_2 T, \sigma_2^2 T)\right] \\
&\quad - h^2 \log e^{-g\frac{(y-m)^2}{2h^2} + g\frac{m^2}{2h^2}} \\
&= -h^2 \log\left[e^{-g\frac{(y-m)^2}{2h^2} + g\frac{m^2}{2h^2}}\left((1-a)\phi(y|\mu_1 T, \sigma_1^2 T) + a\phi(y|\mu_2 T, \sigma_2^2 T)\right)\right] \\
&= -h^2 \log\left[(1-\bar{a})\phi(y|\bar{\mu}_1 T, \bar{\sigma}_1^2 T) + \bar{a}\phi(y|\bar{\mu}_2 T, \bar{\sigma}_2^2 T)\right] - h^2 \hat{V}(m),
\end{aligned} \quad (45)$$

where $\bar{\mu}_1, \bar{\mu}_2, \bar{\sigma}_1, \bar{\sigma}_2, \bar{a}$ are new "renormalized" parameters for a single-particle NES potential, and $\hat{V}(m)$ stands for terms that depend on m but not on y. They can be easily computed using the well-known relation expressing a product of two Gaussian densities as a rescaled third Gaussian density:

$$\phi(y|\mu_1,\sigma_1^2)\phi(y|\mu_2,\sigma_2^2) = \frac{S_{12}}{\sqrt{2\pi\sigma_{12}^2}} \exp\left[-\frac{(y-\mu_{12})^2}{2\sigma_{12}^2}\right] = S_{12}\phi(y|\mu_{12},\sigma_{12}^2), \tag{46}$$

where

$$\sigma_{12}^2 = \frac{\sigma_1^2\sigma_2^2}{\sigma_1^2+\sigma_2^2}, \quad \mu_{12} = \left(\frac{\mu_1}{\sigma_1^2}+\frac{\mu_2}{\sigma_2^2}\right)\sigma_{12}^2,$$

$$S_{12} = \frac{1}{\sqrt{2\pi(\sigma_1^2+\sigma_2^2)}} \exp\left[-\frac{(\mu_1-\mu_2)^2}{2(\sigma_1^2+\sigma_2^2)}\right]. \tag{47}$$

Using these relations, we obtain the renormalized parameters and the function $\hat{V}(m)$:

$$\bar{\mu}_k = \frac{\mu_k + \frac{g}{h^2}\sigma_k^2 m}{1+\frac{g}{h^2}\sigma_k^2 T}, \quad \bar{\sigma}_k^2 = \frac{\sigma_k^2}{1+\frac{g}{h^2}\sigma_k^2 T}, \quad k=1,2,$$

$$\bar{a} = \left[1+\frac{1-a}{a}\sqrt{\frac{h^2+g\sigma_2^2 T}{h^2+g\sigma_1^2 T}}e^{\frac{g(m-\mu_2 T)^2}{2(h^2+g\sigma_2^2 T)}-\frac{g(m-\mu_1 T)^2}{2(h^2+g\sigma_1^2 T)}}\right]^{-1}, \tag{48}$$

$$\hat{V}(m) = \frac{gm^2}{2h^2} + \log\left(\frac{(1-a)h}{\sqrt{h^2+g\sigma_1^2 T}}e^{-\frac{g(m-\mu_1 T)^2}{2(h^2+g\sigma_1^2 T)}} + \frac{ah}{\sqrt{h^2+g\sigma_2^2 T}}e^{-\frac{g(m-\mu_2 T)^2}{2(h^2+g\sigma_2^2 T)}}\right).$$

These formulae show that additional quadratic and linear terms in equation (44) can be re-absorbed into rescaled parameters of the Gaussian mixture, where the Gaussian means become linear functions of the mean field m, while the mixing coefficient depends on m nonlinearly. This implies that within the mean field approximation, the dynamics of market log-returns can be modeled as the dynamics of a fictitious single stock, with initial single-stock model's parameters being "dressed" by interactions according to equation (48).

Phase Transitions of MANES

For a confining potential $V(y)$, the original finite dimensional Langevin equation (18) or (20) produces *ergodic and reversible* dynamics under the Gibbs measure

$$\mu_N(dy) = \frac{1}{Z_N} e^{-U(y_1,\ldots,y_N)/h^2} dy_1 \cdots dy_N, \tag{49}$$

where Z_N is a normalization factor and $U(y_1, \ldots, y_N)$ is the potential (19), as a consequence of the linearity of the corresponding FPE, and uniqueness of the stationary state of this FPE. On the other hand, the McKean–Vlasov with a confining but non-convex self-interaction potential $V(y)$ can lead to *violations of ergodicity and phase transitions* [1].

To explore the phase structure in our setting with the NES potential (23), first note that using equation (24), the integral entering the self-consistency condition (43) can be expressed in terms of an integral that involves Ψ_0^2:

$$m = \frac{1}{Z(m)} \int dy\, y\, e^{-\frac{2}{h^2}\left[V(y) + g\left(\frac{y^2}{2} - my\right)\right]} = \frac{1}{Z(m)} \int dy\, y\, \Psi_0^2(y) e^{-\frac{2g}{h^2}\left(\frac{y^2}{2} - my\right)}. \tag{50}$$

Analysis of the phase structure of the model under variations of parameters is based on equation (50) where m is viewed as an order parameter. This is similar to the classical Ising model where the mean magnetization is analogously used as an order parameter, and a second order phase transition occurs at a critical temperature T_c at which the self-consistency equation $m = \tanh(m/T)$ of the Ising model bifurcates and produces a solution $m \neq 0$, in addition to the "trivial" solution $m = 0$ which describes a high-temperature regime $T \to \infty$.

A similar pattern of bifurcations and phase transitions can also be found, for certain types of the potential $V(y)$, for the self-consistency equation (50) that deals with continuous random variables rather than binary spins of the Ising model. In particular, in the Desai–Zwangiz model, equation (50) is solved with the quartic double-well potential $V(y) = -ay^2 + \lambda y^4$ with $a, \lambda > 0$. However, some general properties of the resulting stochastic system are determined by general properties of the potential, such as the number of local minima, and the asymptotic behavior at $|y| \to \infty$, rather than its particular functional form. In this work, I will use the Log-GM NES potential (24) which produces a tractable double-well potential for certain values of parameters, however the general analysis in the following holds for an arbitrary confining potential $V(y)$.

If the potential $V(y)$ is symmetric, $V(-y) = V(y)$, then $m = 0$ is always a solution to the self-consistency equation. This is similar to how the state with zero magnetization is the only stable solution for the Ising model in the high-temperature limit (which corresponds to the limit $h \to \infty$ in our conventions). However, for lower temperatures, the Ising spins becomes aligned with the mean magnetization $m = \pm 1$, producing a bifurcation of the state equation in the parameter space. The bifurcation provides a mean field approximation description of the second order phase transition in the Ising model. Similarly in the current setting, for certain shapes of the potential $V(y)$ and sufficiently low volatilities h, the self-consistency equation (50) can produce multiple solutions below some critical volatility h_c. The analysis of the solution of this equation as a function of the "temperature" parameter h should be performed at different values of parameter g that drives interactions in the system, as well as other parameters of the WF (24). The equilibrium second-order phase transition describing the order-disorder transitions for the Desai–Zwanzig model was established in [1]. As is shown next, a similar second-order phase transition can also occur for the MANES model when the potential $V(y)$ is symmetric.

Self-consistency equation: Second-order phase transition for a symmetric potential

While for many choice of interesting non-convex potentials the analysis requires numerical integration that needs some extra care in the presence of multiple solutions (see [6]), in our case the integral involved in the self-consistency relation (50) can be easily computed analytically for the WF Ψ_0 defined in equation (24), with its square given by equation (25). Using equation (46), we obtain

$$m = \frac{\int dy\, y\, \Psi_0^2(y) e^{-\frac{2g}{h^2}\left(\frac{y^2}{2}-my\right)}}{\int dy\, \Psi_0^2(y) e^{-\frac{2g}{h^2}\left(\frac{y^2}{2}-my\right)}} = \frac{h^2}{2g}\frac{\frac{\partial Z(m)}{\partial m}}{Z(m)} = \frac{h^2}{2g}\frac{\partial}{\partial m}\log Z(m). \quad (51)$$

The partition function $Z(m)$ that enters this expression can be computed using equations (44) and (45):

$$Z(m) = \int dy\, e^{-\frac{2g}{h^2} V_{\text{eff}}(y)}$$

$$= e^{2\hat{V}(m)}\left[\frac{(1-\bar{a})^2}{\bar{\sigma}_1} + \frac{\bar{a}^2}{\bar{\sigma}_2} + \frac{2\bar{a}(1-\bar{a})}{\sqrt{(\bar{\sigma}_1^2+\bar{\sigma}_2^2)/2}} e^{-\frac{(\bar{\mu}_1-\bar{\mu}_2)^2 T}{2(\bar{\sigma}_1^2+\bar{\sigma}_2^2)}}\right] \quad (52)$$

where $\hat{V}(m)$ and parameters $\bar{a}, \bar{\mu}_1, \bar{\mu}_2$, etc. are defined in equation (48). For a symmetric NES potential with $\mu_1 = -\mu_2 = \mu$, $\sigma_1 = \sigma_2 = \sigma$, and $a = 1/2$, this expression is further simplified:

$$Z(m) = \frac{1}{\sigma} e^{2\hat{V}_s(m)}\left[1 - \frac{1 - e^{-\frac{h^2\mu^2 T}{\sigma^2(h^2+g\sigma^2 T)}}}{2\cosh^2\left(\frac{g\mu T m}{h^2+g\sigma^2 T}\right)}\right] \quad (53)$$

where $\hat{V}_s(m)$ stands for the symmetric version of $\hat{V}(m)$ in equation (48):

$$\hat{V}_s(m) = \frac{gm^2}{2h^2} - \frac{g(m^2 + \mu^2 T^2)}{2(h^2 + g\sigma^2 T)} + \log \cosh \frac{g\mu T m}{h^2 + g\sigma^2 T} + \log \frac{h}{\sqrt{h^2 + g\sigma^2 T}}. \tag{54}$$

The derivative $\partial \log Z / \partial m$ with the symmetric potential is therefore as follows:

$$\frac{\partial}{\partial m} \log Z(m) = \frac{2g^2 \sigma^2 T m}{h^2(h^2 + g\sigma^2 T)} + \frac{2g\mu T}{h^2 + g\sigma^2 T} \frac{\sinh \frac{2g\mu T m}{h^2 + g\sigma^2 T}}{\cosh \frac{2g\mu T m}{h^2 + g\sigma^2 T} + e^{-\frac{h^2 \mu^2 T}{\sigma^2(h^2 + g\sigma^2 T)}}}. \tag{55}$$

Substituting this expression into (51), we obtain the MANES self-consistency equation for a symmetric potential:

$$m = \mu T \frac{\sinh\left(\frac{2g\mu T m}{h^2 + g\sigma^2 T}\right)}{\cosh\left(\frac{2g\mu T m}{h^2 + g\sigma^2 T}\right) + e^{-\frac{h^2 \mu^2 T}{\sigma^2(h^2 + g\sigma^2 T)}}} \tag{56}$$

which looks similar to the self-consistency equation for the Ising model, see Figure 3. Note that a symmetric potential assumed in equation (56) may not necessarily match the market data, and actual self-interaction potentials $V(y)$ implied by market prices are typically *not* symmetric (see section "Fitting Model Parameters Using Option Data"). Nevertheless, analysis of symmetric potentials is of interest because it connects with the theory of second-order phase transitions. As shown in section "MANES bifurcation

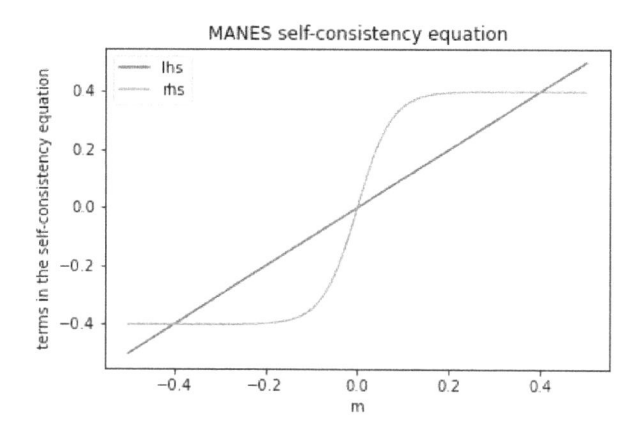

Figure 3. The self-consistency equation (56). The presence of three solutions for the expected log-return m at $m = 0$ and $m = \pm m_c$ with $m_c \neq 0$ is illustrated for the following choice of parameters: $\mu_1 = -\mu_2 = 0.4$, $\sigma_1 = \sigma_2 = 0.1$, $a = 0.5$, $T = 1.0$, $g = 0.2$, $h = 0.1$.

diagrams and phase transitions", for a symmetric potential, the mean field vanishes if the volatility h exceeds a certain critical value h_c, and becomes non-zero, $m = \pm m_0$ for some value m_0, for yet lower values $h < h_c$, with a continuous change from $m = 0$ for $h > h_c$ to non-vanishing values of m for $h < h_c$. This describes a continuous (second-order) phase transition, similar to the one obtained for the Ising model with the vanishing magnetic field field ($B = 0$).

On the other hand, a non-symmetric self-interaction potential $V(y)$ can be approximated by a symmetric potential $V^{(s)}(y)$ with a fictitious "magnetic field" B_0, see equation (32). Driven by the analogy with the Ising model, we could expect that this setting would produce scenarios for a first-order phase transition describing a decay of a metastable state, rather than a second-order phase transition. As we see next, this is indeed the case in the present model.

Non-symmetric potential: A first-order phase transition

When the potential $V(y)$ is not symmetric, the self-consistent mean field m should be computed using the general formula (51). Such analysis can be further simplified using a linear approximation to an asymmetric potential by adding a fictitious external field B_0 to a symmetric potential $V^{(s)}(y)$, see equations (32) and (34). This is equivalent to replacing $m \to m + \frac{1}{g}B_0$ and $V(y) \to V^{(s)}(y)$ in equation (44) that defines the effective potential $V_{\text{eff}}(y)$. Therefore, the generalization of equation (53) to the case of a slightly asymmetric potential $V(y)$ can be obtained by the same replacement $m \to m + \frac{1}{g}B_0$:

$$Z(m) = \frac{1}{\sigma} e^{2\hat{V}^{(s)}\left(m+\frac{1}{g}B_0\right)} \left[1 - \frac{1 - e^{-\frac{h^2\mu^2 T}{\sigma^2(h^2+g\sigma^2 T)}}}{2\cosh^2\left(\frac{g\mu T\left(m+\frac{1}{g}B_0\right)}{h^2+g\sigma^2 T}\right)} \right]. \quad (57)$$

Using this expression, we can obtain a generalization of equation (56) for the case $B_0 \neq 0$:

$$m = \frac{\sigma^2 T}{h^2}B_0 + \mu T \frac{\sinh\left(\frac{2g\mu T}{h^2+g\sigma^2 T}\left(m+\frac{1}{g}B_0\right)\right)}{\cosh\left(\frac{2g\mu T}{h^2+g\sigma^2 T}\left(m+\frac{1}{g}B_0\right)\right) + e^{-\frac{h^2\mu^2 T}{\sigma^2(h^2+g\sigma^2 T)}}}. \quad (58)$$

As implied by this equation, when $B_0 \neq 0$, the mean field m does not vanish for any value of h, and therefore there is no second-order phase transition when $B_0 \neq 0$. Instead, in this case, we obtain a first-order phase transition under variations of the mean field B_0. When B_0 is very small but still non-vanishing, its sign determines whether the negative or positive solution with $m = -m_0$ or $m = m_0$ will have the lowest energy and thus will be the true ground state, where $\pm m_0$ are two degenerate mean field solutions obtained in the strict limit $B_0 = 0$. The discrete $\mathbb{Z}_2$ symmetry $y \leftrightarrow -y$ is explicitly broken when $B_0 \neq 0$. If an initial state with, e.g. $m = +m_0$ is observed at time $t = 0$ and the "magnetic field" $B_0 > 0$, then it is the left-well state $m = -m_0$ that would be the true ground state, while $m = +m_0$ will be a metastable state. Vice versa, for $B_0 < 0$, the state $m = m_0$ would be the true ground state, while $m = -m_0$ could be a metastable state released at $t = 0$. A decay of such a metastable state is described as a first-order phase transition, which amounts to a sudden discontinuous jump from $m = -m_0$ to $m = m_0$ (or vice versa, depending on the sign of B_0 and the initial state). An example of an asymmetric potential leading to such a first-order phase transition is shown in the following section. The model thus suggests an interplay between the first- and second-order phase transitions in different regimes of parameters (h, B_0) which is similar to the phase transitions pattern in the Ising model.

MANES bifurcation diagrams and phase transitions

Analysis of the phase structure of the model is performed using the traditional approach similar to the mean field analysis of the Ising model. We solve the self-consistency equation to compute a function $m = m(h)$ while keeping other model parameters fixed. The same exercise can be repeated for different values of another important model parameter g that controls interactions between individual particles (assets).

The phase diagram obtained with this method for a symmetric potential assumed in equation (56) is shown in Figure 4. The order parameter m obtained as a function of a continuously varying volatility parameter h bifurcates at the critical value h_c. The transition from a zero to a non-zero mean field at the bifurcation point $h = h_c$ is continuous, as it should be for a second-order phase transition.

To identify stable versus unstable solutions of the self-consistency equation for either a symmetric or asymmetric potential $V(y)$ (see, respectively, equations (56) and (58)), we need to compute the Gibbs free energy $\mathcal{F}$ which is given by the following expression [6]:

$$\mathcal{F} = \frac{h^2}{2}\int p(x) \log p(x) dx + \int V(x) p(x) dx + \frac{g}{2} \int\int V_{\text{int}}(x-y) p(x) p(y) dx dy. \tag{59}$$

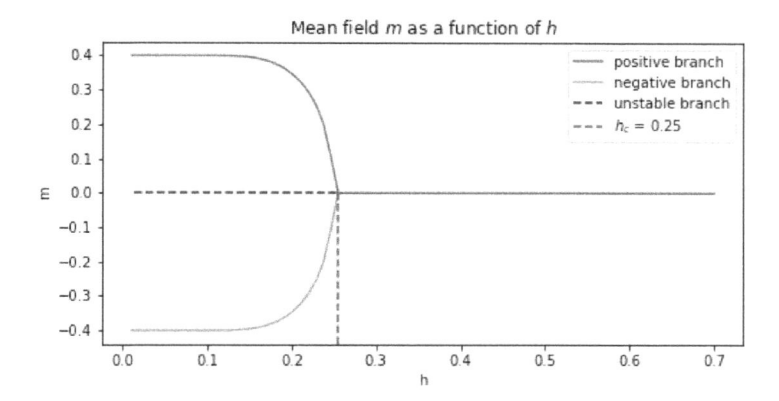

Figure 4. Bifurcation diagram of the order parameter m (the expected market log-return) as a function of the noise level h, obtained for a symmetric potential with the following parameters: $\mu_1 = -\mu_2 = 0.4$, $\sigma_1 = \sigma_2 = 0.1$, $a = 0.5$, $T = 1.0$, $g = 0.2$. The bifurcation occurs at the critical value $h_c = 0.25$.

For the stationary density (42) and the quadratic Curie–Weiss interaction potential, the free energy $\mathcal{F} = \mathcal{F}(m)$ can be computed in a more explicit form:

$$\mathcal{F}(m) = -\frac{h^2}{2}\log Z(m) + \frac{g}{2}m^2, \quad Z(m) = \int dy\, \Psi_0^2(y) e^{-\frac{2g}{h^2}\left(\frac{y^2}{2} - my\right)}. \quad (60)$$

The gradient of the free energy with respect to m is therefore as follows:

$$\frac{\partial \mathcal{F}}{\partial m} = -g \frac{\int dy\, \Psi_0^2(y) y\, e^{-\frac{2g}{h^2}\left(\frac{y^2}{2} - my\right)}}{\int dy\, \Psi_0^2(y) e^{-\frac{2g}{h^2}\left(\frac{y^2}{2} - my\right)}} + gm. \quad (61)$$

The stationary points of the free energy are obtained by setting this expression to zero, which again produces the self-consistency equation (51).

While the relations (59) and (61) are general and apply for the generalized Desai–Zwanzig model with an arbitrary self-interaction potential, in this work that uses the log-GM NES potential (23), the free energy can be computed in closed form for either symmetric or asymmetric potentials $V(y)$ using, respectively equation (53) or equation (52). To have even more compact formulae, it is convenient to use the partition function (57) corresponding to a weakly asymmetric potential whose asymmetry is controlled by a fictitious external field B_0. This produces the following expression:

$$\mathcal{F}(m) = -\frac{g\sigma^2 T}{h^2 + g\sigma^2 T} B_0 m + \frac{gh^2}{2(h^2 + g\sigma^2 T)} m^2$$

$$-\frac{h^2}{2}\log\left[b + \cosh\left(\frac{2g\mu T\left(m + \frac{1}{g}B_0\right)}{h^2 + g\sigma^2 T}\right)\right] + \cdots \quad (62)$$

where the ellipses stand for omitted constant terms that do not depend on m, and parameter b is defined as follows:

$$b := e^{-\frac{h^2\mu^2 T}{\sigma^2(h^2+g\sigma^2 T)}}. \tag{63}$$

Clearly, the free energy $\mathcal{F}(m)$ is not symmetric, i.e. $\mathcal{F}(m) \neq \mathcal{F}(-m)$ for $B_0 \neq 0$ as $\mathcal{F}(m) \sim B_0 m + O(m^2)$ for small values of m. More generally, as can be seen from equation (60), the partition function $Z(m)$ and the free energy $\mathcal{F}(m)$ are symmetric as long as $\Psi_0(y)$ is symmetric in the y-space, i.e. $\Psi_0(y) = \Psi_0(-y)$.

For examples of shapes of the free energy leading to scenarios with phase transitions, see Figure 5. Note that for the left graph obtained with a symmetric potential with $\mu_1 = -\mu_2$, $\sigma_1 = \sigma_2$, $a = 0.5$, the two minima at $m = \pm m_c$ with $m_c \simeq 0.4$ are degenerate, and the point $m = 0$ is a local maximum and is therefore unstable. This setting corresponds to the second-order phase transition for $h < h_c$ and spontaneous breaking of the $\mathbb{Z}_2$ symmetry $m \leftrightarrow -m$ of the free energy $\mathcal{F}(m)$ for $B_0 = 0$. On the other hand, on the right graph, the free energy is non-symmetric for a non-symmetric potential obtained with $\sigma_1 \neq \sigma_2$ and $a \neq 0.5$, which can be approximated by having a non-zero field B_0. For a non-zero field B_0, the $\mathbb{Z}_2$ symmetry $m \leftrightarrow -m$ of the free energy $\mathcal{F}(m)$ is explicitly broken. For this case, the potential has a true minimum for a positive value of m and a local minimum for a negative value of m. If the system is released at time $t = 0$ in the state with a negative m, this state is metastable as it has a higher energy than the state with a positive value of m. In this scenario, a transition between

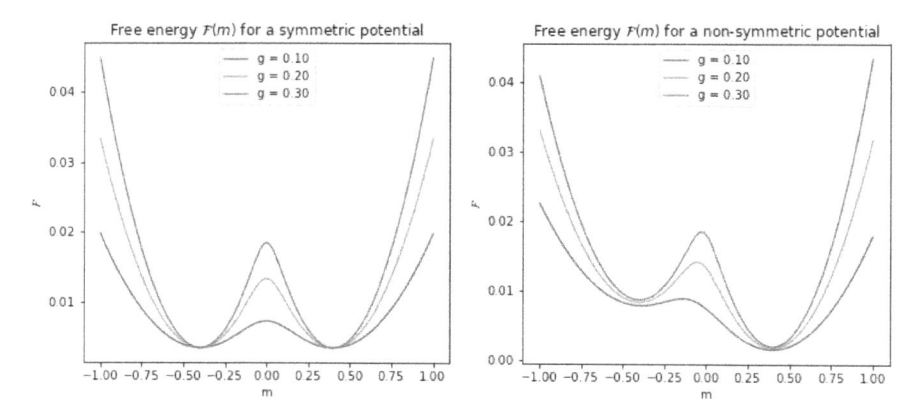

Figure 5. Free energy $\mathcal{F}$ as a function of m. The graph on the left is obtained for a symmetric potential with parameters $\mu_1 = -\mu_2 = 0.4$, $\sigma_1 = \sigma_2 = 0.1$, $a = 0.5$, $T = 1.0$. The graph on the right is obtained for a non-symmetric potential with $\mu_1 = -\mu_2 = 0.4$, $\sigma_1 = 0.1$, $\sigma_2 = 0.15$ and $a = 0.4$.

the state with $m < 0$ and the true ground state with $m > 0$ happens very quickly at a random time, and corresponds to a first-order phase transition. This behavior is again similar to the Ising model, where adding a non-zero magnetic field breaks the $\mathbb{Z}_2$ symmetry of the free energy as a function of magnetization, and produces a first-order phase transition under variations of the temperature.

Critical exponents: α and β

Critical behavior in the present model is defined in a similar way to the Ising model. For the latter, the critical behavior and bifurcation diagram are usually considered with the temperature being the control parameter and average magnetization being the order parameter, while the external magnetic field is used as an additional control parameter. Similarly, in the framework considered here, the volatility parameter h serves in a similar way to the temperature parameter T in the Ising model, while the coupling constant g is used as an additional degree of freedom similar to the magnetic field H in the Ising model.

To investigate the critical behavior of the model in the vicinity of the phase transition at $h = h_c$, the free energy (62) is expanded into a Taylor series around $m = 0$ to the fourth order in m and second order in B_0:

$$\mathcal{F}(h^2) = -f_1 B_0 m + f_2 m^2 + f_4 m^4 + \cdots \tag{64}$$

where constant terms and higher order terms in B_0, m are omitted, and parameters f_1, f_2, f_4 are defined in terms of parameter b introduced in equation (63) and other model parameters as follows:

$$f_1 = \frac{g\sigma^2 T}{h^2 + g\sigma^2 T}\left(1 + \frac{h^2 \mu^2 T^2 (1+b)^{-1}}{h^2 + g\sigma^2 T}\right),$$

$$f_2 = \frac{gh^2}{2(h^2 + g\sigma^2 T)}\left(1 - \frac{2g\mu^2 T^2 (1+b)^{-1}}{h^2 + g\sigma^2 T}\right), \tag{65}$$

$$f_4 = \frac{h^2(2-b)}{3(1+b)^2}\left(\frac{g\mu T}{h^2 + g\sigma^2 T}\right)^4.$$

The free energy $\mathcal{F}$ is written in equation (64) as a function of the noise variance h^2 rather than of the mean field m because in this section we want to explore its dependence on h^2.

Note that equation (63) implies that $0 \leq b \leq 1$, therefore the coefficient in from of m^4 is always positive, ensuring stability of any approximate solution that would be based on the small-m expansion (64). On the other hand, one can see that the coefficient f_2 can change the sign depending on the value of h^2. A bifurcation point $h^2 = h_c^2$ corresponds to the value of h at which the coefficient in front of m^2 in equation (64) vanishes and then becomes negative for yet smaller values $h^2 < h_c^2$. This produces the following relation for the critical volatility parameter h_c:

$$h_c = \sqrt{2g\mu^2 T^2 (1+b)^{-1} - g\sigma^2 T}. \tag{66}$$

To produce a real-valued parameter h_c, the expression under the square root should be positive, producing a constraint on parameter combinations that may lead to bifurcation scenarios:

$$\frac{2\mu^2 T}{\sigma^2} \geq 1 + b. \tag{67}$$

The coefficient f_2 in the expansion (64) can therefore be written in a more suggestive form:

$$f_2 = \frac{gh^2}{2(h^2 + g\sigma^2 T)^2} \left(h^2 - h_c^2\right). \tag{68}$$

In the vicinity of the bifurcation point $h = h_c$, the solution for m in the limit $B_0 \to 0$ can be well approximated by a solution to equation (64), which reads

$$m^2 = -\frac{f_2}{2f_4} = \frac{3g(h^2 + g\sigma^2 T)^2 (1+b)^2}{4(g\mu T)^4 (2-b)} \left(h_c^2 - h^2\right), \quad h \leq h_c. \tag{69}$$

This produces the following expression for the mean field m in the vicinity of the critical point $h = h_c$:

$$m = \pm \frac{h^2 + g\sigma^2 T}{2(g\mu T)^2} \sqrt{\frac{3g(1+b)^2}{2-b}} \left(h_c^2 - h^2\right)^\beta \quad \beta = \frac{1}{2}, \quad \text{if } 0 \leq \frac{h_c - h}{h_c} \ll 1, \tag{70}$$

which again looks very similar to the relation $m \sim (T_c - T)^{1/2}$ with the same critical exponent $\beta = 1/2$ arising for the Ising model.[11] As in the Ising model, the order parameter m is continuous across the critical point $h^2 = h_c^2$, indicating that we deal here with a second-order (continuous) phase transition.

Substituting equation (69) back into equation (64) gives an approximate expression for the free energy for values of h that are slightly below the critical value h_c:

$$\mathcal{F}(h^2) = -\frac{f_2^2}{4f_4} = -\frac{3h^2}{16g^2\mu^4 T^4} \left(h_c^2 - h^2\right)^2 \quad 0 \leq \frac{h_c - h}{h_c} \ll 1. \tag{71}$$

---

[11] Here T stands for the temperature, not the time interval as in the previous formulas.

While this expression approximates the free energy $\mathcal{F}(h)$ for $B_0 = 0$ for values of h that are approaching h_c from below, for values of h that are above h_c, the value of $\mathcal{F}(h)$ in equation (64) will be zero. Further following the analogy with the Ising model, we next define the "specific heat" C_H to be proportional to the second derivative of the free energy with respect to noise variance parameter h^2:

$$C_H = -h^2 \frac{\partial^2 \mathcal{F}(h^2)}{\partial (h^2)^2}. \tag{72}$$

Because the expression in equation (71) arises only for $h^2 < h_c^2$ but vanishes for $h^2 > h_c^2$, it is clear that the first derivative $\partial \mathcal{F}(h^2)/\partial h^2$ is continuous at the critical point $h^2 = h_c^2$, but its second derivative $\frac{\partial^2 \mathcal{F}(h^2)}{\partial (h^2)^2}$ has a finite jump at $h^2 = h_c^2$, translating into a finite jump of the specific heat at the critical point:

$$\begin{aligned}\Delta C_H &:= -\lim_{\varepsilon \to 0} \left(h^2 \frac{\partial^2 \mathcal{F}(h^2)}{\partial (h^2)^2}\bigg|_{h^2=h_c^2-\varepsilon} - h^2 \frac{\partial^2 \mathcal{F}(h^2)}{\partial (h^2)^2}\bigg|_{h^2=h_c^2+\varepsilon} \right) \\ &= \frac{3(1+b)^2}{8} \frac{h_c^4}{2-b} \frac{h_c^4}{g^2 \mu^4 T^4}.\end{aligned} \tag{73}$$

This is again the behavior characterizing a second-order (continuous) phase transition which is similar to the second-order phase transition of the Ising model. Due to a finite jump, we obtain a vanishing value for the critical exponent α entering the formula for the specific heat $C_H \propto |h - h_c|^\alpha$, i.e. $\alpha = 0$.

Partition function and mean field without homogeneity

To consider other properties of the model such as pairwise correlations, in this section we depart from the approximation of a homogeneous market portfolio that was used above, and consider a heterogeneous market where different stocks may have different parameters μ_i, σ_i. The partition function Z relevant for this setting is given by the following expression:

$$Z[B] = \int \prod_{i=1}^{N} dy_i e^{-\frac{2}{h^2}\left[\sum_{i=1}^{N} V_i(y_i) - \sum_{i=1}^{N} y_i B_i + \frac{g}{2N}\sum_{i,j=1}^{N} \frac{1}{2}(y_i-y_j)^2\right]}. \tag{74}$$

Here control parameters B_i are introduced to facilitate calculations of various expectations and correlation functions and are similar in their meaning to an external magnetic field used in the Ising model and other models of phase transitions for similar purposes.

With the Curie–Weiss quadratic interaction potential and MANES self-interaction potential, the partition function can be computed analytically in the limit $N \to \infty$. To this end, first note that the interaction term in the potential can be written as follows:

$$e^{-\frac{2}{h^2}\frac{g}{2N}\sum_{i\neq j}^{N}\frac{1}{2}(y_i-y_j)^2} = e^{-\frac{2}{h^2}\left(\frac{N-1}{N}\sum_{i=1}^{N}\frac{1}{2}gy_i^2 - \frac{g}{2N}\sum_{i\neq j}^{N}y_iy_j\right)}$$
$$\simeq e^{-\frac{2}{h^2}\sum_{i=1}^{N}\frac{1}{2}gy_i^2 + \frac{1}{2}\sum_{i,j=1}^{N}y_iJ_{ij}y_j}, \quad (75)$$

where in the last step I replaced $(N-1)/N \to 1$ assuming that $N \gg 1$ and introduced matrix $\mathbf{J}$ with matrix coefficients $J_{ij} = \frac{2g}{Nh^2}(1-\delta_{ij})$. This can also be written in the matrix form as follows:

$$\mathbf{J} = \frac{2g}{Nh^2}\left(-\mathbf{I} + \mathbf{1}\mathbf{1}^T\right), \quad (76)$$

where $\mathbf{I}$ is a unit $N \times N$ matrix, and $\mathbf{1}$ is a vector of ones of size N. The interaction term can now be represented using integration over auxiliary variables ϕ_i with $i = 1, \ldots, N$ using the Hubbard–Stratonovich transformation:

$$e^{\frac{1}{2}\sum_{i,j=1}^{N}y_iJ_{ij}y_j} = \frac{1}{\sqrt{\det J}}\int_{-\infty}^{\infty}\prod_{i=1}^{N}\frac{d\phi_i}{\sqrt{2\pi}}e^{-\frac{1}{2}\sum_{i,j=1}^{N}\phi_iJ_{ij}^{-1}\phi_j + \sum_{i=1}^{N}y_i\phi_i} \quad (77)$$

which holds for any real symmetric and invertible matrix $\mathbf{J}$, whose inverse matrix has matrix elements J_{ij}^{-1}. The inverse of matrix $\mathbf{J}$ defined in equation (76) is computed using the Sherman–Morrison formula[12]:

$$\mathbf{J}^{-1} = \frac{Nh^2}{2g}\left(-\mathbf{I} + \frac{\mathbf{1}\mathbf{1}^T}{N-1}\right). \quad (78)$$

Using equation (77), the partition function (74) can be written as follows:

$$Z[B] = \frac{1}{\sqrt{\det J}}\int_{-\infty}^{\infty}\prod_{i=1}^{N}\frac{d\phi_i}{\sqrt{2\pi}}e^{-\frac{1}{2}\sum_{i,j=1}^{N}\phi_iJ_{ij}^{-1}\phi_j}$$
$$\times \prod_{i=1}^{N}\int dy_i e^{-\frac{2}{h^2}\left[V_i(y_i) + \frac{1}{2}gy_i^2 - \left(\frac{h^2}{2}\phi_i + B_i\right)y_i\right]}. \quad (79)$$

---

[12]The Sherman–Morrison formula

$$\left(\mathbf{A} + \mathbf{b}\mathbf{c}^T\right)^{-1} = \mathbf{A}^{-1} - \frac{1}{1 + \mathbf{c}^T\mathbf{A}^{-1}\mathbf{b}}\mathbf{A}^{-1}\mathbf{b}\mathbf{c}^T\mathbf{A}^{-1}$$

holds for a non-singular matrix $\mathbf{A}$ and column vectors $\mathbf{b}$, $\mathbf{c}$ such that the combination $\mathbf{A} + \mathbf{b}\mathbf{c}^T$ is non-singular.

The last expression contains a product of one-dimensional integrals, which can be evaluated in close form by noting that terms proportional to y_i and y_i^2 in the exponential can be combined with the potential $V_i(y_i)$ into a new effective potential similarly to equations (44) and (45):

$$V_i^{\text{eff}}(y_i) := V_i(y_i) + g\left(\frac{1}{2}y_i^2 - \psi_i y_i\right), \quad \psi_i := \frac{1}{g}\left(\frac{h^2}{2}\phi_i + B_i\right). \tag{80}$$

Using equation (45) with m replaced by ψ_i, we obtain

$$\prod_{i=1}^{N} \int dy_i e^{-\frac{2}{h^2}[V_i(y_i) + g(\frac{1}{2}y_i^2 - \psi_i y_i)]} = \prod_{i=1}^{N} \int dy_i e^{-\frac{2}{h^2}V_i^{\text{eff}}(y_i)}$$

$$= \prod_{i=1}^{N} e^{2\hat{V}_i(\psi_i) + \log \Omega_i(\psi_i)} \tag{81}$$

where function $\hat{V}_i(\psi_i)$ is defined as in equation (48), and

$$\Omega_i(\psi_i) = \frac{(1 - \bar{a}_i(\psi_i))^2}{\bar{\sigma}_{i1}} + \frac{\bar{a}_i^2(\psi_i)}{\bar{\sigma}_{i2}} + \frac{2\bar{a}_i(\psi_i)(1 - \bar{a}_i(\psi_i))}{\sqrt{(\bar{\sigma}_{i1}^2 + \bar{\sigma}_{i2}^2)/2}} e^{-\frac{(\bar{\mu}_{i1}(\psi_i) - \bar{\mu}_{i2}(\psi_i))^2 T}{2(\bar{\sigma}_{i1}^2 + \bar{\sigma}_{i2}^2)}} \tag{82}$$

and parameters $\bar{\mu}_{i1}, \bar{\mu}_{i2}, \bar{\sigma}_{i1}, \bar{\sigma}_{i2}, \bar{a}$ are defined as in equation (48) for each stock i (I write them here as $\bar{\mu}_{i1}(\psi_i)$, etc. to emphasizes their dependence on parameters ψ_i).

Substituting (81) into equation (79) and changing for convenience the integration variables from ϕ_i to ψ_i according to equation (80), the latter can be written as follows:

$$Z[B] = \frac{1}{\sqrt{(2\pi)^N \det J}} \left(\frac{2g}{h^2}\right)^N \int_{-\infty}^{\infty} \prod_{i=1}^{N} d\psi_i e^{-\frac{2}{h^2}\mathcal{H}(\psi, B)}, \tag{83}$$

where

$$\mathcal{H}(\psi, B) = \frac{1}{h^2} \sum_{i,j=1}^{N} (g\psi_i - B_i) J_{ij}^{-1} (g\psi_j - B_j)$$

$$- h^2 \sum_{i=1}^{N} \left(\hat{V}_i(\psi_i) + \frac{1}{2}\log \Omega_i(\psi_i)\right) \tag{84}$$

is the effective Hamiltonian for variables ψ_i. The multi-dimensional integral with respect to ψ_i can be well approximated in the large-N limit

$N \to \infty$ by a saddle point solution, i.e. a solution of the variational equation $\delta \mathcal{H}(\psi, B)/\delta \psi_i = 0$. The saddle point equation is therefore

$$\frac{2g}{h^2} \sum_{j=1}^{N} J_{ij}^{-1} (g\psi_j - B_j) - h^2 \left(\frac{\partial \hat{V}(\psi_i)}{\partial \psi_i} + \frac{1}{2} \frac{\partial \log \Omega(\psi_i)}{\partial \psi_i} \right) = 0. \quad (85)$$

The solution $\psi = \bar{\psi}$ of this equation hence satisfies the following equation:

$$\bar{\psi}_i = \frac{B_i}{g} + \frac{h^4}{2g^2} \sum_{j=1}^{N} J_{ij} \left[\frac{\partial \hat{V}(\bar{\psi}_j)}{\partial \bar{\psi}_j} + \frac{1}{2} \frac{\partial \log \Omega(\bar{\psi}_j)}{\partial \bar{\psi}_j} \right]. \quad (86)$$

The partition function (83) with the saddle point approximation then reads

$$Z[B] = \frac{1}{\sqrt{(2\pi)^N \det J}} \left(\frac{2}{h^2} \right)^N e^{-\frac{2}{h^2} \mathcal{L}(\bar{\psi}, B)} = e^{-\frac{2}{h^2} F(B)} \quad (87)$$

where $\bar{\psi}$ is a solution to equation (86), and $F(B) = \mathcal{H}(\bar{\psi}, B)$ is the free energy. The local mean field $m_i = \langle y_i \rangle$ is defined as a partial derivative of F:

$$m_i = \frac{h^2}{2} \frac{\partial \log Z}{\partial B_i} = -\frac{\partial F}{\partial B_i} = \frac{2}{h^2} \sum_{j=1}^{N} J_{ij}^{-1} (g\bar{\psi}_j - B_j). \quad (88)$$

This equation can be inverted to express variables $\bar{\psi}_i$ in terms of local mean fields m_i:

$$\bar{\psi}_i = \frac{1}{g} \left[\frac{h^2}{2} \sum_{j=1}^{N} J_{ij} m_j + B_i \right] = \frac{1}{N} \sum_{j \neq i}^{N} m_j + \frac{1}{g} B_i. \quad (89)$$

Using equations (88) and (89), we can now write the saddle point equation (85) in terms of local mean fields m_i:

$$m_i = \frac{h^2}{g} \left[\frac{\partial \hat{V}(\bar{\psi}_i)}{\partial \bar{\psi}_i} + \frac{1}{2} \frac{\partial \log \Omega(\bar{\psi}_i)}{\partial \bar{\psi}_i} \right] \bigg|_{\bar{\psi}_i = \frac{1}{N} \sum_{j \neq i} m_j + \frac{1}{g} B_i}. \quad (90)$$

The last equation is a general mean field self-consistency equation for the MANES model that defines the local mean field m_i (i.e. the expectation of the log-return y_i for the i-th stock) in terms of the expected log-returns for other stocks. Other versions of the self-consistency equation can be obtained from equation (90) if we make further assumptions. For example, if we use it with a symmetric single-stock NES potential $V(y)$, we obtain

$$\frac{\partial \hat{V}(\bar{\psi}_i)}{\partial \bar{\psi}_i} + \frac{1}{2}\frac{\partial \log \Omega(\bar{\psi}_i)}{\partial \bar{\psi}_i} = \frac{g^2 \sigma_i^2 T \psi_i}{h^2(h^2 + g\sigma_i^2 T)} + \frac{g\mu_i T}{h^2 + g\sigma_i^2 T}$$

$$\times \frac{\sinh \frac{2g\mu_i T \psi_i}{h^2 + g\sigma_i^2 T}}{\cosh \frac{2g\mu_i T \psi_i}{h^2 + g\sigma_i^2 T} + e^{-\frac{h^2 \mu_i^2 T}{\sigma_i^2(h^2 + g\sigma_i^2 T)}}}. \quad (91)$$

For a homogeneous version of this self-consistency equation, we set $m_i \to m$ and also remove indices from model parameters $\mu_i \to \mu$, $\sigma_i \to \sigma$. Furthermore, we have in this case $\psi_i \to \psi = \frac{N-1}{N}m + \frac{1}{g}B$ which is well approximated by $\psi = m + \frac{1}{g}B$ in the limit of large N. This produces the self-consistency equation for the homogeneous portfolio:

$$m = \frac{\sigma^2 T}{h^2}B + \mu T \frac{\sinh\left[\frac{2g\mu T}{h^2 + g\sigma^2 T}\left(m + \frac{1}{g}B\right)\right]}{\cosh\left(\frac{2g\mu T}{h^2 + g\sigma^2 T}\left(m + \frac{1}{g}B\right)\right) + e^{-\frac{h^2 \mu^2 T}{\sigma^2(h^2 + g\sigma^2 T)}}}. \quad (92)$$

This coincides with equation (58) provided we identify the external control field B introduced in equation (74) with the fictitious field B_0 introduced in equation (34) to mimic slightly asymmetric potentials.

Susceptibilities

Formulas for the local mean field approximation developed above enable computing susceptibilities for both the homogeneous and heterogeneous market portfolio settings. Starting with a homogeneous setting, the susceptibility χ defined in a similar way to the Ising model:

$$\chi = \left.\frac{\partial m}{\partial B}\right|_{B \to 0}. \quad (93)$$

While in the Ising model such expression computes the sensitivity of the mean magnetization to changes of an external magnetic field, in the present setting, χ is the sensitivity of the expected market log-return to the amount of asymmetry in the the single-stock self-interaction potential.

To compute the susceptibility χ, assume a small but non-vanishing value of B, and expand the right hand side of equation (92) to the first order in B. After re-grouping terms, this gives

$$m = \frac{1}{gh^2} \frac{h^2 h_c^2 + 2gh^2\sigma^2 T + g^2\sigma^4 T^2}{h^2 - h_c^2} B + O\left(B^2\right) \tag{94}$$

where the critical volatility parameter h_c is defined in equation (66). This produces the following result for the susceptibility χ:

$$\chi = \frac{1}{gh^2} \frac{h^2 h_c^2 + 2gh^2\sigma^2 T + g^2\sigma^4 T^2}{h^2 - h_c^2}. \tag{95}$$

Therefore, at the critical volatility value $h^2 = h_c^2$, ξ diverges as $(h_c^2 - h^2)^{-1}$, similar to the Ising model behavior.

A similar analysis can be performed without using the homogeneous portfolio setting but rather working with equation (90) defined for the local mean fields (i.e. expected values) m_i. Again, for small external fields $B_i \to 0$, we also expect the local fields m_i to be small. In this regime, one can retain only the leading linear term in ψ_i in the expansion of the second term in (91) and write

$$\frac{h^2}{g}\left[\frac{\partial \hat{V}(\bar{\psi}_i)}{\partial \bar{\psi}_i} + \frac{1}{2}\frac{\partial \log \Omega(\bar{\psi}_i)}{\partial \bar{\psi}_i}\right]\bigg|_{\psi = \frac{1}{N}\sum_{j\neq i} m_j + \frac{1}{g}B_i} = A_i\left(\frac{1}{N}\sum_{j\neq i} m_j + \frac{1}{g}B_i\right) \tag{96}$$

where

$$A_i := \frac{h^2 h_c^2 + 2gh^2\sigma_i^2 T + g^2\sigma_i^4 T^2}{\left(h^2 + g\sigma_i^2 T\right)^2}. \tag{97}$$

Plugging equation (96) back into equation (90), the latter can be written as a linear system of equations

$$\sum_j \mathbf{G}_{ij} m_j = \frac{1}{g} A_i B_i \tag{98}$$

where $\mathbf{G}$ is a matrix with matrix elements

$$G_{ij} = \delta_{ij}\left(1 + \frac{A_i}{N}\right) - \frac{A_i}{N}. \tag{99}$$

The solution of (98) is

$$\mathbf{m} = \frac{1}{g}\mathbf{G}^{-1}\cdot(\mathbf{A}\circ\mathbf{B}) \tag{100}$$

where $\mathbf{A} \circ \mathbf{B}$ stands for a direct product of vectors $\mathbf{A}$ and $\mathbf{B}$. This produces the following result for local susceptibilities

$$\chi_{ij} := \frac{\partial m_i}{\partial B_j} = \frac{1}{g} G_{ij}^{-1} A_j. \qquad (101)$$

The inverse of matrix $\mathbf{G}$ can be found using the Sherman–Morrison formula:

$$G_{ij}^{-1} = \frac{1}{1 + \frac{A_j}{N}} \delta_{ij} + \frac{1}{1 - \frac{1}{N} \sum_i A_i} \frac{A_i}{N + A_i} \frac{N}{N + A_j}. \qquad (102)$$

In the limit of large N, this can be well approximated by a simpler expression

$$G_{ij}^{-1} = \delta_{ij} + \frac{1}{1 - \langle A \rangle} \frac{A_i}{N}, \quad \langle A \rangle := \frac{1}{N} \sum_{i=1}^{N} A_i. \qquad (103)$$

Log-return covariances and fluctuation-response relations

The covariance of log-returns for two assets i and j is defined as follows:

$$C_{ij} := \langle y_i y_j \rangle - \langle y_i \rangle \langle y_j \rangle = \langle (y_i - \langle y_i \rangle)(y_j - \langle y_j \rangle) \rangle. \qquad (104)$$

This can be computed from the the partition function as follows:

$$C_{ij} = \left(\frac{h^2}{2}\right)^2 \frac{\partial^2}{\partial B_j \partial B_j} \log Z[B] \bigg|_{B \to 0} = \frac{h^2}{2} \frac{\partial m_i}{\partial B_j} \bigg|_{B \to 0}, \qquad (105)$$

where m_i is the local mean field defined in equation (88). Using equations (101) and (103), we obtain

$$C_{ij} = \begin{cases} \dfrac{h^2}{2gN} \dfrac{A_i A_j}{1 - \langle A \rangle}, & \text{if } i \neq j \\[6pt] \dfrac{h^2}{2g} A_i, & \text{if } i = j. \end{cases} \qquad (106)$$

The first relation here shows that in the mean field approach, we obtain $C_{ij} \propto A_i A_j$ implying that the covariance matrix $\mathbf{C}$ has rank one. This is similar to the relation $C_{ij} \propto \beta_i \beta_j$ that arises in a one-factor model with a common "market" factor for all stocks, with β_i being the regression coefficient of stock i's return on the market return.

A link with the calculations performed for the homogeneous setting is provided by considering a special case of the partition function (74) for a homogeneous external field which is the same for all stocks, i.e. $B_i \to B$. For this case, the first derivative of $\log Z$ reads

$$\frac{\partial \log Z}{\partial B} = \frac{2}{h^2} \frac{1}{Z} \int \prod_{i=1}^{N} dy_i \sum_{i=1}^{N} y_i e^{-\frac{2}{h^2}\left[\sum_{i=1}^{N} V_i(y_i) - \sum_{i=1}^{N} y_i B + \frac{g}{2N} \sum_{i,j=1}^{N} \frac{1}{2}(y_i - y_j)^2\right]}$$

$$= \frac{2N}{h^2} \langle m \rangle. \tag{107}$$

Differentiating once more, one obtains

$$\frac{\partial^2 \log Z}{\partial B^2} = \left(\frac{2}{h^2}\right)^2 \left[\left\langle \sum_i y_i \sum_j y_j \right\rangle - \left\langle \sum_i y_i \right\rangle \left\langle \sum_j y_j \right\rangle\right]$$

$$= \left(\frac{2}{h^2}\right)^2 \sum_{i,j} \text{Cov}(y_i, y_j) = \frac{2N}{h^2} \frac{\partial \langle m \rangle}{\partial B} \tag{108}$$

where equation (107) is used at the last step. Re-arranging the last equation here, we obtain the following relation between the "magnetic susceptibility" χ and the average covariance:

$$\frac{1}{N^2} \sum_{i,j} C_{ij} = \frac{h^2}{2N} \frac{\partial m}{\partial B} = \frac{h^2}{2N} \chi \tag{109}$$

where χ is computed in equation (95). This relation shows that the susceptibility is driven by the fluctuations in the system. Such relations are known in statistical physics as *fluctuation-response* formulae.

For a homogeneous setting with $A_i = A$, the mean covariance entering equation (109) can be computed as follows:

$$\bar{C} = \frac{1}{N^2} \sum_{i,j} C_{ij} = \frac{1}{N^2} \left(\frac{Nh^2}{2g} \frac{A^2}{1-A} + \frac{Nh^2}{2g} A\right) = \frac{h^2}{2gN} \frac{A}{1-A}. \tag{110}$$

Using here equation (97) with A_i replaced by A, we finally obtain

$$\bar{C} = \frac{1}{2gN} \frac{h^2 h_c^2 + 2gh^2\sigma^2 T + g^2\sigma^4 T^2}{h^2 - h_c^2} = \frac{h^2}{2N} \chi \tag{111}$$

where χ is the susceptibility for the homogeneous market computed in equation (95). We have therefore verified that covariances for a heterogeneous market defined in equation (106) reduce for a homogeneous market portfolio to the average covariance $\bar{C}$ which is proportional to the susceptibility χ.

In addition to establishing a correspondence with the homogeneous market setting, the local mean field approach of this section leads to the following observation. While the susceptibility χ diverges as $(h^2 - h_c^2)^{-1}$ as $h \to h_c$, this divergence originates in off-diagonal elements C_{ij} with $i \neq j$, and arises due to the factor $1 - \langle A \rangle$ in the denominator in the first relation in equation (106). Therefore, while off-diagonal covariances are proportional to $1/N$ and are hence parametrically small in the limit of large N, their values are increased as the volatility parameter h approaches its critical value h_c. This implies that, for certain combinations of parameters that produce scenarios with $h^2 \simeq h_c^2$, covariances and/or correlations between different stocks obtained in this modeling framework can be made comparable with the average correlation between stocks in the real market, which is around 0.4 for stocks in the S&P 500 universe.

Fitting Model Parameters Using Option Data

As shown in section "Renormalization of the single-stock NES potential by interactions", with the mean field approximation, the dynamics of the market index can be represented as a single-stock dynamics with renormalized parameters $\bar{a}$ and $\bar{\mu}_k, \bar{\sigma}_k$ (with $k = 1, 2$) given by equation (48) which I repeat here for relations involving $\bar{\mu}_k, \bar{\sigma}_k$:

$$\bar{\mu}_k = \frac{h^2 \mu_k + g\sigma_k^2 m}{h^2 + g\sigma_k^2 T}, \quad \bar{\sigma}_k^2 = \frac{h^2 \sigma_k^2}{h^2 + g\sigma_k^2 T}, \quad k = 1, 2. \tag{112}$$

Therefore, the resulting effective single-stock model has six parameters $\bar{\mu}_1, \bar{\mu}_2, \bar{\sigma}_1, \bar{\sigma}_2, a, h$. In [8], this model specification was further reduced to five parameters by setting $\bar{\mu}_1 = -\bar{\mu}_2 = \bar{\mu}$. In this chapter, such constraint will not be imposed.

Calibration of the model to market prices of S&P 500 options (SPX options) or other index options therefore produces six parameters $\bar{\mu}_1, \bar{\mu}_2, \bar{\sigma}_1, \bar{\sigma}_2, \bar{a}, h$. On the other hand, the full set of model parameters in the original multi-asset version of the model involves seven parameters $\mu_1, \mu_2, \sigma_1, \sigma_2, a, h, g$, plus one more unknown value of the expected log-return m, which thus effectively serves as the eighth parameter. Clearly, as eight parameters cannot be uniquely recovered from six parameters $\bar{\mu}_1, \bar{\mu}_2, \bar{\sigma}_1, \bar{\sigma}_2, a, h$ that could be found by calibration to SPX options, we need additional constraints to fix their values. The next few sub-sections develop such constraints on model parameters.

Fixing the coupling constant g and volatilities σ_k

To estimate the coupling constant g that controls interactions in the system, one can try to fix it by fitting the single-stock volatility and pair-wise return correlation obtained in our homogeneous portfolio setting to the average stock vol and correlations obtained in the real market. This can be readily done using the homogeneous version of equation (106). It gives the mean single-stock volatility

$$\bar{\sigma}_M = \frac{h}{\sqrt{T}}\sqrt{\frac{A}{2g}} = \frac{h}{\sqrt{T}}\sqrt{\frac{h^2 h_c^2 + 2gh^2\sigma^2 T + g^2\sigma^4 T^2}{2g\left(h^2 + g\sigma^2 T\right)^2}} \qquad (113)$$

where equation (97) was used at the last step. Note that the factor $1/\sqrt{T}$ in the right-hand side of this relation arises because $\bar{\sigma}_M$ is defined in annualized terms. The mean pairwise correlation is obtained from equation (106) as the ratio of the first row to the second row:

$$\bar{\rho}_M = \frac{1}{N}\frac{A}{1-A}. \qquad (114)$$

Note that $\bar{\rho}_M \sim 1/N$, which implies that correlations die off in the strict thermodynamic limit $N \to \infty$. The mean correlation ρ_M can also be estimated differently by computing the variance of the mean field m in terms of ρ_M and the mean single-stock volatility σ_M:

$$\mathrm{Var}(m) = \left(\frac{1}{N}\bar{\sigma}_M^2 + \bar{\rho}_M\bar{\sigma}_M^2\right)\Delta t. \qquad (115)$$

Neglecting the first term in the right hand side, this relation gives an approximation for ρ_M in terms of the ratio of annualized variance of the market index' log-return σ_m^2 to the variance $\bar{\sigma}_M^2$ of the representative stock in the portfolio:

$$\bar{\rho}_M \simeq \frac{\sigma_m^2}{\bar{\sigma}_M^2}, \qquad (116)$$

where $\sigma_m^2 = \mathrm{Var}(m)/\Delta t$ is the annualized variance of the market log-return. This produces an estimate $\bar{\rho}_M \simeq 0.3$–0.4 which can be used to produce further estimates for parameters in the model. In particular, inverting equation (114) to find A in terms of $\bar{\rho}_M$, and then inverting equation (113) to compute g, we obtain

$$A = \frac{N\bar{\rho}_M}{1 + N\bar{\rho}_M}, \quad g = \frac{h^2 A}{2\bar{\sigma}_M^2 T}. \qquad (117)$$

This suggests that A should be very close to one, approaching it from below, and respectively $g \simeq h^2/(2\bar{\sigma}_M^2 T)$. Next, we can invert the second relation in equation (112) to obtain

$$\sigma_k^2 T = \frac{\bar{\sigma}_k^2 T}{1 - \frac{g}{h^2}\bar{\sigma}_k^2 T}. \qquad (118)$$

This relation shows that g should be such that $\frac{g}{h^2}\bar{\sigma}_k^2 T < 1$ in order to keep the single-stock volatility real-valued and finite, g should satisfy the following constraint:

$$g < \frac{h^2}{\bar{\sigma}_k^2 T}. \tag{119}$$

Using the second relation in (117), the constraint (119) can also be re-stated as the following constraint on the volatility σ_M of the "representative" stock:

$$\bar{\sigma}_M^2 \geq \frac{\bar{\sigma}_k^2}{2} A \simeq \frac{\bar{\sigma}_k^2}{2}, \quad k = 1, 2, \tag{120}$$

where the second approximate form follows as long as $A \simeq 1$ as suggested by equation (117). The second observation with equation (118) is that the single-stock variance $\sigma_k^2 T$ is higher than the variance of the mean log-return $\bar{\sigma}_k^2 T$ in both states $k = 1, 2$. Equivalently, it means that the variance of the mean $\bar{\sigma}_k^2 T$ is *smaller* than the individual variance $\sigma_k^2 T$, which is as expected from the central limit theorem.

How far are we from the critical value $h = h_c$ of the volatility parameter?

If we use the values $\bar{\rho}_M = 0.3 - 0.4$ and $N = 500$, then equation (117) imply that A should approach one from below, and $g \sim h^2/(2\bar{\sigma}_M^2 T)$. On the other hand, using equation (97), we can write the expression for A as follows:

$$A = \frac{h^2(h^2 + h_c^2 - h^2) + 2gh^2\sigma^2 T + g^2\sigma^4 T^2}{(h^2 + g\sigma^2 T)^2} = 1 - \frac{h^2(h^2 - h_c^2)}{(h^2 + g\sigma^2 T)^2}. \tag{121}$$

Given that A should be slightly below one, this implies that h^2 should be slightly above the critical value h_c^2. This means that the model is in a high-temperature phase, yet close to the bifurcation point $h^2 = h_c^2$. In this regime, the non-vanishing mean field m is solely due to the explicit breaking of $\mathbb{Z}_2$ symmetry due to asymmetry of the potential, which can be mimicked by adding the fictitious field B_0, see equation (34).

This behavior of the model appears to be quite reasonable. Indeed, a benign market regime is typically associated with a steadily growing market which, while exhibiting some volatility, has a positive trend. While it is not easy to estimate this trend accurately, a market with a positive market trend is certainly very different from a market with a negative trend. In a world with a symmetric potential and a second-order phase transition, a non-vanishing expected market return can only arise due to a spontaneous breaking of the $\mathbb{Z}_2$ symmetry, which implies that both choices are "physically" equivalent, i.e. equally preferable, in contradiction with the common sense. This suggests that the scenario in which the $\mathbb{Z}_2$ symmetry is broken *explicitly* due to asymmetry of the potential, while the sign of the the mean field m is fixed by the potential, appears more plausible in the present context. The following section shows how the mean field m can be computed for this scenario using the MANES model calibrated to option prices.

Expected market return from self-consistency equation

As was discussed above, the equilibrium expected market log-return is given by the mean field parameter m in the McKean–Vlasov equation (40), and it should satisfy the self-consistency equation (50). Here I present an explicit solution of this equation, assuming that the model is calibrated to option prices, so that renormalized parameters $\bar{\mu}_1, \bar{\mu}_2, \bar{\sigma}_1, \bar{\sigma}_2, a, h$ are known.

Using equation (118) in equation (112), the relations between the "bare" and "renormalized" parameters, resp. μ_k and $\bar{\mu}_k$, can be written as follows:

$$\bar{\mu}_1 = \mu_1 \left(1 - \frac{g}{h^2}\bar{\sigma}_1^2 T\right) + \frac{g}{h^2}\bar{\sigma}_1^2 m, \quad \bar{\mu}_2 = \mu_2 \left(1 - \frac{g}{h^2}\bar{\sigma}_2^2 T\right) + \frac{g}{h^2}\bar{\sigma}_2^2 m. \quad (122)$$

According to equation (26), the mean of the third Gaussian component in equation (25) is as follows:

$$\bar{\mu}_3 = \frac{\bar{\mu}_1 \bar{\sigma}_2^2}{\bar{\sigma}_1^2 + \bar{\sigma}_2^2} + \frac{\bar{\mu}_2 \bar{\sigma}_1^2}{\bar{\sigma}_1^2 + \bar{\sigma}_2^2}. \quad (123)$$

The self-consistency condition (50) is equivalent to the constraint on the expected value of y obtained in the model with the effective potential V_{eff} with the interaction-dressed parameters defined in equations (122) and (123). Therefore, the self-consistency condition (50) now takes the following simple form:

$$m = \bar{\omega}_1 \bar{\mu}_1 T + \bar{\omega}_2 \bar{\mu}_2 T + \bar{\omega}_3 \bar{\mu}_3 T, \quad (124)$$

where weights $\bar{\omega}_k$ are defined as in equation (28) using the "interaction-dressed" parameters $\bar{\mu}_1, \bar{\mu}_2, \bar{\sigma}_1, \bar{\sigma}_2, h, \bar{a}$. Using here equation (123) and re-grouping terms, we obtain

$$m = \left(\bar{\omega}_1 + \frac{\bar{\omega}_3 \bar{\sigma}_2^2}{\bar{\sigma}_1^2 + \bar{\sigma}_2^2}\right) \bar{\mu}_1 T + \left(\bar{\omega}_2 + \frac{\bar{\omega}_3 \bar{\sigma}_1^2}{\bar{\sigma}_1^2 + \bar{\sigma}_2^2}\right) \bar{\mu}_2 T. \quad (125)$$

This relation provides a closed-form solution of the self-consistency condition (50) in terms of renormalized parameters that are directly calibrated to option prices. Recall that equation (48) imply that renormalized weights $\bar{\omega}_k$ depend on "bare" weights ω_k and the mean field m. Therefore, have we used the bare weights ω_k instead of renormalized weights $\bar{\omega}_k$ and expressed $\bar{\mu}_1, \bar{\mu}_2$ in terms of μ_1, μ_2 and m according to equation (122), equation (125) would amount to a nonlinear equation similar to the self-consistency equation (58). Instead, by working with the MANES effective potential (45) and renormalized parameters (122), the original self-consistency condition (43) of the McKean–Vlasov equation (40) is resolved here analytically rather than numerically. Interestingly, when expressed in terms of renormalized parameters in equation (125), the expected market log-return m does not explicitly depend on the coupling constant g, and is found in terms of parameters directly calibrated to option prices.

On the other hand, the "bare" model parameters μ_1, μ_2, etc. *do* depend on the value of g. In particular, after the mean field m is computed from equation (125), the "bare" model parameters μ_1, μ_2 can be obtained from equation (122), volatilities σ_1, σ_2 are obtained using equation (118), and the bare mixing coefficient a can be computed by inverting the formula for $\bar{a}$ in equation (48):

$$a = \left[1 + \frac{1-\bar{a}}{\bar{a}} \sqrt{\frac{h^2 + g\sigma_1^2 T}{h^2 + g\sigma_2^2 T}} e^{\frac{g(m-\mu_1 T)^2}{2(h^2+g\sigma_1^2 T)} - \frac{g(m-\mu_2 T)^2}{2(h^2+g\sigma_2^2 T)}}\right]^{-1}. \qquad (126)$$

Examples of calibration to SPX options

In this section, I present three sets of examples of calibration to market quotes on European options on SPX (the S&P 500 index). I use the same set of option quotes that were used in [8] to illustrate the working of the NES model. In all three sets of experiments, I calibrate to market quotes on 10 call options and 10 put options. The strikes are chosen among available market quotes to cover the the range of option deltas between 0.02 and 0.5, in absolute terms, so that the model is calibrated to both ATM strikes and deep OTM strikes. Details of the loss function are described in [8]. Optimization of the loss function is done using the `shgo` algorithm available in the Python scientific computing package scipy.

In all experiments presented in the following, parameters $\bar{\mu}_1, \bar{\mu}_2, \bar{\sigma}_1, \bar{\sigma}_2, \bar{a}, h$ are found by calibration to SPX options. Inferred parameters are different from those reported in [8] because here I do *not* enforce the constraint $\bar{\mu}_2 = -\bar{\mu}_1$ that was used in [8]. In addition, I display estimated parameters g and m. The coupling constant g is roughly estimated according to the following formula that is obtained by combining equations (117) and (120):

$$g \simeq \frac{h^2}{2\bar{\sigma}_M^2 T}, \quad \bar{\sigma}_M^2 = \max_k \left(\frac{\bar{\sigma}_k^2}{2}\right) + \Delta\sigma^2, \tag{127}$$

where $\Delta\sigma^2$ is a margin that controls the strength of the inequality (120). For numerical examples, the value $\Delta\sigma^2 = 0.05$ is used in the following examples. Furthermore, the equilibrium log-return (mean field) m is computed according to equation (125).

In the first example, I consider SPX 1M options on 07/12/2021 with maturity on 08/09/2021. The results of calibration are presented in Table 1 and Figures 6 and 7. Potentials shown in these figures and in the examples to follow are effective potentials (45) and are computed as a single-stock NES potential with parameters $\bar{\mu}_1, \bar{\mu}_2, \bar{\sigma}_1, \bar{\sigma}_2, \bar{a}, h$ inferred from calibration to index options. Note the difference in implied potentials for puts and calls, as well as different values of inferred model parameters. For both potentials, the current value of the log-return y_0, as shown by the vertical lines, is

Table 1. NES parameters obtained by calibration to 10 put and 10 call options on 1M SPX options with expiry 08/09/2021 on 07/12/2021. The last column shows the mean absolute pricing errors (MAPE).

	$\bar{\mu}_1$	$\bar{\mu}_2$	$\bar{\sigma}_1$	$\bar{\sigma}_2$	$\bar{a}$	h	g	m	MAPE
Puts	0.190	0.118	0.449	0.093	0.525	0.159	1.23	0.01	3.45%
Calls	0.136	-0.592	0.089	0.669	0.342	0.163	0.63	0.01	0.293%

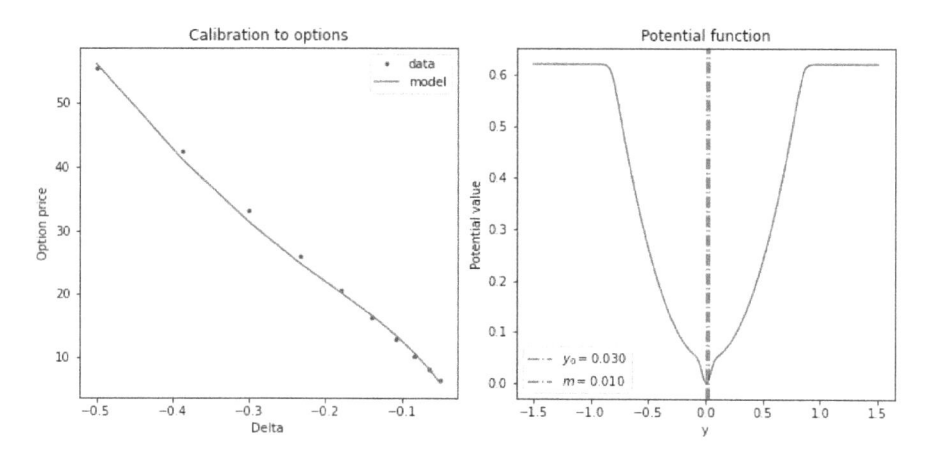

Figure 6. Calibration to 1M SPX put options with expiry 08/09/2021 on 07/12/2021 and the corresponding effective potentials (45). The vertical lines correspond to the current value of log-return y_0 and the equilibrium log-return m.

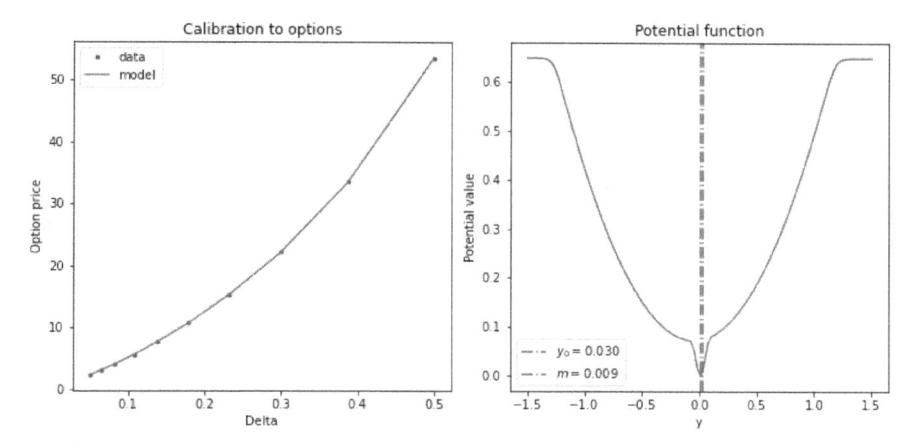

Figure 7. Calibration to 1M SPX call options with expiry 08/09/2021 on 07/12/2021 and the corresponding effective potentials (45). The vertical lines correspond to the current value of log-return y_0 and the equilibrium log-return m.

Table 2. NES parameters obtained by calibration to 10 put and 10 call options on 1Y SPX options with expiry 09/21/2021 on 11/06/2020. The last column shows the mean absolute pricing errors (MAPE).

	$\bar{\mu}_1$	$\bar{\mu}_2$	$\bar{\sigma}_1$	$\bar{\sigma}_2$	$\bar{a}$	h	g	m	MAPE
Puts	1.200	0.269	0.906	0.303	0.438	0.198	0.05	0.455	3.49%
Calls	0.047	−1.866	0.168	1.423	0.467	0.221	0.03	−0.085	2.80%

located near the bottom of the potential well. Furthermore, the equilibrium log-return m is also located near the bottom of the potential for for potentials. This suggests that the price dynamics on this date correspond to an equilibrium regime of small fluctuations around a stable minimum.

In the second example, I look at longer option tenors and consider SPX 1Y options on 11/06/2020 with maturity on 09/21/2021. The results of calibration are presented in Table 2 and Figures 8 and 9. Again, we can note the difference in implied effective potentials for puts and calls, as well as different values of inferred model parameters. Also as in the previous example, for both potentials, the current value of the log-return y_0 and equilibrium expected log-return m are located near the bottom of the potential wells, suggesting an equilibrium regime of small price fluctuations on this date.

Finally, the last example considers 6M options with the expiry on 09/18/2020 on 03/16/2020, at the peak of COVID-19 crisis where the SPX index had the largest drop. This example thus illustrates the model behavior for a severely distressed market. The results of calibration are presented in

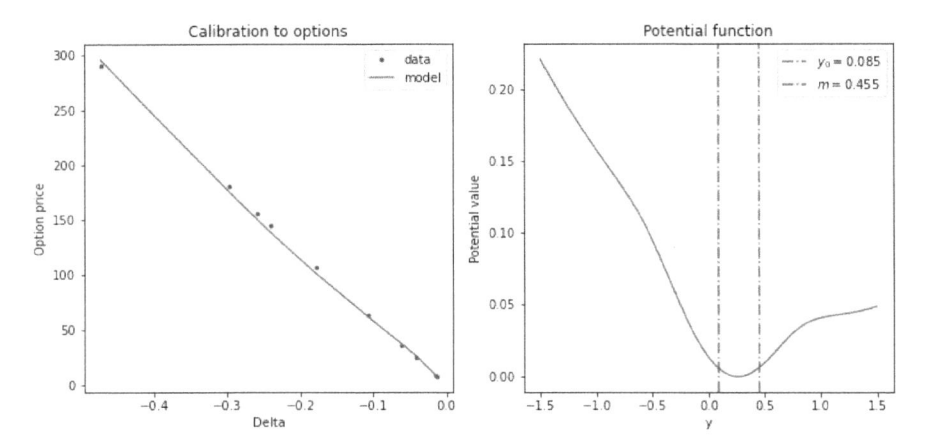

Figure 8. Calibration to 1Y SPX put options with expiry 09/21/2021 on 11/06/2020 and the corresponding effective potentials (45). The vertical lines correspond to the current value of log-return y_0 and the equilibrium log-return m.

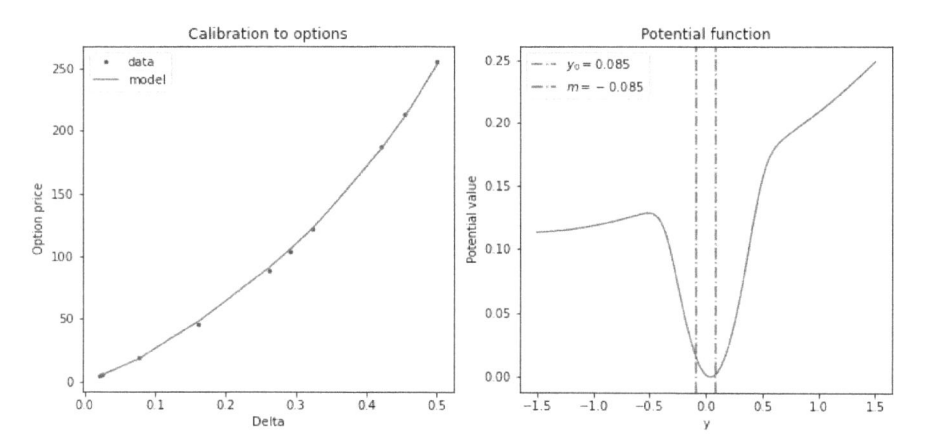

Figure 9. Calibration to 1Y SPX call options with expiry 09/21/2021 on 11/06/2020 and the corresponding effective potentials (45). The vertical lines correspond to the current value of log-return y_0 and the equilibrium log-return m.

Table 3 and Figures 10 and 11. As in the previous examples, again note the difference in implied effective potentials for puts and calls, as well as different values of inferred model parameters.

Unlike the previous examples, for the present case of a distressed market, the initial log-return y_0 on 03/16/2020 is located far away from the global minima for both effective potentials, indicating that this initial state is strongly non-equilibrium, consistently with a prevailing market sentiment on that date.

Table 3. NES parameters obtained by calibration to 10 put and 10 call options on 6M SPX options with expiry 09/18/2020 on 03/16/2020. The last column shows the mean absolute pricing errors (MAPE).

	$\bar{\mu}_1$	$\bar{\mu}_2$	$\bar{\sigma}_1$	$\bar{\sigma}_2$	$\bar{a}$	h	g	m	MAPE
Puts	0.234	−0.585	0.229	1.10	0.757	0.630	0.57	−0.07	0.956%
Calls	0.487	−0.864	0.146	0.700	0.624	0.831	2.28	0.04	1.664%

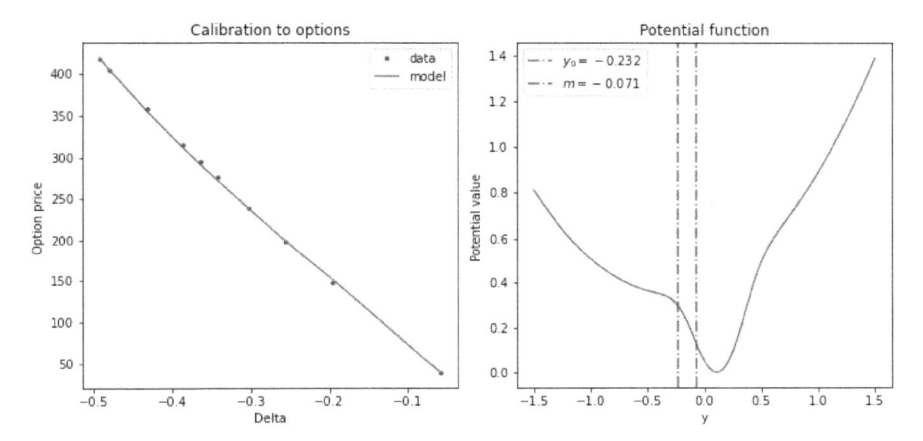

Figure 10. Calibration to 6M SPX put options with expiry 09/18/2020 on 03/16/2020 and the corresponding effective potentials (45). The vertical lines correspond to the current value of log-return y_0 and the equilibrium log-return m.

Interestingly, while both implied effective potentials suggest that the current value y_0 is far from the global minimum of the potential and hence describes a non-equilibrium scenario, they differ in the character of a subsequent relaxation mechanism for this initial state. The implied potential for puts in Figure 10 is a single-well potential, therefore the initial state y_0 is unstable. If the potential itself remains constant through time, this initial state would eventually relax into its global minimum. This would happen even in the limit of zero volatility (zero noise). When the noise is present with $h > 0$, it produces an uncertainty for the time needed to reach the minimum of the potential, as well as small fluctuations around this minimum.

Different from the puts, the call options seem to suggest a different type of relaxation dynamics, as the implied potential for the calls in Figure 11 is a double-well potential quite similar to the one shown in the left column in Figure 2. The initial location is in the vicinity of the local minimum corresponding to the left well, while the right well corresponds to the global minimum. As they are separated by the barrier, the potential in

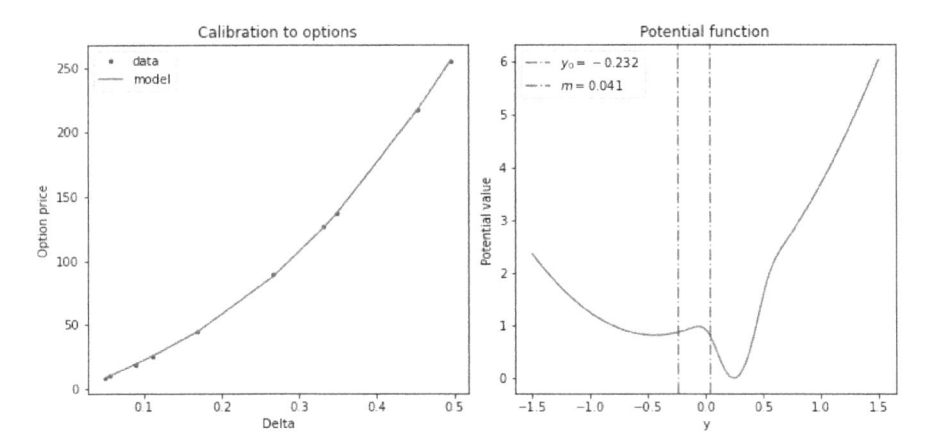

Figure 11. Calibration to 6M SPX call options with expiry 09/18/2020 on 03/16/2020 and the corresponding effective potentials (45). The vertical lines correspond to the current value of log-return y_0 and the equilibrium log-return m.

Figure 11 describes a scenario of metastability. The relaxation to the global minimum (the true ground state) proceeds via instanton transitions as discussed in [7,8]. Instanton transitions are only possibly when the volatility parameter is non-zero, $h > 0$, however small it can be in practice. This suggests that double-well potentials implying metastable dynamics may occur when markets are in distress or during periods of a high market uncertainty, e.g. during general crises such as the 2008 crisis or the COVID-19 crisis of 2020 or during general elections.[13]

Summary and Outlook

This chapter proposed a tractable nonlinear model of interacting and non-equilibrium market, formulated as statistical mechanics of interacting nonlinear oscillators where individual stocks' log-returns y_i are viewed as coordinates of "particles" describing these stocks. Both self-interactions of oscillators y_i and their pairwise interactions are explained in terms of money flows into the market from external investors. In particular, correlations between log-returns of individual stocks originate from the dependence of money flow a_i into stock i on the average previous performance of all other stocks.

---

[13]In particular, [5] found a bimodal implied distribution during British elections of 1987 (though not through other British elections in 1992) and suggested that option prices can be used to monitor the market sentiment during elections.

The main idea of this chapter was to approximate the return of a market index in a heterogeneous market by a mean field of a *homogeneous* market made of N replicas of the same "representative" stock with a self-interaction potential $V(y)$. This enables modeling the dynamics of the market log-return using the mean field approximation that produces the McKean–Vlasov equation as the equation governing the dynamics of the system. As the McKean–Vlasov equation is a nonlinear equation corresponding to the thermodynamic limit $N \to \infty$, it produces far richer dynamics than the original multi-particle Fokker–Planck equation (35), giving rise to ergodicity breaking and first- and second-order phase transitions in different parameter regimes. The resulting dynamics of the mean field resembles the Desai–Zwanzig model of interacting nonlinear oscillators [1,2] and its generalized version in [6].

Furthermore, while the exploration of the phase structure in the generalized Desai–Zwanzig model requires dedicated numerical methods to identify stable and unstable solutions [6], the analysis in this chapter is considerably simplified due to its reliance on the Non-Equilibrium Skew (NES) potential (24). The NES model with the NES potential (24) was introduced in [8] as a flexible and highly tractable nonlinear model of single-stock dynamics that is capable of describing either a benign or stressed market regime, and tracks the pre-asymptotic dynamic behavior of transition probabilities. In particular, non-equilibrium, pre-asymptotic corrections to the moments of a stationary (asymptotic) distribution of log-returns are explicitly controlled in the NES model in terms of the model parameters entering the NES potential (24).

In this chapter, the NES potential from [8] is used to model a multi-asset market. In section "NES and MANES: Multi-Asset dynamics with the Non-Equilibrium Skew potential", the NES potential is introduced as a tractable approximation to a nonlinear self-interaction potential $V(y)$ that originates in money flows and their impact effect on market returns. In addition, the quadratic Curie–Weiss interaction potential (21) is obtained as an approximate interaction potential following the same lines of analysis. This results in a multi-asset extension of the previous single-stock NES model, referred to as the Multi-Asset NES (MANES) model in this chapter.

As shown in this chapter, the new multi-asset MANES model is as tractable as the previous single-stock NES model. This is made possible due to the fact that with the NES self-interaction potential $V(y)$, the new effective potential $V_{\text{eff}}(y)$ that incorporates interactions in the system can be expressed in terms of a single-stock NES potential $V(y)$ with renormalized parameters, see equations (45) and (48). This suggests that for the purpose of calibration to market prices of index options, the multi-asset MANES model is computationally equivalent to a single-stock NES model applied to the "representative" stock. Also due to this property of the model, the self-consistency equation of the mean field approximation that is normally expressed as a nonlinear equation (43) is resolved here analytically in terms of renormalized parameters that are directly calibrated to market prices of index options. Furthermore, estimates made in section "How far are we from the critical value $h = h_c$ of the volatility parameter?" suggest that the model operates in a "high-temperature" phase close to the criticality point, where the volatility parameter h is slightly higher than the critical value $h = h_c$ that corresponds to the bifurcation point of the order parameter m.

Just as the single-stock NES model, the MANES model demonstrates that a *single volatility parameter* is sufficient to accurately match available market prices of index options. These results stand in stark contrast to the most of other option pricing models such as local, stochastic, or rough volatility models that need more complex specifications of noise to fit the market data. Alongside a single volatility parameter h, the effective single-stock potential $V_{\text{eff}}(y)$ (45) has parameters $\bar{\mu}_1, \bar{\mu}_2, \bar{\sigma}_1, \bar{\sigma}_2, \bar{a}$ that can be calibrated to market prices of index options. By fitting these parameters to the market data, we produce implied potentials which replace implied volatility smiles as a way to fit market data. If desired, the calibrated model parameters $\bar{\mu}_1, \bar{\mu}_2, \bar{\sigma}_1, \bar{\sigma}_2, \bar{a}$ can be approximately mapped onto the parameters that enter equation (22), and thus could be interpreted as as implied money flow and market impact parameters. Clearly, it does not preclude one from using more sophisticated models of noise such as stochastic or rough volatility models but only if needed beyond the need to explain market prices of vanilla index options.

The MANES model can be used for several applications or practical interest. In particular, option-implied moments of future returns can be used as predictors for actual future returns, volatilities, and skewness, and employed for portfolio trading, see, e.g. [5,15]. While such analyses typically use risk-neutral moments implied by option prices, the MANES model enables extracting both risk-neutral and real-measure moments and thus enriches the set of predictors for such tasks. The model can also use other market data for model calibration. In particular, in addition to using market prices of options, we could incorporate open interest data in the model calibration. The MANES model could also be jointly calibrated to the equity and credit markets data, by adding fitting to credit indices such as CDX as proxies to probabilities of large market drops. Such applications and extensions are left here for a future research.

Acknowledgments

I would like to thank John Dance, Sebastian Jaimungal, Lisa Huang, and Yinsen Miao for comments and helpful remarks.

References

[1] D. A. Dawson. (1983). Critical dynamics and fluctuations for a mean-field model of cooperative behavior. *Journal of Statistical Physics* 31(1), 29–85.

[2] R. C. Desai, and R. Zwanzig. (1978). Statistical mechanics of a nonlinear stochastic model. *Journal of Statistical Physics* 19(1), 1–24.

[3] T. D. Frank. (2005). *Nonlinear Fokker-Planck Equations*. Springer Series in Synergetics. Springer-Verlag, Berlin.

[4] A. Frazzini, and O. A. Lamont. (2008). Dumb money: Mutual fund flows and the cross-section of stock returns. *Journal of Financial Economics* (Elsevier) 88(2), 299–322.

[5] G. Gemmill, and A. Saflekos. (2000). How useful are implied distributions? Evidence from stock-index options. *The Journal of Derivatives* 7(3), 83–91 (2000). Available at https://www.bis.org/publ/bisp06e.pdf.

[6] S. N. Gomes, S. Kalliadis, G. A. Pavlotis, and P. Yatsyshin. (2019). Dynamics of the Desai-Zwanzig model in multi-well and random energy landscapes. https://arxiv.org/abs/1810.06371.

[7] I. Halperin, and M. F. Dixon. (2020). Quantum equilibrium-disequilibrium: Asset price dynamics, symmetry breaking, and defaults as dissipative instantons. *Physica A* 537, 122187, https://doi.org/10.1016/j.physa.2019.122187.

[8] I. Halperin. (2022). Non-equilibrium skewness, market crises, and option pricing: Non-linear Langevin model of markets with supersymmetry. *Physica A* 594, 127065. https://doi.org/10.1016/j.physa.2022.127065.

[9] I. Halperin. (2020). The inverted world of classical quantitative finance: A non-equilibrium and non-perturbative finance perspective. https://arxiv.org/abs/2008.03623.

[10] L. D. Landau, and E. M. Lifschitz. (1980). *Quantum Mechanics*. Elsevier, Pergamon Press, London, 3rd edition, 1977.

[11] P. Langevin. (1908). Sur la Théorie du Mouvement Brownien. *Comptes Rendus de l'Académie des Sciences* (Paris) 146, 530–533.

[12] N. Martzel, and C. Aslangul. (2001). Mean-field treatment of the many-body Fokker-Planck equation. https://arxiv.org/abs/cond-mat/0106101.

[13] H. P. McKean Jr. (1966). A class of Markov processes associated with nonlinear parabolic equations. *Proceedings of the National Academy of Sciences of the United States of America* 56, 1907–1911.

[14] D. A. McQuarrie. (1973). *Statistical Mechanics*. Harper & Row, New York.

[15] P. S. Stilger, A. Kostakis, and S. H. Poon. (2016). What does risk-neutral skewness tell us about future stock returns? *Management Science* 63(3), 1657–2048.

[16] C. Zinn-Justin. (2002). *Quantum Field Theory and Critical Phenomena*, 4th edn. Clarendon Press, Oxford.

Chapter 3

Mathematics of Embeddings: Spillover of Polarities over Financial Texts

Mengda Li[*] and Charles-Albert Lehalle[†,‡,§]

[*]*Independent Researcher*
[†]*Abu Dhabi Investment Authority (ADIA), UAE*
[‡]*Imperial College London, London, UK*
[§]*charles.lehalle@adia.ae*

Abstract

In this chapter, we perform a mathematical analysis of the word2vec model. This sheds light on how the decision to use such a model makes implicit assumptions on the structure of the language. Beside, under Markovian assumptions that we discuss, we provide a very clear theoretical understanding of the formation of embeddings and, in particular, the way it captures what we call *frequentist synonyms*. These assumptions allow to conduct an explicit analysis of the loss function commonly used by these NLP techniques that asymptotically reaches a cross-entropy between the language model and the underlying true generative model.

Moreover, we produce synthetic corpora with different levels of structures and show empirically how the word2vec algorithm succeed, or not, to learn them. It leads us to empirically assess the capability of such models to capture structures on a corpus of around 42 millions of financial news covering 12 years. And, we rely on the Loughran–McDonald Sentiment Polarity Word Lists and we show that embeddings are exposed to mixing terms with opposite polarity because of the way they treat antonyms as frequentist synonyms. Besides, we study the non-stationarity of such a financial corpus that has surprisingly not be documented in the literature.

Keywords: Quantitative finance, natural language processing, word embeddings, financial news, statistical learning.

Positioning of This Work

Recent advances on Natural Language Processing (NLP) are largely based on the use of embeddings. It started with performing SVD on vectorial representation of words (see, for instance, [17]) before turning to less linear analysis like the use of Self-Organizing Maps [16]. More recently, the term "embeddings" emerged with the idea of learning a lookup table while performing downstream tasks [24]; later on, downstream tasks used LSTM (Long Short-Term Memory) networks to account for the sequential aspects of the natural languages [11]. Now Large Language Models (like BERT, GPT, or LLaMA) propose to plug embeddings in sequences of local LSTM structures, to obtain more contextual (i.e. bi-directional) Euclidean representations of terms [6,33,34] or [8] for a more critical presentation.

Despite the apparent successes of this series of improvements, very few mathematical modeling has been proposed. Some interesting papers are descriptive, exposing the formulas behind the mechanisms of embeddings, like Levy and Goldberg [19] that investigate an SVD viewpoint on embeddings, but they do not provide a probabilistic model corresponding to the implicit assumptions made by the considered NLP modeling algorithm. As a comparison, the literature on topic extraction naturally provides this kind of generative models [3].

For financial applications, different proposals have been made to adapt the models; because of the cost of a full training, they are essentially based on an existing model that is fine-tuned on specific financial texts or tasks. See [13] or [4]. When they are fully trained from scratch (like in [33]), they are trained only once with current texts and then used in the context of baktests to demonstrate their "efficiency". As shown in Figure 8, there is *a risk of containing future information*, since the language model is learnt on text containing description of scandals "before they took place" in the backtest. This is the topic behind this chapter: most financial applications attempt to associate polarities to sentences or segments of sentences, paragraphes, sections, titles, or full body of texts; what process takes place when language models are used to produce "sentiments' polarities"? Can this "coloring of embeddings" spillover to undesirable concepts, like company names? If it is the case, the exact same sentence on Apple or Silicon Valley Bank can exhibit a different sentiment.

The goal of this chapter is to provide a mathematical understanding of how and what embeddings are learning. Since it is a complicated task, we are limiting ourselves to the word2vec model that is paradigmatic of this family of language models. The sophistication of subsequent approaches (like LSTM or BERT) is "simply" to add some sequential memory and some "locality" (since they use an ensemble of embeddings, each of them specializing locally, i.e. in the neighborhood of a syntactic context). Writing the formulas for such models is probably possible but would require a lot of sophisticated notations. We clearly restrict our analysis to global embeddings of the word2vec family but cover both Skip-gram and CBOW modeling approaches.

For readers focused on deeper language models, it is of importance to understand that embeddings of word2vec can be considered as the first term of a Taylor expansion of the embeddings of higher dimensional models. Take the example of the multiple attention heads perfectly described in Vaswani et al. [34] and formally compare it with formula (1): the word2vec formula is made of a product of the transformed (essentially compressed) representation of the word x_i (at the left) with one of the words x_j (at the right). These representations, formally $W^T x$ in the formula, are called "embeddings" since they lift a word (that should be in dimension of the order of the number of different words in English) in a lower-dimensional one. To understand how this matrix W is obtained in multi-head models, one should first focus on one *scaled dot-product attention* architecture: in context, the word X_k is surrounded by words $X_{k-C}, \ldots, X_{k+C}$; these words are used as *keys* K and *queries* Q to form $\mathcal{W} := \mathsf{softmax}(QK^T/\sqrt{d})$, where d is the dimension of the embeddings. This $\mathcal{W}$ is then multiplied by X_k to obtain an equivalent of $W^T X_k$, in which the embeddings $\mathcal{W}$ are now *contextual*: where for word2vec, the embedding of one word is the same across all the documents of the corpus, and for a *scaled dot-product attention* head, it is context-dependent. It is richer of course, but it means also that *locally*, i.e. within a context, the embedding can be considered as constant. If one would like to perform a Taylor expansion of such a head, he or she would find as a first term something that can be read as a word2vec embedding. The second subtlety of deeper models is that they have multiple heads: instead of using one contextual embedding for each word, the architecture has access to a series of embeddings, that are randomly initialised, and then evolve during the learning. Some papers like [27] have shown that the main effect of these multiple layers is that these contextual embeddings specialize: some focus on grammatical aspects, others on adjectives, etc. Again it is nothing more than a contextual effect that would boil down to word2vec embeddings for the first term of a Taylor expansion. Last but not least, the second vector x_j has exactly the same treatment, and finally they are mixed not with the simple $\mathsf{softmax}((W^T x_i)^T W') x_j$ for word2vec, but with an apparently more complex formula, where the left-hand side W^T is replaced by a key and query of the embedding of x_i (that could be named $K(W_k^T X_k)$ and $Q(\tilde{W}_k^T X_k)$) and the right-hand side W' is the embedding of the right word $X_{k'}$. One obtains

$$\mathsf{softmax}(\underbrace{Q(\tilde{W}_k^T X_k)\, K(W_k^T X_k)^T/\sqrt{d}}_{\mathcal{W}})\, \underbrace{\bar{W}_k^T X_{k'}}_{\bar{X}_{k'}}$$

that is locally similar to formula (1) if one follows the underbraced expressions that can be considered as "locally constant" in a Taylor expansion view of the multiple attention heads.[1]

This should be enough to convince the reader that all the analysis in this chapter is locally valid for more complex and deeper models that are mathematically intractable if considered globally. The conclusions of this chapter can hence be considered as "locally valid" for more complex models. The notion of "frequentist synonym" that is usefully developed in this chapter has to be read as "local frequentist synonym" for more complex models.

The theoretical analysis of this chapter is centered around the notion of *Reference Model*. Qualitatively speaking, it is a very large stochastic matrix collecting for each word of the vocabulary the probability of occurrence of all words in "its neighborhood". Given this Reference Model, no embedding is needed (or one can consider that the associated embedding is the identity matrix). It allows us to define a word2vec representation of a corpus as a compression of its Reference Model in two components: one the one hand, an embedding, qualitatively mapping "similar words" (in the sense of the Reference Model) together, and hence defining "frequentist synonyms", and on the other hand, a context matrix, allowing to recover the Reference Model once it is multiplied by the embedding (and after a slightly nonlinear operation, i.e. a softmax layer).

We are here in the spirit of other papers presenting embeddings as a compression (see, for example, [1,22,25]), along with the initial papers on SVD, except that we can prove, under some assumptions, that the minimized criterion is a cross-entropy between the Reference Model and the compressed representation made of the embedding and the context matrices.

In the second stage, we explore further the concept of frequentist synonyms (that "should be put in the same neighborhood by the compression") in association with the identifiability of word2vec embeddings. For that we generate synthetic corpora with know properties thanks to the control of

---

[1] The feed-forward layers of attention heads are not commented since they are made of one ReLu layer and one output layer: again they have a local effect that becomes constant in the (formal) context of a Taylor expansion.

their Reference Model. It allows us to test empirically that if the word2vec model can recover part of the structure we inject in the synthetic texts. The results are below standard expectations: the embeddings of synonyms do not exhibit strong cosine similarities, despite the fact that we use generative models satisfying strong mathematical assumptions (that is not the case of natural language).

Last but not least, we use this understanding to train embeddings on a corpus of 42 millions of financial news from 2008 to 2020. The literature on the use of NLP on financial data has been pushed forward by the work of Loughran and McDonald who analyzed the polarity of 10-K fillings of listed company [21]. One of the outcomes of their work is a dictionary of polarized words in categories (Positive, Negative, Litigious, etc.) that have been later used as a supervised dataset [15], as a benchmark [14], or as an input [20] to analyze other corpora (financial news being one of the most used). It is clear that NLP is now part of the toolbox to predict stock returns [2,35], but to authors' knowledge, no systematic study analyzing the capability of language models to capture financial semantics has been proposed. In this chapter, we do not try to predict stock returns, but we rather focus on providing insight about the structure of financial news from the perspective of embeddings, with a focus on semantic antonyms and non-stationarity.

This chapter is structured as follows: second section starts by defining properly word2vec model and its Skip-gram version and then explains the relationship between the internal representation of such an language model and the underlying "truth" of a Reference Model. Since a Reference Model is target that embeddings try to learn, it is pivotal in our analysis. We then make additional Markovian assumptions to ease the process of proving how the standard Skip-gram loss function asymptotically behaves like a cross-entropy between the learned model and the Reference Model. This allow us to comment the influence of synonyms on the structure of embeddings. Third section exhibits experiments on synthetic corpus generated under restrictive assumptions, allowing us to control what the embeddings should learn. It shows the importance of the structure of the underlying model, and then it shows that the similarity between the representation of synonyms is not as large as expected but significant when compared to other groups of words. Then fourth section exploits a corpus made of 42 millions of financial News in conjunction with the Loughran–McDonald Lexicon. The latter is providing the structure of polarized words (mainly, Positive, Negative, and Litigious) needed for our analysis. It allows us to explore the way embeddings can mix synonyms and antonyms, and we find and explain different results obtained on headlines only versus on the full text of financial News. Moreover, we study the non-stationarity of this structure, in particular around one example of a company name that is listen under Wikipedia's *List of corporate collapses and scandals*.

Theoretical Analysis: Learning Embeddings and Its Relation with Generative Models

An embedding is an vectorial representation of terms of a document, that is learned jointly with other NLP tasks that are learned to model a language or more recently jointly with different downstream task (like part of speech tagging or entity recognition). The underlying idea is that the learning phase positions the terms in the space of embeddings such a way that the downstream tasks are easier. Thus it is expected that the embeddings will position in the neighbourhood of terms having similar roles in sentences. Academic papers working of the topology in the space of embedding identified some interesting algebra properties in the space of embeddings, like "*King - man + woman = Queen*". In this chapter, we focus on the word2vec algorithm which is proposed in [23,24], especially on its Skip-gram version.

The word2vec model

To be compatible with the standard word2vec Skip-gram notations of [29], let us assume that we face a set of documents or a very long document that is a sequence of words belonging to a vocabulary $\mathcal{V} := \{x_1, \ldots, x_V\}$. The word2vec proposed in [12] can be written as a neural network with one hidden layer of size N, that is called the *embedding size* and that is usually far lower than V. Typically, English vocabulary needs at least 70,000 words and the embedding size used by practitioners is between 150 and 400.

To feed to the neural network, the i-th word of the vocabulary is encoded using a one-hot vector x_i, i.e. $x_i \in \mathbb{R}^V$, that is a vector with zeros everywhere but only one 1 at its i-th coordinate. The weights between the input layer and the output layer can hence be represented by a $V \times N$ matrix W. Each row of W is the N-dimension vector representation e_i of the associated word of the input layer. Formally, row i of W can be identified with the representation of i-th word and noted as $x_i^T W$, given as the context word one-hot vector x_i. The activation of the hidden layer that is linear $h_{x_i}^T = x_i^T W = W_{(i,\cdot)}$, which is essentially copying the i-th row of W to h_i. In this chapter, we use row vector notation. We call W the *word-embedding matrix*, and its i-th row is commonly called the *embedding of the i-th word of the vocabulary*.

From the hidden layer to the output layer, there is a matrix of weight commonly noted W', which is an $N \times V$ matrix, associating an embedding to a long row of size V. $W'x_j$ the jth column of the *context matrix* W'. Its output layer as a softmax activation, mapping a vector $(z_i)_i$ to another vector $(\exp(z_i)/\sum_j \exp(z_j))_i$ that is positive and which coordinates sum to one. The output of such a neural network is thus homogeneous to a vector of probabilities. As a consequence, the word2vec allows us to compute the expected probability of occurrence of any j-th word of the vocabulary conditioned on the input word i that reads $\mathsf{softmax}(x_i^T WW')x_j$.

Moreover, the word2vec Skip-gram model has an hyperparameter C that is the number of words that is explicitly considered by the model. In this chapter, for the simplicity of notation, the considered words are the C words following the input word.[2] Thus the variable

$$y_{c,j}^i = \mathsf{softmax}(x_i^T WW')x_j, \quad \forall c \in \{1,..,C\}, \tag{1}$$

computes the estimate of the probability of occurrence of the j-th word of the vocabulary among the C words following the i-th word of the vocabulary in the considered corpus of documents. Note that each output $y_{c,j}^i$ is independent and only depends on the input word.

Definition 1 (Parameters of a word2vec model). *A word2vec model is defined by the triplet of parameters: embeddings, contexts, and width, i.e. (W, W', C).*

The Skip-Gram Loss Function

Assuming the existence of an "underlying" model of the text: In the following sections, we use $(X_k)_k$ to denote the sequence of words in a corpus. To be able to define a probabilistic description of the embeddings, one needs to first make the weak assumption that the sequence of the K words constituting the corpus is drawn from a probability distribution $\mathcal{L}_\mathcal{T}$ that exists. You can think that this distribution allows us to generate the whole sample at once, with a lot of bi-directional dependencies, or you can think that $(X_k)_k$ is a stochastic process (meaning that there is an "information arrow" pointing from the left to the right in the text). It may be generated by a Markov chain if you believe in very short term memory in the text or even be i.i.d. realizations of a random variable if you think the words are generated according to their histogram of frequencies in the English language. The existence of this law will enable us to write probabilities.

---

[2] It is not needed in general; their positions can be chosen by the user.

Since equality (1) defines the likelihood of the occurrence of word j in the neighborhood of word i, the word2vec model implies a specific structure of this probability $\mathcal{L}_T$. Or at least it assume that the loss function it minimizes pushes this likelihood to be compatible with the underlying distribution. This is typically what we will explore in this chapter: which assumptions on $\mathcal{L}_T$ are naturally compatible with the way the word2vec model build its likelihoods? For that we need to start with a good understanding of the Skip-gram loss function.

Following our notations: X_k is the k-th word in the document while x_i is the i-th word in the vocabulary. x_i denotes the one hot encoding of the i-th word in the vocabulary (word index). Hence the event "$X_k = x_j$" means that the k-th word observed in the corpus is the j-th word of the vocabulary.

Writing the Skip-gram loss function: The part of the loss function associated with the k-th word of the document corresponds to the estimate by the word2vec model of likelihood to observe the C following words of the corpus. It reads

$$\ell(X_k, \ldots, X_{k+C}) = -\log \hat{\mathbb{P}}_{W,W'}(X_{k+1} = x_{j_1}, \ldots, X_{k+C} = x_{j_C} | X_k = x_i)$$

$$= -\log \prod_{c=1}^{C} \mathsf{softmax}(x_i^T W W') x_{j_c}$$

$$= -\log \prod_{c=1}^{C} \frac{\exp(x_i^T W W' x_{j_c})}{\sum_{j'=1}^{V} \exp(x_i^T W W' x_{j'})}$$

$$= -\sum_{c=1}^{C} x_i^T W W' x_{j_c} + C \cdot \log \sum_{j'=1}^{V} \exp(x_i^T W W' x_{j'}), \quad (2)$$

where $x_{j_1}, \ldots, x_{j_C}$ are the word occurring after the k-th one in the text. During the learning phase, the Skip-gram word2vec hence deals with sequences of $C+1$ words that can be found in the text.

The full loss function is the average of $\ell(\cdot)$ over all the words of the text that can be expressed as $\frac{1}{K}\sum_k \ell(X_k, \ldots, X_{k+C})$, i.e. the empirical average over the words of the text.

Defining a Reference Model: First define what we will call a "*Reference Model*": qualitatively it is a word2vec model with a trivial embedding, i.e. the size of the embedding is the size of the vocabulary: $N = V$, and hence the embedding matrix W is the identity. Since the rows of the context matrix of a Reference Model do not interact and the softmax layer is not needed, it is enough to arbitrary scale its rows to sum to one. This nonlinearity can thus be removed from the word2vec of a Reference Model.

Definition 2 (Reference Model). *A Reference Model RM is a triplet* (Id, W_0', C) *that can be identified to the parameters of a word2vec neural network. With V the size of the vocabulary, W_0' is a $V \times V$ stochastic matrix which element (i, j) records the probability to see the j-th word of the vocabulary in the C words following the i-th word.*

Different kinds of Reference Models can be built:

(i) The Reference Model corresponding to a given word2vec model (W, W', C) that is (Id, W_0', C) where for any i, $x_i^T W_0' :=$ softmax $(x_i^T W W')$; we use the notation $\mathsf{RM}(W, W')$.

(ii) The Reference Model corresponding to a corpus of document that is (Id, W_0', C) where W_0' is this time made of the averaged empirical probabilities of occurrence of the C words following each possible word; we use the notation $\mathsf{RM}(X_1, \ldots, X_K)$.

(iii) The Reference Model corresponding to underlying probability distribution $\mathcal{L}_T$ that generated the corpus; we use the notation $\mathsf{RM}(\mathcal{L}_T)$.

Note that for the same corpus:

- Given $(X_1, \ldots, X_K)$ has been generated by $\mathcal{L}_T$ (and provided that K is large enough), we expect to have $\mathsf{RM}(X_1, \ldots, X_K) \simeq \mathsf{RM}(\mathcal{L}_T)$.
- If a word2vec is a "good model" for this corpus, the corresponding Reference Model $\mathsf{RM}(W, W')$ should be "close" to the third one $\mathsf{RM}(\mathcal{L}_T)$.

The question of a *generative model* can be formulated this way: *What are the assumptions over $\mathcal{L}_T$ (i.e. the true underlying distribution) such that the Reference Model of the word2vec learned on a (very long) sample generated by $\mathcal{L}_T$ is the same as the Reference Model of $\mathcal{L}_T$?*

Probabilistic weaknesses of the word2vec embeddings

It is now clear that the information contained in a word2vec model on the learned language is captured by a factorization over $V \times N + N \times V$ coefficients (i.e. the embeddings W and the context W') of the joined probability of occurrences of $C + 1$ words at a given distance that could be described by $(V \times V)^C$ parameters. The key quantity that is manipulated by the word2vec is an estimate of $\mathbb{P}_{\mathcal{L}_T}(X_{k+1} = x_{j_1}, \ldots, X_{k+C} = x_{j_C} | X_k = x_i)$ for any word position k in the observed text. We will use the notation $\hat{\mathbb{P}}_{W,W'}(X_{k+1} = x_{j_1}, \ldots, X_{k+C} = x_{j_C} | X_k = x_i)$ to underline that this probability is estimated using matrices W and W' only.

We will list here the main weaknesses of this model and give pointers and intuitions to how more recent models answered to these points.

Weakness 1: The input is made of only one word. The first potential issue of the word2vec Skip-gram approach is that there is only one word as input. It implies that once the model parameters W and W' are chosen, there is no difference between $\hat{\mathbb{P}}_{W,W'}(X_{k+1} = x_{j_1}, \ldots, X_{k+C} = x_{j_C} | X_k = x_i, X_{k-1} = x_{i'})$ and $\hat{\mathbb{P}}_{W,W'}(X_{k+1} = x_{j_1}, \ldots, X_{k+C} = x_{j_C} | X_k = x_i, X_{k-1} = x_{i''})$.[3] There are a lot of different ways to explicitly fix this issue about a potential fragility of the input:

- the *CBOW* version of the word2vec algorithm is one of them. The CBOW uses the same weights W and W' but instead of minimizing a loss that is the likelihood of occurrence of a list of words observed "after" X_k, it takes a list of word occurring "before" X_k and averages their representations to target the $k+1$ word of the text:

$$\hat{\mathbb{P}}^{CBOW}_{W,W'}(X_{k+C} = x_{j_C} | X_{k+C-1} = x_{j_{C-1}}, \ldots, X_{k+1} = x_{j_1}, X_k = x_i)$$

$$= \mathsf{softmax}\left(\frac{1}{C}(x_{j_{C-1}} + \cdots + x_{k+1} + x_k)^T W W'\right) x_{j_C}.$$

It means that the internal representations of the $C-1$ input vectors are averaged and that this average is used to predict the likelihood of the next word.

- Since Skip-gram is trying to estimate the averaged probability of occurrence of words occurring after the pivotal k-th word, and the CBOW is doing the reverse (estimate the probability of occurrence of the $k+1$-th word given an average of the word that are before X_k), one could imagine a mix of these two approaches: using a weighted average of word "before" X_k to predict a weighted averaged probability of words positioned after X_k.

It is not very far away of what the self-attention mechanism is doing [34] except that weights for the average are not chosen *a priori*, but they are learned and they are a function of words surrounding each weighted word: it is the *attention* associated with the occurrence of the sequence of words $(X_{k-1} = x_{i'}, X_k = x_i, X_{k+1} = x_{j_1}, \ldots, X_{k+C-1} = x_{j_{C-1}})$.

---

[3]If we consider the learning process, it is clear that the loss function had to cope with the following two sequences: $(X_{k-1} = x_{i'}, X_k = x_i, X_{k+1} = x_{j_1}, \ldots, X_{k+C-1} = x_{j_{C-1}})$ and $(X_{k-1} = x_{i''}, X_k = x_i, X_{k+1} = x_{j_1}, \ldots, X_{k+C-1} = x_{j_{C-1}})$. The information is inside W and W', but at this stage, it is difficult to know how, and it is sure that it is averaged using as weights the relative number of occurrences of these sequences in the text.

- Another way to have more information about the words before X_k would be to increase the size of the state space and to concatenate X_k and X_{k-1}. But it is considered to be far too demanding in number of parameters. NLP methods prefer to use averages.

The take-away if this weakness is that it prevents a word2vec to capture some subtleties of the learned corpus. One solution is to average over more words in such a way that their weights in the average (i.e. attention) are conditioned by the joined distribution of words. It certainly produces *regularization* and *localization* (in the sense that the embeddings are weighted differently according to their position in the very high dimensional space of observed sequences of words). It means that the conclusions we will obtain on word2vec will probably be only valid locally for attention mechanisms.

Weakness 2: The ordering of words is not taken into account. From the previous section, it is clear that for a Skip-gram word2vec (and we know it is the same for a CBOW one), once the parameters W and W' are chosen, there is no difference between the word2vec estimate of $\hat{\mathbb{P}}_{W,W'}(X_{k+1} = x_{j_1}, \ldots, X_{k+C} = x_{j_C} | X_k = x_i)$ and $\hat{\mathbb{P}}_{W,W'}(X_{k+1} = x_{\sigma(j_1)}, \ldots, X_{k+C} = x_{\sigma(j_C)} | X_k = x_i)$, for any C-permutation σ.

Nevertheless, if we consider the learning process, only sequences that have been seen in the text are considered in the minimization of the loss function, hence W and W' incorporate a trace of the sequences. But it would be interesting to explicitly inject the positions of the words in the probabilistic model.

This is what BERT is doing, by concatenating the embeddings of the words (i.e. the W matrix of the word2vec) with a positional embedding [6]. It is clear that it adds the positional information that is missing in word2vec. As an illustration, just have a look at the expected effect on the estimate of the likelihood if this idea of "positional embeddings" is transposed to the word2vec mechanism: say that we augment the two matrices W and W' by a positional one P that is used the standard word2vec way. Now we can replace (with $C = 2$, to keep it simple, and with the notation p_u to encode that a word is in position u)

$$\hat{\mathbb{P}}_{W,W'}(X_{k+1} = x_j, X_{k+2} = x_{j'} | X_k = x_i)$$
$$= \frac{1}{2}\left(\mathsf{softmax}(x_i^T WW')x_j + \mathsf{softmax}(x_i^T WW')x_{j'}\right)$$

by

$$\hat{\mathbb{P}}_{W,W',P}(X_{k+1} = x_j, X_{k+2} = x_{j'}|X_k = x_i)$$
$$= \frac{1}{2}\left(\mathsf{softmax}(x_i^T WW' + p_1^T PP' p_2) x_j \right.$$
$$\left. + \mathsf{softmax}(x_i^T WW' + p_1^T PP' p_3) x_{j'}\right).$$

This produces different shifts inside the softmax depending on the relative positions of the words. It implies that using such a mechanism, a word2vec would consider two words to be interchangeable with respect to a given x_i, not only if they have the same $x_i^T WW'$ but also if they are positioned at the same distance if i.

Markovian generative models

In this section, we explore the theoretical properties of the word2vec embeddings under the assumption that the underlying language text is generated by a Markov model. We restrict ourselves to the Skip-gram approach: the loss function that is minimized is derived form $\ell(k)$ defined by equality (2).

Definition 3 (Markov generative model for texts). *A Markov generative model for a text over a vocabulary $\mathcal{V}$ of size V is defined by a stochastic transition matrix K of size $V \times V$ such that $K_{i,j}$ is the probability that the j-th word of the vocabulary follows the i-th word of the vocabulary.*

Thanks to K, and provided that an initial distribution m_0 over words is defined, a text of size T is a stochastic process $(X_k)_{1 \leq k \leq T}$ such that $X_0 \sim m_0$ and

$$\mathbb{P}(X_{k+1} = x_j | X_k = x_i) = K_{i,j}. \tag{3}$$

Markov generative models are restricted to irreducible Markov chains.

It is natural to restrict Markov generative models to irreducible Markov chain since it is realistic to think that a language can link two arbitrary words of the vocabulary by a finite text.

Texts generated by Markov models are of course poorer that standard English texts. Nevertheless, we could easily imagine "multiple inputs" or "local" versions of Markov text generation that would correspond the the formerly identified probabilistic weaknesses of the Skip-gram word2vec. Moreover, one could think about extensions of the Markovian framework to more subtle ones like the one chosen for the Latent Dirichlet Allocation [3] without killing most of the theoretical mechanisms that will be used in this section, but this is out of the scope of this chapter.

Since a Markovian text satisfies $\mathbb{P}(X_{(k+c)} = x_j | X_k = x_i) = x_i^T K^c x_j$, it is straightforward to see that the main quantity of interest for a Skip-gram word2vec is linked to the Reference Model RM(K) corresponding to the Markov kernel K since

$$W'_0(\text{RM}(K)) = \frac{1}{C} \sum_{i=1}^{C} K^i. \tag{4}$$

This comes from the fact that the language model is trained on a sequence of C words without any consideration of ordering. Hence, given that $\mathcal{I}$ is a uniform random variable over $\{1, \ldots, C\}$, the Skip-gram word2vec tries to estimate

$$\mathbb{P}(X_{k+\mathcal{I}} | X_k) = \sum_{i=1}^{C} \mathbb{P}(X_{k+i} | X_k) \mathbb{P}(\mathcal{I} = i)$$

$$= \frac{1}{C} \sum_{i=1}^{C} \mathbb{P}(X_{k+i} | X_k). \tag{5}$$

Markov chain properties

In this section, we formulate well-known properties of Markov chains [26] that we will need thereafter within our models and notations. Let $\mathcal{D}(T) = (X_1, X_2, \ldots, X_T)$ be a document made of T words generated using a Markov chain with a transition matrix K on a vocabulary of V words, and denote by μ the stationary distribution of the Markov chain. The state space of our Markov chain is the vocabulary $\mathcal{V}$ which is finite. Hence our Markov chain is positive recurrent; we will largely rely on the ergodic theorem.

Property 4 (Convergence of empirical distribution).

$$\lim_{T \to \infty} \frac{1}{T} \sum_{k=1}^{T} \mathbb{1}(X_k = x_1) = \mu(x_i). \tag{6}$$

Note that μ is a vector of size V, and the ith coordinate $\mu_i = \mu(x_i)$ is the probability of appearance of work i in the stationary distribution.

Proof. Direct consequence of Ergodic theorem. See Theorem 4.16 in [18] or Corollary 13.6.2 in [9]. □

Then we state a property for the convergence of empirical conditional distribution:

Property 5 (Convergence of empirical conditional distribution). *Let $\mathcal{I}$ be an independent uniformly random index from 1 to C (each index has equal probability $\frac{1}{C}$),*

$$\lim_{T\to\infty} \frac{\sum_{k=1}^{T} \mathbb{1}(X_k = x_i, X_{k+\mathcal{I}} = x_j)}{\sum_{k=1}^{T} \mathbb{1}(X_k = x_i)} = \frac{1}{C}\sum_{l=1}^{C} K^l(x_i, x_j).$$

Proof. We only stage the proof in the simplest $C = 1$ case to keep it short; when $C > 1$, the proof is similar. Just observe that $(X_k, X_{k+1})_{k\in\mathbb{N}}$ is an irreducible Markov chain too over the set of states $\{(x,y)|K(x,y) > 0\}$ with kernel $Q((a,b),(c,d)) = K(c,d)\mathbb{1}(b=c)$. Its unique invariant probability measure is $\pi(x,y) = \mu(x)K(x,y)$. Applying the ergodic theorem to $\mathbb{1}(X_k = x_i, X_{k+1} = x_j)$ reads

$$\frac{\frac{1}{T}\sum_{k=1}^{T}\mathbb{1}(X_k=x_i, X_{k+1}=x_j)}{\frac{1}{T}\sum_{k=1}^{T}\mathbb{1}(X_k=x_i)} \xrightarrow[T\to\infty]{} \frac{\mu(x_i)K(x_i,x_j)}{\mu(x_i)} = K(x_i,x_j).$$

□

From a Reference Model to a Markov kernel

Given a Reference Model with $C = 1$, i.e. $(\mathsf{Id}, W_0', 1)$, a Markov chain with $K := W_0'$ is the corresponding Markov generative model: By construction, the two Reference Models coincide. But when $C > 1$, there does not always exist a K satisfying (4) for a given W_0'.

Theorem 6 (Representative generative model). *Given a reference model (Id, W_0', C) on a vocabulary of V words, there exists a Markov chain with a transition matrix K verifying equality (4) if one of these two conditions is verified:*

1. *$C = 1$.*
2. *W_0' is symmetric (or diagonalizable) and all its eigenvalues are in $[0,1]$.*

If our kernel K is diagonalizable (for example, if it is reversible), then $K = P\Delta P^{-1}$ (and if it reversible, P is a orthonormal basis, i.e. $P^{-1} = P^T$) where Δ is a diagonal matrix. Following the spectral properties of transition kernel (Lemma 9 in [30]), if K is ergodic, then the maximal value in the diagonal Δ is 1 and it is unique. The other diagonal elements have absolute value less than 1. We can deduce the relation linking W_0' and C with the diagonal decomposition of K:

$$P^{-1}W_0'P = \frac{1}{C}\sum_{i=1}^{C}\Delta^i. \tag{7}$$

The right-hand side of the equation can be calculated exactly by the formula of geometric sum: It is a diagonal matrix whose terms are $\frac{1}{C}\sum_{i=1}^{C} \lambda^i$ for all $\lambda \neq 1$.

Now we can show the equivalence:

- Given a stochastic matrix K, we can always find W'_0 such that W'_0 is also a stochastic matrix and satisfying (4).
  The matrix constructed in this way $K := P\Delta P^{-1}$ is stochastic. $v = (1,1,...,1)$ is the eigenvector of W'_0 of eigenvalue 1 such that $W'_0 v = v$ and the first column of P is v. So $P^{-1}v = (1, 0, ..., 0)$ hence $P\Delta P^{-1}v = P(1, 0, ..., 0) = v$. $Kv = v$ means that each row of K sums up to 1 which prove that K is stochastic.
- Inversely, given a *diagonalizable* stochastic matrix W'_0 with eigenvalues in $[0, 1]$, we can also find such a matrix K. In fact, if $W'_0 = PDP^{-1}$ where D is a diagonal matrix with value between 0 and 1, we can find a diagonal matrix Δ such that $\frac{1}{C}\sum_{i=1}^{C} \Delta^i = D$ because the function

$$x \mapsto \frac{1}{C}\sum_{i=1}^{C} x^i = \frac{1}{C}\left(\frac{1 - x^{C+1}}{1 - x} - 1\right)$$

is bijective from $[0, 1]$ to $[0, 1]$. □

Understanding Word2Vec as a Compression of a Reference Model

How to compress a Reference Model

Assume that we start with the Reference Model (Id, W'_0, C) of a corpus and try to qualitatively understand what one can expect from a word2vec model of the same corpus when N, the dimension of the word embeddings, is $V - 1$:

- On the one hand, the likelihood associated with the word2vec model to the pair of words (x_i, x_j) reads

$$\mathsf{softmax}(x_i^T W W')x_j.$$

- On the other hand, the empirical occurrences in the corpus says that this likelihood should be

$$x_i^T W'_0 x_j.$$

If a pair of words $(x_i, x_{i'})$ is such that $x_i^T W'_0 x_j = x_{i'}^T W'_0 x_j$, then the rows i and i' of W'_0 are identical. The best way to compress W'_0 is hence

1. to use the embedding matrix to map rows i and i' on the same embedding. We will use the notation $E[i' \mapsto i]$ for this embedding matrix that is the $V \times V$ identity matrix with column i' that is removed and with 1 in place of 0 at its element (i', i),
2. to remove the row i' of W_0'. We use the notation $W_0'[i']^\ominus$ for a $(V-1) \times V$ matrix corresponding to W_0' once its i'-th row is deleted.

Then $E[i' \mapsto i] \cdot W_0'[i']^\ominus = W_0'$. As a consequence:

Property 7 (Trivial compression of a Reference Model by a word2vec). *With the upper notations, the word2vec model* $(E[i' \mapsto i], W_0'[i']^\ominus, C)$ *is an* exact compression *of the Reference Model* (Id, W_0', C) *such that the i-th and i'-th words of W_0' are identical, i.e.*

$$\forall (k, k') : \mathsf{softmax}(x_k^T E[i' \mapsto i] W_0'[i']^\ominus) x_{k'} = x_k^T W_0' x_{k'}. \tag{8}$$

We can have a qualitative look at $E[i' \mapsto i]$, the matrix of embeddings of this exact compression, and note that it mapped i and i' on the same embedding vector. And, qualitatively once more, the nature of words xi and $x_{i'}$ is that they have the same "probability vector", i.e. the same probability of occurrence of surrounding words. See Levy and Goldberg [19] and references therein to different perspectives on compression.

A frequentist (and confusing) viewpoint on synonyms: From a frequentist viewpoint, x_i and $x_{i'}$ are interchangeable; this is the reason why the training algorithm will map them on the same embeddings. One could call them "synonyms from a word2vec viewpoint" or *frequentist synonyms*. For two words being frequentist synonyms, it is enough that in the corpus all sentences containing x_i have a similar sentence where it is replaced by $x_{i'}$. This is clearly not the definition of semantic synonyms, for instance, if all sentences containing the word "*bad*" have an copy with the word "*good*" in place, like

$$\begin{vmatrix} \text{look at this } \boxed{bad} \text{ guy} & \leftrightarrow & \text{look at this } \boxed{good} \text{ guy,} \\ \text{this is } \boxed{bad} \text{ English} & \leftrightarrow & \text{this is } \boxed{good} \text{ English,} \\ \text{etc.} \end{vmatrix}$$

and these semantic antonyms will become synonyms in embeddings learned on this corpus. Elaborating a little more on application, one can bet that on a corpus of cars, colors will probably be frequentist synonyms, since most cars can be seen in any color, hence such a corpus may have as many sentences like "a *gray* car had an accident" than "a *blue* car had an accident". But on a corpus made of cooking recipes, colors will not be frequentist synonyms because there is no "*red* bananas" but "*red* apples" and "*red* peppers".

We will empirically see in the fourth section that if frequentist synonyms provide structure to a word2vec mode, they can confuse a task related to exploit polarity of sentiments of financial texts.

A formal definition for compression of Markov chains

The notion of Reference Model enables to define clearly what kind of compression can be expected from a text embedding, in particular for the word2vec class of models. The vocabulary size being V, and the embedding dimension being N, compressing *linearly* a Reference Model (Id, W_0', C) using a word2vec will end up with a $V \times N$ matrix R and a $N \times V$ matrix $R'W_0'$, such that the probabilities defined by $(R, R'W_0, C)$ are as close as possible from the original ones. Qualitatively,

$$\forall x_i, x_j : \mathsf{softmax}(x_i^T R R' W_0') x_j \text{ "close to" } x_i^T W_0' x_j.$$

The main point is to choose a quantitative criterion that has a sense to define this desired proximity. Qualitatively, the following is clear:

- The rank of RR' needs to be maximal (taking into account that the more frequentist synonyms in the language the lower the rank of W_0') to recover the space spanned by W_0'.
- Without the softmax nonlinearity and if the SVD (Singular Value Decomposition) of W_0' is $U\Sigma V^T$, one could expect that $R \simeq U\Sigma$ and $R' \simeq V^T$ (see [19, Section 4.2] for more details). Nevertheless the softmax changes the setting. Moreover, the SVD compression criterion is the minimization of the unexplained variance whereas in word embedding, the proximity of conditional probabilities seems to be a better criterion.

Definition 8 (Compression of a reference model (Id, W_0', C)). *We define the compression to dimension N of a reference model of size V using two mappings $\phi(W_0')$ and $\phi'(W_0')$. The compressed model is the $(\phi(W_0'), \phi'(W_0'), C)$ word2vec model. For simplicity of the notation we set*

$$\Phi(W_0') := \phi(W_0')\phi'(W_0')$$

and we define $\mathfrak{C}$ the class of function Φ that can be written this way.

If we consider the sequence of words X_k of a text of length T as a stochastic process, then we may take any distance $d(\cdot, \cdot)$ between probability vectors (i.e. $x_k^t W_0'$ or $\mathsf{softmax}(x_k^T WW')$) and write our optimization problem as

$$\min_{R,R'} \frac{1}{T} \sum_{k=1}^{T} d(\mathsf{softmax}(X_k^T \phi(W_0')\phi'(W_0')), X_k^T W_0'). \qquad (9)$$

For linear compression, just note that $RR'W_0' := \Phi(W_0')$. A last qualitatively remark: $\mathsf{softmax}(X_k^T \Phi(W_0'))$ and $X_k^T W_0'$ are both row vectors whose coordinate j' represents the probability of occurrence of the j'-th word of the vocabulary (in the neighborhood of $X_k = x_i$, i.e. if the word $x_{j'}$ occurs at least once in the next C words after $X_k = x_i$). Hence the former will be a "good compression" of the latter if the these two probability distributions are "close".

A natural choice is to consider the *cross-entropy* (that is a shifted version of Kullback–Leiber divergence) between these two distributions:

$$\mathbf{H}_{CE}(\mathbb{P}_{W_0'|X_k}, \hat{\mathbb{P}}_{\Phi(W_0')|X_k}) := -\mathbb{E}_{X_k^T W_0'} \log \mathsf{softmax}(X_k^T \Phi(W_0')) \qquad (10)$$

$$= -\sum_{x_{j'} \in \mathcal{V}} \log \mathsf{softmax}(X_k^T \Phi(W_0')) x_{j'} \cdot X_k^T W_0' x_{j'}.$$

Thanks to the ergodic theorem[4] that can be used on the stochastic process made of the sequence of words when the length of the corpus goes to infinity, it is now possible to state this definition of a compression:

Definition 9 (Compression criterion on a Markov chain generated text). *If a text $\mathfrak{X} = X_1, \ldots, X_k, \ldots$ stems from a Markov generative model which Reference Model is (Id, W_0', C), we define a compression Φ, taken in the class of functions $\mathfrak{C}$, of its Markov kernel W_0' thanks to the expectation (according to the invariant distribution μ of W_0') of the Cross-Entropy between the output vectors of a word2vec. It reads the following:*

$$\min_{\phi \in \Phi} \mathbb{E}_{X_k \sim \mu} \mathbf{H}_{CE}\left(\mathbb{P}_{W_0'|X_k}, \mathsf{softmax}(X_k^T \Phi(W_0'))\right). \qquad (11)$$

Note that with the already defined notation $\hat{\mathbb{P}}_{\Phi(W_0')|X_k}$ for $\mathsf{softmax}(X_k^T \Phi(W_0'))$, we recover a minimization that is compatible with (10).

Moreover, this definition goes beyond the Skip-gram word2vec embeddings, since the compression Φ is quite generic at this stage. It is possible to extend the definition using to the natural filtration $\mathfrak{X}_k$ associated with $\mathfrak{X}$ and replacing $\mathsf{softmax}(X_k^T \phi(W_0'))$ by $\mathsf{softmax}(X_k^T \phi(W_0', \mathfrak{X}_k))$ in formula (11). It would need to change the writing of the expectation, but it allows the compression to use all the words up to the k-th word of the text and hence to embed a memory or local metrics, addressing, for instance, part of the Weakness 1 exposed in "Probabilistic weaknesses of the word2vec embeddings" section.

---

[4] That is $\lim_{T \to \infty} \frac{1}{T} \sum_t f(X_k) = \mathbb{E}_\mu f(X_i)$.

Convergence of the Skip-gram word2vec loss function to a cross-entropy

Show that the mean of Skip-gram loss function converges to the criterion when the size of the corpus size goes to infinity, i.e. replacing it by something link $\mathbb{E}_t \ell(X_t)$ that would be close to (11).

Theorem 10 (Correspondence between Skip-gram loss function and compression criterion). *Assume the words $X_1, \ldots, X_T$ of a document of length T are generated thanks to an ergodic stochastic process according to a Reference Model $(\mathsf{Id}, W_0', 1)$ and an initial probability distribution m_0 over the vocabulary. Then the loss function of a Skip-gram word2vec model $(W, W', 1)$ over this corpus converges toward the expectation of cross-entropy between $\mathbb{P}_{W_0'|X_k}$ and $\hat{\mathbb{P}}_{W,W'|X_k}$*

$$\lim_{T \to \infty} \frac{1}{T} \sum_{k=1}^{T} -\log \mathsf{softmax}(X_k^T W W') X_{k+1} = \mathbb{E}_{X_k \sim \mu} \mathbf{H}_{CE}(\mathbb{P}_{W_0'|X_k}, \hat{\mathbb{P}}_{W,W'|X_k}). \tag{12}$$

Proof. For the sake of notations, we will restrict the theorem and its proof to $C = 1$. Note μ as the ergodic measure of the stochastic process $(X_k)_k$, i.e. $\mu(x_i) = \lim_{T \to \infty} \frac{1}{T} \sum_{k=1}^{T} \mathbb{1}(X_k = x_i)$. Then we can write the loss function as follows:

$$-\frac{1}{T} \sum_{k=1}^{T} \log \mathsf{softmax}(X_k^T W W') X_{k+1}$$

$$= -\sum_{i=1}^{V} \frac{\sum_{k=1}^{T} \mathbb{1}(X_k = x_i)}{T}$$

$$\times \sum_{j=1}^{V} \frac{\sum_{k=1}^{T} \mathbb{1}(X_k = x_i, X_{k+1} = x_j)}{\sum_{k=1}^{T} \mathbb{1}(X_k = x_i)} \log \mathsf{softmax}(x_i^T W W') x_j.$$

With the notations $\mathbb{P}_{W_0'|x_i}$ for the probability distribution of words given the occurrence of x_i (i.e. $x_i^T W_0'$) and $\hat{\mathbb{P}}_{WW'|x_i}$ for the probability of the same events modeled by the Skip-gram word2vec (i.e. $\mathsf{softmax}(x_i^T W W')$). Then the limit when the number of words goes to infinity of the Skip-gram loss reads

$$-\lim_{T \to \infty} \frac{1}{T} \sum_{k=1}^{T} \log \mathsf{softmax}(X_k^T W W') X_{k+1}$$

$$= -\sum_{i=1}^{V} \mu(x_i) \sum_{j=1}^{V} \frac{\mu(x_i) x_i^T W_0' x_j}{\mu(x_i)} \log \mathsf{softmax}(x_i^T W W') x_j$$

$$= \sum_{i=1}^{V} \mu(x_i) \, \mathbf{H}_{CE}(\mathbb{P}_{W'_0|x_i}, \mathsf{softmax}(x_i^T W W'))$$

$$= \mathbb{E}_{X_k \sim \mu} \mathbf{H}_{CE}(\mathbb{P}_{W'_0|X_k}, \hat{\mathbb{P}}_{WW'|X_k}).$$

□

When $C > 1$, the deduction is similar.

Synthetic Experiments: Empirical Study of the Role of Structures

In this section, we leverage on different elements of our theoretical analysis to perform numerical explorations around the identifiability of word embedding models. We mainly leverage on these two elements

- Under restrictive assumption, we can use a Markov chain with a kernel K to build a Reference Model (Id, W'_0, C) thanks to equation (4).
- The existence of *frequentist synonyms* influences the capability to compress efficiently a Reference Model.

We will hence generate different Reference Models having more or less structure (here structure means having blocks of frequentist synonyms or not), on vocabularies of different sizes, and observe how easy or difficult it is to recover them from a generated corpus. If the compression performed by a word2vec is not a principal component analysis (since it is more "low rank" than "low variance" driven, because of the nonlinearity introduced by the softmax and because the minimized criterion in a cross-entropy with an unknown Reference Model via a sample of text), it can nevertheless be expected to find commonalities with the pitfalls identified long ago by Random Matrix Theory (see [32] for an overview) that are playing an important role in the theoretical understanding of the limits of deep learning [5].

Experimental conditions: We generate different Reference Models on vocabularies of size V that will be "compressed" via embeddings of dimension N (for illustration purposes, we even consider some configurations for which $N > V$). We use a Markov chain to build the reference model so that it is "fully random" (uniformly generated) or has a structure. This structure is made of "blocks" whose rows are identical to each other (that is the definition of frequentist synonyms).

We train a Skip-gram word2vec during a large number of epochs (enough to stabilize the learning) via a standard SGD (Stochastic Gradient Descent, generally Adam with a fixed rate of 10^{-4}) implemented in pyTorch and running on Google Colab or AWS using CUDA acceleration.

Types of structure: We experiment different level of structure:

- *No structure* when K is a dense matrix generated randomly and uniformly in the space of stochastic matrices each of its rows is sampled by the uniform distribution in the space of simplex using Dirichlet distribution with $\alpha = 1$.
- *Structure* when K is made of blocks of duplicated rows, when the number of block varies, the overall size of the structure is always kept constant. For instance, when $V = 1000$ we can take either 160 blocks of 5 rows or 40 block of 20 rows because $150 \times 5 = 40 \times 20$.

The expected role of blocks is the following: the intrinsic dimension of a Reference Model made of B blocks of size S plus "noise" on $V - B \times S$ components is between $B + 1$ and $B + (V - B \times S) = V - B(S - 1)$. It is expected to be closer to $B + 1$ in a "low overfitting" configurations and closer to $V - B(S - 1)$ when the model considers that it is equivalent to learn one row or S similar rows. Our theoretical analysis suggests that the Skip-gram word2vec should be more in the first configuration that in the second, since attributing one vector of embedding to a block is far more rewarding (in terms of the loss function) than attributing it to one isolated row of W'_0.

Criteria to monitor: We focus on two criteria:

- The loss function during the learning, to analyze the performance of the compression. It is interesting to note that the loss function (2) being compatible with the cross-entropy (11), and since we are generating synthetic dataset, in some case, one can really expect a full success of the "compression" that is in these cases a simple identification of the generating model.
- The distance between the compressed vectors and the original ones. Once again it is a way to quantify the success of the identification of the properties of the generative model.
  For that, we focus on the (mean) cosine similarity between two words x_i and $x_{i'}$ defined as the scalar product between the two probability vectors:

$$C_S^P(x_i, x_{i'}) := \sum_{j=1}^{V} \mathsf{softmax}(x_i^T WW')x_j \cdot x_{i'}^T W'_0 x_j.$$

To formulate a criterion that is minimal when two groups G_1 and G_2 of words are close, we will use

$$d_C(G_1, G_2) := 1 - \frac{1}{\#G_1 \cdot \#G_2} \sum_{x_i \in G_1} \sum_{x_{i'} \in G_2} \mathbf{C}_S^P(x_i, x_{i'}). \qquad (13)$$

This metric will allow us to understand if words belonging to the same group of frequentist synonyms are closer between themselves rather than to other words.

First experimentation: Low vocabulary size

Figure 1 shows that the Skip-gram word2vect has difficulties to find the generative W_0' when the dimension of the word2vec is larger than the original dimension. This exhibits a clear identifiability issue. Of course in general the size of the embeddings is lower than the vocabulary size, so for word2vec this configuration should never occur in practice. Nevertheless, not that the relative size of the embeddings vs. the one of the vocabulary is not low for embedding-driven language models like BERT.

When the dimension of the embeddings is low, Figure 2 suggests that the structure can be captured by the compression. Typically, one could expect that an embedding of dimension 10 has more chances to capture a Reference Model made of 8 blocks, than one made of 32 blocks, and it is verified in our experiments.

In any case, Figure 3 underlines the fact that higher dimension of embeddings, even if it is lower than the vocabulary size, is better in the presence of a structure. For instance, when the Reference Model is made of 160 blocks of 5 rows (i.e. 800 rows over 1000 of the Reference Model are exhibiting structure): embedding sizes of 200 and 500 perform similarly (with more noise for $N = 500$), whereas $n = 800$ performs very poorly. This underlines the fact that the structure has to be taken into account to choose the targeted compression dimension; in a highly structure language (with a lot of frequentists synonyms), a low dimension can be a proper choice. We can expect that very repetitive sentences (from a semantic perspective), like the headlines of financial News, should exhibit identifiability issues compared to the body of the same news, that are made of more semantically diverse sentences.

Beyond the loss function: Assessing the quality of the captured structure

Figure 4 exhibits how the compressed model WW' recovers the blocks of the Reference Model W_0'. The standard deviations inside a block or between

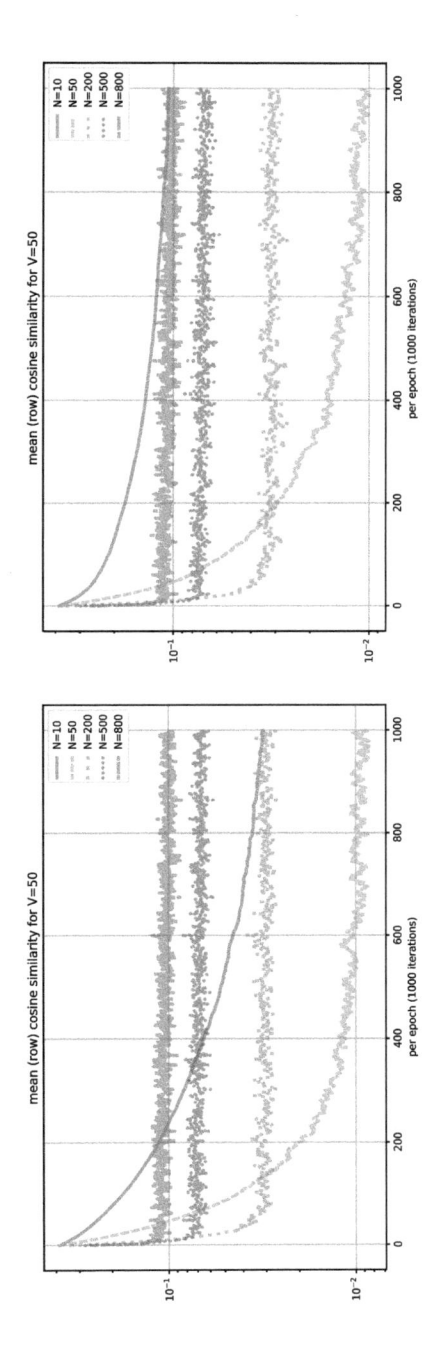

Figure 1. The role of structure: cosine distance between W'_0 and WW' during the learning (x-axis), for a vocabulary size $V = 50$ and for 8 blocks of 5 (left) and no block (right).

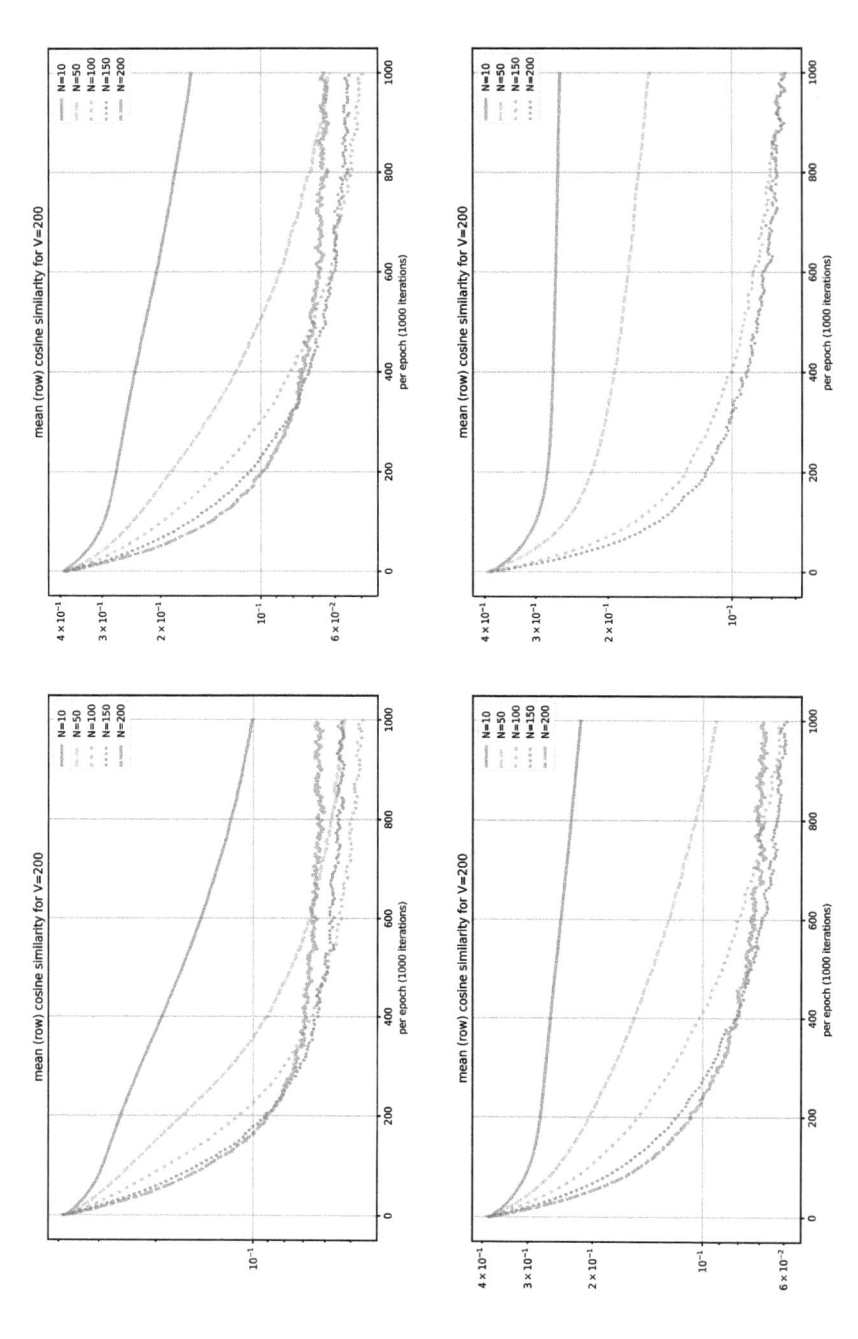

Figure 2. The role of structure: cosine distance between W'_0 and WW' during the learning (x-axis), for a vocabulary size $V = 200$ and for 8 blocks of 20 (top left), 16 blocks of 10 (top right), 32 blocks of 5 (bottom left) and no block (bottom right).

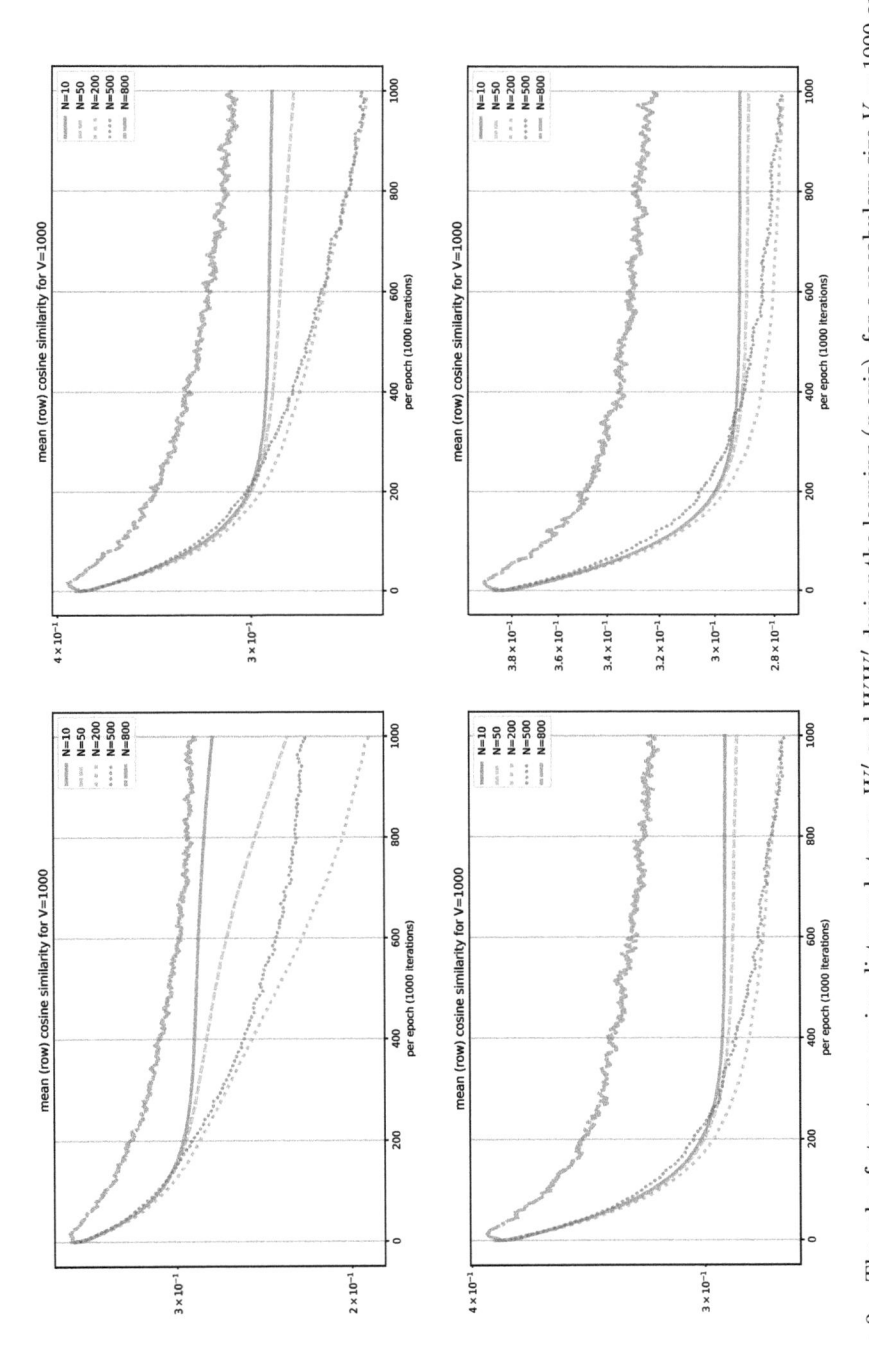

Figure 3. The role of structure: cosine distance between W'_0 and WW' during the learning (x-axis), for a vocabulary size $V = 1000$ and for 10 blocks of 80 (top left), 40 blocks of 20 (top right), 160 blocks of 5 (bottom left), and no block (bottom right).

8 block of size 5 for a vocabulary size of 50.

40 blocks of size 20 for a vocabulary size of 1000.

160 blocks of size 5 for a vocabulary size of 1000.

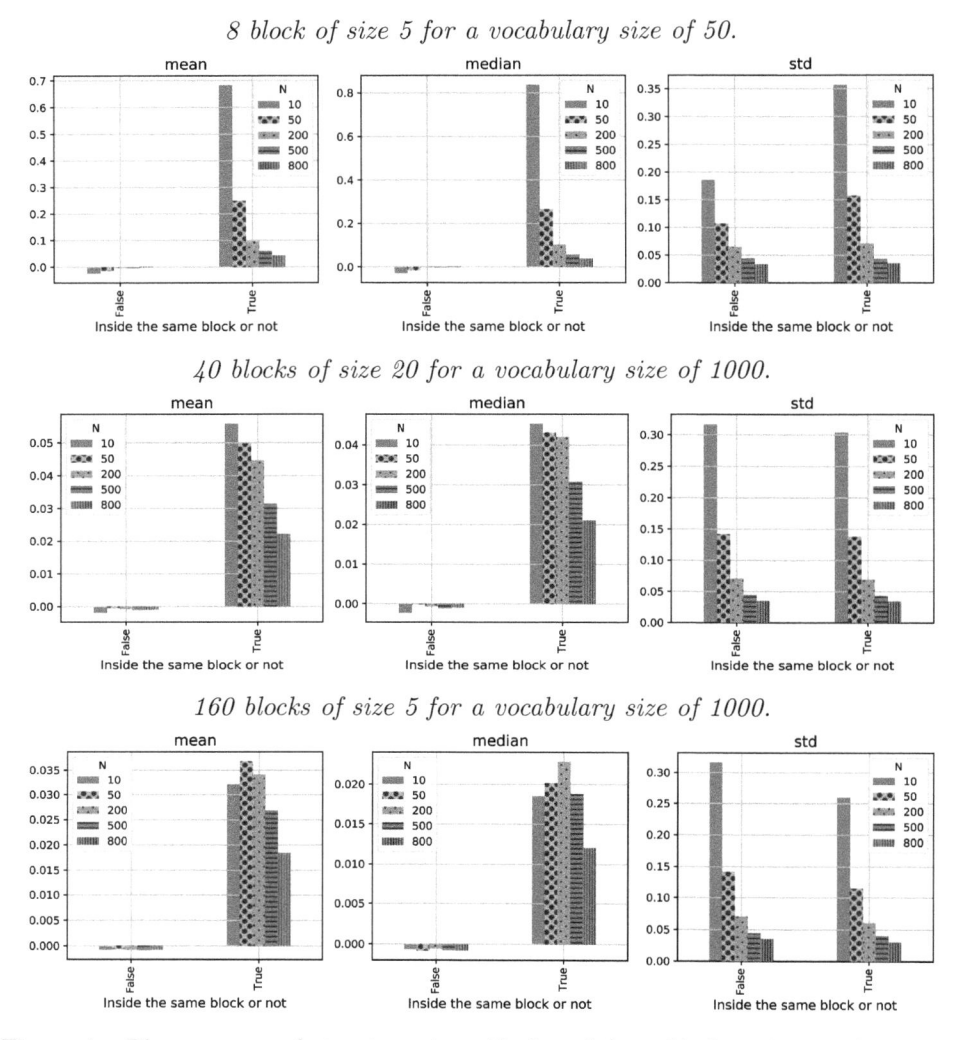

Figure 4. The recovery of structure: inter-block and intra-block cosine similarities of embeddings.

words of another block are similar, and as expected the cosine similarity between words belonging to the same block is higher than the cosine between words of two different blocks. Nevertheless, if this cosine is large in low dimension (0.8 in the best case of $V = 50$), it is far lower in high dimension. It is 0.02 when $V = 1000$ for $N \leq 500$, but it is halves when $N = 800$, showing that overfitting is really present in such a case.

The qualitative conclusions to this empirical study for synthetic data are as follows:

- The dimension of the embedding has to cope with the structure of the Reference Model, with a highly structured Reference Model, a too large embedding dimension is detrimental to the performances.
- The skip-gram word2vec faces identifiability issues in general, and we suspect it is not specific to this model.
- Surprisingly, even if the performances of the model is not good, it succeeds in putting words closer to frequentist synonyms (i.e. within the same block) than to other words.

Our experiments on the dimensionality of word embeddings seem to be compatible with [36] that is in favor of a dimension close to $N = 300$.

Experiments on a Corpus of Financial News: What can be Learned?

This section is an empirical study learning embeddings on a financial corpus. Our goal is to explore how a particular type of structure, that is usually named *sentiment polarity* in NLP, is preserved or not by embeddings.

The usage of NLP in finance often targets to make the difference between "good news" and "bad news" on listed company, as a pre-requisit to build investment strategies (see [10] and [35] for details). The "Loughran–McDonlad Lexicon" tuned by human experts on 10-K regulatory fillings by US corporates (see [21] for details) is commonly used to quantify the polarity of text. Since it lists positive, negative, and litigious words, we will use these lists of words as blocks of synonyms or antonyms. They study play a similar role as the "blocks" of the generative models studied in the previous section.

Sentiments as a source of structure: Does financial news understand better Loughran–McDonald's polarity than Wikipedia?

We use a corpus of financial news provided by a large provider of professional financial news covering years from 2008 to 2020. We will do experiments on the headlines only or on the whole body of the news (i.e. the headline followed by the text). We used different *ad hoc* filters to prevent the repetition of

Table 1. Descriptive statistics of the processed news full text and headlines (*avg. LM* is the average number of words from the Longhran–McDonald lexicon in a document of the corpus, *avg. size* is the average number of words, and *nbe* is the number of news).

	News full text			News headlines		
	avg. LM	avg. size	nbe	avg. LM	avg. size	nbe
2008	0.26	156.18	415,524	0.24	10.14	406,315
2009	0.25	146.84	2,780,525	0.23	10.20	2,497,544
2010	0.22	173.07	2,911,113	0.21	10.17	2,651,780
2011	0.20	354.10	4,038,762	0.19	9.65	1,937,799
2012	0.19	387.70	4,650,571	0.23	10.23	132,126
2013	0.19	370.67	4,797,842	0.21	9.96	187,842
2014	0.19	360.51	4,785,053	0.18	9.90	195,455
2015	0.20	374.79	4,763,002	0.16	9.12	1,303,531
2016	0.24	242.99	3,716,652	0.18	9.54	2,763,163
2017	0.22	133.33	3,401,207	0.15	9.89	2,191,320
2018	0.21	129.60	3,357,355	0.14	9.75	1,980,199
2019	0.17	160.64	1,634,809	0.15	10.00	1,545,239
2020	0.17	175.42	1,388,059	0.15	9.97	1,449,560

news (sometimes a news is repeated, in such a case we only keep its first appearance, and some times a header does not have enough words, in such a case we remove it from the headline corpus). Table 1 provides descriptive statistics on these corpora; it is restricted to the news that will be processed by our word2vec: the headlines contain on average 10 words where the body of the news contains on average around 250 words. Our filtering reduced the number of headlines years 2008, 2012, 2013, and 2014. Due to this, we will restrict some of our analysis to other years only.

As a reference, we use embedding pre-trained in Wikipedia (see Appendix for details).

The content of the Loughran–McDonald Sentiment Word Lists (2018) is described in Table 2. This lexicon contains more negative words than positive words; those are the two categories we will mainly focus on. Not all these words appear in the considered corpora, for instance, embeddings trained on Wikipedia contains only 349 of the 353 positive words where the headlines of our financial News from 2008 to 2012 contains 343 of them.

Here are examples of the first words of this lexicon:

- Negative: *abandon, abandoned, abandoning, abandonment, abandonments, abandons, abdicated, abdicates,* etc.
- Positive: *able, abundance, abundant, acclaimed, accomplish, accomplished, accomplishes, accomplishing, accomplishment,* etc.

Table 2. Number of word in each category present in each corpus.

	LM (ref)	In Wikipedia	In fin. news headlines	In fin. news full text
Negative words	2,354	2,097	2,190	2,335
Positive words	353	349	343	354
Uncertainty words	396	270	273	294
Litigious words	903	590	637	820
StrongModal words	19	18	19	19
WeakModal words	27	26	26	27
Constraining words	183	172	178	181

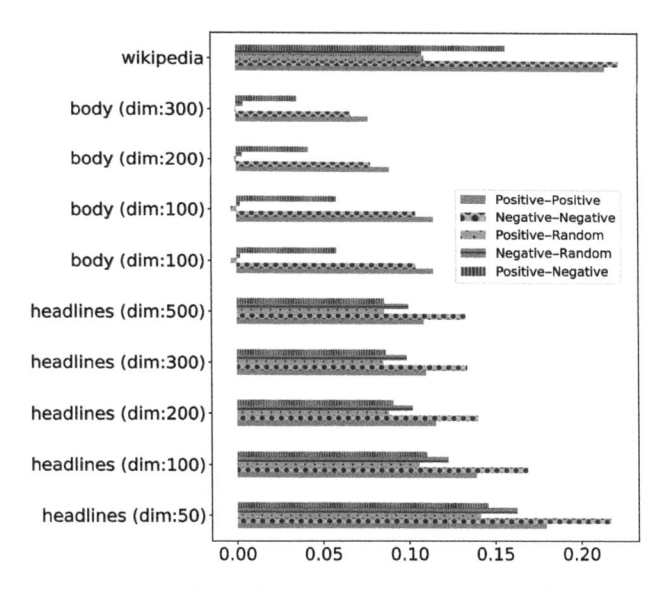

Figure 5. Average cosine similarity between embeddings of different groups of polarities trained on financial news or on Wikipedia (*random* means either random words — for financial news — or all the other words — for Wikipedia).

Figure 5 exhibits the cosine similarity between different sections of polarized words (as it is a scalar product between two vectors of norm 1, its maximum value is 1). The exact computation process is the following: (1) we trained Skip-gram word2vec embedding on each year, (2) for each year we compute the cosine similarity between any two words of the considered lists (for instance between any two positive words, or between any positive word and any negative word), (3) we average over all the obtained cosine similarities (we plot the obtained time series in Figure 6), and (4) we average the obtained numbers excluding years 2008, 2012, 2013 and 2014 (Figure 6 shows that average cosine similarity is too different over these years

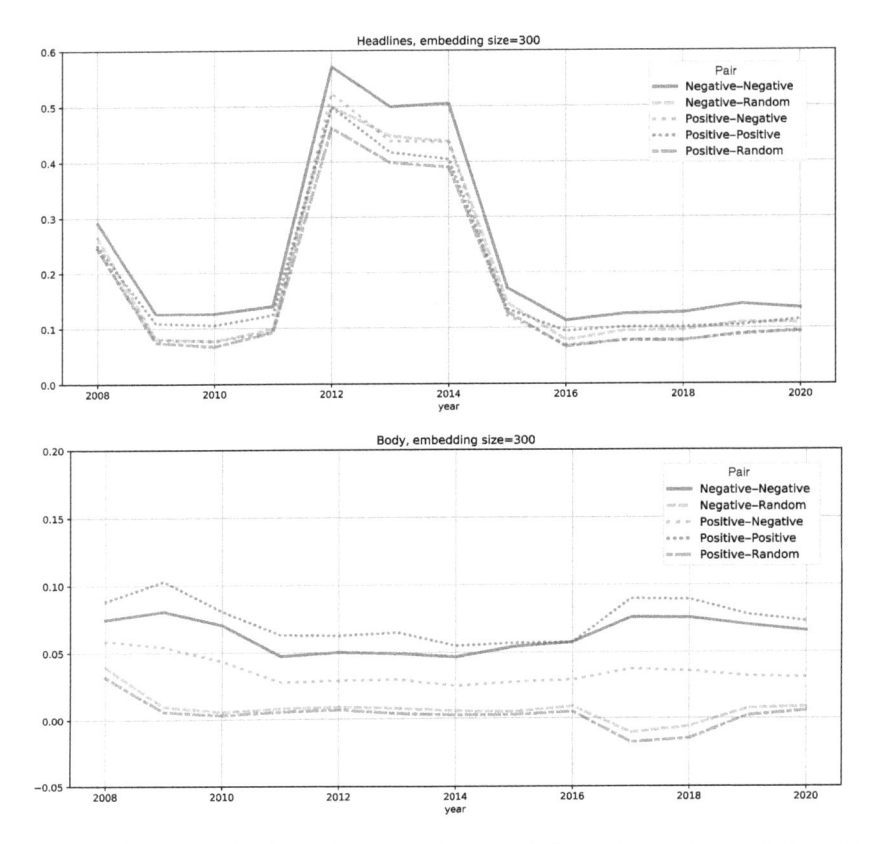

Figure 6. Yearly cosine similarity between the embeddings of Loughran–McDonald Positive and Negative lexicon, for a skip-gram word2vec model learned on the headlines only (bottom) or on the full text (top) of financial news.

for embeddings learned on headlines, probably because of the low number of headlines available).

If the learned embeddings would have been compatible with the polarities expressed by the lexicon, we should observe the following:

- the highest average similarity for groups *Positive–Positive* and *Negative–Negative* bars,
- the lowest average similarity for the *Positive–Negative* bar,
- medium similarities for groups *Positive–Random* and *Negative–Random* bars.

It is not what we observe:

1. If Positive–Positive and Negative–Negative bars are the highest, it is not by far, especially for embeddings learned on the headlines.

2. *Positive–Negative bar similarity is never the lowest*, it is comparable to distance to random words for embeddings.
3. Positive–Random and Negative–Random bars are not very low *except for embeddings learned on the full text* of the news.

Conjectures on frequentist synonyms among polarized words in finance: The first empirical conclusion we can make at this stage is that *it is difficult for embeddings to separate financial polarized antonyms*. The fact that the Positive–Negative bars are never lower than the Positive or Negative vs. Random bars is an evidence of this difficulty. Nevertheless, *it is easier to separate financial polarized words with embeddings learned on the full text of news rather than on headlines only*. We can conjecture that it is because financial headlines are written to be quickly understood by humans, and that for they contain similar sentences with positive or negative words in the same environments. Indeed, when a corpus is structure with short sentences of similar structure, the chances that semantic antonyms becomes frequentist synonyms are high. This configuration disappear when the corpus of full text (body) of financial news is used. Last but not least, Positive–Random and Negative–Random similarity is particularly close to zero when embeddings are learned on the full body of the news. This similarity is even lower for embeddings trained on the full text of financial news rather than on the Wikipedia corpus.

Our theoretical analysis shed light on this: It is difficult for the Skip-gram word2vec to make the difference between frequentist synonyms. From a corpus of short and very structured sentences, antonyms have good chances to become frequentist synonyms.

As a conclusion of this first analysis, we can conclude that using embeddings to discriminate polarity will be more difficult on a the corpus of headlines, that is not diverse enough, and financial enough (since it does not work is well on Wikipedia).

Stationarity of embeddings associated to polarized words

The cosine similarity between polarities is stationary

Table 1 shows the number of document in our corpus each year; we have clearly less headlines in 2008, 2012, 2013, and 2014, and it is reflected in the average cosine similarities of Figure 6 (top panel). There is not enough documents, hence not enough diversity in the sentences, and as a consequence words are not been seen in differentiating enough contexts: the global cosine similarity is higher.

That being put aside, we observe that the averaged results of Figure 5 are in line with their time series representation: the ordering of cosine similarities between groups of words of same polarity are the same every year. Despite a widening of the difference between words of the same polarity (Positive–Positive and Negative–Negative lines) and the cosine between polarized words and random words (Negative–Random and Positive–Random lines) in 2017 and 2018, *the embeddings seem to be stationary* in the sense that there is no real change in their relative ranking from one year to another.

Influence of the embedding size on the polarities

Figure 7 Shows the influence of embedding size on the cosine similarity between embeddings of Positive words with other classes (Positive, Negative, and Random) words.[5] The empirical results are mixed since on the one hand when the dimension decreases, the similarity inside the class of positive words increase (that is good), but the similarity between positive and negative words increases too. The similarity to random words stays close to zero.

Conjecture on the influence of embeddings, size on frequentist synonyms: This observation is compatible with the idea that the lower the dimension to represent the language, the more difficult to memorize the differences in contexts of words. As a consequence, words with the same polarity will become more similar, but antonyms will become more similar too.

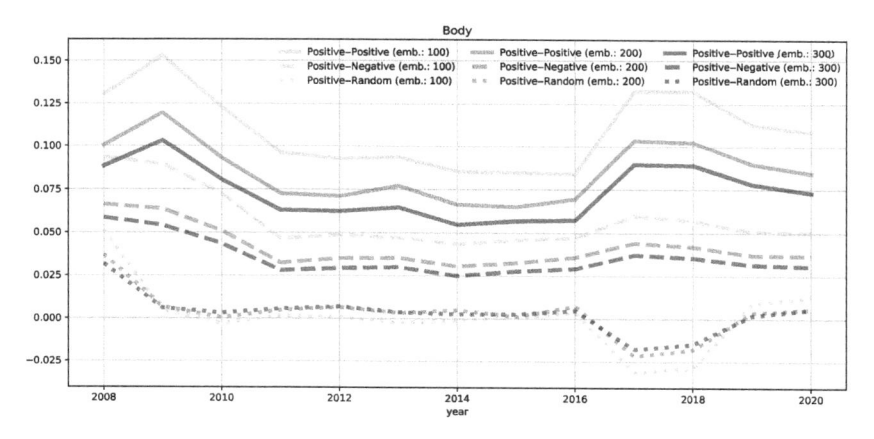

Figure 7. Change of the cosine similarity between positive words and random words as a function of the size of embeddings.

---

[5]The effects are the same for Negative words.

When a company name becomes a frequentist synonym

Figure 8 document an interesting effect that can have an influence on using embeddings to predict returns of listed companies. The important question to ask is if we want that a sentence like *"Microsoft printed booming results"*[6] is understood the same way as *"Google printed booming results"* by an embedding-based system. It would require that *Microsoft* and *Google* have no polarity. There is no semantic reason for a company name to be positively or negatively biased, i.e. to have a cosine similarity different from zero to a group of polarized words.

This analysis is not systematic, we simply took the only company name of Wikipedia's *List of corporate collapses and scandals* that is in our database, and computed its cosine similarity with the five most numerous lists of the Loughran–McDonald lexicon (Positive, Negative, Litigious, and Uncertain).

Before commenting the results, have a look at the quick summary made by Wikipedia[7] of the Theranos case:

> In March 2018 the US Securities and Exchange Commission charged Theranos, its CEO Elizabeth Holmes and former president Ramesh "Sunny" Balwani, claiming they had engaged in an "elaborate, years-long fraud" wherein they "deceived investors into believing that its key product — a portable blood analyzer — could conduct comprehensive blood tests from finger drops of blood".

Figure 8. Yearly cosine similarity between the embeddings of Theranos and the Loughran–McDonald lexicon (it is measured using the `model.wv.n_similarity` function of gensim).

---

[6] "Booming" is part of the Loughran–McDonald list of positively polarized words.
[7] On Theranos Wikipedia page as of the 10th of March 2020.

On Figure 8, we see the change in the polarity of the embedding of the term *Theranos* in the embeddings. This means that in 2018 and 2019, sentences with the name of this company is tinted with negativity and litigation compared to the same sentence concerning another company. With an exaggerated anthropomorphism, we could say that the embedding now "believes" that most sentences concerning this company are negative or litigious. In fact this company name became a frequentist synonym of negative and litigious terms. Probably because from the viewpoint of skip-gram word2vec loss function, it was "easier" to get this company name closer to negative and litigious terms than to keep it away from them (like any company name should semantically be).

The goal of this paper is not to investigate further on this kind of polarization of entities, that is in fact covered in a completely different context by the literature on *fairness of NLP*, see [28] for an example and [31] for an overview.

Conclusion

In this chapter, we presented some theoretical understanding of word embeddings, essentially using the skip-gram word2vec model. Moreover, we explain why more sophisticated model should inherit, at least locally, of some of these properties. It allows us the define the concept of Reference Model (that is the uncompressed version of an embedding model) and to show that asymptotically the loss function of such a learning algorithm is a cross-entropy between the representation of the model and the distribution of the Reference Model. Moreover, it lead us to define frequentists synonyms, i.e. words that have the same context in the considered corpus. It is impossible for embeddings to make the difference between exact frequentist synonyms and difficult for approximate synonyms.

Then we test these concepts on synthetic corpora generated using controlled Markovian models, so that we can focus on the identifiability of skip-gram word2vec embeddings. We observe that if their identifiability is poor, the cosine similarity between embeddings makes sense, even when it is low: frequentist synonyms are closer to word from their class than to words of another group of synonyms.

Last but not least, we provide empirical observations on a financial corpus: we use the Loughran–McDonald lexicon to obtain semantic synonyms and antonyms: lists of polarized words (Positive vs. negative words essentially). And we compare the polarity of embeddings trained on headlines of news with other trained on the full text of the same news. We observe that on news headlines, that are short and structured sentences, semantic antonyms are often frequentist synonyms, and hence it is difficult for embeddings learned on such headlines to make the difference between positive and negative words. On the opposite, embeddings learned on the full body of the news are more reflecting the polarities of the considered lexicon. In fact they are better reflecting financial polarities than embeddings trained on Wikipedia. It seems that the dimension of the embeddings has an influence on the cosine similarities between polarities: the lower the dimension, the more difficult to make the difference, in the space of embeddings, between positive and negative words. We moreover observe that names of companies can be tinted with polarity. The structure of the loss function of embeddings can lead them to accept to represent a company name close to a polarized word if it appears a lot in negative news: this company name is thus becoming a frequentist synonym of negative and litigious words.

Acknowledgments

Authors would like to thank Sylvain Champonnois for deep discussions about the nature of text polarity and biases of embeddings, Jean-Charles Nigretto for preliminary work on biases in doc2vec models, Elise Tellier for long discussions about the optimal size of embeddings, Laurent El Ghaoui for challenging discussions about the representation of language models by embeddings, and Gérard Ben Arous for discussions on the convergence of criteria and the choice of *observables* in the empirical analysis of word embeddings.

Appendix: Technical Details

We use Python 3.7.6 and Gensim 3.8.3 to train our models on AWS with financial news. On financial news headlines, we set `min_count=1` to include every word in headlines while we let `min_count=5` as the default value of Gensim on financial news which means we only consider the words count more than 5 times. Our `epoque =1`, i.e. each training sample is used only 1 time. As the model is initialized by training sentences, our `total_examples=model.corpus_count`.

Training skip-gram word2vec on one year of News takes around 2h and on one year of headlines it takes around 1h30. We use `ml.p2.xlarge` as of March 2021, i.e. 1 NVIDIA K80 GPU, 1 vCPU, and 64GB of RAM.

The model trained on English Wikipedia Dump of February 2017 uses Gensim Continuous skip-gram with no lemmatization. It is provided by Language Technology Group at the University of Oslo in NLPL word embeddings repository [7].

References

[1] A. Acharya, R. Goel, A. Metallinou, and I. Dhillon. (2019). Online embedding compression for text classification using low rank matrix factorization. In *Proceedings of the AAAI Conference on Artificial Intelligence*, Vol. 33, pp. 6196–6203.

[2] D. Araci. (2019). Finbert: Financial sentiment analysis with pre-trained language models. *arXiv preprint arXiv:1908.10063*.

[3] D. M. Blei, A. Y. Ng, and M. I. Jordan. (2003). Latent dirichlet allocation. *The Journal of Machine Learning Research* 3, 993–1022.

[4] Q. Chen. (2021). Stock movement prediction with financial news using contextualized embedding from bert. *arXiv preprint arXiv:2107.08721*.

[5] A. Choromanska, M. Henaff, M. Mathieu, G. B. Arous, and Y. LeCun. (2015). The loss surfaces of multilayer networks. In *Artificial Intelligence and Statistics*, pp. 192–204. PMLR.

[6] J. Devlin, M.-W. Chang, K. Lee, and K. Toutanova. (2019). BERT: Pre-training of deep bidirectional transformers for language understanding. In *Proceedings of the 2019 Conference of the North American Chapter of the Association for Computational Linguistics: Human Language Technologies, Volume 1 (Long and Short Papers)*, pp. 4171–4186, Minneapolis, Minnesota, June 2019. Association for Computational Linguistics.

[7] M. Fares, A. Kutuzov, S. Oepen, and E. Velldal. (2017). Word vectors, reuse, and replicability: Towards a community repository of large-text resources. In *Proceedings of the 21st Nordic Conference on Computational Linguistics*, pp. 271–276, Gothenburg, Sweden, May 2017. Association for Computational Linguistics.

[8] L. Floridi and M. Chiriatti. (2020). Gpt-3: Its nature, scope, limits, and consequences. *Minds and Machines* 30, 681–694.

[9] J.-F. Le Gall. (2006). Intégration, probabilités et processus aléatoires.

[10] M. Gentzkow, B. Kelly, and M. Taddy. (2019). Text as data. *Journal of Economic Literature* 57(3), 535–74.

[11] S. Ghosh, O. Vinyals, B. Strope, S. Roy, T. Dean, and L. Heck. (2016). Contextual lstm (clstm) models for large scale nlp tasks. *arXiv preprint arXiv:1602.06291*. Presented at KDD 2016.

[12] I. Goodfellow, Y. Bengio, and A. Courville. (2016). *Deep Learning*. MIT Press. http://www.deeplearningbook.org.

[13] A. Gutiérrez-Fandiño, P. N. Kolm, M. Noguer i Alonso, and J. Armengol-Estapé. (2022). Fineas: Financial embedding analysis of sentiment. *The Journal of Financial Data Science* 4(3), 45–53.

[14] Z. T. Ke, B. T. Kelly, and D. Xiu. (2019). Predicting returns with text data. Technical report, National Bureau of Economic Research.

[15] A. Kumar, A. Sethi, M. S. Akhtar, A. Ekbal, C. Biemann, and P. Bhattacharyya. (2017). Iitpb at semeval-2017 task 5: Sentiment prediction in financial text. In *Proceedings of the 11th International Workshop on Semantic Evaluation (SemEval-2017)*, pp. 894–898.

[16] K. Lagus, T. Honkela, S. Kaski, and T. Kohonen. (1999). Websom for textual data mining. *Artificial Intelligence Review* 13(5), 345–364.
[17] T. K. Landauer, D. Laham, and P. Foltz. (1998). Learning human-like knowledge by singular value decomposition: A progress report. *Advances in Neural Information Processing Systems* 10, 45.
[18] D. A. Levin, Y. Peres, and E. L. Wilmer. (2006). *Markov Chains and Mixing Times*. American Mathematical Society.
[19] O. Levy, and Y. Goldberg. (2014). Neural word embedding as implicit matrix factorization. In Z. Ghahramani, M. Welling, C. Cortes, N. Lawrence, and K. Q. Weinberger (eds.) *Advances in Neural Information Processing Systems*, Vol. 27, pp. 2177–2185. Curran Associates, Inc..
[20] X. Li, H. Xie, L. Chen, J. Wang, and X. Deng. (2014). News impact on stock price return via sentiment analysis. *Knowledge-Based Systems* 69, 14–23.
[21] T. Loughran and B. McDonald. When is a liability not a liability? Textual analysis, dictionaries, and 10-ks (2011). *The Journal of Finance* 66(1), 35–65.
[22] A. May, J. Zhang, T. Dao, and C. Ré. (2019). On the downstream performance of compressed word embeddings. *Advances in Neural Information Processing Systems* 32, 11782.
[23] T. Mikolov, K. Chen, G. Corrado, and J. Dean. (2013). Efficient estimation of word representations in vector space. *arXiv e-prints, arXiv:1301.3781* (January).
[24] T. Mikolov, I. Sutskever, K. Chen, G. S. Corrado, and J. Dean. (2013). Distributed representations of words and phrases and their compositionality. In C. J. C. Burges, L. Bottou, M. Welling, Z. Ghahramani, and K. Q. Weinberger (eds.) *Advances in Neural Information Processing Systems*, Vol. 26, pp. 3111–3119. Curran Associates, Inc.
[25] V. Raunak, V. Gupta, and F. Metze. (2019). Effective dimensionality reduction for word embeddings. In *Proceedings of the 4th Workshop on Representation Learning for NLP (RepL4NLP-2019)*, pp. 235–243.
[26] D. Revuz. (2008). *Markov Chains*. Elsevier.
[27] A. Rogers, O. Kovaleva, and A. Rumshisky. (2021). A primer in bertology: What we know about how bert works. *Transactions of the Association for Computational Linguistics* 8, 842–866.
[28] A. Romanov, M. De-Arteaga, H. Wallach, J. Chayes, C. Borgs, A. Chouldechova, S. Geyik, K. Kenthapadi, A. Rumshisky, and A. T. Kalai. (2019). What's in a name? Reducing bias in bios without access to protected attributes. *arXiv preprint arXiv:1904.05233*.
[29] X. Rong. (2014). word2vec Parameter Learning Explained. *arXiv e-prints, arXiv: 1411.2738* (November).
[30] J. Salez. Temps de mélange des chaînes de markov.
[31] D. Shah, H. A. Schwartz, and D. Hovy. (2019). Predictive biases in natural language processing models: A conceptual framework and overview. *arXiv preprint arXiv:1912.11078*.
[32] T. Tao. (2012). *Topics in Random Matrix Theory*, Vol. 132. American Mathematical Society.
[33] H. Touvron, T. Lavril, G. Izacard, X. Martinet, M.-A. Lachaux, T. Lacroix, B. Rozière, N. Goyal, E. Hambro, F. Azhar, et al. (2023). Llama: Open and efficient foundation language models. *arXiv preprint arXiv:2302.13971*.
[34] A. Vaswani, N. Shazeer, N. Parmar, J. Uszkoreit, L. Jones, A. N. Gomez, Ł. Kaiser, and I. Polosukhin. (2017). Attention is all you need. In *Proceedings of the 31st International Conference on Neural Information Processing Systems*, pp. 6000–6010.

[35] F. Z. Xing, E. Cambria, and R. E. Welsch. (2018). Natural language based financial forecasting: A survey. *Artificial Intelligence Review* 50(1), 49–73.

[36] Z. Yin, and Y. Shen. (2018). On the dimensionality of word embedding. In S. Bengio, H. Wallach, H. Larochelle, K. Grauman, N. Cesa-Bianchi, and R. Garnett (eds.) *Advances in Neural Information Processing Systems*, Vol. 31, pp. 887–898. Curran Associates, Inc.

Chapter 4

Optimal ESG Portfolios: Which ESG Ratings to Use?

Anatoly Schmidt[*,‡] **and Xu Zhang**[†,§]

*Department of Finance and Risk Engineering,
NYU Tandon School, New York, USA*
[†]*Independent Researcher*
[‡]*as8098@nyu.edu*
[§]*xz2657@nyu.edu*

Abstract

The idea behind the optimal ESG portfolio (OESGP) is to expand the mean–variance theory by adding the portfolio ESG value (PESGV) multiplied by the ESG strength parameter γ (which is the investor's choice) to the minimizing objective function [26,27]. PESGV is assumed to be the sum of portfolio constituents' weighted ESG ratings that are offered by several providers. In this work, we analyzed the sensitivity of the OESGP based on the constituents of the Dow Jones Index to the ESG ratings provided by MSCI, S&P Global, and Sustainalytics. We describe discrepancies among various ESG ratings for the same securities and their effects on the OESGP performance. We found that with growing γ, the OESGP diversity and Sharpe ratio may monotonically decrease. However, the *ESG-tilted Sharpe ratio* has one or two maximums. The 1st maximum exists at moderate values of γ and yields a moderately diversified OESGP, which can serve as a criterion for optimal ESG portfolios. The 2nd maximum at large γ corresponds to highly concentrated OESGPs. It appears as if the portfolio has one or two securities with a lucky combination of high returns and high ESG ratings.

Keywords: Portfolio choice, mean–variance theory, ESG, Dow Jones index.

Introduction

There has been significant interest in environmental, social, and governance (ESG) investing as a viable supplement, if not an alternative, to the traditional, wealth-oriented asset management [2,3,16,24,25]. While there are indications that neglecting the so-called "sin stocks" can decrease portfolio returns [17], many studies point to a positive correlation between the corporate ESG ratings and financial performance [23,32]. This, however, does not guarantee the *outperformance* of the ESG-based portfolios and indexes [20,21]. Also, high performance of the ESG investments can be driven by money inflows due to their increased popularity [31].

A reasonable investing strategy that reflects investors' interest in a socially responsible growth of their wealth is simultaneous mean–variance portfolio (MVP) optimization in terms of portfolio return, risk, and portfolio ESG value (PESGV) [26,27]. PESGV can be calculated using the ESG ratings of the portfolio constituents that are provided by various agencies. Three problems need to be addressed within this approach. First, it is a choice of a relevant portfolio performance measure. Unfortunately, the Sharpe ratio (Sh), being a classical portfolio performance criterion, can monotonically decline with increasing PESGV due to small correlations between returns and PESGV. Schmidt [27] proposed using a synthetic performance measure, the *ESG-tilted Sharpe ratio*

$$Sh\_ESG = Sh(1 + \text{PESGV}) \qquad (1)$$

that may have maximums at some non-zero PESGVs and therefore can serve as a criterion for choosing an optimal ESG portfolio (OESGP).

Alessandrini and Jondeau [33] offered another performance measure with a mixing parameter γ:

$$Eff_p = (1 - \gamma)Sh + \gamma \text{PESGV}/\sigma_p. \qquad (2)$$

In equation (2), σ_p is portfolio volatility. Alessandrini and Jondeau [33] considered equal-weight portfolios rather than optimized Eff_p in terms of portfolio weights. Chen and Mussalli [6] used a combo of financial returns and the ESG asset ratings similar to equation (2) for optimizing the informational ratio of the unconstrained ESG portfolio.

Another problem related to deriving OESGP is that MVPs are highly concentrated [15]. This, along with high estimation errors of portfolio covariance matrix and expected returns [7], often yields inferior MVP performance with respect to equal-weight portfolios (EWPs) [9,12,29].

Nadler and Schmidt [22] offered a way to increase portfolio diversity by replacing Pearson's correlations with partial correlations conditioned on the state of the economy (mimicked for the US equity market by the S&P 500 index). The motivation for introducing partial correlations-based MVP

(PartMVP) as an alternative to Pearson-based correlations MVP (PearMVP) is that according to the capital asset pricing model, most of the individual equity asset returns follow the equity market trend, which increases their Pearson's correlations with each other. On the other hand, partial correlations conditioned on the market returns describe co-movements of the excess asset returns unaffected by the market momentum. As a result, partial correlations are lower than the corresponding Pearson correlations, and PartMVPs are more diversified than PearMVPs. It was found that PartMVPs may outperform PearMVPs and EWPs out of the sample [5,22]. Another advantage of PartMVP is that while the PearMVP weights change frequently and significantly over time, the PartMVP weights outside the bear market effects can be almost constant. It should be noted that the idea of replacing Pearson's correlations with other measures of price co-movements is not novel: DeMiguel et al. [10] and Gerber et al. [14] used implied volatility and a robust co-movement measure, respectively, instead of Pearson's correlations.

Finally, there is a problem with the choice of ESG ratings for individual companies that may differ significantly among their providers [4,8,11]. Avramov et al. [1] proposed treating the uncertainty of the ESG ratings as a noisy measurement problem. Schmidt [28] suggested that since the ESG ratings are not regulated and are vaguely defined as arbitrary combinations of various environmental, social, and corporate growth categories, the ratings' agencies should provide individual estimates of all ESG categories. Then, investors would be able to derive ESG ratings according to their own vision.

In this work, we conduct an analysis of how sensitive the OESGPs may be to the choice of the ESG ratings. Specifically, we compare the OESGPs derived with the ratings available from the MSCI, S&P Global (SPGI), and Sustainalytics (SUST) on their websites in January 2021. We use these ratings for estimating the OESGPs formed with the constituents of Dow Jones Index (DJI) in 2020. Since the SUST ESG ratings in public domains are available only for the current year, we started with the analysis of OESGPs using the data for 2020. In general, a one-year lookback period may be insufficient for the robust estimates of the MVP weights. Therefore, we also compared OESGPs derived with the MSCI and SPGI ratings averaged over the three-year period of 2018–2020.

In this work, we consider long-only portfolios. The optimization problem for these portfolios is not analytically tractable and hence requires numerical methods. The drawback of optimal portfolios with unconstrained asset weights is that they can have extreme (greater than 100%) long and short positions [18]. As a result, optimal unconstrained ESG portfolios can have short positions for underperforming (outperforming) companies with high (low) PESGV, which may be undesirable for investors. We consider both PearMVP and PartMVP for discussing the PESGV effects on portfolio diversity.

The mean–variance optimization problem for the ESG portfolios is formulated in the following section. In the following two sections, we describe the data used in this work and discuss the specifics of the OESGPs derived with various ESG ratings.

The Framework for an Optimal ESG Portfolio

The classical MVP derivation is based on minimizing the objective function [13]:

$$U_{\text{MVP}} = 0.5\lambda\sigma_p^2 - r_p. \tag{3}$$

In equation (3), λ is the risk aversion parameter that is usually chosen for yielding either minimum variance portfolio or maximum Sharpe portfolio. We use the latter in this work. Portfolio variance σ_p^2 equals

$$\sigma_p^2 = \sum_{i,j=1}^{N} w_i w_j \sigma_{ij}, \tag{4}$$

where w_i are portfolio weights, $i = 1, 2, \ldots, N$; $\sum_{i=1}^{N} w_i = 1$, and σ_{ij} are the covariances between asset returns, r_p is the mean portfolio return

$$r_p = \sum_{i=1}^{N} w_i r_i, \tag{5}$$

and r_i are the mean portfolio asset returns. The Sharpe ratio is defined as follows:

$$Sh = (r_p - r_f)/\sigma_p, \tag{6}$$

where r_f is the risk-free rate of return.

For OESGPs, we expand U_{MVP} with PESGV:

$$U_{\text{ESG-MVP}} = U_{\text{MVP}} + \gamma \text{PESGV}, \tag{7}$$

where γ is the ESG strength parameter, which is the investors' choice depending on how important for them is PESGV. We assume that the PESGV is a linear function of the portfolio constituents' ESG ratings δ_i:

$$\text{PESGV} = \sum_{i=1}^{N} w_i \delta_i. \tag{8}$$

Then, the derivation of an optimal long-only ESG portfolio represents the following quadratic programming problem:

$$\min \left[0.5\lambda \sum_{i,j=1}^{N} w_i w_j \sigma_{ij} - \gamma \sum_{i=1}^{N} \delta_i w_i \right], \tag{9}$$

$$\text{s.t.} \sum_{i=1}^{N} w_i r_i = r_p, \tag{10}$$

$$\sum_{i=1}^{N} w_i = 1, \tag{11}$$

$$w_i \geq 0; \quad i = 1, 2, \ldots, N. \tag{12}$$

As was indicated in the introduction, the covariance matrix can be calculated using partial correlations rather than Pearson's correlations. The partial correlation coefficient, $\rho_{ij|k}$, between variables X_i and X_j that is conditioned on variable X_k measures the correlation between residuals of linear regressions of X_i on X_k and X_j on X_k [19]. It can be introduced via partial covariance [30]:

$$\sigma_{ij|k} \equiv \text{cov}(X_i, X_j | X_k) = \text{cov}(X_i, X_j) - \text{cov}(X_i, X_k)\text{cov}(X_j, X_k)/\text{var}(X_k)$$
$$= \rho_{ij|k} \sigma_{i|k} \sigma_{j|k}. \tag{13}$$

In equation (13), partial variance $\sigma_{i|k}^2 \equiv \text{cov}(X_i, X_i | X_k)$ equals

$$\sigma_{i|k}^2 = \sigma_i^2 - \sigma_{ik}^4 / \sigma_k^2. \tag{14}$$

The partial correlation coefficient can be calculated using Pearson's correlations:

$$\rho_{ij|k} = \frac{\rho_{ij} - \rho_{ik}\rho_{jk}}{\sqrt{1 - \rho_{ik}^2}\sqrt{1 - \rho_{jk}^2}}. \tag{15}$$

Portfolio variance in terms of partial covariance equals

$$\sigma_p^2 = \sum_{i,j=1}^{N} w_i w_j \sigma_{ij|k}. \tag{16}$$

Hence, for the derivation of PartMVP, equation (16) should be used instead of equation (4).

The choice of the OESGP is implemented in two steps: First, the maximum Sharpe portfolios are estimated using the minimization protocol (9)–(12) for various values of the ESG strength parameter γ. Then, the OESGP that yields the maximum value of the ESG-tilted Sharpe ratio $Sh\_ESG(1)$ is chosen.

The Data

We considered a portfolio of 28 stocks that constituted DJI in December 2020. Two DJI constituents, Dow Chemical and United Technologies, for which prices prior to their mergers with other companies were not available, were excluded from the list. We used adjusted daily closing prices from finance.yahoo.com. Risk-free rate included in the definition of the Sharpe ratio (6) was neglected in this work. Partial correlations between the portfolio constituents were estimated with respect to returns of the SPDR S&P 500 ETF (SPY).

In order to use similar scaling of the ESG ratings from different providers, we transformed them so that they all were in the same range [0, 1]. Since the original SPGI ratings are in the range [0, 100], they were divided by 100. We mapped the original MSCI seven-level literal ratings on a linear grid and scaled them with the maximum value of seven (see Table 1).

The original SUST ratings define the ESG risk within the range [0, 100] rather than the ESG value. Therefore, we used the inverse SUST risks as the corresponding ESG values.

Figures 1 and 2 illustrate the following specifics of various ESG ratings. First, the low granularity of the MSCI ratings yields the same ratings for multiple companies.

Second, the typical inverse SUST ESG risks are by the order of magnitude lower than the MSCI and SPGI ESG ratings. There are also dramatic differences between the MSCI and SPGI ESG ratings even though they all are in the same range. For example, the MSCI rating for Johnson and Johnson (JNJ) is about twice lower than the SPGI rating. On the other hand, the

Table 1. MSCI ESG ratings.

MSCI ratings	CCC	BBB	BB	B	A	AA	AAA
Linear grid	1	2	3	4	5	6	7
Scaled scores	0.143	0.286	0.429	0.571	0.714	0.857	1

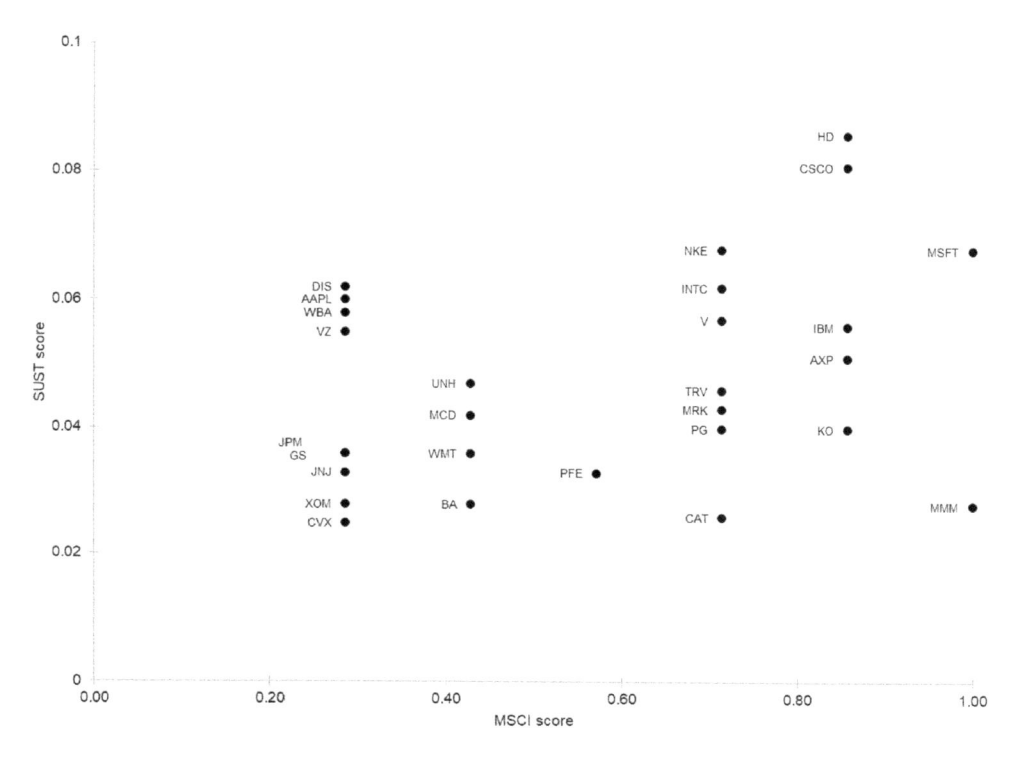

Figure 1. The MSCI and SUST ESG ratings in 2020.

MSCI rating of Home Depot (HD) is about four times higher than the SPGI rating.

Results and Discussion

All weights and performance details of the maximum Sharpe OESGPs formed with the DJI constituents and calibrated with the data for 2020 are listed in Tables A1–A3 for the MSCI, SPGI, and SUST ratings, respectively. The same results for the OESGPs calibrated within the three-year period of 2018–2020 are listed in Tables A4 and A5 for the MSCI and SPGI ratings, respectively.

For zero ESG strength ($\gamma = 0$), PearMVP for 2020 (2018–2020) has only three (five) weights higher than 0.1% out of 28 DJI constituents. On the other hand, PartMVP for 2020 (2018–2020) has 26 (27) weights higher than 0.1% (see Figures 3–7 for the MSCI, SPGI, and SUST ratings, respectively). With increasing ESG strength, both PearMVP and PartMVP become less diversified regardless of the source of the ESG ratings. In fact, the advantage

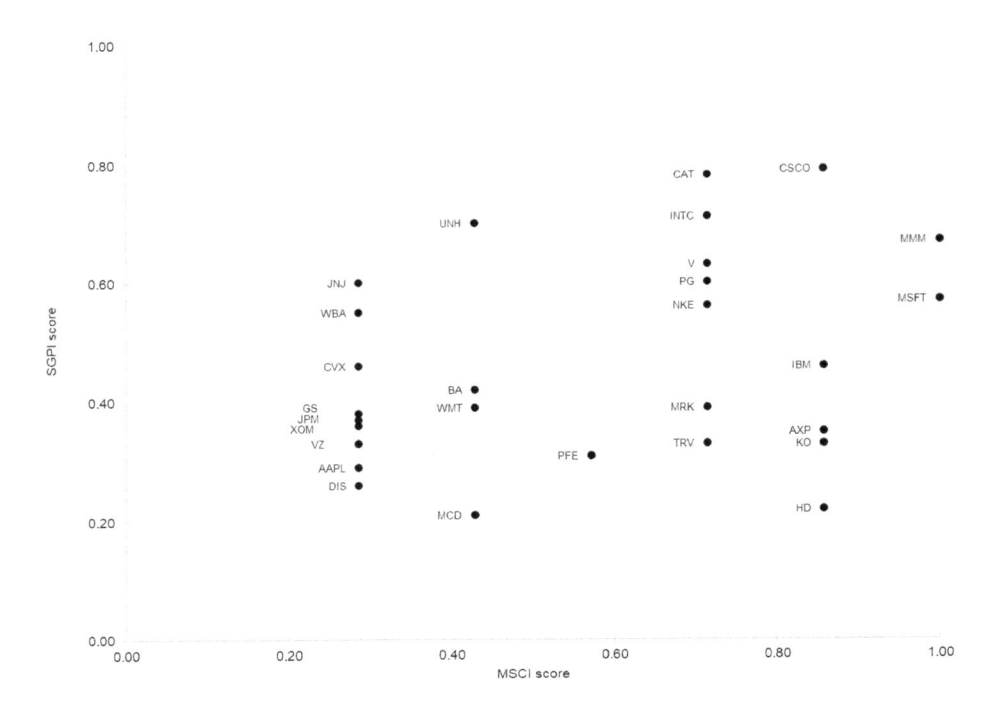

Figure 2. The MSCI and SPGI ESG ratings in 2020.

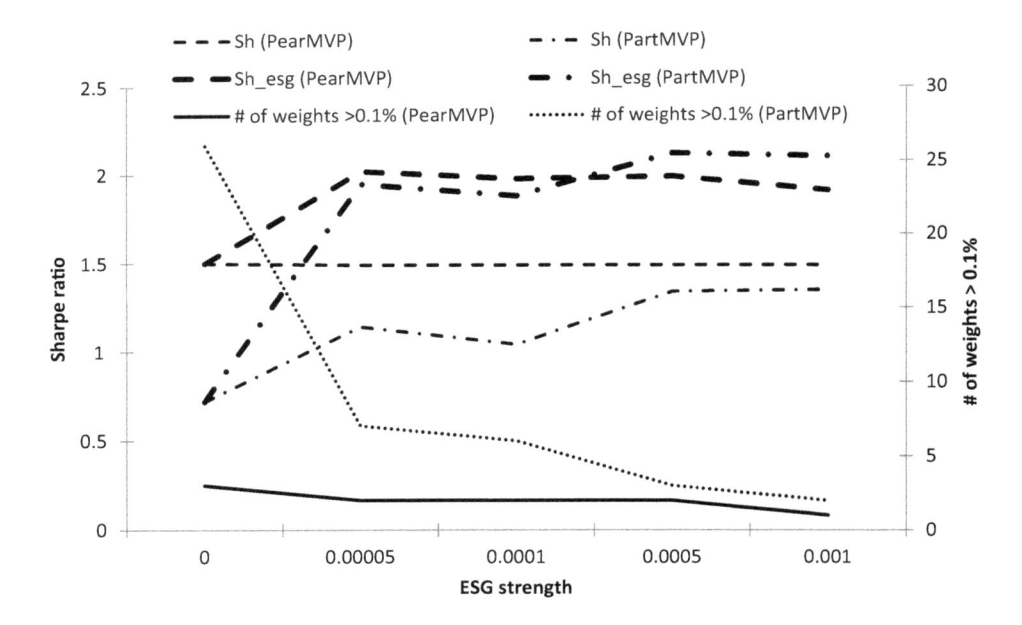

Figure 3. OESGP performance for the MSCI ESG ratings in 2020.

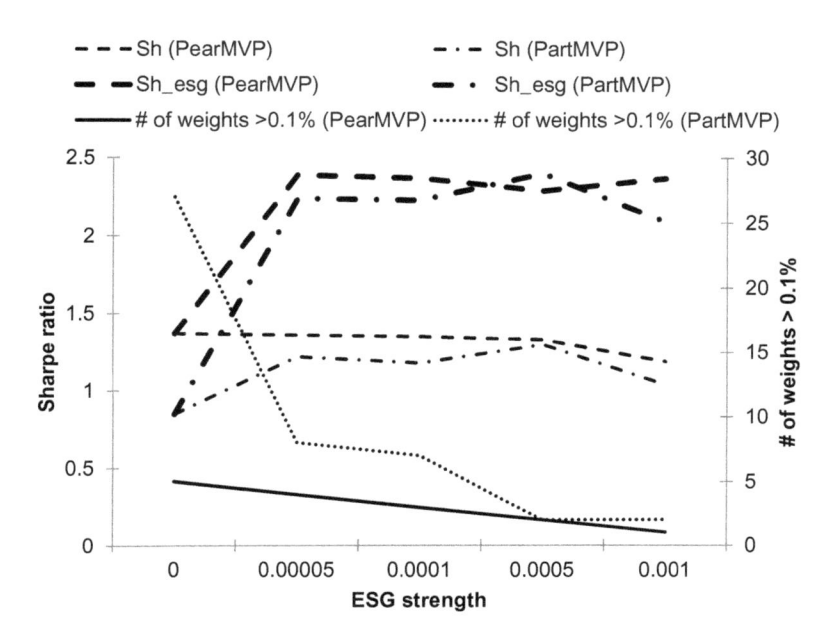

Figure 4. OESGP performance for the MSCI ESG ratings in 2018–2020.

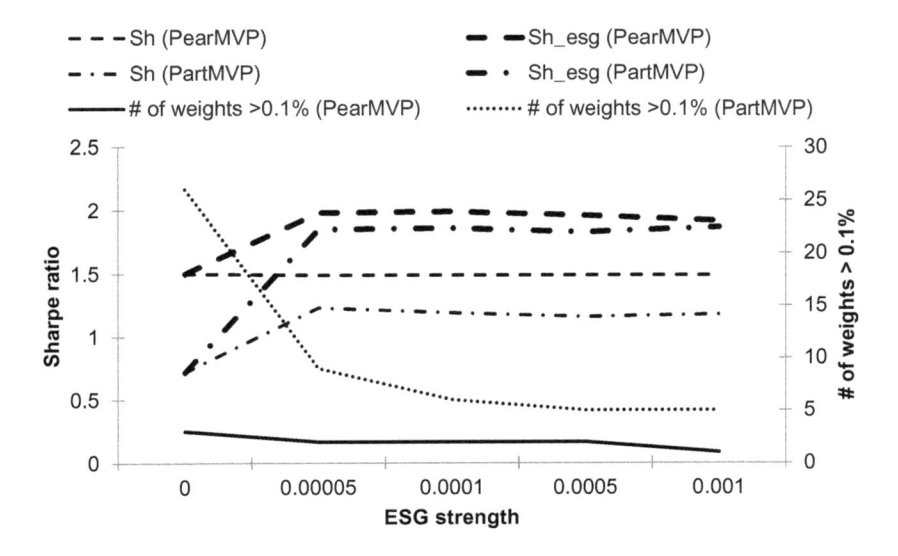

Figure 5. OESGP performance for the SPGI ESG ratings in 2020.

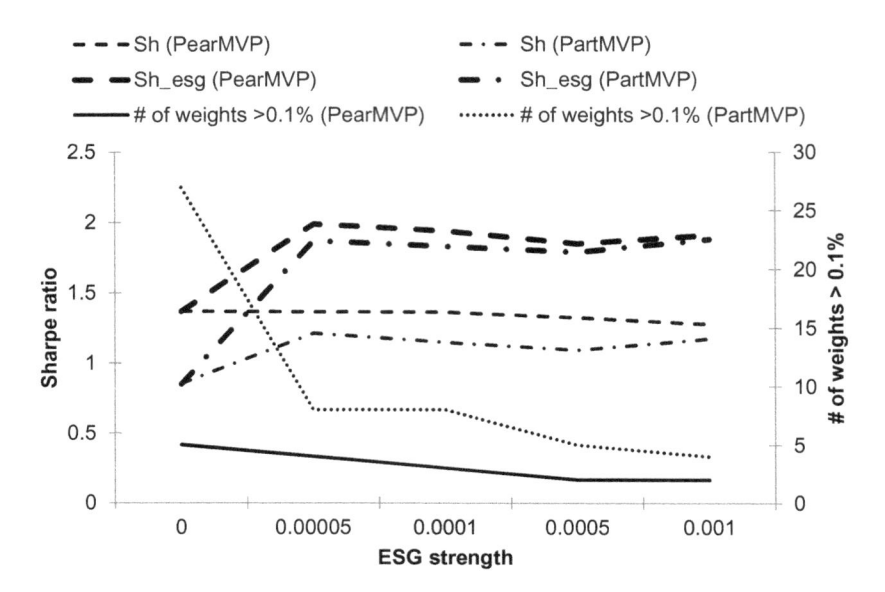

Figure 6. OESGP performance for the SPGI ESG ratings in 2018–2020.

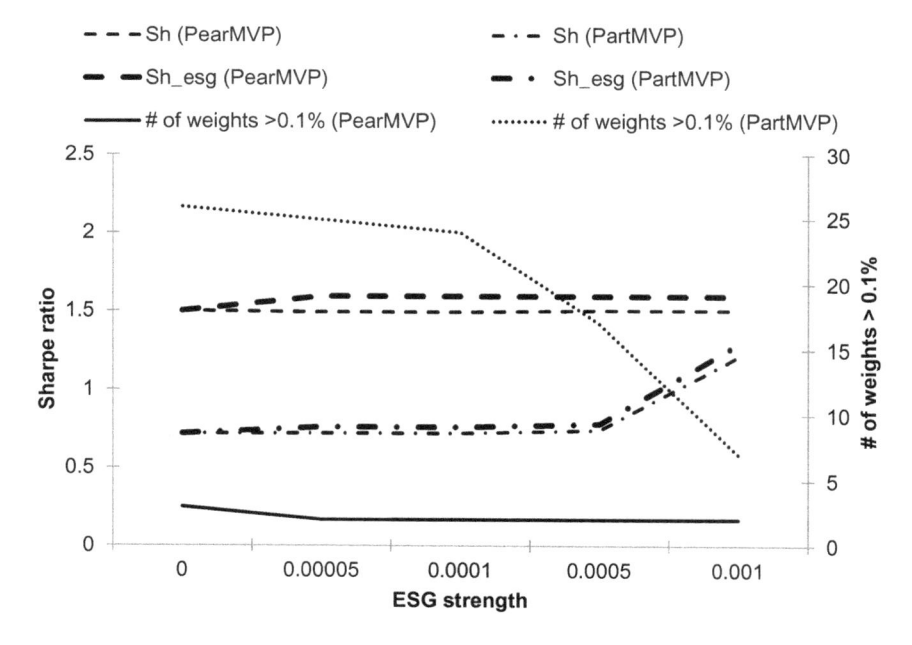

Figure 7. OESGP performance for the SUST ESG ratings in 2020.

of the PartMVP over PearMVP in terms of portfolio diversity that manifests for ESG-neutral portfolios may disappear. The OESGP Sharpe ratios (Sh) declined with increasing ESG strength in 2018–2020 but remained flat in 2020. This is caused by that the correlations between the DJI constituents' total returns and their ESG ratings are very low (see Tables A1–A5). However, the dependencies of the ESG-tilted Sharpe ratio (Sh_ESG) on the ESG strength are non-monotonic. There may be two types of maximums in these dependencies. One maximum corresponds to moderate γ and yields moderately diversified OESGPs. Another maximum may appear at higher values of γ if the portfolio of interest has one or two securities with lucky combinations of high returns and high ESG ratings. This maximum can be higher than the 1st maximum.

In 2020, PearMVPs with the MSCI, SPGI, and SUST ratings had only maximums of the 1st type. On the other hand, PartMVPs with the MSCI and SPGI ratings had maximums of both types while the SUST ratings yielded a maximum of the 2nd type.

In 2018–2020, both PearMVPs and PartMVPs with the MSCI and SPGI ratings had maximums of both types. The relevant portfolio weights are listed in Table 2. While there is some overlap between the OESGP constituents defined with the MSCI and SPGI ratings (particularly, AAPL, MSFT, NKE, and PG), generally their weights vary significantly.

It should be noted that even when the 2nd maximum of Sh_ESG is higher than its 1st maximum, investors may choose a more diversified portfolio for mitigating future market risk. The same rule can be used for selecting between PartMVPs and PearMVPs. Ultimately, it is the performance out of the sample that is a decisive criterion for choosing one OESGP over another.

Unfortunately, our results do not lead to straightforward conclusions over the preference of one source of the ESG ratings over the others. Yet we can offer the following comments for addressing this problem.

Usage of the inverse SUST risks as the proxies to the ESG ratings yields a notably lower sensitivity of the OESGP performance to the ESG strength parameter in comparison with the MSCI and SPGI ratings.

The advantage of the MSCI and SPGI ratings is that they are freely available in public domains for the last five years while the SUST risks are offered only for the current year. Moreover, besides the SPGI ESG ratings, their E, S, and G components are available in public domains too. This permits constructing customized ESG ratings with different weightings of their components.

Table 2. The OESGP weights in 2018–2020.

Portfolio	Scores	Maximum	ESG strength	Sh_esg	AAPL	AXP	CAT	CSCO	HD	JPM	KO	MSFT	NKE	PG	UNH	V
PaMVP	MSCI	1st	5.0E-05	2.24	17.9	12.1	1.4	0.0	2.0	0.0	0.5	45.8	15.2	5.1	0.0	0.0
PeMVP	MSCI	1st	5.0E-05	2.39	37.7	0.0	0.0	0.0	0.0	0.0	0.0	27.4	22.4	12.5	0.0	0.0
PaMVP	MSCI	2nd	5.0E-04	2.40	36.3	0.0	0.0	0.0	0.0	0.0	0.0	63.7	0.0	0.0	0.0	0.0
PeMVP	MSCI	2nd	1.0E-03	2.37	0.0	0.0	0.0	0.0	0.0	0.0	0.0	100.0	0.0	0.0	0.0	0.0
PaMVP	SPGI	1st	5.0E-05	1.87	17.5	0.0	6.8	0.0	0.0	2.5	0.0	32.4	16.2	5.1	7.8	11.5
PeMVP	SPGI	1st	5.0E-05	1.99	39.6	0.0	0.0	0.0	0.0	0.0	0.0	20.9	25.7	13.7	0.0	0.0
PaMVP	SPGI	2nd	1.0E-03	1.88	0.0	0.0	0.0	2.5	0.0	0.0	0.0	73.2	9.3	0.0	14.9	0.0
PeMVP	SPGI	2nd	1.0E-03	1.91	26.7	0.0	0.0	0.0	0.0	0.0	0.0	73.3	0.0	0.0	0.0	0.0

For better or worse, the granularity of the scaled MSCI ratings of $1/7 \approx 0.143$ is much lower than that of the SPGI-scaled ratings of 0.01. We refrain from characterizing the significance of this specific without an analysis of the OESGP sensitivity to small changes in the ESG ratings.

Significant discrepancies among the ESG ratings from different providers for the same securities need to be resolved before using them with confidence for deriving OESGPs that contain securities from various equity sectors. We suppose that these discrepancies may be lower or at least more tractable for securities from the same equity sectors. Hence, deriving the OESGPs based on single equity sectors may be a useful development in this field. However, we believe that the most promising approach in making sense of the ESG ratings is their analysis on the level of the individual ESG categories. The Sustainability Accounting Standards Board (SASB) lists 26 ESG categories, and the ratings' agencies use their proprietary solutions for estimating and mixing these categories into the environmental, social, and corporate growth pillars and weighting the pillars into the composite ESG ratings. Hence, short of strict regulation, one cannot expect that the ESG ratings offered by various providers will converge any time soon. Therefore, investors may want to use the estimates of the individual corporate ESG categories (possibly comparing the data from several providers) and derive the composite ESG scores according to their own vision.

Appendix

Table A1. Maximum Sharpe OESGP for the DJI with the MSCI ratings (2020).

ticker	ESG score	ESG strength Return	0 Pearson	0 Partial	5E-05 Pearson	5E-05 Partial	1E-04 Pearson	1E-04 Partial	5E-04 Pearson	5E-04 Partial	0.001 Pearson	0.001 Partial	0.002 Pearson	0.002 Partial
AAPL	0.29	79.6%	88.3%	12.5%	84.9%	25.0%	90.2%	11.5%	88.8%	56.7%	100.0%	62.0%	0.0%	0.0%
AXP	0.86	−3.7%	0.0%	0.3%	0.0%	7.6%	0.0%	8.6%	0.0%	0.0%	0.0%	0.0%	0.0%	0.0%
BA	0.43	−34.6%	0.0%	1.1%	0.0%	0.0%	0.0%	0.0%	0.0%	0.0%	0.0%	0.0%	0.0%	0.0%
CAT	0.71	23.2%	0.0%	3.3%	0.0%	15.1%	0.0%	13.9%	0.0%	0.0%	0.0%	0.0%	0.0%	0.0%
CSCO	0.86	−5.7%	0.0%	1.7%	0.0%	0.0%	0.0%	0.0%	0.0%	0.0%	0.0%	0.0%	0.0%	0.0%
CVX	0.29	−25.7%	0.0%	0.0%	0.0%	0.0%	0.0%	0.0%	0.0%	0.0%	0.0%	0.0%	0.0%	0.0%
DIS	0.29	22.2%	0.0%	2.2%	0.0%	0.0%	0.0%	0.0%	0.0%	0.0%	0.0%	0.0%	0.0%	0.0%
GS	0.29	13.4%	0.0%	2.8%	0.0%	0.0%	0.0%	0.0%	0.0%	0.0%	0.0%	0.0%	0.0%	0.0%
HD	0.86	23.6%	0.0%	6.4%	0.0%	10.4%	0.0%	11.3%	0.0%	0.0%	0.0%	0.0%	0.0%	0.0%
IBM	0.86	−3.4%	0.0%	6.0%	0.0%	0.0%	0.0%	0.0%	0.0%	0.0%	0.0%	0.0%	0.0%	0.0%
INTC	0.71	−17.9%	0.0%	1.1%	0.0%	0.0%	0.0%	0.0%	0.0%	0.0%	0.0%	0.0%	0.0%	0.0%
JNJ	0.29	9.8%	0.0%	2.0%	0.0%	0.0%	0.0%	0.0%	0.0%	0.0%	0.0%	0.0%	0.0%	0.0%
JPM	0.29	−7.9%	0.0%	7.7%	0.0%	0.0%	0.0%	0.0%	0.0%	0.0%	0.0%	0.0%	0.0%	0.0%
KO	0.86	2.4%	0.0%	3.7%	0.0%	0.0%	0.0%	0.0%	0.0%	0.0%	0.0%	0.0%	0.0%	0.0%
MCD	0.43	8.0%	0.0%	0.5%	0.0%	0.0%	0.0%	0.0%	0.0%	0.0%	0.0%	0.0%	0.0%	0.0%
MMM	1.00	0.3%	0.0%	4.4%	0.0%	0.0%	0.0%	0.0%	0.0%	0.0%	0.0%	0.0%	0.0%	32.8%
MRK	0.71	−9.7%	0.0%	13.9%	0.0%	27.9%	0.0%	41.3%	0.0%	41.2%	0.0%	0.0%	0.0%	0.0%
MSFT	1.00	39.5%	0.0%	2.9%	0.0%	13.8%	0.0%	13.3%	0.0%	2.1%	0.0%	38.0%	100.0%	67.2%
NKE	0.71	39.9%	11.5%	1.2%	15.1%	0.0%	9.8%	0.0%	11.2%	0.0%	0.0%	0.0%	0.0%	0.0%
PFE	0.57	3.0%	0.0%	5.4%	0.0%	0.0%	0.0%	0.0%	0.0%	0.0%	0.0%	0.0%	0.0%	0.0%
PG	0.71	14.4%	0.0%	2.6%	0.0%	0.2%	0.0%	0.0%	0.0%	0.0%	0.0%	0.0%	0.0%	0.0%
TRV	0.71	3.9%	0.0%	3.9%	0.0%	0.0%	0.0%	0.0%	0.0%	0.0%	0.0%	0.0%	0.0%	0.0%
UNH	0.43	19.9%	0.0%	8.2%	0.0%	0.0%	0.0%	0.0%	0.0%	0.0%	0.0%	0.0%	0.0%	0.0%
V	0.71	15.0%	0.0%	1.5%	0.0%	0.0%	0.0%	0.0%	0.0%	0.0%	0.0%	0.0%	0.0%	0.0%
VZ	0.29	−0.6%	0.0%	0.8%	0.0%	0.0%	0.0%	0.0%	0.0%	0.0%	0.0%	0.0%	0.0%	0.0%
WBA	0.29	−30.5%	0.0%	0.8%	0.0%	0.0%	0.0%	0.0%	0.0%	0.0%	0.0%	0.0%	0.0%	0.0%
WMT	0.43	23.2%	0.2%	3.1%	0.0%	0.0%	0.0%	0.0%	0.0%	0.0%	0.0%	0.0%	0.0%	0.0%
XOM	0.29	−36.6%	0.0%		0.0%	0.0%	0.0%	0.0%	0.0%	0.0%	0.0%	0.0%	0.0%	0.0%
# of weights >= 0.1%			3	26	2	7	2	6	2	3	1	2	1	2
Ann volatility			44.3%	35.3%	43.7%	38.9%	44.7%	39.0%	44.5%	43.4%	46.7%	43.9%	43.9%	37.5%
Ann return			66.3%	25.4%	65.4%	44.3%	66.8%	40.6%	66.5%	58.0%	69.5%	59.4%	42.9%	31.2%
Ann Sh			1.50	0.72	1.49	1.14	1.49	1.04	1.49	1.34	1.49	1.35	0.98	0.83
PESGV			0.00	0.00	0.35	0.71	0.33	0.81	0.33	0.59	0.29	0.56	1.00	1.00
Ann Sh_ESG			1.50	0.72	2.02	1.95	1.98	1.88	1.99	2.12	1.91	2.11	1.95	1.66
Return-ESG score corr	0.12													

Table A2. Maximum Sharpe OESGP for the DJI with the SPGI ratings (2020).

ticker	ESG strength score	ESG Return correlation	0 Pearson	0 Partial	5E-05 Pearson	5E-05 Partial	1E-04 Pearson	1E-04 Partial	5E-04 Pearson	5E-04 Partial	0.001 Pearson	0.001 Partial	0.002 Pearson	0.002 Partial	0.003 Pearson	0.003 Partial	0.004 Pearson	0.004 Partial	0.005 Pearson	0.005 Partial
AAPL	0.29	79.6%	88.3%	12.5%	86.1%	32.7%	84.0%	26.4%	92.9%	23.8%	100.0%	31.9%	100.0%	64.7%	52.9%	60.8%	0.0%	0.0%	0.0%	0.0%
AXP	0.35	−3.7%	0.0%	0.3%	0.0%	1.6%	0.0%	0.0%	0.0%	0.0%	0.0%	0.0%	0.0%	0.0%	0.0%	0.0%	0.0%	0.0%	0.0%	0.0%
BA	0.42	−34.6%	0.0%	1.1%	0.0%	0.0%	0.0%	0.0%	0.0%	0.0%	0.0%	0.0%	0.0%	0.0%	0.0%	0.0%	0.0%	0.0%	0.0%	0.0%
CAT	0.78	23.2%	0.0%	3.3%	0.0%	19.9%	0.0%	24.5%	0.0%	33.8%	0.0%	42.8%	0.0%	35.3%	47.1%	39.2%	98.4%	90.5%	100.0%	100.0%
CSCO	0.79	−5.7%	0.0%	1.7%	0.0%	0.0%	0.0%	0.0%	0.0%	0.0%	0.0%	0.0%	0.0%	0.0%	0.0%	0.0%	0.0%	0.0%	0.0%	0.0%
CVX	0.46	−25.7%	0.0%	0.0%	0.0%	0.0%	0.0%	0.0%	0.0%	0.0%	0.0%	0.0%	0.0%	0.0%	0.0%	0.0%	0.0%	0.0%	0.0%	0.0%
DIS	0.26	22.2%	0.0%	2.2%	0.0%	2.8%	0.0%	0.0%	0.0%	0.0%	0.0%	0.0%	0.0%	0.0%	0.0%	0.0%	0.0%	0.0%	0.0%	0.0%
GS	0.38	13.4%	0.0%	2.8%	0.0%	0.0%	0.0%	0.0%	0.0%	0.0%	0.0%	0.0%	0.0%	0.0%	0.0%	0.0%	0.0%	0.0%	0.0%	0.0%
HD	0.22	23.6%	0.0%	6.4%	0.0%	0.3%	0.0%	0.0%	0.0%	0.0%	0.0%	0.0%	0.0%	0.0%	0.0%	0.0%	0.0%	0.0%	0.0%	0.0%
IBM	0.46	−3.4%	0.0%	6.0%	0.0%	0.0%	0.0%	0.0%	0.0%	0.0%	0.0%	0.0%	0.0%	0.0%	0.0%	0.0%	0.0%	0.0%	0.0%	0.0%
INTC	0.71	−17.9%	0.0%	1.1%	0.0%	0.0%	0.0%	0.0%	0.0%	0.0%	0.0%	0.0%	0.0%	0.0%	0.0%	0.0%	0.0%	0.0%	0.0%	0.0%
JNJ	0.60	9.8%	0.0%	0.0%	0.0%	0.0%	0.0%	0.0%	0.0%	0.0%	0.0%	0.0%	0.0%	0.0%	0.0%	0.0%	0.0%	0.0%	0.0%	0.0%
JPM	0.37	−7.9%	0.0%	2.0%	0.0%	0.0%	0.0%	0.0%	0.0%	0.0%	0.0%	0.0%	0.0%	0.0%	0.0%	0.0%	0.0%	0.0%	0.0%	0.0%
KO	0.33	2.4%	0.0%	7.7%	0.0%	0.0%	0.0%	0.0%	0.0%	0.0%	0.0%	0.0%	0.0%	0.0%	0.0%	0.0%	0.0%	0.0%	0.0%	0.0%
MCD	0.21	8.0%	0.0%	3.7%	0.0%	0.0%	0.0%	0.0%	0.0%	0.0%	0.0%	0.0%	0.0%	0.0%	0.0%	0.0%	0.0%	0.0%	0.0%	0.0%
MMM	0.67	0.3%	0.0%	0.5%	0.0%	0.0%	0.0%	0.0%	0.0%	0.0%	0.0%	0.0%	0.0%	0.0%	0.0%	0.0%	0.0%	0.0%	0.0%	0.0%
MRK	0.39	−9.7%	0.0%	4.4%	0.0%	0.0%	0.0%	0.0%	0.0%	0.0%	0.0%	0.0%	0.0%	0.0%	0.0%	0.0%	0.0%	0.0%	0.0%	0.0%
MSFT	0.57	39.5%	0.0%	13.9%	0.0%	19.1%	0.0%	23.0%	0.0%	24.6%	19.3%	0.0%	0.0%	0.0%	0.0%	0.0%	0.0%	0.0%	0.0%	0.0%
NKE	0.56	39.9%	11.5%	2.9%	13.8%	14.4%	16.0%	14.5%	7.1%	8.8%	1.9%	0.0%	0.0%	0.0%	0.0%	0.0%	1.6%	9.5%	0.0%	0.0%
PFE	0.31	3.0%	0.0%	1.2%	0.0%	0.0%	0.0%	0.0%	0.0%	0.0%	0.0%	0.0%	0.0%	0.0%	0.0%	0.0%	0.0%	0.0%	0.0%	0.0%
PG	0.60	14.4%	0.0%	5.4%	0.0%	0.0%	0.0%	0.0%	0.0%	0.0%	0.0%	0.0%	0.0%	0.0%	0.0%	0.0%	0.0%	0.0%	0.0%	0.0%
TRV	0.33	3.9%	0.0%	2.6%	0.0%	2.1%	0.0%	0.0%	0.0%	0.0%	0.0%	0.0%	0.0%	0.0%	0.0%	0.0%	0.0%	0.0%	0.0%	0.0%
UNH	0.70	19.9%	0.0%	3.9%	0.0%	7.1%	0.0%	9.5%	0.0%	8.9%	0.0%	4.1%	0.0%	0.0%	0.0%	0.0%	0.0%	0.0%	0.0%	0.0%
V	0.63	15.0%	0.0%	8.2%	0.0%	0.0%	0.0%	2.0%	0.0%	0.0%	0.0%	0.0%	0.0%	0.0%	0.0%	0.0%	0.0%	0.0%	0.0%	0.0%
VZ	0.33	−0.6%	0.0%	1.5%	0.0%	0.0%	0.0%	0.0%	0.0%	0.0%	0.0%	0.0%	0.0%	0.0%	0.0%	0.0%	0.0%	0.0%	0.0%	0.0%
WBA	0.55	−30.5%	0.0%	0.8%	0.0%	0.0%	0.0%	0.0%	0.0%	0.0%	0.0%	0.0%	0.0%	0.0%	0.0%	0.0%	0.0%	0.0%	0.0%	0.0%
WMT	0.39	23.2%	0.2%	0.8%	0.1%	0.0%	0.0%	0.0%	0.0%	0.0%	0.0%	0.0%	0.0%	0.0%	0.0%	0.0%	0.0%	0.0%	0.0%	0.0%
XOM	0.36	−36.6%	0.0%	3.1%	0.0%	0.0%	0.0%	0.0%	0.0%	0.0%	0.0%	0.0%	0.0%	0.0%	0.0%	0.0%	0.0%	0.0%	0.0%	0.0%
# of weights >= 0.1%			3	26	2	9	2	6	2	5	1	5	1	2	2	2	2	2	1	1
Ann volatility			44.3%	35.3%	43.9%	38.2%	43.6%	37.9%	45.2%	38.0%	46.7%	38.6%	46.7%	40.6%	39.8%	40.2%	44.6%	43.0%	44.9%	44.9%
Ann return			66.3%	25.4%	65.7%	46.9%	65.1%	45.3%	67.6%	44.0%	69.5%	45.8%	69.5%	55.9%	51.4%	54.4%	31.2%	32.2%	31.0%	31.0%
Ann Sh			1.50	0.72	1.49	1.23	1.49	1.19	1.49	1.16	1.49	1.18	1.49	1.38	1.29	1.35	0.70	0.75	0.69	0.69
PESGV			0.00	0.00	0.33	0.51	0.33	0.56	0.31	0.59	0.29	0.58	0.29	0.46	0.52	0.48	0.78	0.76	0.78	0.78
Ann Sh.ESG			1.50	0.72	1.98	1.85	1.99	1.86	1.96	1.83	1.92	1.87	1.92	2.02	1.96	2.01	1.24	1.32	1.23	1.23
Return-ESG score corr	−0.05																			

Table A3. Maximum Sharpe OESGP for the DJI with the SUST ratings (2020).

ESG strength		0		5E-05		1E-04		5E-04		0.001		0.01		0.02		0.03		0.04		0.05		0.06		0.07		0.1	
ticker / correlation	ESG score	Return	Pearson	Partial	Pearson	Partial	Pearson	Partial	Pearson	Partial	Pearson	Partial	Pearson	Partial	Pearson	Partial	Pearson	Partial	Pearson	Partial	Pearson	Partial	Pearson	Partial	Pearson	Partial	Pearson	Partial
AAPL	0.06	79.6%	88.3%	0.0%	12.5%	0.0%	88.3%	0.0%	12.4%	0.0%	87.8%	11.3%	87.2%	0.0%	34.9%	85.3%	0.0%	51.7%	100.0%	61.8%	55.6%	65.1%	58.9%	37.1%	0.0%	0.0%	0.0%	0.0%
AXP	0.05	−3.7%	0.0%	0.3%	1.1%	0.7%	1.0%	0.0%	1.2%	0.8%	1.2%	4.0%	0.0%	4.5%	0.0%	0.0%	0.0%	0.0%	0.0%	0.0%	0.0%	0.0%	0.0%	0.0%	0.0%	0.0%	0.0%	0.0%
BA	0.03	−34.6%	0.0%	1.1%	0.0%	1.0%	0.0%	0.0%	0.8%	0.0%	0.0%	0.0%	0.0%	0.0%	0.0%	0.0%	0.0%	0.0%	0.0%	0.0%	0.0%	0.0%	0.0%	0.0%	0.0%	0.0%	0.0%	0.0%
CAT	0.03	23.2%	0.0%	3.3%	0.0%	2.9%	0.0%	2.4%	0.0%	0.0%	0.0%	0.0%	0.0%	0.0%	0.0%	0.0%	0.0%	0.0%	0.0%	0.0%	0.0%	0.0%	0.0%	0.0%	0.0%	0.0%	0.0%	0.0%
CSCO	0.08	−5.7%	0.0%	1.7%	0.0%	2.5%	0.0%	3.3%	0.0%	7.3%	0.0%	4.6%	0.0%	0.0%	0.0%	0.0%	0.0%	0.0%	0.0%	0.0%	0.0%	0.0%	0.0%	0.0%	0.0%	0.0%	0.0%	0.0%
CVX	0.02	−25.7%	0.0%	0.0%	0.0%	0.0%	0.0%	0.0%	0.0%	0.0%	0.0%	0.0%	0.0%	0.0%	0.0%	0.0%	0.0%	0.0%	0.0%	0.0%	0.0%	0.0%	0.0%	0.0%	0.0%	0.0%	0.0%	0.0%
DIS	0.06	22.2%	0.0%	2.2%	0.0%	2.2%	0.0%	2.3%	0.0%	2.3%	0.0%	6.9%	0.0%	0.0%	0.0%	0.0%	0.0%	0.0%	0.0%	0.0%	0.0%	0.0%	0.0%	0.0%	0.0%	0.0%	0.0%	0.0%
GS	0.04	13.4%	0.0%	2.8%	0.0%	2.4%	0.0%	0.0%	0.0%	2.0%	0.0%	0.0%	0.0%	0.0%	0.0%	0.0%	0.0%	0.0%	0.0%	0.0%	0.0%	0.0%	0.0%	0.0%	0.0%	0.0%	0.0%	0.0%
HD	0.09	23.6%	0.0%	6.4%	0.0%	7.7%	0.0%	8.9%	0.0%	15.6%	0.0%	19.0%	0.0%	46.6%	0.0%	38.2%	0.0%	44.4%	0.0%	34.9%	0.0%	41.1%	0.0%	62.9%	100.0%	100.0%	100.0%	100.0%
IBM	0.06	−3.4%	0.0%	6.0%	0.0%	6.2%	0.0%	6.4%	0.0%	6.3%	0.0%	0.0%	0.0%	0.0%	0.0%	0.0%	0.0%	0.0%	0.0%	0.0%	0.0%	0.0%	0.0%	0.0%	0.0%	0.0%	0.0%	0.0%
INTC	0.06	−17.9%	0.0%	1.1%	0.0%	0.9%	0.0%	0.7%	0.0%	0.0%	0.0%	0.0%	0.0%	0.0%	0.0%	0.0%	0.0%	0.0%	0.0%	0.0%	0.0%	0.0%	0.0%	0.0%	0.0%	0.0%	0.0%	0.0%
JNJ	0.03	9.8%	0.0%	0.0%	0.0%	0.0%	0.0%	0.0%	0.0%	0.0%	0.0%	0.0%	0.0%	0.0%	0.0%	0.0%	0.0%	0.0%	0.0%	0.0%	0.0%	0.0%	0.0%	0.0%	0.0%	0.0%	0.0%	0.0%
JPM	0.04	−7.9%	0.0%	2.0%	0.0%	1.9%	0.0%	1.9%	0.0%	0.7%	0.0%	0.0%	0.0%	0.0%	0.0%	0.0%	0.0%	0.0%	0.0%	0.0%	0.0%	0.0%	0.0%	0.0%	0.0%	0.0%	0.0%	0.0%
KO	0.04	2.4%	0.0%	7.7%	0.0%	7.7%	0.0%	7.7%	0.0%	6.2%	0.0%	0.0%	0.0%	0.0%	0.0%	0.0%	0.0%	0.0%	0.0%	0.0%	0.0%	0.0%	0.0%	0.0%	0.0%	0.0%	0.0%	0.0%
MCD	0.04	8.0%	0.0%	3.7%	0.0%	2.7%	0.0%	1.5%	0.0%	0.0%	0.0%	0.0%	0.0%	0.0%	0.0%	0.0%	0.0%	0.0%	0.0%	0.0%	0.0%	0.0%	0.0%	0.0%	0.0%	0.0%	0.0%	0.0%
MMM	0.03	0.3%	0.0%	0.5%	0.0%	0.0%	0.0%	0.0%	0.0%	0.0%	0.0%	0.0%	0.0%	0.0%	0.0%	0.0%	0.0%	0.0%	0.0%	0.0%	0.0%	0.0%	0.0%	0.0%	0.0%	0.0%	0.0%	0.0%
MRK	0.04	−9.7%	0.0%	4.4%	0.0%	4.3%	0.0%	4.2%	0.0%	1.3%	0.0%	0.0%	0.0%	0.0%	0.0%	0.0%	0.0%	0.0%	0.0%	0.0%	0.0%	0.0%	0.0%	0.0%	0.0%	0.0%	0.0%	0.0%
MSFT	0.07	39.5%	0.0%	13.9%	0.0%	14.0%	0.0%	14.2%	0.0%	15.3%	0.0%	13.5%	0.0%	0.0%	0.0%	0.0%	0.0%	0.0%	0.0%	0.0%	0.0%	0.0%	0.0%	0.0%	0.0%	0.0%	0.0%	0.0%
NKE	0.07	39.9%	11.5%	2.9%	11.7%	3.5%	11.7%	4.1%	12.2%	7.2%	12.8%	16.6%	14.7%	1.6%	0.0%	0.0%	0.0%	0.0%	0.0%	0.0%	0.0%	0.0%	0.0%	0.0%	0.0%	0.0%	0.0%	0.0%
PFE	0.03	3.0%	0.0%	1.2%	0.0%	0.7%	0.0%	0.2%	0.0%	0.0%	0.0%	0.0%	0.0%	0.0%	0.0%	0.0%	0.0%	0.0%	0.0%	0.0%	0.0%	0.0%	0.0%	0.0%	0.0%	0.0%	0.0%	0.0%
PG	0.04	14.4%	0.0%	5.4%	0.0%	4.9%	0.0%	4.3%	0.0%	0.0%	0.0%	0.0%	0.0%	0.0%	0.0%	0.0%	0.0%	0.0%	0.0%	0.0%	0.0%	0.0%	0.0%	0.0%	0.0%	0.0%	0.0%	0.0%
TRV	0.05	3.9%	0.0%	2.6%	0.0%	2.6%	0.0%	2.5%	0.0%	1.6%	0.0%	0.0%	0.0%	0.0%	0.0%	0.0%	0.0%	0.0%	0.0%	0.0%	0.0%	0.0%	0.0%	0.0%	0.0%	0.0%	0.0%	0.0%
UNH	0.05	19.9%	0.0%	3.9%	0.0%	3.8%	0.0%	3.6%	0.0%	1.9%	0.0%	0.0%	0.0%	0.0%	0.0%	0.0%	0.0%	0.0%	0.0%	0.0%	0.0%	0.0%	0.0%	0.0%	0.0%	0.0%	0.0%	0.0%
V	0.06	15.0%	0.0%	8.2%	0.0%	8.4%	0.0%	8.6%	0.0%	8.1%	0.0%	0.0%	0.0%	0.0%	0.0%	0.0%	0.0%	0.0%	0.0%	0.0%	0.0%	0.0%	0.0%	0.0%	0.0%	0.0%	0.0%	0.0%
VZ	0.06	−0.6%	0.0%	1.5%	0.0%	2.2%	0.0%	2.9%	0.0%	6.8%	0.0%	0.0%	0.0%	0.0%	0.0%	0.0%	0.0%	0.0%	0.0%	0.0%	0.0%	0.0%	0.0%	0.0%	0.0%	0.0%	0.0%	0.0%
WBA	0.06	−30.5%	0.0%	0.8%	0.0%	1.0%	0.0%	1.2%	0.0%	2.3%	0.0%	0.0%	0.0%	0.0%	0.0%	0.0%	0.0%	0.0%	0.0%	0.0%	0.0%	0.0%	0.0%	0.0%	0.0%	0.0%	0.0%	0.0%
WMT	0.04	23.2%	0.2%	0.8%	0.0%	0.2%	0.0%	0.0%	0.0%	0.0%	0.0%	0.0%	0.0%	0.0%	0.0%	0.0%	0.0%	0.0%	0.0%	0.0%	0.0%	0.0%	0.0%	0.0%	0.0%	0.0%	0.0%	0.0%
XOM	0.03	−36.6%	0.0%	3.1%	0.0%	3.0%	0.0%	2.9%	0.0%	1.9%	0.0%	0.0%	0.0%	0.0%	0.0%	0.0%	0.0%	0.0%	0.0%	0.0%	0.0%	0.0%	0.0%	0.0%	0.0%	0.0%	0.0%	0.0%
# of weights >= 0.1%			3	26	2	25	2	24	2	17	2	7	2	3	1	2	2	2	2	2	2	2	1	1	1	1		
Ann volatility		44.3%	35.3%	44.4%	35.4%	44.3%	35.5%	44.3%	35.7%	44.1%	39.0%	43.8%	41.5%	46.7%	42.2%	41.9%	42.5%	42.0%	41.4%	43.6%	43.6%							
Ann return		66.3%	25.4%	66.3%	25.5%	66.3%	25.6%	66.2%	26.3%	66.0%	47.3%	65.5%	51.1%	69.5%	54.8%	52.4%	56.1%	53.7%	45.2%	30.9%	30.9%							
Ann Sh		1.50	0.72	1.49	0.72	1.49	0.72	1.50	0.74	1.50	1.21	1.49	1.23	1.49	1.30	1.25	1.32	1.28	1.09	0.71	0.71							
PESGV		0.00	0.00	0.06	0.06	0.06	0.06	0.06	0.06	0.06	0.07	0.06	0.07	0.06	0.07	0.07	0.07	0.07	0.08	0.09	0.09							
Ann Sh_ESG		1.50	0.72	1.59	0.76	1.59	0.76	1.59	0.78	1.59	1.29	1.59	1.32	1.58	1.39	1.34	1.41	1.37	1.18	0.77	0.77							
Return-ESG score corr	0.36																											

Table A4. Maximum Sharpe OESGP for the DJI with the MSCI ratings (2018-2020).

ESG strength ticker/correlation	ESG score	Return	0 Pearson	0 Partial	5E-05 Pearson	5E-05 Partial	1E-04 Pearson	1E-04 Partial	5E-04 Pearson	5E-04 Partial	0.001 Pearson	0.001 Partial	0.002 Pearson	0.002 Partial
AAPL	0.57	222.7%	46.9%	10.4%	37.7%	17.9%	46.7%	8.6%	62.7%	36.3%	0.0%	0.0%	0.0%	0.0%
AXP	0.86	26.5%	0.0%	0.0%	0.0%	12.1%	0.0%	13.4%	0.0%	0.0%	0.0%	0.0%	0.0%	0.0%
BA	0.43	−23.3%	0.0%	1.2%	0.0%	0.0%	0.0%	0.0%	0.0%	0.0%	0.0%	0.0%	0.0%	0.0%
CAT	0.71	24.2%	0.0%	3.8%	0.0%	1.4%	0.0%	0.0%	0.0%	0.0%	0.0%	0.0%	0.0%	0.0%
CSCO	0.76	25.4%	0.0%	2.0%	0.0%	0.0%	0.0%	0.0%	0.0%	0.0%	0.0%	0.0%	0.0%	0.0%
CVX	0.33	−23.6%	0.0%	0.4%	0.0%	0.0%	0.0%	0.0%	0.0%	0.0%	0.0%	0.0%	0.0%	0.0%
DIS	0.29	66.6%	0.0%	2.3%	0.0%	0.0%	0.0%	0.0%	0.0%	0.0%	0.0%	0.0%	0.0%	0.0%
GS	0.29	7.5%	0.0%	3.5%	0.0%	0.0%	0.0%	0.0%	0.0%	0.0%	0.0%	0.0%	0.0%	0.0%
HD	0.86	51.7%	0.0%	6.9%	0.0%	2.0%	0.0%	3.1%	0.0%	0.0%	0.0%	0.0%	0.0%	0.0%
IBM	0.86	−7.1%	0.0%	3.4%	0.0%	0.0%	0.0%	0.0%	0.0%	0.0%	0.0%	0.0%	0.0%	0.0%
INTC	0.71	12.1%	0.0%	1.6%	0.0%	0.0%	0.0%	0.0%	0.0%	0.0%	0.0%	0.0%	0.0%	0.0%
JNJ	0.29	21.6%	0.0%	2.6%	0.0%	0.0%	0.0%	0.0%	0.0%	0.0%	0.0%	0.0%	0.0%	0.0%
JPM	0.38	26.9%	0.0%	3.5%	0.0%	0.0%	0.0%	0.0%	0.0%	0.0%	0.0%	0.0%	0.0%	0.0%
KO	0.81	32.0%	0.0%	4.7%	0.0%	0.5%	0.0%	0.5%	0.0%	0.0%	0.0%	0.0%	0.0%	0.0%
MCD	0.43	31.4%	0.0%	2.8%	0.0%	0.0%	0.0%	0.0%	0.0%	0.0%	0.0%	0.0%	0.0%	0.0%
MMM	1.00	−18.7%	0.0%	1.9%	0.0%	0.0%	0.0%	0.0%	0.0%	0.0%	0.0%	18.2%	0.0%	18.2%
MRK	0.71	56.6%	0.0%	4.1%	0.0%	0.0%	0.0%	0.0%	0.0%	0.0%	0.0%	0.0%	0.0%	0.0%
MSFT	1.00	169.1%	14.3%	13.6%	27.4%	45.8%	38.3%	57.7%	37.3%	63.7%	100.0%	81.8%	100.0%	81.8%
NKE	0.71	130.1%	24.3%	3.7%	22.4%	15.2%	15.0%	12.5%	0.0%	0.0%	0.0%	0.0%	0.0%	0.0%
PFE	0.57	18.8%	0.0%	2.7%	0.0%	0.0%	0.0%	0.0%	0.0%	0.0%	0.0%	0.0%	0.0%	0.0%
PG	0.81	65.8%	8.5%	3.3%	12.5%	5.1%	0.0%	4.2%	0.0%	0.0%	0.0%	0.0%	0.0%	0.0%
TRV	0.71	13.4%	0.0%	1.9%	0.0%	0.0%	0.0%	0.0%	0.0%	0.0%	0.0%	0.0%	0.0%	0.0%
UNH	0.43	63.4%	0.0%	3.5%	0.0%	0.0%	0.0%	0.0%	0.0%	0.0%	0.0%	0.0%	0.0%	0.0%
V	0.71	94.4%	0.0%	6.6%	0.0%	0.0%	0.0%	0.0%	0.0%	0.0%	0.0%	0.0%	0.0%	0.0%
VZ	0.33	23.8%	0.0%	2.4%	0.0%	0.0%	0.0%	0.0%	0.0%	0.0%	0.0%	0.0%	0.0%	0.0%
WBA	0.33	−42.1%	0.0%	0.9%	0.0%	0.0%	0.0%	0.0%	0.0%	0.0%	0.0%	0.0%	0.0%	0.0%
WMT	0.33	55.3%	6.0%	2.2%	0.0%	0.0%	0.0%	0.0%	0.0%	0.0%	0.0%	0.0%	0.0%	0.0%
XOM	0.29	−42.3%	0.0%	4.1%	0.0%	0.0%	0.0%	0.0%	0.0%	0.0%	0.0%	0.0%	0.0%	0.0%
# of weights $\geq$ = 0.1%			5	27	4	8	3	7	2	2	1	2	1	2
Ann volatility			27.3%	23.8%	27.4%	28.1%	30.1%	28.6%	32.2%	31.4%	32.2%	29.5%	32.2%	29.5%
Ann return			37.4%	20.3%	37.4%	34.2%	40.7%	33.7%	42.7%	40.8%	38.3%	30.8%	38.3%	30.8%
Ann Sh			1.37	0.85	1.36	1.22	1.35	1.18	1.33	1.30	1.19	1.04	1.19	1.04
PESGV			0.00	0.00	0.75	0.85	0.76	0.89	0.73	0.84	1.00	1.00	1.00	1.00
Ann Sh.ESG			1.37	0.85	2.39	2.24	2.37	2.23	2.29	2.40	2.37	2.09	2.37	2.09
Return-ESG score corr	0.28													

Table A5. Maximum Sharpe OESGP for the DJI with the SPGI ratings (2018–2020).

ESG strength ticker/correlation	ESG score	Return	0 Pearson	0 Partial	5E-05 Pearson	5E-05 Partial	1E-04 Pearson	1E-04 Partial	5E-04 Pearson	5E-04 Partial	1E-03 Pearson	1E-03 Partial	2E-03 Pearson	2E-03 Partial	3E-03 Pearson	3E-03 Partial	4E-03 Pearson	4E-03 Partial	5E-03 Pearson	5E-03 Partial
AAPL	0.28	222.7%	46.9%	10.4%	39.6%	17.5%	50.9%	8.9%	62.5%	0.0%	26.7%	0.0%	0.0%	0.0%	0.0%	0.0%	0.0%	0.0%	0.0%	0.0%
AXP	0.37	26.5%	0.0%	0.0%	0.0%	0.0%	0.0%	0.0%	0.0%	0.0%	0.0%	0.0%	0.0%	0.0%	0.0%	0.0%	0.0%	0.0%	0.0%	0.0%
BA	0.29	−23.4%	0.0%	1.2%	0.0%	0.0%	0.0%	0.0%	0.0%	0.0%	0.0%	0.0%	0.0%	0.0%	0.0%	0.0%	0.0%	0.0%	0.0%	0.0%
CAT	0.75	24.2%	0.0%	3.8%	0.0%	6.8%	0.0%	10.7%	0.0%	8.0%	0.0%	0.0%	0.0%	0.0%	0.0%	0.0%	0.0%	0.0%	0.0%	0.0%
CSCO	0.81	25.4%	0.0%	2.0%	0.0%	0.0%	0.0%	2.6%	0.0%	9.4%	2.7%	0.0%	0.0%	0.0%	0.0%	0.0%	22.0%	11.6%	82.8%	83.1%
CVX	0.43	−23.6%	0.0%	0.4%	0.0%	0.0%	0.0%	0.0%	0.0%	0.0%	0.0%	0.0%	0.0%	21.0%	0.0%	16.3%	0.0%	0.0%	0.0%	0.0%
DIS	0.29	66.6%	0.0%	2.3%	0.0%	0.0%	0.0%	0.0%	0.0%	0.0%	0.0%	0.0%	0.0%	0.0%	0.0%	0.0%	0.0%	0.0%	0.0%	0.0%
GS	0.38	7.5%	0.0%	3.5%	0.0%	0.0%	0.0%	0.0%	0.0%	0.0%	0.0%	0.0%	0.0%	0.0%	0.0%	0.0%	0.0%	0.0%	0.0%	0.0%
HD	0.28	51.7%	0.0%	6.9%	0.0%	0.0%	0.0%	0.0%	0.0%	0.0%	0.0%	0.0%	0.0%	0.0%	0.0%	0.0%	0.0%	0.0%	0.0%	0.0%
IBM	0.45	−7.1%	0.0%	3.4%	0.0%	0.0%	0.0%	0.0%	0.0%	0.0%	0.0%	0.0%	0.0%	0.0%	0.0%	0.0%	0.0%	0.0%	0.0%	0.0%
INTC	0.57	21.6%	0.0%	1.6%	0.0%	0.0%	0.0%	0.0%	0.0%	0.0%	0.0%	0.0%	0.0%	0.0%	0.0%	0.0%	0.0%	0.0%	0.0%	0.0%
JNJ	0.62	21.6%	0.0%	2.6%	0.0%	0.0%	0.0%	0.0%	0.0%	0.0%	0.0%	0.0%	0.0%	0.0%	0.0%	0.0%	0.0%	0.0%	0.0%	0.0%
JPM	0.40	26.9%	0.0%	3.5%	0.0%	2.5%	0.0%	4.6%	0.0%	0.0%	0.0%	0.0%	0.0%	0.0%	0.0%	0.0%	0.0%	0.0%	0.0%	0.0%
KO	0.33	32.0%	0.0%	4.7%	0.0%	0.0%	0.0%	0.0%	0.0%	0.0%	0.0%	0.0%	0.0%	0.0%	0.0%	0.0%	0.0%	0.0%	0.0%	0.0%
MCD	0.24	31.4%	0.0%	2.8%	0.0%	0.0%	0.0%	0.0%	0.0%	0.0%	0.0%	0.0%	0.0%	0.0%	0.0%	0.0%	0.0%	0.0%	0.0%	0.0%
MMM	0.66	−18.7%	0.0%	1.9%	0.0%	0.0%	0.0%	0.0%	0.0%	0.0%	0.0%	0.0%	0.0%	0.0%	0.0%	0.0%	0.0%	0.0%	0.0%	0.0%
MRK	0.39	56.6%	0.0%	4.1%	0.0%	0.0%	0.0%	0.0%	0.0%	0.0%	0.0%	0.0%	0.0%	0.0%	0.0%	0.0%	0.0%	0.0%	0.0%	0.0%
MSFT	0.57	169.1%	14.3%	13.6%	20.9%	32.4%	26.4%	35.6%	37.5%	53.8%	73.3%	73.2%	90.7%	60.2%	100.0%	68.5%	59.5%	76.8%	0.0%	0.0%
NKE	0.57	130.1%	24.3%	3.7%	25.7%	16.2%	22.7%	15.4%	0.0%	13.0%	0.0%	9.3%	7.1%	0.0%	0.0%	0.0%	0.0%	0.0%	0.0%	0.0%
PFE	0.29	18.8%	0.0%	2.7%	0.0%	0.0%	0.0%	0.0%	0.0%	0.0%	0.0%	0.0%	0.0%	0.0%	0.0%	0.0%	0.0%	0.0%	0.0%	0.0%
PG	0.57	65.8%	8.5%	3.3%	13.7%	5.1%	0.0%	4.6%	0.0%	0.0%	0.0%	0.0%	0.0%	0.0%	0.0%	0.0%	0.0%	0.0%	0.0%	0.0%
TRV	0.40	13.4%	0.0%	1.9%	0.0%	0.0%	0.0%	0.0%	0.0%	0.0%	0.0%	0.0%	0.0%	0.0%	0.0%	0.0%	0.0%	0.0%	0.0%	0.0%
UNH	0.72	63.4%	0.0%	3.5%	0.0%	7.8%	0.0%	11.1%	0.0%	15.8%	14.9%	0.0%	0.0%	0.0%	0.0%	15.2%	18.5%	11.5%	17.2%	16.9%
V	0.58	94.4%	0.0%	6.6%	0.0%	11.5%	0.0%	11.1%	0.0%	0.0%	0.0%	0.0%	2.2%	18.8%	0.0%	0.0%	0.0%	0.0%	0.0%	0.0%
VZ	0.29	23.8%	0.0%	2.4%	0.0%	0.0%	0.0%	0.0%	0.0%	0.0%	0.0%	0.0%	0.0%	0.0%	0.0%	0.0%	0.0%	0.0%	0.0%	0.0%
WBA	0.36	−42.1%	0.0%	0.9%	0.0%	0.0%	0.0%	0.0%	0.0%	0.0%	0.0%	0.0%	0.0%	0.0%	0.0%	0.0%	0.0%	0.0%	0.0%	0.0%
WMT	0.31	55.3%	6.0%	2.2%	0.0%	0.0%	0.0%	0.0%	0.0%	0.0%	0.0%	0.0%	0.0%	0.0%	0.0%	0.0%	0.0%	0.0%	0.0%	0.0%
XOM	0.36	−42.3%	0.0%	4.1%	0.0%	0.0%	0.0%	0.0%	0.0%	0.0%	0.0%	0.0%	0.0%	0.0%	0.0%	0.0%	0.0%	0.0%	0.0%	0.0%
# of weights >= 0.1%			5	27	4	8	3	8	2	5	2	4	3	3	1	3	3	3	2	2
Ann volatility		27.3%	23.8%	20.3%	27.1%	27.0%	29.8%	26.8%	32.2%	28.0%	31.4%	29.5%	31.1%	29.1%	32.2%	29.6%	29.0%	30.1%	29.9%	30.0%
Ann return		37.4%	20.3%	0.85	37.1%	32.8%	40.6%	30.8%	42.7%	30.6%	40.1%	34.7%	37.5%	29.9%	38.3%	31.7%	29.7%	33.4%	14.3%	14.3%
Ann Sh		1.37	0.85	0.00	1.37	1.21	1.36	1.15	1.33	1.09	1.28	1.17	1.03	1.19	1.07	1.02	1.11	0.48	0.48	
PESGV		0.00	0.00	0.46	0.54	0.42	0.59	0.39	0.63	0.50	0.60	0.58	0.65	0.57	0.63	0.65	0.62	0.80	0.80	
Ann Sh_ESG		1.37	0.85	1.99	1.87	1.94	1.83	1.85	1.79	1.91	1.88	1.90	1.70	1.87	1.75	1.69	1.79	0.86	0.86	
Return-ESG score corr	0.05																			

References

[1] D. Avramov, S. Cheng, A. Lioui, and A. Tarelli. (2021). Sustainable investing with ESG rating uncertainty. *Journal of Financial Economics*. Available online: https://www.sciencedirect.com/science/article/abs/pii/S0304405X21003974?via%3Dihub.

[2] D. P. Baron. (2009). A positive theory of moral management, social pressure, and corporate social performance. *Journal of Economics and Management Strategy* 18, 7–43.

[3] R. Benabou, and J. Tirole. (2010). Individual and corporate social responsibility. *Economica* 77 (January), 1–19.

[4] F. Berg, J. Kölbel, and R. Rigobon. (2019). Aggregate confusion: The divergence of ESG ratings. *MIT Sloan Research Paper* No. 5822-19. Available at SSRN: https://ssrn.com/abstract=3438533.

[5] H. Cai, and A. B. Schmidt. (2020). Comparing mean–variance portfolios and equal-weight portfolios for major US equity indexes. *Journal of Asset Management* 21, 326–332.

[6] M. Chen, and Mussalli. G. (2020). An integrated approach to quantitative ESG investing. *The Journal of Portfolio Management* 46, 65–74.

[7] V. K. Chopra, and W. T. Ziemba. (1993). The effect of errors in means, variances, and covariances on optimal portfolio choice. *The Journal of Portfolio Management* 19(2), 6–11.

[8] D. M. Christensen, G. Serafeim, and S. Sikochi (2021). Why is corporate virtue in the eye of thebeholder? The case of ESG ratings. *The Accounting Review*. Available online: https://doi.org/10.2308/TAR2019-0506.

[9] V. DeMiguel, L. Garlappi, and R. Uppal. (2009). Optimal versus Naïve diversification: How inefficient is the 1/N portfolio strategy? *Review of Financial Studies* 22, 1915–1953.

[10] V. DeMiguel, Y. Plyakha, R. Uppal, and G. Vilkov. (2013). Improving portfolio selection using option-implied volatility and skewness. *Journal of Financial and Quantitative Analysis* 48, 1813–1845.

[11] E. Dimson, P. Marsh, and M. Staunton. (2020). Divergent ESG ratings. *The Journal of Portfolio Management* 471, 75–87.

[12] R. Duchin, and H. Levy. (2009). Markowitz versus the Talmudic portfolio diversification strategies. *Journal of Portfolio Management* 35, 71–74.

[13] E. J. Elton, M. J. Gruber, S. J. Brown, and W. N. Goetzmann. (2009). *Modern Portfolio Theory and Investment Analysis*. John Wiley and Sons, Hoboken.

[14] S. Gerber, H. M. Markowitz, P. A. Ernst, Y. Miao, B. Javid, and P. Sargen. (2021). The Gerber statistic: A robust co-movement measure for portfolio optimization. *The Journal of Portfolio Management*. Available online: https://doi.org/10.3905/jpm.2021.1.316.

[15] R. C. Green, and B. Hollifield. (1992). When will mean–variance efficient portfolios be well diversified? *Journal of Finance* 47, 1785–1809.

[16] O. Hart, and L. Zingales. 2017). Companies should maximize shareholder welfare not market value. *Journal of Law, Finance, and Accounting* 2, 247–274.

[17] H. Hong, and M. Kacperczyk. (2009). The price of sin: The effects of social norms on markets. *Journal of Financial Economics* 93(1), 15–36.

[18] H. Jacobs, S. Müller, and M. Weber. (2013). How should individual investors diversify? An empirical evaluation of alternative asset allocation policies. *Journal of Financial Markets* 19, 62–85.

[19] J. Johnston, and J. DiNardo. (1997). *Econometric Methods*. McGraw-Hill, New York.

[20] R. Kumar. (2019). ESG: Alpha or duty? *The Journal of Index Investing* 9(4), 58–66.

[21] M. La Torre, F. Mango, A. Cafaro, and S. Leo. (2020). Does the ESG index affect stock return? Evidence from the Eurostoxx50. *Sustainability* 12(16), 6387.

[22] D. Nadler, and A. B. Schmidt. (2014). Portfolio theory in terms of partial covariance. Available at SSRN: http://ssrn.com/abstract=2436478.

[23] Z. Nagy, A. Kassam, and L. E. Lee. (2016). Can ESG add alpha? An analysis of ESG tilt and momentum strategies. *The Journal of Investing* 25(2), 113–124.

[24] M. Oehmke, and M. M. Opp. (2020). A theory of socially responsible investment. *Swedish House of Finance Research Paper* No. 20-2. Available at SSRN: https://ssrn.com/abstract=3467644.

[25] L. Pastor, R. F. Stambaugh, and L. A. Taylor. (2020). Sustainable investing in equilibrium. Available at SSRN: https://ssrn.com/abstract=3498354.

[26] L. H. Pedersen, S. Fitzgibbons, and L. Pomorski. (2021). Responsible investing: The ESG-efficient frontier. *Journal of Financial Economics* 142, 572–597.

[27] A. B. Schmidt. (2020). Optimal ESG portfolios: An example for the Dow Jones index. *Journal of Sustainable Finance and Investment.* https://doi.org/10.1080/20430795.2020.1783180.

[28] A. B. Schmidt. (2021). The ESG conundrum: An outsider's view. Available online: https://papers.ssrn.com/sol3/papers.cfm?abstract_id=3942572.

[29] J. Tu, and G. Zhou. (2009). Markowitz meets Talmud: A combination of sophisticated and Naïve diversification strategies. *Journal of Financial Economics* 99, 204–215.

[30] J. Whittaker. (1990). *Graphical Models in Applied Multivariate Statistics*. Wiley, Hoboken.

[31] P. van der Beck. (2021). Flow-driven ESG returns. *Swiss Finance Institute Research Paper* No. 21–71. Available online: https://ssrn.com/abstract=3929359.

[32] G. Friede, T. Busch, and A. Bassen. (2015). ESG and financial performance: aggregated evidence from more than 2000 empirical studies, *Journal of Sustainable Finance & Investment* 5(4), 210–233.

[33] F. Alessandrini, and E. Jondeau. (2021). Optimal strategies for ESG portfolios. *The Journal of Portfolio Management* 47(6), 114–138.

© 2024 World Scientific Publishing Company
https://doi.org/10.1142/9789811281747_0005

Chapter 5

Centrality of the Supply Chain Network

Liuren Wu

Zicklin School of Business, Baruch College, CUNY, New York, USA
liuren.wu@baruch.cuny.edu

Abstract

As the supply chain information becomes more readily accessible, researchers are paying increasing attention to information flows and interactions between suppliers and customers. Shocks to a supplier not only impact its immediate customers but also generate ripple effects on the whole economy through the supply chain network. This chapter strives to define the relative importance, or centrality, of a supplier in the whole supply chain network and understand how the most central suppliers interact with the aggregate economy. Analysis shows that supplier-central companies tend to be more volatile, and their stock performance tends to precede the movements of the aggregate market.

Keywords: Supply chain, directed network, degree centrality, eigenvector centrality, Kleinberg centrality.

Introduction

Researchers increasingly appreciate the important roles that structural relations and connections play in understanding the statistical relations of financial securities across different companies. Such structural relations include, for example, similarities in the underlying business, as often captured by industry classification and peer group identification. Stock returns from the same industry/peer group tend to move more closely together.[1]

---

[1] Largely because of this observation, commonly used risk models in the industry such as BARRA treat each industry as a stand-alone risk factor.

Similarly, when analyzing the behavior of a particular company, financial analysts and investors pay close attention to the company's major suppliers and customers, from which they strive to infer the potential risks and opportunities for the company.[2] The closure or production delay of a major supplier can cause significant issues for the company's production, whereas the changing demand of the customer base poses challenges for the company's sales projection. Menzly and Ozbas [12] argue that fully digesting the implication of the information from structurally connected firms can take time. As evidence, they show that stocks that are in economically related supplier and customer industries cross-predict each other's returns. Chen et al. [5] show that the return predictability across supplier–customer industries can be even stronger for the corporate bond market. Shahrur et al. [15] use a sample of equities listed on the exchanges of 22 developed countries to show that equity returns on customer industries lead the returns of supplier industries. At the firm level, Cohen and Frazzini [6] use a data set of firms' principal customers to identify a set of economically related firms and show that the stock price of a firm does not fully incorporate news involving its related firms, thus generating predictable subsequent price moves. These studies highlight the importance of understanding the structural connections across companies and the implications of these connections for cross relations.

Understanding the structural connections is important not only for enhancing the identification of pair-wise statistical relations (both contemporaneous and across different leads and lags) but also for identifying the key drivers of the business cycle and aggregate market fluctuations. For this purpose, this chapter proposes to examine the supplier–customer relation not in terms of pair-wise connections, but from the perspective of an economic supply chain network, with each relation serving as a directed node of the network. This chapter examines the potentials of a list of centrality measures in capturing the major determinants of the supply chain network. Of these measures, the supplier–customer centrality pair defined based on Kleinberg [10]'s algorithm looks particularly promising. The supplier centrality of a company is defined as the sum of the customer centrality of all its customers, while the customer centrality of a company is defined as the sum of the supplier centrality of all its suppliers. The two types of centralities can

---

[2]Luo and Nagarajan [11] examine the antecedents and consequences of analysts choosing to become supply chain analysts, i.e. analysts following both a supplier and its major customer. Guan et al. [7] find that the likelihood of an analyst following a supplier–customer firm pair increases with the strength of the economic ties along the supply chain, as measured by the percentage of the supplier's sales to the customer.

then be solved as the leading eigenvectors of the products of the supplier–customer network matrix.

Based on the information from the FactSet supply chain relationship database, this chapter builds a supplier–customer network matrix at the beginning of each year from 2004 to 2014, computes the supplier and customer centralities of each company, and constructs supplier and customer central stock portfolios based on the centrality estimates. Empirical analysis shows that supplier central portfolios tend to be more volatile than customer central portfolios. Furthermore, the stock performance of supplier central portfolios tends to predict the movements of the overall stock market.

In related work, Aobdia, Caskey, and Ozel [3] define centrality of industries based on cross-industry trade flows and find that stock returns on central industries depend more on aggregate risks and shocks to central industries propagate more strongly than shocks to other industries. Ahern and Harford [2] represent the economy as a network of industries connected through customer and supplier trade flows and analyze how merger waves propagate through the network. Ahern [1] show that industries that are more central in the network of intersectoral trade earn higher stock returns than industries that are less central.

The remainder of this chapter is organized as follows. The following section describes the FactSet supply chain relationship database. The third section discusses different centrality measures and their economic meanings within the context of the supply-chain network. The fourth section analyzes the supplier centralities of the US economy. The last section concludes.

Data Source and Sample Construction

The firm-level supplier–customer relationship information is from FactSet, which maps the historical supplier–customer relationship for each company based on information from the company's regulatory disclosure reports, annual reports, and other primary sources. Based on this database, this chapter builds the supply chain network among the constituents of the S&P Composite 1500 index, which covers about 90% of the US market capitalization and includes three leading indices: the S&P 500 index, the S&P MidCap 400 index, and the S&P SmallCap 600 Index. From January 2004 to January 2014, the analysis takes snapshots of the database at the beginning of each year and merges the information with price and accounting data from Bloomberg. Over the 11-year period, the number of mapped companies ranges from 1214 to 1340, and the number of identified supplier–customer pairs range from 4169 to 6061.

The FactSet database indicates the applicable time period for each relationship, from which one can in principle extract the relational mapping at any time; nevertheless, this chapter decides to take the snapshots at a relatively low frequency because of the concern that it may take time for the applicable relationship to become fully known to the average investor. The underlying assumption for taking a snapshot once a year is that the structural relation is relatively stable and holds over the coming year.

The US Securities and Exchange Commission (SEC) has a mandate (rule SFAS 131) that calls for companies to disclose a customer if their revenue exposure to the customer is 10% or greater. In addition, some companies choose to disclose more customers and suppliers if they believe the information will help improve their attractiveness and draw more attention from investors. FactSet expands and consolidates such information through cross-validation and a reverse linkage methodology. For example, if company A identifies company B as a supplier, company A becomes the customer of company B. In practices, the relationships are not symmetric due to size differences (among other things). For example, if company A is much smaller than company B, it is possible that although company B is a significant supplier for company A and company A voluntarily discloses A as an important supplier, company A is not a significant customer for B and hence company B does not disclose A as a significant customer. In this case, the cross-validation approach may bring a certain degree of bias to the network construction.

In performing the analysis, the author is conscious of the fact that a company can have supplier/customer relationships, possibly with lesser significance, with many more companies than reported in the database. Furthermore, by confining the universe to the constituents of the S&P Composite 1500 index, the constructed network is necessarily a truncated one, where important foreign and/or private companies outside the index universe are trimmed from the actual supply-chain network.

Important questions also arise on how to define the relative significance of a relationship. When highlighting the significance of a supplier to a company, the company may cite how much it spends on this supplier, for example, in percentage of its total cost of goods sold. On the other hand, when highlighting the significance of a customer to a company, the company can cite how much percentage of its sales come from this customer. While companies sometimes report these numbers and the FactSet database captures them, the information is still far too sparse for systematic analysis. Absent from more detailed information, this chapter treats each identified supplier as equally significant to its customer, and vice versa. Depending on the specific applications, one can also consider altering the significance of each relationship according to factors, such as industry proximity, direction of disclosure, and size of firms. Furthermore, measures that capture the irreplaceability and/or value added of a supplier can also be used as importance metrics.

Supply-Chain Centrality Measures

Instead of focusing on the pair-wise relationship between a company and its suppliers, this chapter examines the relationship from the perspective of the aggregate economy and the economy-wide network of the supply chain. In particular, this chapter strives to define appropriate measures that capture the relative importance of each supplier in terms of its contribution to the networked economy.

Formally, the supplier network at a given point in time can be represented by an $N \times N$ supplier matrix A, with N denoting the number of companies in the network. We set the (i,j)-th element of the matrix, $A_{i,j}$, to 1 if company i is the supplier of company j and set it to 0 otherwise. Under this construction, each column j of the supplier matrix represents an index of the suppliers for company j, and each row i of the matrix is an index of the customers that the supplier i has.

In the language of networks, the supply-chain network is a directed network, where one can imagine drawing an arrow from each customer to each of its suppliers. The literature on networks, e.g. Newman [13], has proposed a long list of centrality measures to represent the relative importance of a vertex in the network. This chapter discusses the applicability of these concepts to the supply-chain network and explores their economic meaning in capturing the relative importance of a supplier in the supply-chain network.

Degree centrality: The number of companies that the supplier serves

Intuitively, a supplier company is more important to the economy if it is the supplier of many companies instead of just one company. Thus, a simple measure of centrality for a supplier, often referred to as the **degree centrality**, can be defined as the number of companies that this supplier serves. From the perspective of the network topology, if one draws an arrow from a company to each of its suppliers, the degree centrality simply measures the degree of a vertex or the number of arrows pointed to a particular supplier.

Given the construction of a supplier matrix A, the degree centrality vector, $\mathbf{c}$, can be computed as the simple sum of each row,

$$\mathbf{c} = A\mathbf{e}, \qquad (1)$$

where $\mathbf{e}$ denotes a vector of ones.

Different disclosure practice and data construction can potentially have important impacts on the measure and what the measure captures. For example, if companies were not to disclose any suppliers but were just to disclose important customers that account for no less than 10% of their sales, this measure for a company would only capture the number of customers this company discloses. Based on the 10% threshold, each company would have a maximum of 10 important customers, and the numbers would be smaller if the sales distribution is less uniform and more concentrated on a few customers. In this case, the measure would have very little to do with how important the company is to the overall economy, but it would capture more of how concentrated or spread out its customer base is. A larger measure would suggest that the company's customer base is more evenly spread out.

On the other hand, if companies were to only report their significant suppliers, based on, for example, the percentage of costs of goods sold the supply accounts for, the degree centrality measure for a company would capture the number of companies that regard this company as an important suppliers and thus would be closer to a measure of importance for this company, in the eyes of its customers. Furthermore, if the supplier is a large company and makes significant supplies to many customer companies, even though the supply to each company may look small to the supplier itself, it is significant to the customer. The number of customer companies who regard this company as a significant supplier can indeed be a measure of the company's importance.

In practice, the SEC rule only asks that a company reports its significant customers. Such reports are not particularly helpful to the meaning of the degree centrality measure. It is their self voluntary report on their important suppliers that gives significance to the degree centrality measure as a measure of supplier importance to the economy.

On the other side, suppose we also attempt to define an analogous customer importance degree centrality measure for each company as the sum of the company's suppliers. In this case, again the self-reported number of significant suppliers may not really measure the company's importance as a customer but more reflects whether the company's production needs equally costly parts or parts that have vastly different costs. A company with more evenly costly parts will report more significant suppliers. For each part, a company can also decide to choose more than one suppliers, but research has shown that the variation for the number of suppliers per part is small.[3] On the other hand, if companies only report their significant customers, then the number of companies that regard this particular company as a significant customer more appropriately reflects the significance of the company as a customer. Thus, in a sense, the importance of a company to the economy comes from the perception from the other side. The cross validation approach of FactSet allows one to collect the information from the other side.

Figure 1 plots the top ten suppliers as of January 2014 with the most number of customers. The top two on the list BFAM and BRK/B have more than 50 identified customers and the next five suppliers have 40 or more customers. The top list spans across different industries from child-care provider (BFAM), multinational conglomerate (BRK/B), manufacturers of industrial goods (CR, GE), to large technology companies (IBM, MSFT).

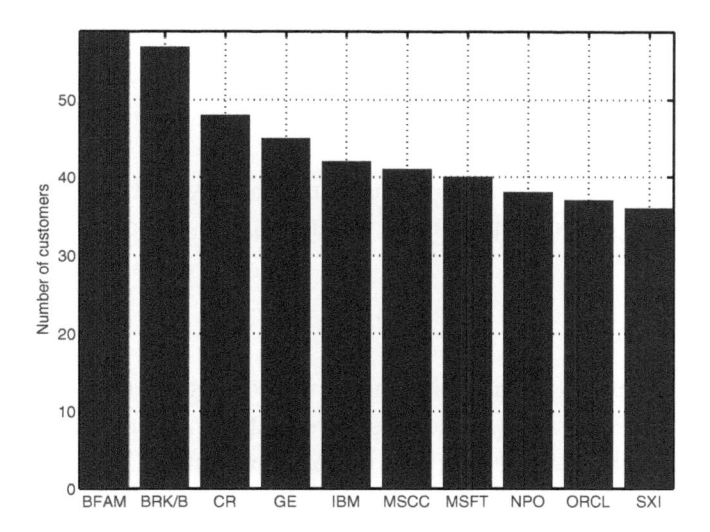

Figure 1. Top ten suppliers with the most number of customers. The plot lists the top ten suppliers with the most number of identified customers as of January 2014, with the bar higher for each company representing the number of identified customers.

---

[3]O'Connor et al. [14] find that the repeated nature of the procurement process typically makes it optimal for the customer to award business to several firms to encourage competition and to preclude a "winner take all" outcome. On the other hand, they find that awarding the contracts to too many suppliers create logistic issues regarding monitoring and quality control. So most companies choose two to three suppliers at most for the same product.

Since the centrality measures are meant to be a relative measure capturing the relative importance of a company within the network economy, one can perform various normalizations to the measures. The definition in (1) uses the raw total number of customers without normalization. Alternatively, one can normalize the measures so that they sum to N, in which case the measures average to one and the magnitude does not vary with the size of the economy, or one can normalize the measures to sum to one (or 100%) so that the estimate for each company represents the percentage contribution of that company to the overall economy. Henceforth, we take the latter normalization and interpret each estimate as a percentage contribution.

Eigenvector centrality: The importance of customers as suppliers

Degree centrality treats all customers equally and awards one "centrality point" for each customer; however, not all customers are equivalent. A customer can be a supplier to other companies, and its importance to the economy increases if the customer itself is an important supplier to many other companies. Instead of awarding suppliers just one point for each customer, **eigenvector centrality** gives each supplier a score proportional to the sum of the scores of its customers.

Let c_i be the eigenvector centrality for company i. The eigenvector centrality of company i is proportional to the sum of the centralities of all its customers,

$$c_i = (1/\lambda) \sum_j A_{i,j} c_j. \tag{2}$$

Equation (2) defines the centrality measure recursively. Given N companies in the network, equation (2) defines N equations, from which one solves for the N centrality measures for the N companies. The proportionality coefficient λ can be determined as a normalization condition by setting, for example, $\sum_j c_j = 1$. One can solve the centrality measures recursively by starting at some initial guesses (say, assuming that all companies have equal centrality) and applying equation (2) repeatedly until the solution converges.

Representing equation (2) in matrix notation, we have,

$$\lambda \mathbf{c} = A \mathbf{c}. \tag{3}$$

It can be shown that the thus defined eigenvector centrality is proportional to the right leading eigenvector of the supplier matrix A Bonacich [4]. Furthermore, since all elements of the supplier matrix A are non-negative, the Perron–Frobenius theorem states that the leading eigenvector has strictly positive components.

The eigenvector centrality has the nice property that it can be large either because a supplier has many customers or because it has important customers, or both. The importance of the customer is defined based on its importance as a supplier itself.

Kleinberg centrality: The importance of customers as economic hubs

Eigenvector centrality treats a customer as important if it is an important supplier itself, but suppliers and customers can be important in different ways. A supplier is important if it supplies to many different customers, whereas a customer can be important simply because of its role as a hub for many different types of suppliers.

From the perspective of a directed network, Kleinberg [10] proposes to construct two types of centralities, labeled as the **authority centrality** and the **hub centrality**, to quantify each vertex's prominence in the two roles, with authorities corresponding to important suppliers and hubs corresponding to important customers in our application. Specifically, each company i has both a supplier authority centrality c_i and a customer hub centrality h_i, representing its importance in the two different roles as a supplier and a customer, respectively. The defining characteristic of a supplier with high authority centrality is that it is the supplier to many customers with high hub centrality. The defining characteristic of a customer with high hub centrality is that it is the customer of many suppliers with high authority centrality.

Mathematically, the two centrality measures can be connected as,

$$c_i = \alpha \sum_j A_{i,j} h_j, \quad h_i = \beta \sum_j A_{j,i} c_j. \qquad (4)$$

Compared to eigenvector centrality, the above definition gives distinct roles to suppliers and customers. In practice, a company discloses both its important customers and its important suppliers. The importance of a company as a supplier is dictated by the number of customer companies that report it as an important supplier, and the importance further increases if the reporting customers are important customers in others' eyes. The importance of a company as a customer is dictated by the number of supplier companies that report it as an important customer, and the importance further increases if the reporting suppliers are important suppliers in other's eyes.

In matrix notation,

$$\mathbf{c} = \alpha A \mathbf{h}, \quad \mathbf{h} = \beta A^\top \mathbf{c}. \qquad (5)$$

Combining the two,

$$AA^\top \mathbf{c} = \lambda \mathbf{c}, \quad A^\top A \mathbf{h} = \lambda \mathbf{h}, \qquad (6)$$

with $\lambda = (\alpha\beta)^{-1}$. Therefore, the supplier and customer centralities are respectively given by leading eigenvectors of $AA^\top$ and $A^\top A$ with the same eigenvalue.

A crucial condition for this approach to work is that $AA^\top$ and $A^\top A$ have the same leading eigenvalue λ; otherwise, one cannot satisfy both conditions in equation (6). This is indeed the case, and in fact that all eigenvalues are the same for the two matrices. To see this, one can start with $AA^\top \mathbf{c} = \lambda \mathbf{c}$, and multiply both sides by $A^\top$ to give

$$A^\top A (A^\top \mathbf{c}) = \lambda (A^\top \mathbf{c})$$

and hence $A^\top \mathbf{c}$ is the eigenvector of $A^\top A$ with the same eigenvalue λ. Comparing to (6), this means that

$$A^\top \mathbf{c} = \mathbf{h}, \qquad (7)$$

which gives us a fast way of calculating the customer centrality once we have the supplier centrality.

It is possible that a company is not the supplier to any other company. This will lead to a row of zeros in the supplier matrix. Both the degree centrality and eigenvector centrality for this company will be zero. A zero centrality measure for this company sounds intuitive as a company that supplies to nobody should be ranked low in the supply chain. However, imagine another company, which is a supplier to thousands of such companies. The fact that it is a supplier of many companies should make it "important" and this importance shows up in the degree centrality measure. However, since its customers all have zero eigenvector centrality, this company will also end up having zero eigenvector centrality. Taking this argument further, a company may be pointed to by others that themselves are pointed to by many more, and so on through many generations, but if the progression ends up at a company or companies that supply to no other companies, it is all for nothing — The final value of the eigenvector centrality will still be zero. Such a chain effect does not sound quite as intuitive.

In a strict sense, a company must have customers, even though these customers may not take the form of a company. Furthermore, companies do not report all their customers. By regulation, they only need to report customer companies that generate over 10% of their revenue. Some companies may choose to report more. Even so, it is unlikely the database lists all possible customers. For many retail companies, their customers are simply the consumers. Thus, one common solution to this issue is to give each company a small amount of centrality "for free", regardless of its position in the network. This modification was first proposed by Katz [9]. The addition can also be different for different companies if one thinks of it as a "prior" centrality measure estimated from information outside of the network matrix.

The customer and supplier hub and authority centralities circumvent this problem from an alternative angle. Companies not cited by any others as a supplier (that is, companies with no customers) have supplier centrality zero, which is reasonable, but they can still have non-zero customer centrality. Thus, their suppliers can have non-zero supplier centrality by virtue of being a supplier. This is perhaps a more elegant solution to the problems of eigenvector centrality in directed networks than the more *ad hoc* method of introducing an additive constant term in Katz [9]. Defining the importance of a supplier based on the importance of its customers as a customer also sounds more economically intuitive than based on the importance of its customers as a supplier.

Central Suppliers of the US Economy

This section analyzes the aggregate behaviors of the constructed supply chain network, with a particular focus on the behavior of the central suppliers, identified via various centrality measures. All centrality measures are normalized to sum to 100% so that the estimate for each company represents its percentage contribution to the total economy.

Distributional behaviors

To understand whether a few companies dominate the behavior of the supply chain network or the importance of companies is relatively evenly distributed, Figure 2 sorts the centrality estimates of different companies at the beginning of at year from large to small and then plots the cumulative summation of these estimates. By normalization, the total sum is equal to 100%. The speed to which the cumulative sum increases reveals the distributional behavior of the centrality measure: A straight line would suggest that all companies are equally central to the economy, whereas a highly concave plot would indicate that a small number of companies contribute to a large proportion to the network behavior.

The four panels in Figure 2 represent the estimates at four selected years. The patterns look similar at other time periods. Within each panel, the solid line denotes the behavior of the degree centrality estimates, the dashed line denotes the behavior of the eigenvector centrality, the dash-dotted line represents the behavior of the Kleinberg supplier (authority) centrality, and the dotted line represents the behavior of the Kleinberg customer

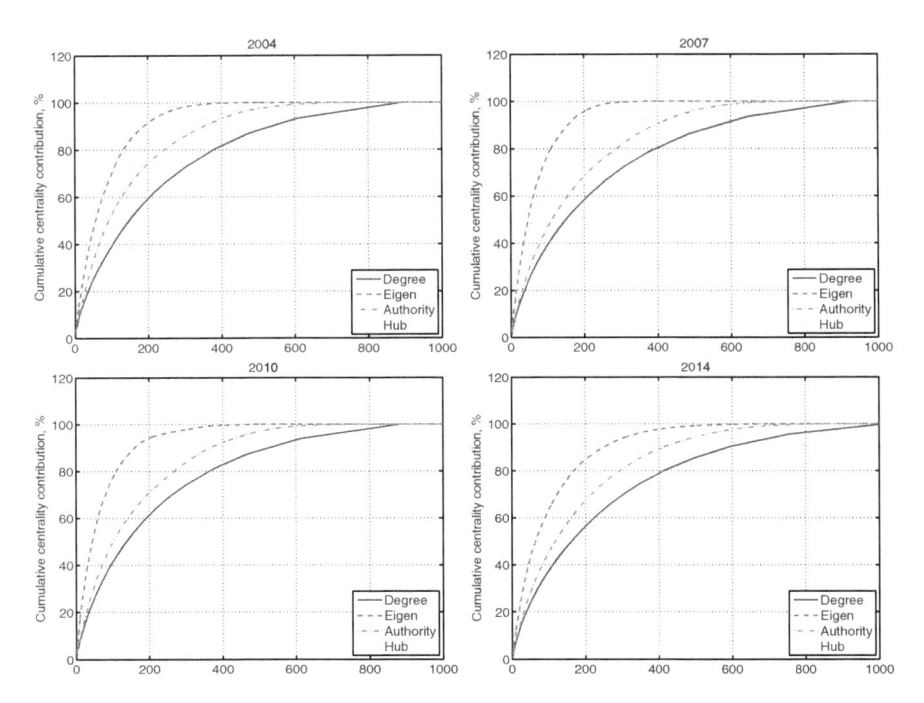

Figure 2. Cumulative distribution of centrality estimates. The four panels plot the cumulative distribution at the beginning of four selected years. The behaviors at other time periods are similar. With each panel, the four lines denote the distribution of the four centrality measures: degree centrality (solid line), eigenvector centrality (dashed line), Kleinberg authority supplier centrality (dash-dotted line), and Kleinberg hub customer centrality (dotted line). At the beginning of each year and for each centrality measure, the line is produced by first sorting the centrality estimates of different companies from large to small and then computing the cumulative summation of the sorted series. By normalization, the centrality estimates across the whole universe sum to 100%.

(hub) centrality. All the lines are concave, highlighting the fact that different companies have different degrees of importance in the network. The least concave and hence the most uniform measure comes from the simple degree centrality measure. The most concave, on the other hand, comes from the eigenvector centrality measure, in which case the recursive definition magnifies the contribution of a few companies while diminishing the contribution of the rest, to the extent that the top 200 companies capture most of the contribution to the aggregate economy.

The Kleinberg centrality measures jointly define the relative importance of a company as a supplier and as a customer via the authority and hub centrality. The distributional behaviors of the two centrality measures fall between the behaviors of the degree centrality and the eigenvector centrality. It is interesting to observe that across all the years, the hub customer

Table 1. Average rank correlation between different centrality measures. Entries report the historical average of the Spearman rank correlation estimates between the different centrality measures estimated at the beginning of each year from 2004 to 2014 on US S&P 1500 companies.

	Degree	Eigenvector	Supplier	Customer
Degree	1.000	0.776	0.891	−0.086
Eigenvector	0.776	1.000	0.805	−0.045
Supplier	0.891	0.805	1.000	−0.111
Customer	−0.086	−0.045	−0.111	1.000

centrality estimates show a steeper curvature than the authority supplier centrality estimates, suggesting that the important customer hubs are more concentrated than important supplier authorities.

To gauge the similarities and differences of the different centrality measures, Table 1 estimates the Spearman rank correlation of the four centrality measures each year and reports the historical average of the cross-correlation estimates. The first three centrality measures are all about the relative importance as a supplier. They show highly positive correlations. The customer centrality, on the other hand, measures the relative importance of a company as a customer and shows negative correlations with all four supplier centrality measures. The most negative correlation is between the Kleinberg customer centrality and its counterpart Kleinberg suppler centrality at −0.111.

Average stock behaviors across different quantiles of centralities

To understand how stocks with different centralities behave differently, Table 2 reports the average stock behaviors within each quantile of the centrality estimates. Columns under TR, ER, Beta, IV, RV, and Corr report the average annualized stock return, average excess stock return, average beta estimates, average three-month at-the-money implied volatility, realized volatility, and average correlation with the stock index. From the three supplier centrality measures, one can observe that stocks with high supper centralities tend to have lower return, higher beta, higher implied and realized volatility but lower correlation with the market. Hence, central supplier companies tend to be more volatile and more idiosyncratic. The customer central stocks, on the other hand, tend to have lower beta, lower volatility but higher correlation with the market.

Table 2. Average stock behaviors under different centrality quantiles. Entries report the average statistics of stocks within each quantile of the centrality estimates. Columns under TR, ER, Beta, IV, RV, and Cor denote respectively the annualized stock return, the annualized excess return, the stock beta computed against the S&P 500 index, the three-month at-the-money option implied volatility, realized volatility, and correlation with the S&P 500 index.

Centrality	TR	ER	Beta	IV	RV	Corr
Quantile	Panel A. Degree centrality					
1	0.104	0.019	1.259	0.377	0.334	0.514
2	0.137	0.035	1.235	0.370	0.322	0.494
3	0.125	0.047	1.267	0.386	0.341	0.499
4	0.133	0.027	1.290	0.385	0.338	0.508
5	0.119	0.030	1.324	0.378	0.337	0.529
Quantile	Panel B. Eigenvector centrality					
1	0.138	0.041	1.223	0.372	0.325	0.514
2	0.149	0.052	1.234	0.371	0.328	0.511
3	0.135	0.040	1.257	0.373	0.330	0.509
4	0.111	−0.000	1.342	0.395	0.343	0.507
5	0.102	−0.001	1.414	0.412	0.363	0.507
Quantile	Panel C. Supplier authority centrality					
1	0.131	0.026	1.223	0.352	0.319	0.540
2	0.117	0.032	1.262	0.378	0.330	0.515
3	0.122	0.021	1.285	0.396	0.342	0.496
4	0.121	0.041	1.274	0.381	0.335	0.499
5	0.108	0.022	1.338	0.391	0.350	0.511
Quantile	Panel D. Customer hub centrality					
1	0.123	0.060	1.240	0.388	0.334	0.503
2	0.112	0.031	1.210	0.368	0.337	0.506
3	0.096	0.028	1.251	0.365	0.325	0.517
4	0.105	0.006	1.243	0.365	0.328	0.515
5	0.112	0.020	1.193	0.344	0.312	0.532

The top ten list

This section ranks the companies based on each centrality measure and lists the top ten companies with the highest centrality estimates for each measure. Panel A of Table 3 provides the ticker of the top ten suppliers with the most customers (and hence the highest degree centrality) at the beginning of each year. Technology companies are the most prominent top suppliers during the earlier period of the sample, but the list becomes more diverse in the more recent years with some companies from the industrial goods and services sectors.

Table 3. Top ten suppliers with the most number of customers and eigenvector centrality. Entries list the tickers of the top ten companies with the most customers and hence the highest degree centrality estimates in panel A and with the highest eigenvector centrality in panel B, at the beginning of each year from 2004 to 2014.

	Panel A. Top ten suppliers with the most number of customers									
2004	GE	HPQ	IBM	INFA	INTC	IWOV	MSFT	ORCL	ROK	WBSN
2005	A	ACN	CNQR	GE	HPQ	IBM	INTC	IWOV	MSFT	ORCL
2006	ACN	GE	HPQ	IBM	INFA	IWOV	MANH	MSFT	ORCL	ROVI
2007	EMC	GE	HPQ	IBM	IWOV	MSFT	ORCL	PRGS	QTM	ROVI
2008	DDR	EMC	HPQ	IBM	IWOV	JDAS	MSFT	ORCL	PRGS	ROVI
2009	DDR	EMC	HPQ	IBM	JDAS	LXP	MSFT	ORCL	PRGS	ROVI
2010	DDR	EMC	HPQ	IBM	JDAS	LXP	MSFT	ORCL	PRGS	ROVI
2011	ACN	DIS	FRT	HPQ	IBM	INTC	JDSU	LXP	MSFT	PTC
2012	ADC	BFS	EP	FRT	HPQ	IBM	INTC	LFUS	MSFT	PWR
2013	BRK/B	CR	HON	HPQ	IBM	IRC	LFUS	MSFT	NPO	ORCL
2014	BFAM	BRK/B	CR	GE	IBM	MSCC	MSFT	NPO	ORCL	SXI
	Panel B. Top ten suppliers with the highest eigenvector centrality									
2004	CYMI	IBM	INTC	KEM	NTAP	ORCL	PSEM	SMTC	TTMI	XLNX
2005	A	AMCC	CYMI	IBM	IRF	KEM	NTAP	PSEM	SMTC	TTMI
2006	A	ATML	IBM	IRF	MSFT	ORCL	PSEM	ROVI	TTMI	XLNX
2007	ATML	BRKS	CA	IBM	JBL	JDSU	KEM	PSEM	QTM	SNIC
2008	ATML	AVT	COGT	CYMI	IBM	KEM	ORCL	PSEM	ROVI	SNIC
2009	ATML	CYMI	EMC	HPQ	IBM	JAVA	MSFT	QLGC	ROVI	SNIC
2010	ATML	EMC	HPQ	IBM	JAVA	MSFT	QLGC	QTM	ROVI	SNIC
2011	ADBE	ATML	CDNS	HPQ	IBM	INTC	KEM	MSFT	QTM	SNIC
2012	ATML	CDNS	HPQ	IBM	INTC	IRF	KEM	LFUS	XCRA	MSFT
2013	IBM	INTC	IRF	JBL	KEM	LFUS	LSI	NVDA	ORCL	XLNX
2014	AKAM	COHU	EXAR	IBM	INTC	LFUS	LSI	MMM	NVDA	XLNX

Panel B of Table 3 provides the top ten suppliers with the highest eigenvector centrality. Once the feedback effect of the network is accounted for, technology companies dominate the top ten lists for most years. In particular, ATML, which designs, develops, and manufactures semiconductor integrated circuit products, becomes the top supplier from 2006 to 2012.

Table 4 provides the top ten list of Kleinberg supplier authority centrality in panel A. By changing the definition of customer importance, the top ten supplier list has a completely new makeover in recent years. In particular, the 2014 top ten list includes six REIT companies, but none from the technology sector. The reasoning behind this shift needs further exploration.

Panel B of Table 4 provides the top ten list of Kleinberg customer hub centrality. The list is quite distinctive and includes many large retail stores. In 2014, this list includes BestBuy, Costco, Home Depot, Lowe's, Sears, Target, and Walmart.

Table 4. Top ten companies with the highest Kleinberg supplier authority centralities and customer hub centralities. Entries list the tickers of the top ten companies with the highest Kleinberg supplier authority centrality (panel A) and customer hub centrality (panel B) at the beginning of each year from 2004 to 2014.

				Panel A. Kleinberg supplier authority centrality					
2004	IBM	INFA	INTC	IRF	KEM	MSFT	ORCL	SANM	WBSN	XLNX
2005	A	EMC	GE	HPQ	IBM	IRF	KEM	MSFT	ORCL	SANM
2006	A	HPQ	IBM	INFA	IRF	IWOV	MSFT	ORCL	PRGS	ROVI
2007	ATML	EQIX	IBM	IRF	IWOV	JCI	MSFT	PRGS	QTM	ROVI
2008	BCSI	DDR	EQY	FICO	FRT	IWOV	MAC	PEI	PRGS	ROVI
2009	DDR	EMC	EQY	HPQ	IBM	MAC	MSFT	PEI	PTC	PRGS
2010	DDR	EMC	EQY	HPQ	IBM	KRG	MAC	MSFT	PEI	PRGS
2011	DDR	FRT	HPQ	IBM	JAH	KRG	MAC	MSFT	PEI	SPG
2012	ADC	AKR	DDR	EQY	FRT	KIM	KRG	MAC	PEI	SPG
2013	ADC	CR	DDR	FRT	HON	IRC	KIM	KRG	LEG	MAC
2014	ADC	CR	DDR	FRT	IRC	KRG	LEG	MAC	MMM	SXI
				Panel B. Kleinberg customer hub centrality					
2004	AVT	BA	CSCO	DELL	GE	HPQ	IBM	INTC	LMT	WMT
2005	BA	CSCO	DELL	F	GE	HPQ	IBM	INTC	LMT	WMT
2006	BA	CSCO	F	GE	HPQ	IBM	INTC	LMT	MSFT	WMT
2007	BA	CSCO	F	GE	HPQ	IBM	LMT	SHLD	T	WMT
2008	BA	CSCO	HD	HPQ	IBM	LMT	SHLD	T	TGT	WMT
2009	BA	HD	HPQ	IBM	KSS	MSFT	SHLD	T	TGT	WMT
2010	BA	BBY	HD	HPQ	IBM	KSS	SHLD	T	TGT	WMT
2011	BBY	HD	HPQ	IBM	JCP	KSS	M	SHLD	TGT	WMT
2012	BBY	COST	HD	JCP	KSS	LOW	M	SHLD	TGT	WMT
2013	BA	GE	HD	KSS	LMT	LOW	NOC	SHLD	TGT	WMT
2014	BA	BBY	COST	HD	LMT	LOW	NOC	SHLD	TGT	WMT

The statistical behavior of central stock portfolios

While the paper defines several centrality measures to identify the central suppliers and customers of an economy, it is unclear how the central vertices of the network behave and how their behaviors impact the aggregate economy. To explore the historical behavior of "central" companies, this section forms equal-weighted stock portfolios using the top ten companies[4] defined by each centrality measure and analyze their statistical behaviors in terms of their relations with the aggregate market, proxied by the S&P 500 index. The compositions of the portfolios are updated at the beginning

---

[4] An alternative is to form stock portfolios with weights proportional to the centrality estimates. The results are qualitatively similar. Varying the number of companies included in the portfolio does not alter the qualitative conclusion, either.

Table 5. Statistical behaviors of central company stock portfolios. I form equal weighted stock portfolios using the top ten companies with the highest centrality measures. The compositions of the portfolios are updated at the beginning of each year based on the renewed centrality measure estimates. Entries report the relations between these portfolios and the S&P 500 index. β measures the contemporaneous daily return regression slope against the index, ρ measures the contemporaneous daily return correlation with the S&P 500 index. The remaining columns report the predictive correlation of quarterly total portfolio returns and quarterly excess portfolio returns on future index returns over three and six month horizons.

Year	β	ρ	Quarterly total return		Quarterly excess return	
			3-month	6-month	3-month	6-month
Degree	1.075	0.879	0.177	0.043	0.088	0.015
Eigenvector	1.245	0.795	0.185	0.079	0.115	0.093
Authority	1.208	0.840	0.232	0.103	0.232	0.177
Hub	0.984	0.872	0.122	0.016	0.010	−0.002

of each year based on the renewed centrality measure estimates. Table 5 summarizes these statistical behaviors.

The column titled "β" in Table 5 reports the daily return beta estimates against the S&P 500 index return. The Kleinberg customer central portfolio has a beta estimate slightly less than one at 0.984, whereas the Kleinberg supplier centrality portfolio and the eigenvector supplier central portfolio both have beta estimates higher than 1.2. Thus, top suppliers companies show larger systemic risk than do the top customer companies. The column "ρ" reports the correlation between the daily central portfolio returns and the S&P 500 index returns. The estimates are very high, all around 0.8 or higher for both supplier and customer central portfolios.

To examine whether the stock performance of top suppliers and customers provide any precursor for the overall market movements, the table computes quarterly returns on the central portfolios and measures their forecasting correlations with the S&P 500 index returns over the next three- and six-month horizons. When using total return on the central portfolios as the predictor, the forecasting correlations are all positive for all central portfolios at both horizons. The predictions are the most positive for the Kleinberg supplier authority central portfolio at both three- and six-month horizons.

Part of the positive correlation prediction is due to the momentum effect identified by, for example, Jegadeesh and Titman [8]. To control for this, the table also generates excess stock returns by regressing each stock return against the S&P 500 index return and then computes the excess returns on the central portfolios. With the excess returns as a predictor of future index returns, the forecasting correlation is virtually zero for the customer central portfolio but remain positive for all supplier central portfolios. In particular, excess return on the Kleinberg supplier authority central portfolio generates the strongest predictive correlation at both three and six month horizons. The evidence seems to suggest that the stock performance of central suppliers predict the following performance of the overall market, but the stock performance of central customers does not lead the market performance.

Comparing the relative forecasting power of the different centrality measures, one can also gain some basic understanding on the effectiveness of different measures. The quarterly forecasting correlation is 8.8% for the simple degree centrality portfolio. By contrast, the eigenvector centrality portfolio generates stronger forecasting correlations at 11.5%, highlighting the importance of the feedback effect. Finally, defining the importance of the supplier based on the importance of the customer as a hub using Kleinberg's [10] algorithm generates the strongest and most persistent forecasting power for the supplier central portfolio. Thus, Kleinberg's joint construction of customer and supplier centrality not only provides an elegant and economically intuitive algorithm for handling singularity issues in the eigenvector centrality definition but also identifies the most predictive central suppliers.

Concluding Remarks

With the increasing availability of supply-chain information, researchers are paying increasing attention to information flows and interactions between suppliers and customers. This chapter examines the interactions from the perspective of an economy-wide supply-chain network and proposes a list of network centrality measures to capture the relative importance of each company within this network. Based on information from the FactSet Supply Chain Relationships database, I build a supplier network matrix at the beginning of each year from 2004 to 2014 and compute the supplier centralities of each company. I then construct supplier central stock portfolio based on the top ten companies with the highest centrality estimates and find that supplier central portfolios tend to be more volatile than customer central portfolios. Furthermore, the stock performance of supplier central portfolios tends to predict the movements of the overall stock market.

The idea of analyzing the centrality of the supply-chain network is relatively new. While the analysis in this chapter shows some promising results, there is much to do for future research. First, this chapter has explored several centrality definitions. A lot more research, both theoretical and empirical, is needed to fully understand which measure is the most relevant for what purpose. Furthermore, many more variations can be constructed both from the perspective of building the supplier network matrix and from the perspective of constructing new centrality measures. For the network matrix, one can explore different firm characteristics and data sources to enrich the network and differentiate the relative importance of each supplier to a customer. The importance metrics can be based on how much it supplies to the customer and/or how unique and irreplaceable the supply is. One can also explore different clustering of the network matrix to understand flows from a more aggregate level, for example, from one industry to another industry or from one economic or geographic region to another region. For centrality measures, much theoretical work can be done on the shock and response dynamics, from which one can motivate the definition of an appropriate centrality measure and better understand the flow of the shocks within the supply chain. Second, given the construction of the network matrix and centrality measures, much research can be done in understanding the statistical behavior of financial security prices and trading behaviors across different centrality levels and how they interact. Shocks to the supply chain can generate ripple effects, which can show up potentially as lead-lag predictive relations in security returns, earning surprises, default probabilities, and/or distinctive term structure patterns in realized and option implied volatilities and credit default swaps.

Acknowledgments

The author thanks Andrey Itkin (the editor), Jeremy Bertomeu, Clarence Kwan, Chen Li, Terrence Martell, Joseph Weintrop, Weina Zhang, Dexin Zhou, and seminar participants at the *5th Annual SWUFE Baruch Research Symposium* and Quantbot Technologies for their comments and suggestions. The author is particularly grateful to Jeremy Zhou of FactSet Revere for making the FactSet Revere supply-chain data available for research and for many helpful discussions, Manoj Boolani of FactSet Research Systems for many references and helpful discussions, and Yafeng Yin for technical assistance with data processing.

References

[1] K. R. Ahern. (2015). Network Centrality and the Cross Section of Stock Returns. Working Paper, University of Southern California.
[2] K. R. Ahern, and J. Harford. (2014). The importance of industry links in merger waves. *Journal of Finance* 69(2), 527–576.

[3] D. Aobdia, J. Caskey, and N. B. Ozel. (2014). Inter-industry network structure and the cross-predictability of earnings and stock returns. *Review of Accounting Studies* 19(3), 1191–1224.

[4] P. Bonacich. (1987). Power and centrality: A family of measures. *American Journal of Sociology* 92(5), 1170–1182.

[5] Chen, L., G. Zhang, and W. Zhang. (2014). Return Predictability in Corporate Bond Market along the Supply Chain. Working Paper, Cheung Kong Graduate School of Business, University of Missouri-St. Louis, and National University of Singapore.

[6] L. Cohen, and A. Frazzini. (2008). Economic links and predictable returns. *Journal of Finance* 63(4), 1977–2011.

[7] Y. Guan, F. Wong, and Y. Zhang. (2015). Analyst following along the supply chain. *Review of Accounting Studies* 210(1), 210–241.

[8] N. Jegadeesh, and S. Titman. (2001). Profitability of momentum strategies: An evaluation of alternative explanations. *Journal of Finance* 56(2), 699–720.

[9] L. Katz. (1953). A new status index derived from sociometric analysis. *Psychometrika* 18(1), 39–43.

[10] J. M. Kleinberg. (1999). Authoritative sources in a hyperlinked environment. *Journal of the ACM* 46(5), 604–632.

[11] S. Luo, and N. J. Nagarajan. (2014). Information complementarities and supply chain analysts. *The Accounting Review* (Forthcoming).

[12] L. Menzly, and O. Ozbas. (2010). Market segmentation and cross-predictability of returns. *Journal of Finance* 65(4), 1555–1580.

[13] M. E. J. Newman. (2010). *Networks: An Introduction.* Oxford University Press, Oxford, UK.

[14] N. G. O'Connor, A. Wu, and S. W. Anderson. (2014). Relative Performance Evaluation in Supply Chain Management. Working Paper, Hong Kong Baptist University, National Chengchi University, and University of California, Davis.

[15] H. Shahrur, Y. L. Becker, and D. Rosenfeld. (2010). Return predictability along the supply chain: The international evidence. *Financial Analysts Journal* 66(3), 60–77.

Chapter 6

Are E-mini S&P 500 Futures Prices Random?

Valerii Salov
Personal Project and Research
v7f5a7@comcast.net

Abstract

Chains of the CME Group Time and Sales E-mini S&P 500 futures tick prices and their a-b-c-d-increments are studied. A discrete probability distribution based on the Hurwitz zeta function and Dirichlet series is suggested for the price increments. The randomness of the ticks is discussed using the notions of typicalness, chaoticness, and stochasticness introduced by Kolmogorov and Uspenskii and developed by predecessors, them, and pupils. They define randomness in terms of the theory of algorithms.

> Can we distinguish random chains from non-random ones?
>
> Andrey Nikolaevich Kolmogorov, Vladimir Andreevich Uspenskii, [43, p. 427/390]

> ... the object of statistical methods is the reduction of data.
>
> Ronald Aylmer Fisher, [24, p. 311]

> Are the very mathematical notions of probability of any relevance to actual market quotations?
>
> Paul Anthony Samuelson, [94, p. 41]

Keywords: Randomness, typicalness, chaoticness, stochasticness, E-mini S&P futures, tick prices, empirical distributions of price increments, discrete probability distributions, Dirichlet series, Hurwitz and Riemann Zeta functions, Zipf's law, power laws, Zipf–Mandelbrot law, Kolmogorov complexity, a-b-c-process, maximum profit strategy, optimal trading elements, C++, relationship between sample moments of prices and their increments.

Introduction

The theory of probability [8, pp. 3–11, I. Chance and Natural Laws; pp. 199–211, X. Philosophical Significance of the Laws of Chance], [9, pp. 87–95, IX. Probability Becomes Certainty], [41, pp. 12–14, $2. Relationship to experience data], [77, pp. 121–198, Letters on Probability; pp. 286–312, Gambling and Probability Theory; pp. 313–325, Notes on Teaching Probability Theory], [42, pp. 27–33, 2. Probability and Frequency], [27, pp. 16–20, $1. Intuitive notions of random events], [34, p. 15], [98, pp. 14–17, 1.2. Randomness and Random Variables], [99, pp. 12–25, 1.1. Randomness, Probability, Expectation], statistics [24, 73], testing statistical hypothesis [69], [42, pp. 137–139, Example], [50, 51], statistical estimation [70], [42, pp. 132–136, Example] rely on *randomness* but do not say what is this. '*A random process is a random function $x(t)$ of an independent variable t*' [44, p. 584] — *the same old randomness* not clarifying the essence. '*It is possible to consider a random process as a collection of either process realizations $X(t)$ or random variables $x_1 = x(t_1)$, $x_2 = x(t_2)$, ...*' obeying a joint distribution, [44, p. 584]. Both schemes imply that conditions at t_1 and t_2 may differ. But, at each t_i, the *trials* must be repeated under the same conditions. From now, t is time.

The words *sequence* and *chain* are borrowed from [43, p. 427/391]: "*When saying 'sequence' we shall have in mind an infinite sequence and when saying 'chain' a finite chain*". Each futures contract has the *first listing day* and *termination of trading* yielding a finite number of trading sessions. Each session brings irregular but finite number of Time & Sales ticks: date and time, price, size, a price condition symbol. *Thus, each futures generates a chain of ticks.*

Market conditions vary, creating *the only* realization that makes approaches, requiring *several* realizations, helpless. Schemes investigating each random variable in a chain by a *single* value are not suitable too. Reference [86, pp. 76, 24.1 To "shy criticism"]. A harsher sentence we find in [106, p. 8]: "*... conclusions obtained by applying probability theory in a situation where there is no statistical ensemble of experiments do not possess scientific credibility*". Yes, a chain of futures prices being an individual object is perceived to be random. "*But what do the words 'perceived to be random' mean? Classical probability theory does not give an answer to this important question*" [43, p. 426/390].

Considering individual sequences of zeros and ones, Kolmogorov and Uspenskii explain the following: "*Random sequences possess three fundamental properties. Each of these may be taken as the basis of a definition* of randomness" [43, p. 428/391]. They introduce the terms *typicalness* [43, pp. 428–429/391–392], *chaoticness* [43, pp. 429–430/392–393], and *stochasticness* [43, pp. 434–438/396–398]. Both [43, 107] are not only a development of these concepts but a review of the contributions of predecessors. The author tried to popularize these results for economists and traders [86, pp. 75–81, 24 A Comment on Randomness]. The latter also applies Δ-*randomness* [86, p. 81] to the July 2013 Corn ZCN13 contract traded down to the limit price on Thursday March 28, 2013 [86, p. 40, Figure 18]. However, what Harold Edwards said about his book applies to this attempt too: *"If you read this book without reading the primary sources you are like a man who carries a sack lunch to a banquet"*, [20, p. xi]. The monographs [14, 53, 71, 109] and recollections [110] are a good help.

Reviewing a sequence with twice as many 0s as 1s, the authors of [107, p. 109/125] conclude: *"... the notion of randomness makes sense only with respect to a given probability distribution ..."*. The futures markets are rich in unique chains of transaction times, prices, their increments, and volumes. *The current objective is to analyze them, introduce a discrete distribution for the futures price increments, and discuss what is needed in order to apply to them the notions of typicalness, chaoticness, and stochasticness. The focus is on the popular E-mini S&P 500 futures Time & Sales chains of ticks. Are they random?*

The Futures Trade Tick Chains

The Time & Sales ticks *representing trades* in the CME Group electronic trading platform *Globex* are triplets of time, price, and positive volume. The *non-trade* ticks, such as the *settlement price* ones, are reported with zero volume.

Trading sessions and time ranges

A futures contract tick chain is organized as $1 \leq D$ *trading sessions* indexed here by $d = 1, \ldots, D$. The beginning "b" and end "e" session times are set by the *contract specifications*: $[t_b^1, t_e^1], \ldots, [t_b^d, t_e^d], \ldots, [t_b^D, t_e^D]$. A dth session encloses $1 \leq R_d$ time ranges indexed here by $r = 1, \ldots, R_d$. This creates the segments

$$([t_b^{1,1}, t_e^{1,1}], \ldots, [t_b^{1,r}, t_e^{1,r}], \ldots, [t_b^{1,R_1}, t_e^{1,R_1}]), \ldots,$$
$$([t_b^{d,1}, t_e^{d,1}], \ldots, [t_b^{d,r}, t_e^{d,r}], \ldots, [t_b^{d,R_d}, t_e^{d,R_d}]), \ldots,$$
$$([t_b^{D,1}, t_e^{D,1}], \ldots, [t_b^{D,r}, t_e^{D,r}], \ldots, [t_b^{D,R_D}, t_e^{D,R_D}]), \tag{1}$$

where $t_b^d = t_b^{d,1}$, $t_e^d = t_e^{d,R_d}$. $0 \leq N_d$ and $0 \leq N_{d,r}$ are the numbers of ticks in the dth session and its rth range. N is the number of ticks per contract life

$$N_d = \sum_{r=1}^{r=R_d} N_{d,r}, \; N = \sum_{d=1}^{d=D} N_d = \sum_{d=1}^{d=D} \sum_{r=1}^{r=R_d} N_{d,r}. \qquad (2)$$

Each d associates with a *session date*, a part of t_e^d, allowing to use a date label instead of d. t_b^d may have an earlier date than t_e^d. Each contract produces own ticks, Figure 1, and contract labels can be useful. Central Time, CT, is quoted.

Times (1) can differ for different types of sessions.

(1) **Ordinary:** E-mini S&P 500 December 2019 Futures ESZ19. Thursday December 19, 2019: $R_{20191219}^{ESZ19} = 2$, $[t_b^{20191219,1} = 2019/12/18\ 17{:}00{:}00$, $t_e^{20191219,1} = 2019/12/19\ 15{:}15{:}00]$, $[t_b^{20191219,2} = 2019/12/19\ 15{:}30{:}00$, $t_e^{20191219,2} = 2019/12/19\ 16{:}00{:}00]$, $N_{20191219,1}^{ESZ19} = 37437$, $N_{20191219,2}^{ESZ19} = 232$, $N_{20191219}^{ESZ19} = N_{20191219,1}^{ESZ19} + N_{20191219,2}^{ESZ19} = 37669$.

(2) **Termination of trading:** ESZ19. Friday December 20, 2019: $R_{20191220}^{ESZ19} = 1$, $[t_b^{20191220,1} = 2019/12/19\ 17{:}00{:}0$, $t_e^{20191220,1} = 2019/12/20\ 08{:}30{:}00]$, $N_{20191220,1}^{ESZ19} = N_{20191220}^{ESZ19} = 4681$. https://www.cmegroup.com/trading/equity-index/us-index/e-mini-sandp500_contract_specifications.html: *"the 3d Friday of the contract month"*.

(3) **Pre-holiday:** March 2020 ESH20. Tuesday December 24, 2019: $R_{20191224}^{ESH20} = 1$, $[t_b^{20191224,1} = 2019/12/23\ 17{:}00{:}00$, $t_e^{20191224,1} = 2019/12/24\ 12{:}15{:}00]$, $N_{20191224,1}^{ESH20} = N_{20191224}^{ESH20} = 138805$.

The *intra-range* triplets $(t_i^{d,r}, P_i^{d,r}, V_i^{d,r})$ are indexed here by $i = 1, \ldots, N_{d,r}$. For $R_d = 1$, the r is omitted in (t_i^d, P_i^d, V_i^d) and i becomes the *intra-session* index. For $N_d = 0$ and/or $N_{d,r} = 0$, the i is undefined. Thus, a contract label and three integers $1 \leq d \leq D$, $1 \leq r \leq R_d$, $1 \leq i \leq N_{d,r}$ identify a futures trade tick chain and locate any tick within it.

The a-b-c-d-increments

The first "f" trade may occur not in the first session and not in the first range of a session. The last "l" trade may happen not in the last session and not in the last range of a session. The first and last triplets are $(t_1^{d_f,r_f}, P_1^{d_f,r_f}, V_1^{d_f,r_f})$ and $(t_{N_{d_l,r_l}}^{d_l,r_l}, P_{N_{d_l,r_l}}^{d_l,r_l}, V_{N_{d_l,r_l}}^{d_l,r_l})$, where $1 \leq d_f \leq d_l \leq D$, $1 \leq r_f \leq R_{d_f}$, $1 \leq r_l \leq R_{d_l}$. For classification and convenience, [86, pp. 16–20] introduces the

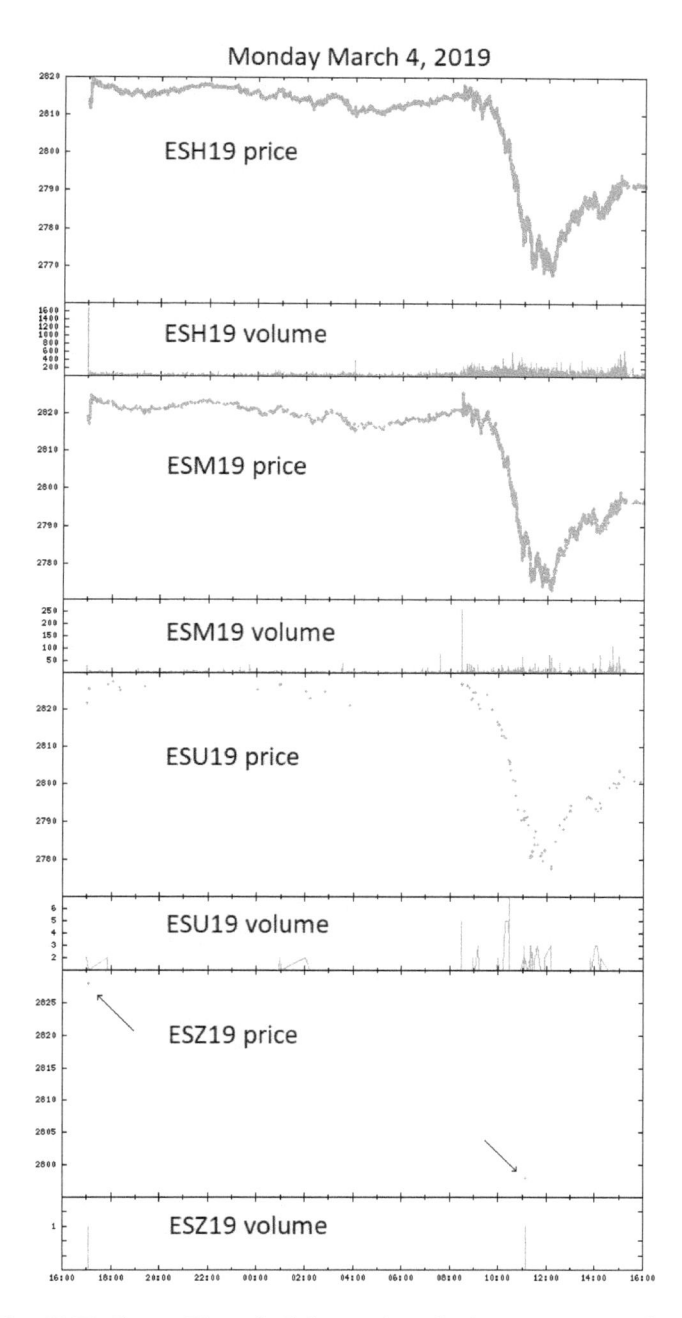

Figure 1. The CME Group Time & Sales trade tick times, prices, and volumes of the E-mini S&P 500 March H, June M, September U, December Z 2019 futures for the session on Monday March 4, 2019 consisting of the ranges ([2019/03/03 17:00:00, 2019/03/04 15:00:00], [2019/03/04 15:15:00, 2019/03/04 16:00:00]).

a-, b-, c-, ci-increments complemented here with the d-, db-, de-, di-, dbi-, dei-increments

a-increment: $\Delta t_i^{d,r} = t_i^{d,r} - t_{i-1}^{d,r}$, $i = 2, \ldots, N_{d,r}$,

b-increment: $\Delta P_i^{d,r} = P_i^{d,r} - P_{i-1}^{d,r}$, $i = 2, \ldots, N_{d,r}$,

c-increment: $\Delta P^d = P_1^d - P_{N_{d-1}}^{d-1}$, $d = 2, \ldots, d_l$,

d-increment: $\Delta t^d = t_1^d - t_{N_{d-1}}^{d-1}$, $d = 2, \ldots, d_l$,

db-increment: $\Delta t_b^d = t_1^d - t_b^d$, $d = d_f, \ldots, d_l$,

de-increment: $\Delta t_e^d = t_e^d - t_{N_d}^d$, $d = d_f, \ldots, d_l$,

ci-increment: $\Delta P^{d,r} = P_1^{d,r} - P_{N_{d,r-1}}^{d,r-1}$, $d = d_f, \ldots, d_l$, $r = 2, \ldots, R_d$,

di-increment: $\Delta t^{d,r} = t_1^{d,r} - t_{N_{d,r-1}}^{d,r-1}$, $d = d_f, \ldots, d_l$, $r = 2, \ldots, R_d$,

dbi-increment: $\Delta t_b^{d,r} = t_1^{d,r} - t_b^{d,r}$, $d = d_f, \ldots, d_l$, $r = 1, \ldots, R_d$,

dei-increment: $\Delta t_e^{d,r} = t_e^{d,r} - t_{N_{d,r}}^{d,r}$, $d = d_f, \ldots, d_l$, $r = 1, \ldots, R_d$. (3)

The a- and b-increments are durations of time and price differences between the neighboring ticks within a session with $R_d = 1$ and $2 \leq N_d = N_{d,1}$ or a range with $2 \leq N_{d,r}$. Both are undefined, if $N_{d,r} < 2$.

The c-increments are price differences between the first and last prices in the current and previous sessions. They associate with the d-increments equal to the sum of the previous session de-increment, a known duration separating two sessions, and the current session db-increment $t_1^d - t_{N_{d-1}}^{d-1} = \Delta t_e^{d-1} + (t_b^d - t_e^{d-1}) + \Delta t_b^d$. The c-, d-, db-, de-increments "do not care" which ranges within a session contain the first and last ticks. The c-, d-increments are undefined, if $N_d = 0$ and/or $N_{d-1} = 0$. The db-, de-increments are undefined, if $N_d = 0$.

The ci-increments are price differences between the first and last prices in the current and previous ranges within a session. They correspond to the the di-increments equal to the sum of the previous range dei-increment, a known duration between two ranges, and the current range dbi-increment $t_1^{d,r} - t_{N_{d,r-1}}^{d,r-1} = \Delta t_e^{d,r-1} + (t_b^{d,r} - t_e^{d,r-1}) + \Delta t_b^{d,r}$. The "i" in "ci", "dbi", and "dei" means *internal* with respect to a session containing several ranges $1 < R_d$. The ci-, di-increments are undefined, if $N_{d,r} = 0$ and/or $N_{d,r-1} = 0$. The dbi-, dei-increments are undefined, if $N_{d,r} = 0$.

For *liquid* contracts, the d- and di-increments are greater than the a-increments and almost constant and equal to the known pauses between sessions and ranges, what often allows to ignore them. Figure 1, ESZ19, and

Table 1 illustrate that d-, db-, de-, di- (not shown), dbi-, dei-increments can be irregular.

Undefined quantities: In Table 1, $N_d = N_{d,1} = N_{d,2} = 0$ for $d = 2019\text{-}03\text{-}05$ making all increments given by equation (3) undefined. $N_d = N_{d,1} = 1$ and $N_{d,2} = 0$ for $d = 2019\text{-}03\text{-}06$ define the db-, de-increments for the session, and dbi-, dei-increments for its first range. In addition, c-, d-increments are defined for $d = 2019\text{-}03\text{-}04$ and $s-1 = 2019\text{-}03\text{-}03$ with $N_d = 3$ and $N_{d-1} = 2$, even, $N_{d-1,2} = 0$. Since for all sessions in Table 1 from two ranges one or both have no ticks, ci-, di-increments are undefined and the corresponding columns are excluded. In Table 1, the c-, d-, db-, de-, dbi-, dei-values are shown on the lines of the first and/or last ticks forming the values.

The a-b-c-d-process

The first $(t_1^{d_f,r_f}, P_1^{d_f,r_f}, V_1^{d_f,r_f})$ and last $(t_{N_{d_l,r_l}}^{d_l,r_l}, P_{N_{d_l,r_l}}^{d_l,r_l}, V_{N_{d_l,r_l}}^{d_l,r_l})$ triplets are the beginning and end of a futures contract trade tick chain — a unique realization of an *a-b-c-d-process*. Reference [86, pp. 16–20] says an *a-b-c-process* ignoring the almost constant and known durations between sessions and ranges supported by a high liquidity assuming big numbers of ticks N_d and volumes V_d. Despite the irregular shocks, the N_d and V_d demonstrate *systematic evolution* throughout the life of the contract [87, p. 13, Figures 5, 6]. The daily volumes are approximated by $V(t) = A(T-t)^B(t-T_0)^C e^{D(t-T_0)}$ [87, pp. 16–20, Figures 9 and 10] with more calibration details in [89, pp. 32–36]. The Kumaraswamy (preferred) and Gama, in contrast with Weibull, distributions fit a specific dependence between the sample skewness and excess kurtosis of the daily volume fluctuations [89, p. 35, Figure 7] and a-increments [86, pp. 23, 24, Figures 8 and 9].

For a while, after listing, the N_d remains small, Figure 1, ESZ19. Prices of the frequently traded, liquid, ESH19, ESM19, ESU19 move *synchronously*. The three (two coincide) ticks of ESZ19 follow them too. *But to move in sync, prices of contracts with infrequent ticks must do greater steps.* The b-increments sample standard deviations S in Table A1 support this conclusion for $d = \text{Date}(t_{N_{d,r}}^{d,r}) = 2019\text{-}03\text{-}04$, $r = 1$: ESH19 0.42614, ESM19 1.2835, ESU19 8.4951, ESZ19 84.853. Figure 2 illustrates this non-stationarity of b-increments and a drop of the range sample distributions in time.

Figure 3 confirms synchronous moves of the ESH19, ESM19, ESU19, ESZ19 prices for two session ranges on March 4, 2019 presenting linear correlations of one minute sample means. In general, ticks arrive at irregular times separated by the a-increments. This complicates regression analysis

Table 1. ESZ19 trade ticks, $\delta = 0.25$, $k = \$50.00$; $r = 1,\ldots,R_d = 2$; $i = 1,\ldots,N_{d,r}$; a-, d-, db-, de-, dbi-, dei-increment in seconds; b-, c-increment in δ; $r = 1$: [Date 1 17:00:00, Date 15:15:00], $r = 2$: [Date 15:30:00, Date 16:00:00]. U means undefined.

d	r	i	$t_i^{d,r}$	$P_i^{d,r}$	$V_i^{d,r}$	a	b	c	d	db	de	dbi	dei
2019-03-01	1	1	2019-03-01 08:49:27	2821.00	1	U	U	U	U	56967	22287	56967	19587
		2	2019-03-01 09:48:33	2810.00	1	3546	−44						
	2	U	U	U	U	U	U	U				U	U
2019-03-04	1	1	2019-03-03 17:04:08	2828.00	1	U	U	72	198935	248		248	
		2	2019-03-04 11:07:41	2798.00	1	65013	−120				17539		14839
		3	2019-03-04 11:07:41	2798.00	1	0	0						
	2	U	U	U	U	U	U	U				U	U
2019-03-05	1	U	U	U	U	U	U	U	U	U	U	U	U
	2	U	U	U	U	U	U	U					
2019-03-06	1	1	2019-03-05 19:55:10	2803.00	2	U	U	U	U	10510	72290	10510	69590
	2	U	U	U	U	U	U	U			U	U	U

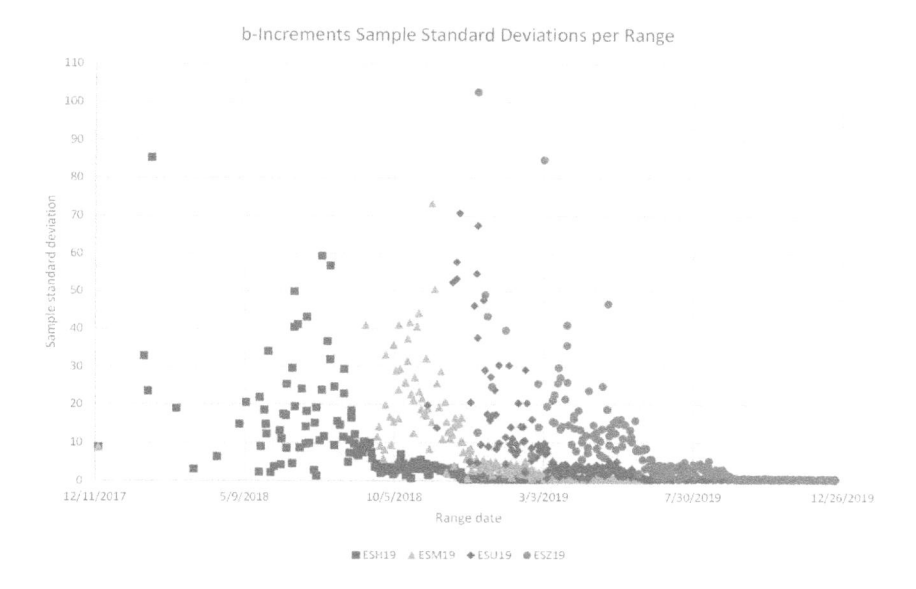

Figure 2. A drop of range sample standard deviations in time, Table A1.

requiring prices observed at one time. That is why, prices from two 1335 and 30 minutes ranges were divided into 1365 one minute intervals and sample mean prices were computed for each interval containing at least one price. A minute interval with prices available for two paired contract months creates one point on Figure 3.

In contrast with the mentioned a-increments dependencies, the b-increments in Table A1 do not show a clear relationship between the sample skewness SK and excess kurtosis EK. The skewness for many ranges is close to zero indicating symmetry of a sample distribution. The excess kurtosis deviates from zero pointing to non-Gaussian properties. Sample distributions, significantly deviating from Gaussian, were reported for futures b-increments [81, p. 35], [86, pp. 41–45, 60, 61], [87, pp. 4–6, 37–40], [92, pp. 23–26].

A formal description of the a-b-c-d-process: For a chain of ranges, each with $2 \leq N_{d,r}$, an *a-property* "generates" at $t_i^{d,r}$ a next a-increment $\Delta t_{i+1}^{d,r}$. If $t_i^{d,r} + \Delta t_{i+1}^{d,r} = t_{i+1}^{d,r} < t_e^{d,r}$, then a *b-property* "generates" a next b-increment $\Delta P_{i+1}^{d,r}$ so that $0 \leq P_{i+1}^{d,r} = P_i^{d,r} + \Delta P_{i+1}^{d,r}$. This is continued as an *a-b-process*, adding to $t_1^{d,r}$ and $P_1^{d,r}$ the sums of the generated a- and b-increments, until a next $t_{i+1}^{d,r} \geq t_e^{d,r}$. The latter terminates the rth range fixing $N_{d,r}$ and $\Delta t_e^{d,r}$. If $r < R_d$, then *dbi-property* "generates" a next dbi-increment $\Delta t_b^{d,r+1}$. This sets $t_1^{d,r+1} = t_b^{d,r+1} + \Delta t_b^{d,r+1}$. The *ci-property*

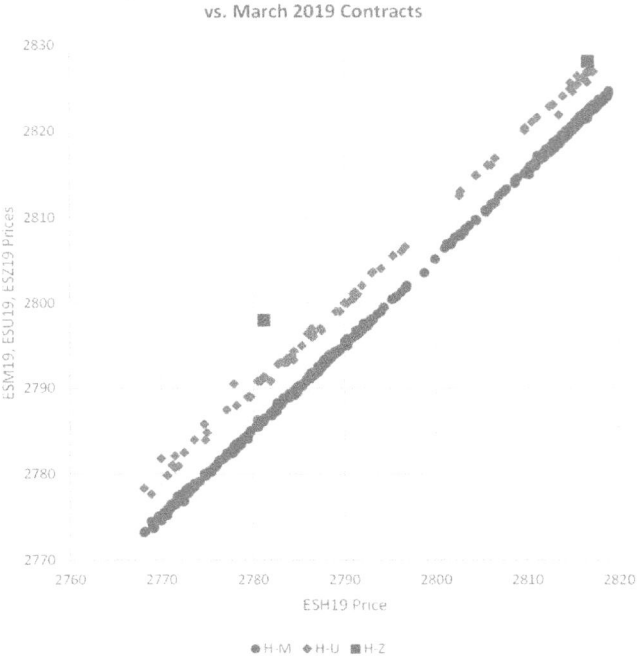

March 4, 2019 Prices of E-Mini June, September, December vs. March 2019 Contracts

Figure 3. Regressions of E-mini S&P 500 one minute mean prices for the pairs of expiration months. H-M: $P^{ESM19} = (1.00836 \pm 0.00077) \times P^{ESH19} + (-18.2 \pm 2.2)$, observations 881, coefficient of linear correlation 0.99993, confidence level 95%; H-U: $P^{ESU19} = (1.0024 \pm 0.0078) \times P^{ESH19} + (3.15 \pm 21.8)$, 93, 0.9993, 95%; H-Z: $P^{ESZ19} = (0.85 \pm n/a) \times P^{ESH19} + (441.5 \pm n/a)$, 2, n/a, n/a. Microsoft Excel, the Analysis ToolPak, Regression.

generates a ci-increment and "concatenates by price" the previous rth and now current $(r+1)$-th ranges setting $0 \leq P_1^{d,r+1} = P_{N_{d,r}}^{d,r} +$ ci-increment. This is continued until the last R_dth range inclusively and terminates the session. After that a *db-property* "creates" a db-increment for the $(s+1)$-th session establishing $t_1^{d+1,1}$. A *c-property* "concatenates by price" the previous dth and now current $(d+1)$-th sessions finalizing $0 \leq P_1^{d+1,r} = P_{N_{d,R_d}}^{d,R_d} +$ c-increment. All repeats until the last session terminates.

On Sample Statistics of Values and Increments

A price chain $P_1^{d,r}, \ldots, P_{N_{d,r}}^{d,r}$ is also a sample of $N_{d,r}$ prices. A b-increments chain $\Delta P_2^{d,r}, \ldots, \Delta P_{N_{d,r}}^{d,r}$ is also a sample of $N_{d,r} - 1$ b-increments, where

$$P_i^{d,r} = P_1^{d,r} + \sum_{j=2}^{j=i}(P_j^{d,r} - P_{j-1}^{d,r}) = P_1^{d,r} + \sum_{j=2}^{j=i}\Delta P_j^{d,r},\ i = 1, \ldots, N_{d,r}, \quad (4)$$

In (4) and elsewhere, the sum is zero, if the top index is less than the bottom. Properties of a sample of values and a sample of increments formed from these values relate. With regard to prices, both the prices and their increments are *"perceived to be random"*. Distributions of prices are at the heart of *Bollinger Bands* [7], *Market Profile* and *Value Area* [81]. Continuous prices and/or rates increments and increments relative to prices are modeled [94] for pricing financial derivatives. What to study [86, pp. 34–35, The chicken and egg question]: fluctuations or their *algebraic* sums — prices? For instance, the sample moments can be computed for both.

The sample beginning and central moments of all orders s in equation (A1) are *arithmetic averages*. It maybe a coincidence, but the character s for the moment order in [26, p. 68] corresponds to the Greek σειρά — order. *"All known types of averages such as arithmetic, quadratic, geometric, harmonic, i.e."* [40, p. 136/388] are *symmetric functions* [40, p. 137/389, II. *"est une fonction symmetrique"*], [68, p. 71, (i) *"eine symmetrische Function"*] — independent on the permutations of arguments. Then, the same is true for the sample standard deviation S, skewness SK, excess kurtosis EK in equation (A1).

Permuting prices does not affect their sample moments but changes increments and their sample moments. Permuting prices increments with the fixed $P_1^{d,r}$ keeps their sample moments but changes prices and corresponding sample moments. From equation (A1), for the $1 \leq q \leq N_{d,r}$ adjacent ticks, ending at $i = q, \ldots, N_{d,r}$, the *moving sample moments of the order s* are

$$a_s^{P^{d,r}}(i) = \frac{\sum_{j=1}^{j=q}(P_{i-q+j}^{d,r})^s}{q}; \quad m_s^{P^{d,r}}(i) = \frac{\sum_{j=1}^{j=q}(P_{i-q+j}^{d,r} - a_1^{P^{d,r}}(i))^s}{q}. \quad (5)$$

Sample moments estimate theoretical *general population moments*. For a *random variable* ξ with the *distribution function* $F_\xi(x)$ [26, p. 28], the theoretical moments α_s, absolute moments β_s, central moments μ_s, central absolute moments ν_s of the *order s* are the *mathematical expectations*, $\mathbf{M}$, [26, p. 68]

$$\alpha_s = \mathbf{M}\xi^s = \int x^s dF_\xi(x), \quad \alpha_0 = \mathbf{M}1 = \int dF_\xi(x) = 1, \quad \alpha_1 = \mathbf{M}\xi,$$

$$\beta_s = \mathbf{M}|\xi|^s = \int |x|^s dF_\xi(x), \quad \beta_0 = 1,$$

$$\mu_s = \mathbf{M}(\xi - \mathbf{M}\xi)^s = \int (x - \alpha_1)^s dF_\xi(x), \quad \mu_0 = 1,$$

$$\nu_s = \mathbf{M}|\xi - \mathbf{M}\xi|^s = \int |x - \alpha_1|^s dF_\xi(x), \quad \nu_0 = 1. \quad (6)$$

Gnedenko and Kolmogorov clarify [26, p. 68] (VS's translation): "... *it is easy to derive that the existence of μ_s and ν_s is equivalent to the existence of α_s and β_s. Therefore, we shall speak further about the s-order moments existence, not mentioning which ones*".

Equations (5) imply that $P_{i-q+1}^{d,r}, \ldots, P_i^{d,r}$ are from a general prices population. Then, the *sample mean* $a_1^{P^{d,r}}(i)$ is an *unbiased*, $\mathbf{M}a_1^{P^{d,r}}(i) = \alpha_1^{P^{d,r}}(i)$, *consistent*, $\lim_{q \to \infty} a_1^{P^{d,r}}(i) = \alpha_1^{P^{d,r}}(i)$, estimator of the *population mean*. The $m_2^{P^{d,r}}(i)$ is a *biased*, $\mathbf{M}m_2^{P^{d,r}}(i) \neq \mu_2^{P^{d,r}}(i)$, consistent, $\lim_{q \to \infty} m_2^{P^{d,r}}(i) = \mu_2^{P^{d,r}}(i)$, estimator of the *population variance*. The *sample variance* $(S^{P^{d,r}}(i))^2 = \frac{q}{q-1} m_2^{P^{d,r}}(i)$, $1 < q$, is an unbiased, $\mathbf{M}(S^{P^{d,r}}(i))^2 = \mu_2^{P^{d,r}}(i)$, consistent, $\lim_{q \to \infty}(S^{P^{d,r}}(i))^2 = \mu_2^{P^{d,r}}(i)$, estimator of the population variance, [44, p. 611]. Bollinger Bands [7, pp. 50–59 Construction; inset page "... FORMULAS"] attracting traders [35, pp. 294–299, 754–755] and analysts [15, 47, 76] apply samples of *daily closing prices* with $q = 20$, and, in other denominations, $a_1^P(i)$, $\sqrt{m_2^P(i)}$.

Others assume that price increments are the primary random variables, while prices are just their sums added to the first price, equation (4). Then, for $2 < q$

$$a_s^{\Delta P^{d,r}}(i) = \frac{\sum_{j=2}^{j=q}(\Delta P_{i-q+j}^{d,r})^s}{q-1}; \quad (S^{\Delta P^{d,r}}(i))^2 = \frac{q-1}{q-2} m_2^{\Delta P^{d,r}}(i),$$

$$m_s^{\Delta P^{d,r}}(i) = \frac{\sum_{j=2}^{j=q}(\Delta P_{i-q+j}^{d,r} - a_1^{\Delta P^{d,r}}(i))^s}{q-1}. \tag{7}$$

While statistics (5), (7) do not depend on the order of values, it cannot be ignored, if prices and increments are studied together. All complicates, if from $i - q + 1$ to i the distributions change. *Even, if the type of the distribution of whether prices or increments remains intact but the moments change, causing non-stationary properties, the application of sample mean and variance rises concerns.*

Relationship between $a_1^{P^{d,r}}(i)$ and $a_1^{\Delta P^{d,r}}(i)$

Changing denominations in (A2)–(A4) for $1 < q \leq N_{d,r}$, $i = q, \ldots, N_{d,r}$ yields

$$a_1^{*\Delta P^{d,r}}(i) = \frac{\sum_{j=2}^{j=q} j \Delta P_{i-q+j}^{d,r}}{\sum_{j=2}^{j=q} j}, \quad a_2^{*\Delta P^{d,r}}(i) = \frac{\sum_{j=2}^{j=q} j(\Delta P_{i-q+j}^{d,r})^2}{\sum_{j=2}^{j=q} j},$$

$$a_1^{P^{d,r}}(i) = P_{i-q+1}^{d,r} + \frac{q^2-1}{q} a_1^{\Delta P^{d,r}}(i) - \frac{(q-1)(q+2)}{2q} a_1^{*\Delta P^{d,r}}(i),$$

$$a_1^{\Delta P^{d,r}}(i) = \frac{q}{q^2-1}\left(a_1^{P^{d,r}}(i) - P_{i-q+1}^{d,r} + P_i^{d,r} - \frac{P_{i-q+1}^{d,r} + \sum_{j=1}^{j=q-1} P_{i-q+j}^{d,r}}{q}\right)$$

$$= \frac{P_i^{d,r} - P_{i-q+1}^{d,r}}{q-1}. \tag{8}$$

Separately, neither $a_1^{P^{d,r}}(i)$ nor $a_1^{\Delta P^{d,r}}(i)$ depend on the order of sample values. Together, they may due to $j\Delta P_{i-q+j}^{d,r}$ in equation (8). From this point of view, the non-moving [88, pp. 28–29] and *moving* sample means are similar.

Relationship between $m_2^{P^{d,r}}(i)$ and $m_2^{\Delta P^{d,r}}(i)$

From equations (5), (7), (8), (A3), (A9),

$$m_2^{P^{d,r}}(i) = \frac{q^2-1}{q}m_2^{\Delta P^{d,r}}(i) - \frac{(q^2-1)(q^2-q-1)(a_1^{\Delta P^{d,r}}(i))^2}{q^2}$$

$$- \frac{(q-1)(q+2)a_2^{*\Delta P^{d,r}}(i)}{2q}$$

$$+ \frac{(q^2-1)(q-1)(q+2)a_1^{\Delta P^{d,r}}(i)a_1^{*\Delta P^{d,r}}(i)}{q^2}$$

$$- \left[\frac{(q-1)(q+2)a_1^{*\Delta P^{d,r}}(i)}{2q}\right]^2$$

$$+ \frac{2}{q}\sum_{j=3}^{j=q}(q-j+1)\Delta P_{i-q+j}^{d,r}\sum_{l=2}^{l=j-1}\Delta P_{i-q+l}^{d,r}. \tag{9}$$

Relationship between $(S^{P^{d,r}}(i))^2$ and $(S^{\Delta P^{d,r}}(i))^2$

From equations (7), (9),

$$(S^{P^{d,r}}(i))^2 = \frac{q^2-q-2}{q-1}(S^{\Delta P^{d,r}}(i))^2 - \frac{(q+1)(q^2-q-1)(a_1^{\Delta P^{d,r}}(i))^2}{q}$$

$$- \frac{(q+2)a_2^{*\Delta P^{d,r}}(i)}{2} + \frac{(q+1)(q^2+q-2)a_1^{\Delta P^{d,r}}(i)a_1^{*\Delta P^{d,r}}(i)}{q}$$

$$- \frac{(q-1)(q+2)^2(a_1^{*\Delta P^{d,r}}(i))^2}{4q^2} +$$

$$+ \frac{2}{q-1}\sum_{j=3}^{j=q}(q-j+1)\Delta P_{i-q+j}^{d,r}\sum_{l=2}^{l=j-1}\Delta P_{i-q+l}^{d,r}. \tag{10}$$

Daily closing prices and b-increments

The difference between today and yesterday closing prices is the sum of the b-c-ci-increments. The number of summands in sessions and ranges varies, Table A1. If it is assumed that the tick price increments are *independent identically distributed, i.i.d.*, random variables, then the sums with the different number of summands, in general, cannot be i.i.d. random variables [26]. If it is assumed that the daily price increments are i.i.d., then the tick increments cannot be anymore recognized i.i.d. The same relates to the so-called *log-returns*. From [88, p. 2], *"Ignorance of intra-day ticks yields controversial daily price applications"*.

Empirical Distributions of b-Increments

Table A1 shows that ES-mini futures sample moments of b-increments change from session to session. Table A2 confirms that the sample distributions of b-increments from several sessions differ. Table A3 proves that these distributions vary within a session or range. Since prices and their increments are integer multiples of the minimal non-zero absolute price change $\delta_{ES} = 0.25$ assigned by the futures contract specifications [89, p. 5, Figure 1], the empirical distributions are discrete. *A discrete distribution is characterized by values associated with probabilities summing to one. The countable number of values can be infinite.*

Maximum futures prices

The Corn and some other futures have daily trading price limits [87, pp. 11–12, 5 Price limits]. This guarantees that tomorrow a futures price cannot be unlimited. Then, a possible maximum positive b-increment is limited too and known in advance. However, it is unknown, if it will be realized tomorrow. The up limit was set for ESM20 on Friday and Tuesday March 13 and 24, 2020. The down limit was set for ESM20 on Mondays March 16 and 23, 2020 Figure A1. *While a futures price is never infinite, absence of a limit makes the potential maximum price and positive increment arbitrary big and uncertain. This implies that a next drop to zero can be unlimited too. Moreover, prices can be negative.*

Minimum and negative futures prices

In [87, pp. 1–2], a zero stock price associates with bankruptcy and a negative price with a garbage removal cost. On Monday April 20, 2020, the West

Texas Intermediate, WTI, crude oil for May delivery fell more than 100% to settle at negative $37.63 per barrel. The oil producers paid traders to take the product off their hands. At futures expiration, the future and spot prices must coincide to avoid arbitrage and the May 2020 Crude Oil futures contract CLK20 expiring on Tuesday April 21, 2020 "dived to the negative bottom". We expect that by the nature of S&P 500 Index, an ES-mini futures price should not drop below zero. *In general, for a discrete distribution of futures prices and increments, negative prices imply that a low distribution value can be uncertain.*

Empirical frequencies of b-increments

The δ futures specification simplifies selection of bins for building empirical discrete distributions of the b-increments, Tables 7 and A1–A3. A bin's frequency divided by the sum of frequencies for all bins estimates a probability associated with the bin. The estimate assumes that the distribution remains the same in time intervals from which the b-increments are collected. Tables A1–A3 indicate that for long intervals this is not so. But too short intervals lose statistical meaning. *It is striking that, despite the non-stationarity, the empirical frequencies depend on the b-increment serving as the bin's rank and form straight lines in specially constructed bi-logarithmic coordinates, Figure 8.*

The author presented dependencies of the frequencies of b-increments on their size, rank (g from "grade"), in [86, pp. 46–47, Figures 20 and 21], [87, pp. 33–34, Figures 24 and 25] and for the *extreme* b-increments in [86, pp. 51–52, Figures 23 and 24], [87, pp. 55–56, Figures 34 and 35]. Figures 4–7 plot selective *empirical quantities Q1–Q8* defined in terms of a *probability mass function*, PMF:

$$Q1(g) = PMF(g), \quad \forall g,$$

$$Q2(g) = \ln(PMF(g)), \quad \forall\, 0 < PMF(g),$$

$$Q3(g < 0) = \frac{PMF(g)}{PMF(g-1)}, \quad Q3(0 < g) = \frac{PMF(g)}{PMF(g+1)},$$

$$Q3^-(g = 0) = \frac{PMF(0)}{PMF(-1)}, \quad Q3^+(g = 0) = \frac{PMF(0)}{PMF(1)},$$

$$Q4(g < 0) = \ln\left(\frac{PMF(g)}{PMF(g-1)}\right), \quad Q4(0 < g) = \ln\left(\frac{PMF(g)}{PMF(g+1)}\right),$$

$$Q4^-(g = 0) = \ln\left(\frac{PMF(0)}{PMF(-1)}\right), \quad Q4^+(g = 0) = \ln\left(\frac{PMF(0)}{PMF(1)}\right),$$

$$Q5(g<0) = \frac{\ln\left(\frac{PMF(g)}{PMF(g-1)}\right)}{\ln\left(\frac{PMF(g-1)}{PMF(g-2)}\right)}, \quad Q5(0<g) = \frac{\ln\left(\frac{PMF(g)}{PMF(g+1)}\right)}{\ln\left(\frac{PMF(g+1)}{PMF(g+2)}\right)},$$

$$Q5^-(g=0) = \frac{\ln\left(\frac{PMF(0)}{PMF(-1)}\right)}{\ln\left(\frac{PMF(-1)}{PMF(-2)}\right)}, \quad Q5^+(g=0) = \frac{\ln\left(\frac{PMF(0)}{PMF(1)}\right)}{\ln\left(\frac{PMF(1)}{PMF(2)}\right)},$$

$$Q6(g\le -1) = \frac{PMF(-1)}{PMF(g)}, \quad Q6(1\le g) = \frac{PMF(1)}{PMF(g)},$$

$$Q7(g\le -1) = \ln\left(\frac{PMF(-1)}{PMF(g)}\right), \quad Q7(1\le g) = \ln\left(\frac{PMF(1)}{PMF(g)}\right),$$

$$Q8(g<-1) = \frac{\ln\left(\frac{PMF(-1)}{PMF(g-1)}\right)}{\ln\left(\frac{PMF(-1)}{PMF(g)}\right)}, \quad Q8(1<g) = \frac{\ln\left(\frac{PMF(1)}{PMF(g+1)}\right)}{\ln\left(\frac{PMF(1)}{PMF(g)}\right)}, \quad (11)$$

where $Q6(-1) = Q6(1) = 1$, $Q7(-1) = Q7(1) = 0$, $Q7(g) = Q2(-1|1) - Q2(g)$. Properties for choosing theoretical distributions for b-increments:

(A) Plots 4–7 b, h hint at the underlying *empirical power laws*.
(B) 4–7 a, c, d point to the *King effect* for zero b-increments.
(C) Relative fluctuations of quantities increase at higher $|g|$ with lower frequencies. 4 c, d, i, j resemble an open *fan* or a *peacock's tail*. This also "bends" the right side of the straight lines on 4 b, h.
(D) At high $|g|$ on 4 c, d, $Q4$ "dances" around zero.
(E) Similarly, on 4 i, j, $Q8$ "dances" around one.
(F) "Circles-Diamonds" on 4 b, d, f, h, j indicate a symmetrical distribution.
(G) But 5, 7 b, d, f, h, j, and less 6, tell about a loss of symmetry.
(H) Table A2 has $11\,Q5(-1)$, $13\,Q5(1) < \frac{\ln(2)}{\ln(\frac{3}{2})}$; $10\,Q5(-1)$, $8\,Q5(1) >= \frac{\ln(2)}{\ln(\frac{3}{2})}$. Table 7 has $5\,Q5(-1), 0\,Q5(1) < \frac{\ln(2)}{\ln(\frac{3}{2})}$; $9\,Q5(-1), 10\,Q5(1) >= \frac{\ln(2)}{\ln(\frac{3}{2})}$.

Theoretical Distributions for b-Increments

The Zipf–Mandelbrot distribution and PMFs constructed from the Riemann and Hurwitz Zeta functions were first time, to the best of the author's knowledge, applied to fit the observed dependencies of futures b-increments on size in [86, pp. 45–47], [87, pp. 30–37, 10 Randomness of Price Increments]. A review of terminology is in [87, pp. 36–37, "Riemann Zeta", "Zeta", "Hurwitz Zeta" distributions]. The Zipf $PMF_Z(g)$, $S = 1$, $Q = 0$, and Zipf–Mandelbrot

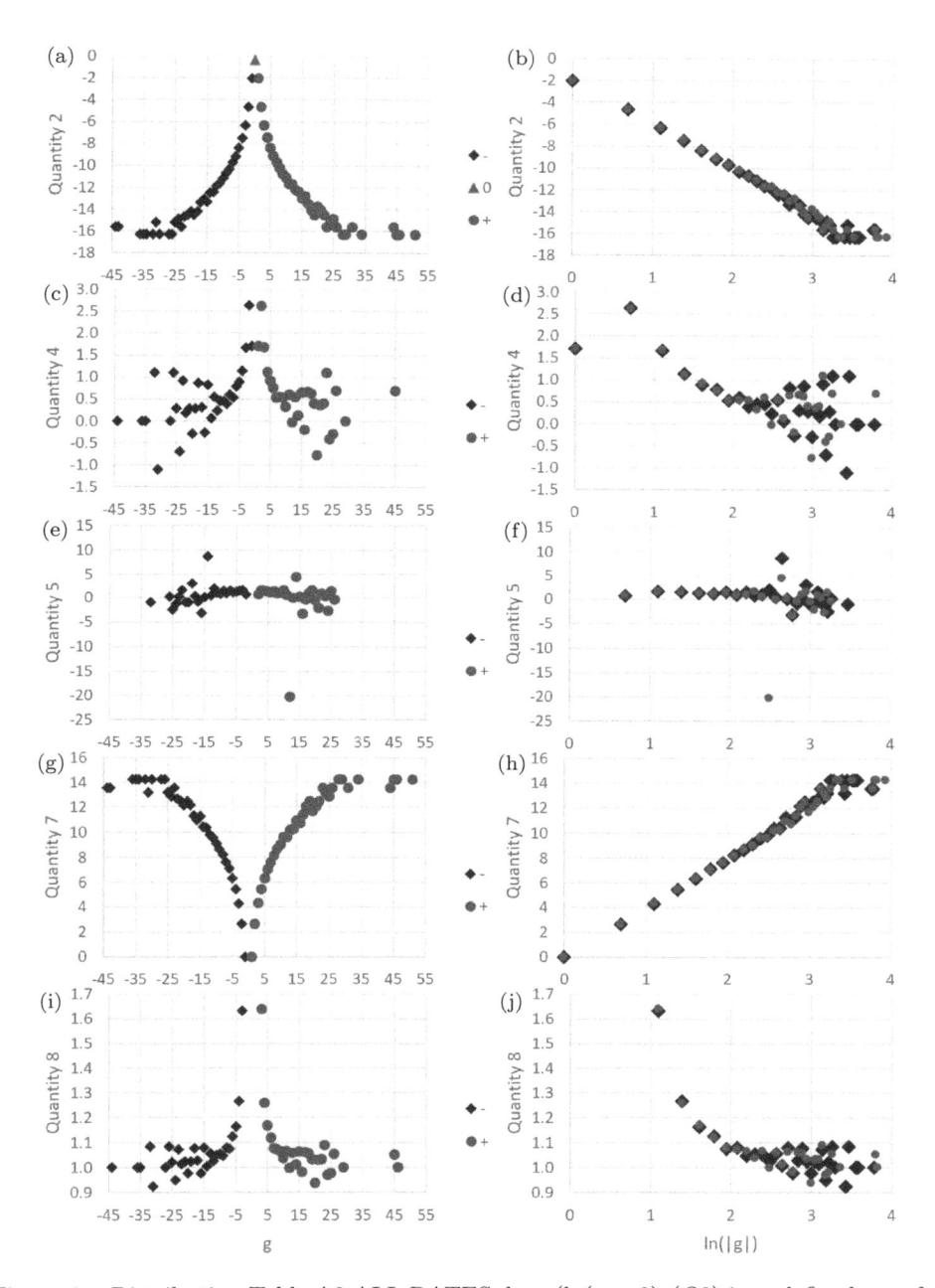

Figure 4. Distribution Table A2 ALL DATES: b — $(\ln(g = 0), (Q2)$ is undefined; c — for $-, +$ branches, Q4 is plotted against $g - 1$, $g + 1$; d — Q4 is plotted against $\ln(|g| + 1)$; e — for $-, +$ branches Q5 is plotted against $g - 2$, $g + 2$; f — Q5 is plotted against $\ln(|g| + 2)$; i — for $-, +$ branches Q8 is plotted against $g - 2$, $g + 2$; j — Q8 is plotted against $\ln(|g| + 2)$.

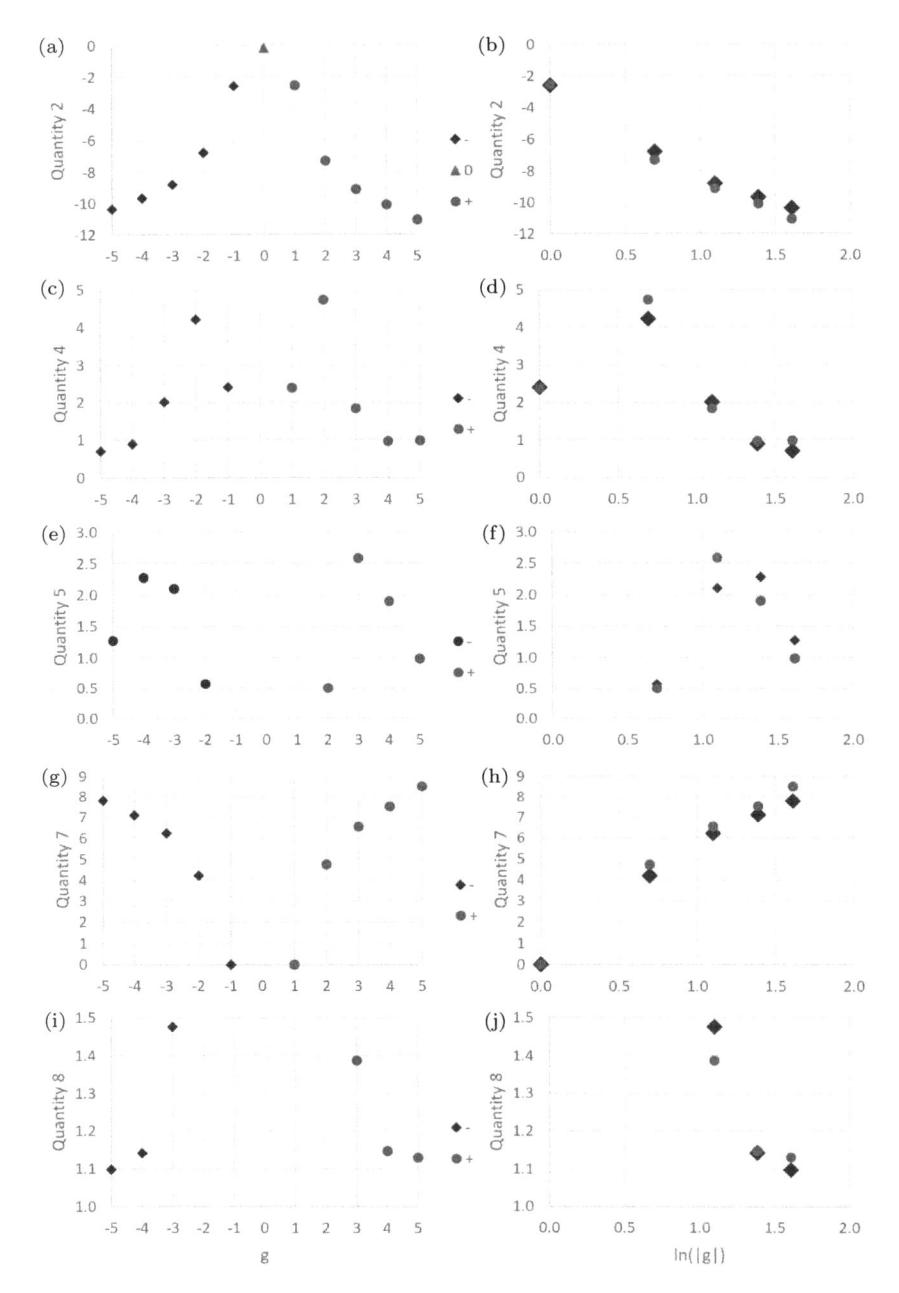

Figure 5. Distribution Table 7 ALL: b — $(\ln(g = 0), Q2)$ is undefined; c — for $-$, $+$ branches, Q4 is plotted against $g - 1$, $g + 1$; d — Q4 is plotted against $\ln(|g| + 1)$; e — for $-$, $+$ branches Q5 is plotted against $g - 2$, $g + 2$; f — Q5 is plotted against $\ln(|g| + 2)$; i — for $-$, $+$ branches Q8 is plotted against $g - 2$, $g + 2$; j — Q8 is plotted against $\ln(|g| + 2)$.

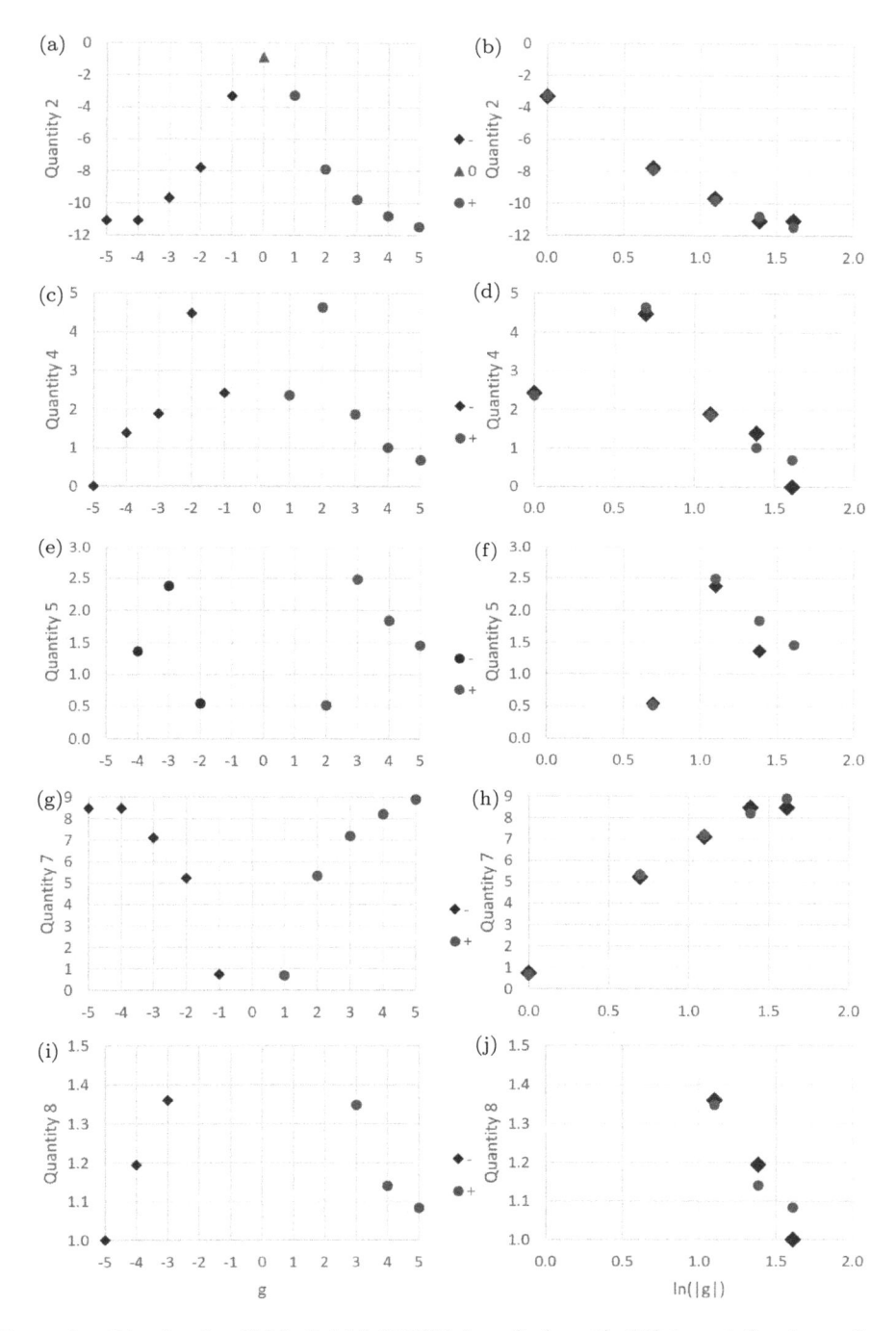

Figure 6. Distribution Table 7 ALL-BOTE: b — $(\ln(g=0), Q2)$ is undefined; c — for $-$, $+$ branches, Q4 is plotted against $g-1$, $g+1$; d — Q4 is plotted against $\ln(|g|+1)$; e — for $-$, $+$ branches Q5 is plotted against $g-2$, $g+2$; f — Q5 is plotted against $\ln(|g|+2)$; i — for $-$, $+$ branches Q8 is plotted against $g-2$, $g+2$; j — Q8 is plotted against $\ln(|g|+2)$.

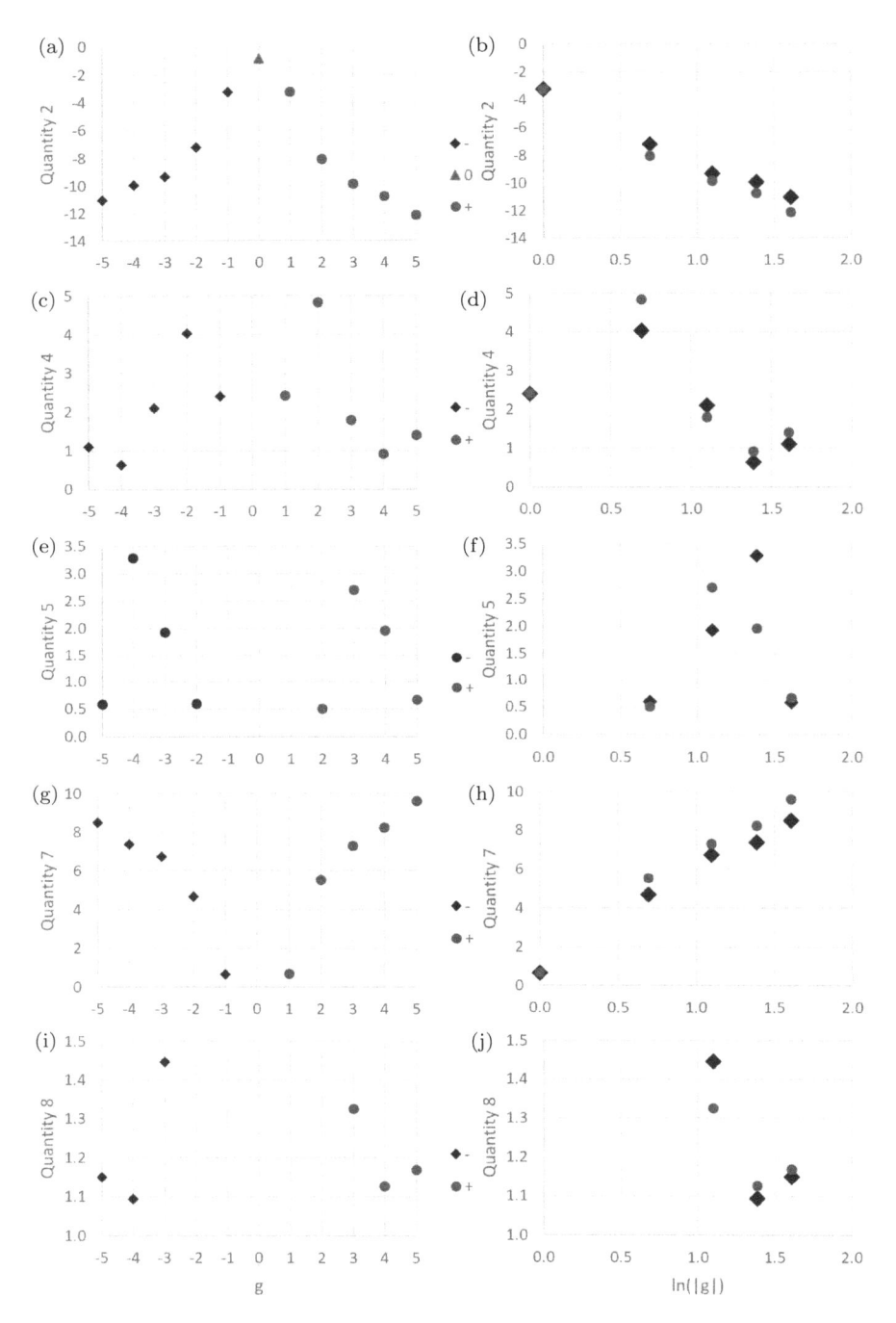

Figure 7. Distribution Table 7 ALL-SOTE: b — $(\ln(g=0), Q2)$ is undefined; c — for $-$, $+$ branches, Q4 is plotted against $g-1$, $g+1$; d — Q4 is plotted against $\ln(|g|+1)$; e — for $-$, $+$ branches Q5 is plotted against $g-2$, $g+2$; f — Q5 is plotted against $\ln(|g|+2)$; i — for $-$, $+$ branches Q8 is plotted against $g-2$, $g+2$; j — Q8 is plotted against $\ln(|g|+2)$.

$PMF_{ZM}(g)$ laws imply a finite number of ranks N_g, extreme b-increments limited by N_g,

$$PMF_{ZM}(g) = \frac{(g+Q)^{-S}}{\sum_{i=1}^{i=N_g}(i+Q)^{-S}}, \quad 0 < S, \ 0 < Q, \ 1 \leq g \leq N_g. \quad (12)$$

A price limit is specified as a change L counted from a previous *settlement price* P_s, where $0 < L < P_s$. This ensures that at any futures tick in a next session $P_s - L \leq P \leq P_s + L$. Thus, for any last price P in a next session with a trading limit: $N_g = g_{\max} - g_{\min} + 1 = \frac{2L}{\delta} + 1$, where

$$g_{\min}(P, P_s, L) = \frac{P_s - L - P}{\delta} \leq 0, \quad g_{\max}(P, P_s, L) = \frac{P_s + L - P}{\delta} \geq 0. \quad (13)$$

These $PMF_{ZM}(g)$ functions can split for negative and positive g and branches of a distribution with zero b-increment included into any of them. For symmetrical distributions the frequencies can be combined using $|g|$. Sometimes, to avoid $\ln(g=0)$ in bi-logarithmic plots, it is needed to shift the rank g. This makes the $|$b-increment$|$ not equal to g and requires a recalibration.

The Riemann Zeta $PMF_R(g)$, $1 < S$, $Q = 1$, does not fit well b-increments [86, p. 45]. The Hurwitz Zeta $PMF_H(g)$ is interesting [86, p. 48, Table 6]

$$PMF_H(g) = \frac{(g+Q)^{-S}}{\sum_{i=0}^{i=\infty}(i+Q)^{-S}} = \frac{(g+Q)^{-S}}{\zeta(Q,S)}, \quad 1 < S, \ 0 < Q, \ 0 \leq g. \quad (14)$$

For $PMF_{R,H}$, g is unlimited. For real $S, Q \in \mathbb{R}$, the Hurwitz Zeta function $\zeta(S, Q)$ is computed using the formula from [86, p. 48, 15.2 Euler–Maclaurin formula for Hurwitz Zeta]. A 64 bits Microsoft Excel XLOPER Excel4 and Excel12 Add-In was written by the author in C++ [32, 101] for the type double, accurate for our purposes, and used for fitting in conjunction with the Microsoft Excel's Solver. The site https://www.wolframalpha.com/examples/mathematics/mathematical-functions/special-functions/zeta-functions/ has shown agreement up to 14–16 meaningful decimal digits in wide ranges of S, Q. Working on [86], the author did not know about [33] researching accurate evaluations of the Hurwitz Zeta. Later, the Hurwitz Zeta was applied for construction of a family of continuous distributions [46]. *But we need a discrete one.*

Normalization, infinite b-increments, asymmetry

The b-increments measured in δ are negative, zero, and positive integers. But probabilities in equation (14) are defined for ranks $0 \leq g$. They can serve a single branch or a symmetrical distribution. Even, this requires a normalization summing probabilities to one for $g < 0$, $g = 0$, $0 < g$. Since the extreme absolute b-increments are, in general, unknown, the expressions should allow $|g| \to \infty$. Truncating the tails of practically negligible probabilities looks a "simpler" task than predicting exactly extreme b-increments. The expressions should support non-symmetrical negative and positive branches. These properties and discreteness suggest to turn to *Dirichlet series* [52, 58].

Dirichlet series

Let $\Lambda = \{\lambda_n\}$ is a *sequence* of positive strictly increasing real numbers $0 < \lambda_1 < \lambda_2 < \cdots$ and $\lim_{n \to \infty} \lambda_n = \infty$. Let $A = \{a_n\}$ is a *sequence* of *complex* numbers, $a_n \in \mathbb{C}$, and $z = \sigma + it \in \mathbb{C}$, where $s, t \in \mathbb{R}$ and $i = \sqrt{-1}$. Then, $(A, \Lambda) = \sum_{n=1}^{n=\infty} a_n e^{-\lambda_n z}$ is a Dirichlet series [58, p. 14], [52, p. 111]. Our interest is *absolutely convergent* series, with $a_n, z \in \mathbb{R}$, ensuring normalization.

Example 1. For $a_n = 1$, $\lambda_n = \ln(n)$, $z = s$, $\zeta(s) = \sum_{n=1}^{n=\infty} a_n e^{-\lambda_n z} = \sum_{n=1}^{n=\infty} \frac{1}{n^s}$ is the Riemann Zeta function [58, p. 15, equation (I.4)]. For $\sigma + it = s \in \mathbb{C}$, this "Dirichlet series ... is convergent for $\sigma > 1$, and uniformly convergent in any finite region in which $\sigma \geq 1 + \delta$, $\delta > 0$. It therefore defines an analytic function $\zeta(s)$, regular for $\sigma > 1$" [105, p. 1]. A particular case $\zeta(S)$, $1 < S \in \mathbb{R}$, $t = 0$ and $it = 0$, supports these properties. Here, $t \in \mathbb{R}$ is not time.

Example 2. For $a_n = 1$, $\lambda_n = \ln(n+q)$, $1 < q \in \mathbb{R}$, $z = s \in \mathbb{C}$, $\zeta(s, q) = \frac{1}{q^s} + \sum_{n=1}^{n=\infty} a_n e^{-\lambda_n z} = \frac{1}{q^s} + \sum_{n=1}^{n=\infty} \frac{1}{(n+q)^s} = \sum_{n=0}^{n=\infty} \frac{1}{(n+q)^s}$ is the Hurwitz Zeta function convergent absolutely and uniformly with respect to s with real parts $\mathrm{Re}(s) > 1$, $\mathrm{Re}(q) > 0$ [3, p. 40], [1, pp. 249, 251, Theorem 12.1].

A zeta function $\zeta(S, Q, H, K)$

Consider real quantities introducing $\zeta(S, Q, H, K)$

$$1 < S,\ 0 < Q,\ 0 < K,\ 1 < S(1+H) \text{ or } \frac{1}{S} - 1 < H,$$

$$a_n = 1,\ \lambda_n = \ln((n+Q)^{1+He^{-Kn}}),\ z = S,$$

$$\zeta(S, Q, H, K) = \frac{1}{Q^{S(1+H)}} + \sum_{n=1}^{n=\infty} a_n e^{-\lambda_n z} = \sum_{n=0}^{n=\infty} \frac{1}{(n+Q)^{S(1+He^{-Kn})}}.$$

(15)

For $H = 0$, the latter is the Hurwitz Zeta $\zeta(S,Q)$ with $0 < Q$, $1 < S \in \mathbb{R}$. For $H = 0$, $Q = 1$, it is the Riemann Zeta $\zeta(S)$ with $1 < S \in \mathbb{R}$. Let us notice that

$$n \in \mathbb{N}_0; \lim_{n \to \infty}(S(1 + He^{-Kn})) = \lim_{x \to \infty}(S(1 + He^{-Kx})) = S; \ m \in \mathbb{N},$$

$$\frac{d^m[S(1+He^{-Kx})]}{dx^m} = (-1)^m SHK^m e^{-Kx},$$

$$\frac{d[S(1+He^{-Kx})]}{dx} = -SHKe^{-Kx},$$

$$1 < S(1+H) \leq S(1+He^{-Kn}) < S, \quad \text{if } \frac{1}{S} - 1 < H < 0,$$

$$1 < S = S(1+He^{-Kn}), \quad \text{if } H = 0,$$

$$1 < S < S(1+He^{-Kn}) \leq S(1+H), \quad \text{if } 0 < H.$$

$S(1+He^{-Kx})$ is bound for $0 \leq x$ and with $\frac{1}{S} - 1 < H < 0$, $H = 0$, or $0 < H$ it is monotonically increasing, constant, or monotonically decreasing. With all said, $\zeta(S,Q,H,K)$ converges absolutely because (see [23, p. 264, Theorem 1])

$$\forall_{1 \leq n, 0 < Q, 1 < S, 0 \leq H} \frac{1}{(n+Q)^{S(1+He^{-Kn})}} \leq \frac{1}{(n+Q)^S},$$

$$\forall_{1 \leq n, 0 < Q, 1 < S, \frac{1}{S}-1 < H \leq 0} \frac{1}{(n+Q)^{S(1+He^{-Kn})}} \leq \frac{1}{(n+Q)^{S(1+H)}}.$$

Later, a bound but non-constant $K(n)$ is considered. Then, the mth derivative above cannot be applied.

A Probability Mass Function based on $\zeta(S,Q,H,K)$

$$1 < S, 0 < Q, \frac{1}{S} - 1 < H, 0 < K \in \mathbb{R}, \ 0 \leq i, m \in \mathbb{N}_0,$$

$$\zeta_m(S,Q,H,K) = \sum_{i=m}^{i=\infty} \frac{1}{(i+Q)^{S(1+He^{-Ki})}}; \ \zeta(S,Q,H,K)$$

$$= \zeta_0(S,Q,H,K); \ \zeta(S,Q,H,K) - \zeta_1(S,Q,H,K) = \frac{1}{Q^{S(1+H)}},$$

$$g \in \mathbb{Z}; \ 1 < S_-, 0 < Q_-, \frac{1}{S_-} - 1 < H_-, 0 < K_- \in \mathbb{R}; \ 1 < S_+, 0 < Q_+,$$

$$\frac{1}{S_+} - 1 < H_+, 0 < K_+ \in \mathbb{R}; \ -\left(\frac{1}{Q_-^{S_-(1+H_-)}} + \frac{1}{Q_+^{S_+(1+H_+)}}\right) < V \in \mathbb{R},$$

$$D = \zeta(S_-, Q_-, H_-, K_-) + \zeta(S_+, Q_+, H_+, K_+) + V,$$

$$Q_-^{-S_-(1+H_-)} + Q_+^{-S_+(1+H_+)} + V = D - \zeta_1(S_-, Q_-, H_-, K_-)$$
$$-\zeta_1(S_+, Q_+, H_+, K_+),$$

$$PMF(g) = \begin{cases} \frac{(-g+Q_-)^{-S_-(1+H_- e^{gK_-})}}{D} & \text{if } g < 0, \\ \frac{Q_-^{-S_-(1+H_-)} + Q_+^{-S_+(1+H_+)} + V}{D} & \text{if } g = 0, \\ \frac{(g+Q_+)^{-S_+(1+H_+ e^{-gK_+})}}{D} & \text{if } 0 < g. \end{cases} \quad (16)$$

Convergence of $\zeta(S, Q, H, K)$ supports $\sum_{g=-\infty}^{g=\infty} PMF(g) = 1$. The subscripts $-$ and $+$ denote two branches of the asymmetric probability distribution. *For a better understanding, let us consider a particular simpler case, $H_- = H_+ = 0$, data, where the simplification is sufficient, and where it does not work.*

Symmetric and asymmetric distributions: $H_- = H_+ = 0$

Probability Mass Functions: From Example 2 and equation (16)

$$1 < S_-, \; 0 < Q_-, \; 1 < S_+, \; 0 < Q_+, \; -\left(\frac{1}{Q_-^{S_-}} + \frac{1}{Q_+^{S_+}}\right) < V,$$

$$D = \zeta(S_-, Q_-) + \zeta(S_+, Q_+) + V,$$

$$PMF(g) = \begin{cases} \frac{(-g+Q_-)^{-S_-}}{D} & \text{if } g < 0, \\ \frac{Q_-^{-S_-} + Q_+^{-S_+} + V}{D} = \frac{D - \zeta_1(S_-, Q_-) - \zeta_1(S_+, Q_+)}{D} & \text{if } g = 0, \\ \frac{(g+Q_+)^{-S_+}}{D} & \text{if } 0 < g. \end{cases} \quad (17)$$

It is symmetric, if $0 < Q_- = Q_+ = Q$, $1 < S_- = S_+ = S$, $-\frac{2}{Q^S} < V$

$$PMF(g) = \begin{cases} \frac{(|g|+Q)^{-S}}{D} = \frac{(|g|+Q)^{-S}}{2\zeta(S,Q)+V} & \text{if } g \neq 0, \\ \frac{2Q^{-S}+V}{D} = \frac{2Q^{-S}+V}{2\zeta(S,Q)+V} = \frac{D - 2\zeta_1(S,Q)}{D} & \text{if } g = 0. \end{cases} \quad (18)$$

If $V = -\frac{1}{Q^S}$, then $PMF(g) = \frac{(|g|+Q)^{-S}}{D} = \frac{(|g|+Q)^{-S}}{2\zeta(S,Q)-Q^{-S}}$ describes all ranks.

Distribution functions. Introducing $\varepsilon(x) = \begin{cases} 0 & \text{if } x < 0 \\ 1 & \text{if } x \geq 0 \end{cases}$, [55, p. 5], we get

$$F(x) = \sum_{g=-\infty}^{g=\infty} PMF(g) \times \varepsilon(x-g) = \sum_{g=-\infty}^{g=\lfloor x \rfloor} PMF(g) \times \varepsilon(x-g). \quad (19)$$

Then, for the asymmetric (17), and symmetric (18) distributions

$$F(x) = \begin{cases} \frac{\zeta_{\lfloor |x| \rfloor}(S_-,Q_+)}{\zeta(S_-,Q_-)+\zeta(S_+,Q_+)+V} & \text{if } \lfloor x \rfloor < 0, \\ \frac{\zeta(S_-,Q_-)+Q_+^{-S_+}+V}{\zeta(S_-,Q_-)+\zeta(S_+,Q_+)+V} & \text{if } \lfloor x \rfloor = 0, \\ \frac{\zeta(S_-,Q_-)+\zeta(S_+,Q_+)+V-\zeta_{\lfloor x \rfloor+1}(S_+,Q_+)}{\zeta(S_-,Q_-)+\zeta(S_+,Q_+)+V} & \text{if } \lfloor x \rfloor > 0, \end{cases}$$

$$F(x) = \begin{cases} \frac{\zeta_{\lfloor |x| \rfloor}(S,Q)}{2\zeta(S,Q)+V} & \text{if } \lfloor x \rfloor < 0, \\ \frac{\zeta(S,Q)+Q^{-S}+V}{2\zeta(S,Q)+V} & \text{if } \lfloor x \rfloor = 0, \\ \frac{2\zeta(S,Q)+V-\zeta_{\lfloor x \rfloor+1}(S,Q)}{2\zeta(S,Q)+V} & \text{if } \lfloor x \rfloor > 0. \end{cases} \quad (20)$$

Characteristic functions. $f(t) = \int_{-\infty}^{\infty} e^{itx} dF(x)$, $i = \sqrt{-1}$, for the purely discrete distribution, [55, p. 17], [88, p. 17], given by (19), reduces to the sum $f(t) = \sum_{g=-\infty}^{g=\infty} PMF(g) e^{itg}$. The symmetric case from (18) is

$$f(t) = \sum_{g=-\infty}^{g=-1} \frac{(-g+Q)^{-S} e^{itg}}{2\zeta(S,Q)+V} + \frac{2Q^{-S}+V}{2\zeta(S,Q)+V} + \sum_{g=1}^{g=\infty} \frac{(g+Q)^{-S} e^{itg}}{2\zeta(S,Q)+V}$$

$$= \frac{2Q^{-S}+V+\sum_{g=1}^{g=\infty} \frac{e^{-itg}+e^{itg}}{(g+Q)^S}}{2\zeta(S,Q)+V} = \frac{2Q^{-S}+V+2\sum_{g=1}^{g=\infty} \frac{\cos(tg)}{(g+Q)^S}}{2\zeta(S,Q)+V}. \quad (21)$$

Since for $1 < S, 0 < Q$, $|\frac{\cos(tg)}{(g+Q)^S}| \leq \frac{1}{(g+Q)^S}$, the series $\sum_{g=1}^{g=\infty} \frac{\cos(tg)}{(g+Q)^S}$ converges. The asymmetric case from (17) is

$$f(t) = \frac{Q_-^{-S_-}+Q_+^{-S_+}+V+\sum_{g=1}^{g=\infty} \left[\frac{e^{-itg}}{(g+Q_-)^{S_-}} + \frac{e^{itg}}{(g+Q_+)^{S_+}}\right]}{\zeta(S_-,Q_-)+\zeta(S_+,Q_+)+V}. \quad (22)$$

Beginning moments of integer order $0 \leq s$. From (6), (17), and (19), where the integrals are replaced with the sums, and the term for $g = 0$ vanishes

$$\alpha_s = \frac{\sum_{g=-\infty}^{g=-1} \frac{g^s}{(g+Q_-)^{S_-}} + \sum_{g=1}^{g=\infty} \frac{g^s}{(g+Q_+)^{S_+}}}{\zeta(S_-,Q_-)+\zeta(S_+,Q_+)+V}$$

$$= \frac{\sum_{g=1}^{g=\infty} g^s \left[\frac{(-1)^s}{(g+Q_-)^{S_-}} + \frac{1}{(g+Q_+)^{S_+}}\right]}{\zeta(S_-,Q_-)+\zeta(S_+,Q_+)+V}. \quad (23)$$

From (18), for $m \in \mathbb{N}$, $0 < Q, 2m+1 < S \in \mathbb{R}$, the symmetric case is

$$\alpha_s = \frac{\sum_{g=1}^{g=\infty} \frac{g^s(1+(-1)^s)}{(g+Q)^S}}{2\zeta(S,Q)+V}; \quad \alpha_{s=2m-1} = 0;$$

$$\alpha_{s=2m} = \frac{2\sum_{g=1}^{g=\infty} \frac{g^{2m}}{(g+Q)^S}}{2\zeta(S,Q)+V} < \frac{2\zeta(S-2m,Q)}{2\zeta(S,Q)+V}. \tag{24}$$

Indeed, $\sum_{g=1}^{g=\infty} \frac{g^{2m}}{(g+Q)^S} = \sum_{g=0}^{g=\infty} \frac{g^{2m}}{(g+Q)^S} < \sum_{g=0}^{g=\infty} \frac{(g+Q)^{2m}}{(g+Q)^S} = \zeta(S-2m,Q)$. The latter converges for $2m+1 < S$. Thus, Inequality (24) and convergence of α_{2m} follow from the convergence of the Hurwitz Zeta function and [23, pp. 264–265, Theorem 1]. Alternatively, we can apply the peculiar *Ermakov's criterion* [23, pp. 285–287] (Vasily Petrovich Ermakov, 03/11/1845 [02/27/1845 — old calendar] — 03/16/1922, [17, p. 88]; on the history of criterion see [17, pp. 29–33]). For $1 \leq x$, the function $f(x) = \frac{x^{2m}}{(x+Q)^S}$ is positive, continuous, and has an extremum at $f'(x) = \frac{2mx^{2m-1}(x+Q)^S - x^{2m}S(x+Q)^{S-1}}{(x+Q)^{2S}} = 0$ or $2m(x+Q) - xS = 0$. $x^* = \frac{2mQ}{S-2m}$ corresponds to a maximum because $0 < f'(x)$ for $x < x^*$ and $f'(x) < 0$ for $x^* < x$. Thus, $f(x)$ is monotonically decreasing for $x^* < x$. Under these conditions, the criterion suggests to check if $\lim_{x\to\infty} \frac{f(e^x)e^x}{f(x)} < 1$ for $1 \leq x$ or $x^* \leq x$, if $1 \leq x^*$. We get $\lim_{x\to\infty} \frac{e^{(2m+1)x}}{(e^x+Q)^S} \frac{(x+Q)^S}{x^{2m}} = \lim_{x\to\infty} \frac{x^{S-2m}}{e^{(S-2m-1)x}} < 1$, if $\frac{S-2m}{S-2m-1} < \frac{x}{\ln(x)}$ for $1 < x$. Then, $\sum_{n=1}^{n=\infty} f(n)$ converges.

In Tables 2–5, without exceptions, $4 < S^*$ and sometimes $5 < S^*$ supporting existence of the mean $\alpha_1 = 0$, α_2, $\alpha_3 = 0$, the variance $\mu_2 = \alpha_2 - \alpha_1^2 = \alpha_2$, $\mu_3 = \alpha_3 - 3\alpha_1\alpha_2 + 2\alpha_1^3 = 0$ for the symmetric case. To compute α_s using (23), (24), we evaluate $\sum_{g=1}^{g=\infty} \frac{g^s}{(g+Q)^S}$ for different $s \in \mathbb{N}$, $S, Q \in \mathbb{R}$, Appendix E.

Central moments of integer order $0 \leq s$: Once theoretical α_s are evaluated using equations (23), (24), and Appendix E, the central moments μ_s are obtained using Formulas (A6) valid for both theoretical and sample moments.

Entropies: Shannon's entropy [96, p. 393, Theorem 2], [38, p. 4] $H = -K\sum_{i=1}^{i=n} p_i \log p_i$ for $n = \infty$ with the omitted constant K and denoted h, since H is reserved, is for asymmetric and symmetric cases

$$h = -\sum_{g=-\infty}^{g=-1} \frac{(-g+Q_-)^{-S_-}}{D} \log_2\left(\frac{(-g+Q_-)^{-S_-}}{D}\right) - \frac{Q_-^{-S_-} + Q_+^{-S_+} + V}{D}$$

$$\times \log_2\left(\frac{Q_-^{-S_-} + Q_+^{-S_+} + V}{D}\right) - \sum_{g=1}^{g=\infty} \frac{(g+Q_+)^{-S_+}}{D} \log_2\left(\frac{(g+Q_+)^{-S_+}}{D}\right)$$

$$= \frac{\log_2(D)}{D}\zeta_1(S_-, Q_-) + \frac{S_-}{D}\sum_{g=1}^{g=\infty}\frac{\log_2(g+Q_-)}{(g+Q_-)^{S_-}} + \frac{\log_2(D)}{D}\zeta_1(S_+, Q_+)$$

$$+ \frac{S_+}{D}\sum_{g=1}^{g=\infty}\frac{\log_2(g+Q_+)}{(g+Q_+)^{S_+}} - \frac{Q_-^{-S_-} + Q_+^{-S_+} + V}{D}$$

$$\times \log_2\left(\frac{Q_-^{-S_-} + Q_+^{-S_+} + V}{D}\right),$$

$$h = \frac{2\log_2(D)}{D}\zeta_1(S, Q) + \frac{2S}{D}\sum_{g=1}^{g=\infty}\frac{\log_2(g+Q)}{(g+Q)^S}$$

$$- \frac{2Q^{-S} + V}{D}\log_2\left(\frac{2Q^{-S} + V}{D}\right). \tag{25}$$

For $1 \le g$, $0 < Q$, $0 < \log_2(g+Q) < g+Q$, $\sum_{g=1}^{g=\infty}\frac{\log_2(g+Q)}{(g+Q)^S} < \zeta_1(S-1, Q)$

$$h < \frac{\log_2(D)}{D}\zeta_1(S_-, Q_-) + \frac{S_-}{D}\zeta_1(S_- - 1, Q_-) + \frac{\log_2(D)}{D}\zeta_1(S_+, Q_+)$$

$$+ \frac{S_+}{D}\zeta_1(S_+ - 1, Q_+) - \frac{Q_-^{-S_-} + Q_+^{-S_+} + V}{D}\log_2\left(\frac{Q_-^{-S_-} + Q_+^{-S_+} + V}{D}\right),$$

$$h < \frac{2\log_2(D)}{D}\zeta_1(S, Q) + \frac{2S}{D}\zeta_1(S-1, Q) - \frac{2Q^{-S} + V}{D}\log_2\left(\frac{2Q^{-S} + V}{D}\right).$$

Both converge for $2 < S_-$, S_+, S. See Appendix F for computing $\sum_{g=1}^{g=\infty}\frac{\log_2(g+Q)}{(g+Q)^S}$.

Relationships between symmetric probabilities

Let us define, similar to $Q7(g)$ in equations (11),

$$f(g) = \begin{cases} \ln\left(\frac{PMF(0)}{PMF(g)}\right) = \ln(2Q^{-S} + V) + S\ln(|g| + Q) & \text{if } g \ne 0, \\ 0 & \text{if } g = 0. \end{cases} \tag{26}$$

The Hurwitz Zeta $\zeta(S, Q)$ is excluded from $f(g)$. For $g \ne 0$, $f(g)$ is the linear function $f(x) = \ln(2Q^{-S} + V) + S \times x$ of $x = \ln(|g| + Q)$ with the slope S and intercept $\ln(2Q^{-S} + V)$. The distance between neighboring gs is ± 1 but between corresponding xs with gs $\ne 0$, it is $\Delta x = \frac{\ln(|g|+1+Q)}{\ln(|g|+Q)}$. If Q increases, then points on a chart of $f(x)$ are getting closer each to other along the x-axis as well as y-axis because $\Delta f(x) = S\frac{\ln(|g|+1+Q)}{\ln(|g|+Q)}$. The $\frac{\Delta f(x)}{\Delta x} = S$ remains constant.

In general, in contrast with the points $g \neq 0$, the point $(x = \ln(|g = 0| + Q) = \ln(Q), y = f(g = 0) = 0)$ is out of the line intersecting the x-axis at $(-\frac{\ln(2Q^{-S}+V)}{S}, 0)$. It gets to the line, if $V = -\frac{1}{Q^S}$. Equations (17), (18) can model the King effect, Property (B), for a single highest, $g = 0$, rank outlier from a line plotted in such bi-logarithmic coordinates [86, p. 49, 16.2 Parabola's shortcomings].

Symmetric distribution, estimation of (Q, S, V) *or* (Q, S, D)

We estimate parameters of the distribution minimizing Pearson's χ^2-square quantity [73,97] for the N_G selected rank classes. Each class $\mathcal{G}_i$, $i = 1, \ldots, N_\mathcal{G}$, is a set of $0 < n_{\mathcal{G}_i}$ ranks $\mathcal{G}_i = \{g_1^{\mathcal{G}_i}, \ldots, g_{n_{\mathcal{G}_i}}^{\mathcal{G}_i}\}$. The sets are pairwise disjoint.

$$PMF(\mathcal{G}_i) = \sum_{g \in \mathcal{G}_i} PMF(g); \quad n_{\mathcal{G}_i} = \sum_{g \in \mathcal{G}_i} n(g); \quad N = \sum_{i=1}^{i=N_\mathcal{G}} n_{\mathcal{G}_i},$$

$$\chi^2(Q, S, V) = \sum_{i=1}^{i=N_G} \frac{(n_{\mathcal{G}_i} - PMF(\mathcal{G}_i) \times N)^2}{PMF(\mathcal{G}_i) \times N} \to \min.$$

Ideally, if for each class $\mathcal{G}_i$ the expected frequency is $5 \leq PMF(\mathcal{G}_i) \times N$.

Example 3. Table A3, two hours interval [09:00:00, 11:00:00]. $N_\mathcal{G} = 7$. $\mathcal{G}_1 = \{-\infty, \ldots, -3\}$, $n_{\mathcal{G}_1} = 1_{-7} + 1_{-6} + 1_{-5} + 5_{-4} + 24_{-3} = 32$; $\mathcal{G}_2 = \{-2\}$, $n_{\mathcal{G}_2} = 221_{-2}$; $\mathcal{G}_3 = \{-1\}$, $n_{\mathcal{G}_3} = 9954_{-1}$; $\mathcal{G}_4 = \{0\}$, $n_{\mathcal{G}_4} = 85500_0$; $\mathcal{G}_5 = \{1\}$, $n_{\mathcal{G}_5} = 9885_1$; $\mathcal{G}_6 = \{2\}$, $n_{\mathcal{G}_2} = 213_2$; $\mathcal{G}_7 = \{3, \ldots, \infty\}$, $n_{\mathcal{G}_7} = 20_3 + 8_4 + 1_5 + 3_6 + 2_7 = 34$. $N = 32_{\mathcal{G}_1} + 221_{\mathcal{G}_2} + 9954_{\mathcal{G}_3} + 85500_{\mathcal{G}_4} + 9885_{\mathcal{G}_5} + 213_{\mathcal{G}_6} + 34_{\mathcal{G}_7} = 105839$. Fitting results are in Table 2.

Table 2. Minimization of the χ^2 quantity for Example 3. $Q^* = 1.93146468557664\text{E-}02$, $S^* = 5.59105010010000\text{E+}00$, $\zeta(S^*, Q^*) = 3.83436043730026\text{E+}09$, $V^* = -7.66872086501599\text{E+}09$, $D^* = 9.58453750610351\text{E+}00$, $Q^{*-S^*} = 3.83436043637933\text{E+}09$. Expected frequencies are greater than five.

Class	Set	Observed	$PMF(\mathcal{G}_i)$	Expected	χ^2
$\mathcal{G}_1$	$\{-\infty, \ldots, -3\}$	32	2.821360E-04	29.86	1.53221E-01
$\mathcal{G}_2$	$\{-2\}$	221	2.051247E-03	217.10	6.99890E-02
$\mathcal{G}_3$	$\{-1\}$	9954	9.375122E-02	9922.54	9.97772E-02
$\mathcal{G}_4$	$\{0\}$	85500	8.078308E-01	85500.00	2.37685E-11
$\mathcal{G}_5$	$\{1\}$	9885	9.375122E-02	9922.54	1.41988E-01
$\mathcal{G}_6$	$\{2\}$	213	2.051247E-03	217.10	7.75030E-02
$\mathcal{G}_7$	$\{3, \ldots, \infty\}$	34	2.821360E-04	29.86	5.73703E-01
Total		105839	1.000000E+00	105839.00	1.11618E+00

Minimizing the χ^2 quantity, we want to find a value less than a tabulated one but not necessarily the least. This adds a constraint to the already constrained optimization. If (Q^*, S^*, V^*) or (Q^*, S^*, D^*) is found, then the null hypothesis H_0: the observed and expected frequencies are statistically indistinguishable — should be accepted. For $g \neq 0$, an individual symmetric contribution $\chi^2(\mathcal{G}) = \frac{(n_-(\mathcal{G})-N\times PMF(\mathcal{G}))^2}{N\times PMF(\mathcal{G})} + \frac{(n_+(\mathcal{G})-N\times PMF(\mathcal{G}))^2}{N\times PMF(\mathcal{G})} = \frac{(n_-(\mathcal{G})-q)^2}{q} + \frac{(n_+(\mathcal{G})-q)^2}{q}$ is the least, if $\frac{d\chi^2(\mathcal{G})}{dq} = \frac{-(n_-^2(\mathcal{G})-q^2)-(n_+^2(\mathcal{G})-q^2)}{q^2} = 0$.
Then, $q = \sqrt{\frac{n_-^2(\mathcal{G})+n_+^2(\mathcal{G})}{2}} = N \times PMF(\mathcal{G})$. For $g = 0$, the least $\chi^2(g = 0) = \frac{(n(0)-N\times PMF(0))^2}{N\times PMF(0)} = 0$ is at $PMF(0) = \frac{n(0)}{N} = \frac{2Q^{-S}+V}{2\zeta(S,Q)+V} = \frac{D-2\zeta_1(S,Q)}{D}$.
If S and Q are fixed, we still can vary V or D to fit better the empirical $PMF(0)$ and model the King effect. Thus, for the rank zero and two other ranks $0 < g_i < g_j$

$$R(g_i, g_j) = \frac{PMF(g_i)}{PMF(g_j)} = \left(\frac{g_i+Q}{g_j+Q}\right)^{-S} = \sqrt{\frac{n^2(-g_i)+n^2(g_i)}{n^2(-g_j)+n^2(g_j)}},$$

$$1 < S = \frac{\ln(R(g_i,g_j))}{\ln(\frac{g_j+Q}{g_i+Q})} \quad \text{or} \quad 0 < Q = \frac{g_j - g_i \times (R(g_i,g_j))^{\frac{1}{S}}}{(R(g_i,g_j))^{\frac{1}{S}} - 1},$$

$$-\frac{2}{Q^S} < V = \frac{2 \times (n(0) \times \zeta(S,Q) - N \times Q^{-S})}{N - n(0)},$$

$$0 < D = \frac{2\zeta_1(S,Q)}{1-PMF(0)}. \tag{27}$$

Equations (27) binding Q, S, V, or D can be useful for getting initial values for a solver. They do not reduce the number of the degrees of freedom $df = N_\mathcal{G} - 1 - 3$. But if they serve as additional equality constraints, then df must be subtracted two. $Q = \frac{2-1\times(\frac{9954^2+9885^2}{221^2+213^2})^{\frac{1}{2S}}}{(\frac{9954^2+9885^2}{221^2+213^2})^{\frac{1}{2S}}-1} = \frac{2-2088.901^{-2S}}{2088.901^{-2S}-1}$, $V = \frac{2\times(85500\times\zeta(S,Q)-105839\times Q^{-S})}{105839-85500}$ were used in Example 3 to get $\chi^2 = 1.116$, $df = 3$ in Table 2. This is less than theoretical $\chi^2(0.01, df = 3) = 11.345$. It is important that parameters are estimated by frequencies grouped using the same classes, where χ^2 is minimized [37, pp. 558–621, Chapter 30, Tests of Fit], [49–51]. If one would like to bind S, Q, V, or D by equation (27), then the only parameter, let us say Q, would remain independent and Example's $df = 1$. But, even, in that case the number is less than theoretical $\chi^2(0.01, 1) = 6.635$. *The fitting is reasonable.*

Example 4. The interval [09:00:00, 11:00:00] is the same as in Example 3 but $N_\mathcal{G} = 9$, Table 2. Q remains an independent parameter. D is computed by equation (27) to fit $PMF(0)$ exactly. Given Q, S is evaluated for a few pairs of $0 < g_i < g_j$ by (27). The pair with the least χ^2 sum is selected. This reduces df by two, if we keep the optimal parameters without an attempt to relax the constraints of (27). For Table 3, the conservative $df = 9 - 1 - 3 - 2 = 3$. The theoretical values $\chi^2(0.05, 3) = 7.815$, $\chi^2(0.01, 3) = 11.345$, $\chi^2(0.001, 3) = 16.266$ are greater than the experimental 7.66837.

The procedure described in Example 4 is applied to the empirical morning and day time two hours interval distributions in Table A3 leading to Table 4. For [13–15], $\chi^2 = 60.812$ indicates poor fitting. This interval includes the stock market closing minutes, where liquidity of ES-mini futures gets high and deviations from distributions observed within the middle of a session are likely. Appendix C confirms that in two hours intervals distributions change over time. Combining all samples in one may influence on the approximation of the sample by a single theoretical distribution. Table 5 combines all 11893137 b-increments registered in Table A2.

Initial Q, S, D were computed using equation (27). Then, constraints were relaxed and the Microsoft GRG Nonlinear Solver found a slightly smaller solution added to the title of Table 5, $df = 31 - 1 - 3 = 27$. The χ^2 value 162.44532 is greater than theoretical $\chi^2(0.001, 27) = 55.476$. This corresponds to the conclusion of Appendix B that the distributions differ for different dates. Then, it is difficult to expect that one set of parameters can describe the combined sample. Nevertheless, the function $f(g)$, equation (26), plotted on Figure 8 fits well the outlying $PMF(0)$, what can be interpreted as a King effect, and the next 22 frequencies. While we cannot accept the null

Table 3. Minimization of the χ^2 quantity for Example 4. $Q^* = 1.00000000000000$E-04, $S^* = 5.51466188979908$E+00, $\zeta(S^*, Q^*) = 1.14458370622345$E+22, $V^* = -2.289167412$ 446910E+22, $D^* = 1.06611171126830$E+01. Expected frequencies are greater than five.

Class	Set	Observed	$PMF(\mathcal{G}_i)$	Expected	χ^2
$\mathcal{G}_1$	$\{-\infty, \ldots, -4\}$	8	6.7113587e-05	7.10	0.11321
$\mathcal{G}_2$	$\{-3\}$	24	2.1925780e-04	23.21	0.02717
$\mathcal{G}_3$	$\{-2\}$	221	2.0511570e-03	217.09	0.07034
$\mathcal{G}_4$	$\{-1\}$	9954	9.3747091e-02	9922.10	0.10257
$\mathcal{G}_5$	$\{0\}$	85500	8.0783076e-01	85500.00	0.00000
$\mathcal{G}_6$	$\{1\}$	9885	9.3747091e-02	9922.10	0.13871
$\mathcal{G}_7$	$\{2\}$	213	2.0511570e-03	217.09	0.07715
$\mathcal{G}_8$	$\{3\}$	20	2.1925780e-04	23.21	0.44293
$\mathcal{G}_9$	$\{4, \ldots, \infty\}$	14	6.7113587e-05	7.10	6.69630
Total		105839	1.0000000e+00	105839.00	7.66837

Table 4. Parameters of distributions given by equation (18) approximating empirical distributions from Table A3. The b-increments from each interval are grouped in nine classes $\{-\infty, \ldots, -4\}, \{-3\}, \{-2\}, \{-1\}, \{0\}, \{1\}, \{2\}, \{3\}, \{4, \ldots, \infty\}$.

Hours	N	Q	S	D	V	$\varsigma(S,Q)$	χ^2
05–07	30171	0.0001	4.599694843671028	9.582979720055942	−5.009673043023590E+18	2.504836521151180E+18	8.621
07–09	93258	0.0001	4.950293331564227	9.043686785731195	−1.265328407126000E+20	6.326642035630000E+19	13.248
09–11	105839	0.0001	5.514661889798908	10.661111711268300	−2.289167412446910E+22	1.144583706222345E+22	7.668
11–13	67199	0.0001	5.612949206407109	10.476668864239500	−5.660135412019440E+22	2.830067706009720E+22	5.312
13–15	88704	0.0001	4.965376646873370	11.141577117796900	−1.453906851504370E+20	7.269534257521850E+19	60.812

Table 5. Minimization of the χ^2 quantity for Example 3. $Q^* = 1.646484044248660\text{E-}01$, $S^* = 4.256171381652780\text{E+}00$, $\zeta(S^*, Q^*) = 2.160629913097710\text{E+}03$, $\zeta_1(S^*, Q^*) = 5.718003183830510\text{E-}01$, $V^* = -4.317249583490080\text{E+}03$, $D^* = 4.010242705348760\text{E+}00$, $Q^{*-S^*} = 2.160058112779330\text{E+}03$. Expected frequencies are greater than five.

Class	Set	Observed	$PMF(\mathcal{G}_i)$	Expected	χ^2
$\mathcal{G}_1$	$\{-\infty, \ldots, -15\}$	121	1.2172482e-05	144.77	3.90253
$\mathcal{G}_2$	$\{-14\}$	46	3.1412331e-06	37.36	1.99857
$\mathcal{G}_3$	$\{-13\}$	49	4.2897469e-06	51.02	0.07986
$\mathcal{G}_4$	$\{-12\}$	85	6.0042621e-06	71.41	2.58651
$\mathcal{G}_5$	$\{-11\}$	108	8.6501019e-06	102.88	0.25513
$\mathcal{G}_6$	$\{-10\}$	171	1.2896494e-05	153.38	2.02421
$\mathcal{G}_7$	$\{-9\}$	272	2.0040086e-05	238.34	4.75385
$\mathcal{G}_8$	$\{-8\}$	402	3.2769296e-05	389.73	0.38632
$\mathcal{G}_9$	$\{-7\}$	737	5.7144358e-05	679.63	4.84357
$\mathcal{G}_{10}$	$\{-6\}$	1258	1.0835381e-04	1288.67	0.72978
$\mathcal{G}_{11}$	$\{-5\}$	2773	2.3014718e-04	2737.17	0.46897
$\mathcal{G}_{12}$	$\{-4\}$	6772	5.7517200e-04	6840.60	0.68793
$\mathcal{G}_{13}$	$\{-3\}$	21176	1.8507954e-03	22011.76	31.73304
$\mathcal{G}_{14}$	$\{-2\}$	111613	9.3188696e-03	110830.59	5.52339
$\mathcal{G}_{15}$	$\{-1\}$	1550611	1.3034452e-01	1550205.23	0.10621
$\mathcal{G}_{16}$	$\{0\}$	8501630	7.1483007e-01	8501571.94	0.00040
$\mathcal{G}_{17}$	$\{1\}$	1549421	1.3034452e-01	1550205.23	0.39673
$\mathcal{G}_{18}$	$\{2\}$	111983	9.3188696e-03	110830.59	11.98263
$\mathcal{G}_{19}$	$\{3\}$	20952	1.8507954e-03	22011.76	51.02263
$\mathcal{G}_{20}$	$\{4\}$	6852	5.7517200e-04	6840.60	0.01900
$\mathcal{G}_{21}$	$\{5\}$	2745	2.3014718e-04	2737.17	0.02239
$\mathcal{G}_{22}$	$\{6\}$	1290	1.0835381e-04	1288.67	0.00138
$\mathcal{G}_{23}$	$\{7\}$	751	5.7144358e-05	679.63	7.49574
$\mathcal{G}_{24}$	$\{8\}$	430	3.2769296e-05	389.73	4.16108
$\mathcal{G}_{25}$	$\{9\}$	252	2.0040086e-05	238.34	0.78296
$\mathcal{G}_{26}$	$\{10\}$	182	1.2896494e-05	153.38	5.34045
$\mathcal{G}_{27}$	$\{11\}$	100	8.6501019e-06	102.88	0.08045
$\mathcal{G}_{28}$	$\{12\}$	103	6.0042621e-06	71.41	13.97515
$\mathcal{G}_{29}$	$\{13\}$	59	4.2897469e-06	51.02	1.24864
$\mathcal{G}_{30}$	$\{14\}$	52	3.1412331e-06	37.36	5.73770
$\mathcal{G}_{31}$	$\{15, \ldots, \infty\}$	141	1.2172482e-05	144.77	0.09812
Total		11893137	1.000000E+00	11893137.00	162.44532

hypothesis, and changing distributions from session to session is a possible explanation, the dots, "sitting" almost on the line, look perspective.

Cancellation errors of $\zeta(S, Q) - Q^{-S}$ can be significant using the C++ type double. Instead of calling HurwitzZeta(S, Q) and subtracting Q^{-S}, the author wrote HurwitzZetaStart(S, Q, start) computing $\zeta_{\text{start}}(S, Q)$, $0 \leq \text{start}$, see ζ_m in (16) for $H = 0$, Appendix G. For start $= 0$, or 1 it is $\zeta(S, Q)$, or $\zeta_1(S, Q)$.

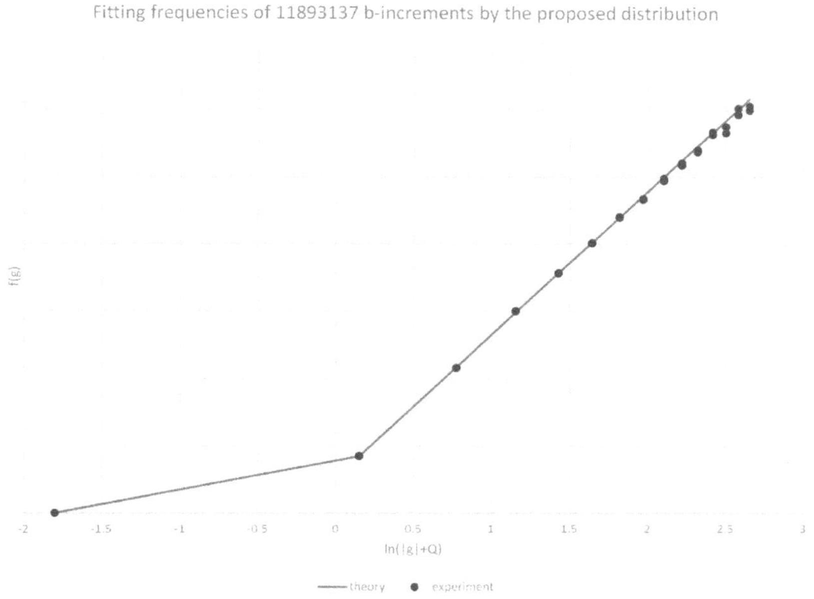

Fitting frequencies of 11893137 b-increments by the proposed distribution

Figure 8. The 11893137 b-increments from 20 ESM20 sessions, see Tables 5, A2, are grouped in 31 classes. The optimal Q, S, V were used to plot the function $f(g)$ from equation (26). The so-called King effect can be simulated.

Example 5. The Microsoft Excel Scientific Formating with 15 decimal places, POWER, HurwitzZeta, HurwitzZetaStart yield: HurwitzZeta $(S = 4.8, Q = 0.0015)$ =3.587252579327390E+13, POWER$(Q, -S)$ = 3.587 252579327280E+13, HurwitzZeta(S, Q) − POWER$(Q, -S)$ = 1.0781 25000000000E+00. The formated difference is 3.9−2.8 = 1.1 but Excel "snapshots" the 10 bytes floating point unit, FPU, registers. The relative error of subtracting POWER would be 4% because HurwitzZetaStart$(S, Q, 1)$ = 1.035834637773560E+00.

The following property is useful and makes $0 < Q < 1$ "special" for $\zeta(S, Q)$.

Proposition 1. Let $0 < Q \in \mathbb{R}$, $\{Q\} = Q - \lfloor Q \rfloor$, then $\zeta(S, Q) = \zeta_{\lfloor Q \rfloor}(S, \{Q\})$.

Proof. $\sum_{n=0}^{n=\infty} \frac{1}{(n+\lfloor Q \rfloor + \{Q\})^S} = \sum_{m=\lfloor Q \rfloor}^{m=\infty} \frac{1}{(m+\{Q\})^S} = \zeta_{\lfloor Q \rfloor}(S, \{Q\})$. ∎

Example 6. $\zeta(S = 4.1, Q = 1.3) = \zeta_1(4.1, 0.3)$=3.863894557863230E−01, $\zeta(S = 4.3, Q = 2.7) = \zeta_1(4.3, 0.7)$=2.014628878506670E−02, $\zeta(S = 5.2, Q = 3.2) = \zeta_1(5.2, 0.2)$=3.280025303811470E−03.

Asymmetric distribution

A big daily price range: The number of trading ticks N in an E-mini S&P 500 futures daily trading session reaches hundreds of thousands. On Thursday September 3, 2020, the December 2020 contract ESZ20 had *Open* 3569.00, *High* 3576.25, *Low* 3414.50, *Close* 3451.25. This is a price range of $3576.25 - 3414.50 = 161.75$ full points or $161.75 / 0.25 = 647 \delta_{ES}$. This session initiated a downturn continued until Thursday September 24, 2020 with Open 3228.00, High 3268.25, Low 3198.00, Close 3238.00. A big daily price change 161.75 points is a relatively rare event. The dollar equivalent per contract is $161.75 \times \$50 = 647 \times \$12.50 = \$8087.50$. This is before *commissions* and *fees*.

Commissions and fees: At the time of this writing, the *National Futures Association*, NFA, fee is \$0.02. The *Exchange fee* is \$1.23. Depending on the broker and the number of contracts traded per month, commissions can be \$0.10 − \$3.50. Depending on the broker, data technology, trading platform, additional routing, and execution fees \$0.05−\$0.50 may be applicable. All are *per contract per side*. The *round trip* cost per contract can be from $2 \times (\$0.02 + \$1.23 + \$0.10 + \$0.05) = \$2.80$ to $2 \times (\$0.02 + \$1.23 + \$3.50 + \$0.50) = \$10.50$. If one would "catch" the exact High and Low, then the profit from trading a single contract with the "pessimistic cost" would be $\$8087.50 - \$10.50 = \$8077.00$.

An old dispute: The same High and Low may happen several times within a session. The volume associated with both can be big. Since each trade involve both the buy and sell sides, somebody trades with the optimal per day prices. Years ago, and only once in life, the author sold on a *Limit order* a Copper futures, HG, which had no a daily price limit, at the daily High, as it became clear *after* the closing. In two days, the trade returned a profit of \$2000 per contract. But it was not accompanied by buying at the Low. Selling or buying at the daily High or Low is more realistic for corn, soybean, wheat futures trading with the daily down or up price limits [92, p. 26, Strategy II].

Dividing, even, a big daily price range by $N - 1$ yields an almost zero sample mean b-increment. This creates an impression that long term trading cannot give a positive result but *people and robots trade* [82, pp. 38–39, Why we speculate], [86, pp. 4–7, Inalienable Property of a Speculative Market].

If we put aside the terms "Trend" and "Volatility" [82, p. 39], to give more time to the financial community to elaborate commonly accepted definitions of these *controversial* notions, and apply, in contrast, the *well-defined* framework of the *maximum profit strategy*, MPS, [80], [81, p. 39, Alternative analysis], [86, pp. 82–101], [87, p. 8], [88, pp. 47–48, MPS studies, 2007–2017], then we get the time intervals of optimal trades with the sets of properties collectively named by the author the *optimal trading elements*, OTE, [82, pp. 39, 43 Optimal trading elements], [84, pp. 26–27, MPS, OTE: A review], [86, pp. 91–100, The Optimal Trading Elements], [87, pp. 39–47]. The chain of the alternating buying, BOTE, and selling, SOTE, elements objectively extracts from the past till today the market offers and supplies a measure of their size and frequency. It looks natural for the author to evaluate possible non-symmetric empirical b-increments distributions and corresponding statistics and apply the new theoretical discrete distribution to BOTE and SOTE.

Estimates with $|g| = 0, 1, 2, 3$. *A criterion for* H_-, $H_+ = 0$

Experimental $PMF(g)$ decrease with increasing $|g|$. Let us assign expressions from (17) to experimental values $Q4$ and $Q5$ from (11)

$$Q4(1) = S_+ \ln\left(\frac{2+Q_+}{1+Q_+}\right) > 0; \quad Q4(2) = S_+ \ln\left(\frac{3+Q_+}{2+Q_+}\right) > 0,$$

$$Q5(1) = \frac{Q4(1)}{Q4(2)} = \frac{\ln\left(\frac{2+Q_+}{1+Q_+}\right)}{\ln\left(\frac{3+Q_+}{2+Q_+}\right)} \Rightarrow Q_+ = \frac{(2+Q_+)^{1+Q5(1)}}{(3+Q_+)^{Q5(1)}} - 1.$$

The last equation has one unknown Q_+. The left is the line $y = Q_+$. The right curve's asymptotic is the line $\lim_{Q_+ \to \infty} \frac{(2+Q_+)^{1+Q5(1)}}{(3+Q_+)^{Q5(1)}} - 1 = Q_+ - 1$ parallel to and *below* $y = Q_+$. The left end of the curve at $Q_+ = 0$ is *above* $y = Q_+ = 0$, if

$$Q_+ = 0 < \frac{(2+0)^{1+Q5(1)}}{(3+0)^{Q5(1)}} - 1 \Rightarrow 0 < Q5(1) < \frac{\ln(2)}{\ln(\frac{3}{2})} \approx 1.709511291351455. \tag{28}$$

Also

$$\frac{d\left[\frac{(2+Q_+)^{1+Q5(1)}}{(3+Q_+)^{Q5(1)}} - 1\right]}{dQ_+}$$

$$= \frac{(2+Q_+)^{Q5(1)}\left[(3+Q_+)(1+Q5(1)) - (2+Q_+)Q5(1)\right]}{(3+Q_+)^{1+Q5(1)}} > 0,$$

$$\lim_{Q_+ \to \infty} \frac{(2+Q_+)^{Q5(1)}\left[(3+Q_+)(1+Q5(1)) - (2+Q_+)Q5(1)\right]}{(3+Q_+)^{1+Q5(1)}} = 1.$$

Under these conditions, the curve and line intersect at some $0 < Q_+^*$, which can be found *numerically* using a *one dimensional solver*. After that,

$$S_+^* = \frac{Q4(1)}{\ln\left(\frac{2+Q_+^*}{1+Q_+^*}\right)} = \frac{Q4(2)}{\ln\left(\frac{3+Q_+^*}{2+Q_+^*}\right)}.$$

Similarly, for the negative distribution branch

$$Q4(-1) = S_- \ln\left(\frac{|-2|+Q_-}{|-1|+Q_-}\right) > 0; \quad Q4(-2) = S_- \ln\left(\frac{|-3|+Q_-}{|-2|+Q_-}\right) > 0,$$

$$0 < Q5(-1) = \frac{Q4(-1)}{Q4(-1)} = \frac{\ln\left(\frac{2+Q_-}{1+Q_-}\right)}{\ln\left(\frac{3+Q_-}{2+Q_-}\right)} \Rightarrow Q_- = \frac{(2+Q_-)^{1+Q5(-1)}}{(3+Q_-)^{Q5(-1)}} - 1,$$

and

$$0 < Q5(-1) < \frac{\ln(2)}{\ln(\frac{3}{2})} \approx 1.709511291351455; \quad S_-^* = \frac{Q4(-1)}{\ln\left(\frac{2+Q_-^*}{1+Q_-^*}\right)} = \frac{Q4(-2)}{\ln\left(\frac{3+Q_-^*}{2+Q_-^*}\right)}.$$

(29)

Using three $PMFs$ each from the negative and positive branches of an empirical b-increments distribution, we obtain Q_-^*, S_-^*, Q_+^*, S_+^*. The six $g \neq 0$ do not have to be $g = -3, -2, -1, 1, 2, 3$ but the latter, after $g = 0$, are statistically the most representative — recollect the "open peacock tail". These values allow to fit *exactly* the applied experimental ratios of $PMF(g)$ but not necessarily the six $PMF(g)$. The latter is achieved with D_+^* and D_-^*

$$D_+^* = \frac{(1+Q_+^*)^{-S_+^*}}{PMF(1)} = \frac{(2+Q_+^*)^{-S_+^*}}{PMF(2)} = \frac{(3+Q_+^*)^{-S_+^*}}{PMF(3)},$$

$$D_-^* = \frac{(|-1|+Q_-)^{-S_-}}{PMF(-1)} = \frac{(|-2|+Q_-)^{-S_-}}{PMF(2)} = \frac{(|-3|+Q_-)^{-S_-}}{PMF(-3)} \quad \text{but also}$$

$$V^* = \frac{PMF(0)(\zeta(S_-^*, Q_-^*) + \zeta(S_+^*, Q_+^*)) - Q_-^{*-S_-^*} - Q_+^{*-S_+^*}}{1 - PMF(0)},$$

$$D^* = \zeta(S_-^*, Q_-^*) + \zeta(S_+^*, Q_+^*) + V^*.$$

$D^* = D_-^* = D_+^*$ would be a coincidence due to all sorts of imperfections including $PMF(0)$ that tends to drop out of both branches. Let us determine the χ^2 sum for seven probabilities, where D^*_-, D^*_+ are replaced with D^*

$$\chi_7^2 = \sum_{g=-3}^{g=-1} \frac{\left[\frac{N(-g+Q^*_-)^{-S^*_-}}{D^*_-} - \frac{N(-g+Q^*_-)^{-S^*_-}}{D^*}\right]^2}{\frac{N(-g+Q^*_-)^{-S^*_-}}{D^*}} + 0$$

$$+ \sum_{g=1}^{g=3} \frac{\left[\frac{N(g+Q^*_+)^{-S^*_+}}{D^*_+} - \frac{N(g+Q^*_+)^{-S^*_+}}{D^*}\right]^2}{\frac{N(g+Q^*_+)^{-S^*_+}}{D^*}}$$

$$= \frac{N}{D^*}\left(\left(\frac{D^*_- - D^*}{D^*_-}\right)^2 \sum_{g=-3}^{g=-1}(-g+Q^*_-)^{-S^*_-}\right.$$

$$\left. + \left(\frac{D^*_+ - D^*}{D^*_+}\right)^2 \sum_{g=1}^{g=3}(g+Q^*_+)^{-S^*_+}\right).$$

For an experiment with the ideally symmetric frequencies, $D^*_- = D^*_+ = D'$, $Q^*_- = Q^*_+ = Q^*$, $S^*_- = S^*_+ = S^*$

$$\chi_7^2 = \frac{2N}{D^*} \times \left(\frac{D' - D^*}{D'}\right)^2 \times \sum_{g=1}^{g=3}(g+Q^*)^{-S^*}.$$

Example 7. In Table 3, $N = 105839$, $Q^* = 0.0001$, $S^* \approx 5.51$, $D^* \approx 10.7$. Then, the symmetric $\chi_7^2 = \frac{2 \times 105839}{10.7} \times (1.0001^{-5.51} + 2.0001^{-5.51} + 3.0001^{-5.51}) \times \left(\frac{D'-10.7}{D'}\right)^2 \approx 20252.58 \times$ (relative deviation of D' from D^*)^2. With the relative deviation 0.01, 1%, $\chi_7^2 \approx 2.03$.

Equations from this section can provide initial values for a multidimensional solver with different optimization criteria or supply estimates when the number of frequencies classes is limited by $|g| = 3$. They give a criterion: if $Q5(-1)$ and/or $Q5(1) \geq 1.709511291351455$, then simplifications $H_- = 0$ and/or $H_+ = 0$ are insufficient and equation (16) are needed.

On "peacocks' tails". A fluctuation in $Q5(-1)$, $Q5(1)$ is estimated as following. If $R = \frac{A}{B}$, then $\frac{\Delta R}{R} \approx \sqrt{[\frac{\Delta A}{A}]^2 + [\frac{\Delta B}{B}]^2}$. $|\Delta \ln(x)| \approx |\frac{\Delta x}{x}|$. $Q5(1) = \frac{\ln(\frac{PMF(1)}{PMF(2)})}{\ln(\frac{PMF(2)}{PMF(3)})}$. Then, $|\frac{\Delta Q5(1)}{Q5(1)}|^2 \approx \left[\frac{\Delta \ln(\frac{PMF(1)}{PMF(2)})}{\ln(\frac{PMF(1)}{PMF(2)})}\right]^2 +$

$$\left[\frac{\Delta \ln(\frac{PMF(2)}{PMF(3)})}{\ln(\frac{PMF(2)}{PMF(3)})}\right]^2 \cdot |\Delta \ln(\frac{PMF(1)}{PMF(2)})|^2 \approx |\frac{\Delta(\frac{PMF(1)}{PMF(2)})}{\frac{PMF(1)}{PMF(2)}}|^2 \approx \left[\frac{\Delta PMF(1)}{PMF(1)}\right]^2 +$$
$$\left[\frac{\Delta PMF(2)}{PMF(2)}\right]^2; |\Delta \ln\left(\frac{PMF(2)}{PMF(3)}\right)|^2 \approx \left[\frac{\Delta PMF(2)}{PMF(2)}\right]^2 + \left[\frac{\Delta PMF(3)}{PMF(3)}\right]^2. \text{ Finally,}$$

$$\left|\frac{\Delta Q5(1)}{Q5(1)}\right| \approx \sqrt{\frac{\left[\frac{\Delta PMF(1)}{PMF(1)}\right]^2 + \left[\frac{\Delta PMF(2)}{PMF(2)}\right]^2}{\ln^2\left(\frac{PMF(1)}{PMF(2)}\right)} + \frac{\left[\frac{\Delta PMF(2)}{PMF(2)}\right]^2 + \left[\frac{\Delta PMF(3)}{PMF(3)}\right]^2}{\ln^2\left(\frac{PMF(2)}{PMF(3)}\right)}}.$$

We shall assume that the absolute fluctuation $\Delta n(g)$ of the frequency $n(g)$ applied for the evaluation of the $PMF(g)$ is proportional to $\sqrt{n(g)}$ and replace $\frac{PMF(g)}{PMF(g+1)}$, independent on the normalizing factor, with $\frac{n(g)}{n(g+1)}$.

Example 8. Table A2 ALL DATES. $n(1) = 1549421$, $n(2) = 111983$, $n(3) = 20952$, $Q5(1) \approx 1.567489$, $\Delta n(1) \approx 1244.8$, $\Delta n(2) \approx 334.6$, $\Delta n(3) \approx 144.7$, $\frac{\Delta n(1)}{n(1)} \approx 0.000803$, $\frac{\Delta n(2)}{n(2)} \approx 0.00299$, $\frac{\Delta n(3)}{n(3)} \approx 0.00691$, $\ln^2(\frac{n(1)}{n(2)}) \approx 6.903$, $\ln^2(\frac{n(2)}{n(3)}) \approx 2.809$, $|\frac{\Delta Q5(1)}{Q5(1)}| \approx 0.00464$. $Q5(1) \approx 1.567 \pm 0.007$.

Example 9. Table 7 ALL. $n(1) = 14848$, $n(2) = 131$, $n(3) = 21$, $Q5(1) \approx 2.583978$, $\Delta n(1) \approx 121.9$, $\Delta n(2) \approx 11.4$, $\Delta n(3) \approx 4.6$, $\frac{\Delta n(1)}{n(1)} \approx 0.00821$, $\frac{\Delta n(2)}{n(2)} \approx 0.0874$, $\frac{\Delta n(3)}{n(3)} \approx 0.218$, $\ln^2(\frac{n(1)}{n(2)}) \approx 22.38$, $\ln^2(\frac{n(2)}{n(3)}) \approx 3.35$, $|\frac{\Delta Q5(1)}{Q5(1)}| \approx 0.130$. $Q5(1) \approx 2.58 \pm 0.34$.

Examples 8, 9 illustrate increasing fluctuations caused by small frequencies. For frequency one the relative error is 100%. This explains "open fans" or "peacocks' tails" on Figure 4 c, d, e, f, i, j and "bending" b, h — Property (C).

Evaluations of $Q5(-1), Q_-, S_-, D_-, Q5(1)$

The ESZ20 Tuesday September 15, 2020 trading session depicted on Figure 9 is "ordinary". For the extraction of BOTE and SOTE, we select the day interval [2020/09/15 06:00:00, 2020/09/15 15:15:00] and apply the *filtering cost* $FC = \$149.99$ and *trading cost*, commissions and fees, $C = \$4.68$ both per contract per transaction. The round trip costs are the doubles $299.98 and $9.36. This yields 15 OTEs, Figure 10 and Table 6.

MPS0 is MPS, which does not reinvest profits and trades a fixed number of contracts — one for simplicity. By definition, MPS0, MPS may not lose: if an interval has no price fluctuations exceeding $2 \times FC$, then the best, optimal from the absolute profit point of view, strategy is "do nothing". The latter always exists. For several MPS and OTE properties and their dependence on FC and C see already presented bibliography.

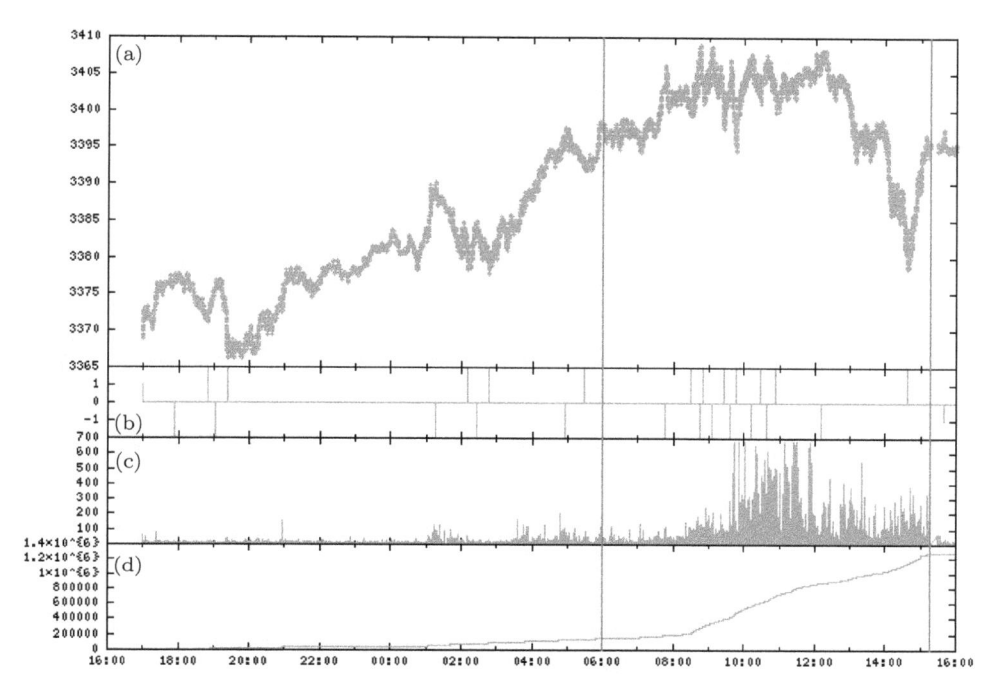

Figure 9. The CME Group Time & Sales trade ticks: prices (a), MPS0 with $FC = \$149.99$ (b), volume (c), and cumulative volume (d) of the E-mini S&P 500 December, Z, 2020 futures, ESZ20, traded on Tuesday September 15, 2020. The two vertical lines indicate the interval from which OTEs are extracted.

It is worth noticing that $\$12.50 \times |a_1^{OTE}| \times \text{Size}^{OTE} - 2 \times C$ taken for an OTE from Table 7 yields the profit from Table 6, where the profits are exact. To check and match the latter, rounding of a_1 is reduced in Table 7.

Example 10. $\$12.50 \times |a_1^{BOTE1}| \times \text{Size}^{BOTE1} - 2 \times C = \$12.50 \times |0.00867522| \times 5533 - 2 \times \$4.68 = \$590.64$; $\$12.50 \times |a_1^{BOTE10} = -0.00493872| \times 53892 - \$9.36 = \$328.14$; $\$12.50 \times |a_1^{BOTE14=-0.00493872}| \times 5467 - \$9.36 = \$1478.14$.

In Table 7, Q_-^* are computed using the GRG Nonlinear Microsoft Solver for $H_- = H_+ = 0$ and negative branches of SOTE2, BOTE3, SOTE6, SOTE8, BOTE9, where $R^- < 1.709511291351455$. $Err_- = \frac{2+Q_-^*}{3+Q_-^*} - Q_-^* - 1$. Some bins are empty preventing a simple evaluation. The bins for $|g| = 3$ have small frequencies causing big relative errors in ratios.

Always $a_1^{BOTE, \text{ b}-\text{increment}} > 0$, $a_1^{SOTE, \text{ b}-\text{increment}} < 0$. At first glance, this bias is "drowning in standard deviation". However, it is namely this bias multiplied by the frequency of trades forms the market offers.

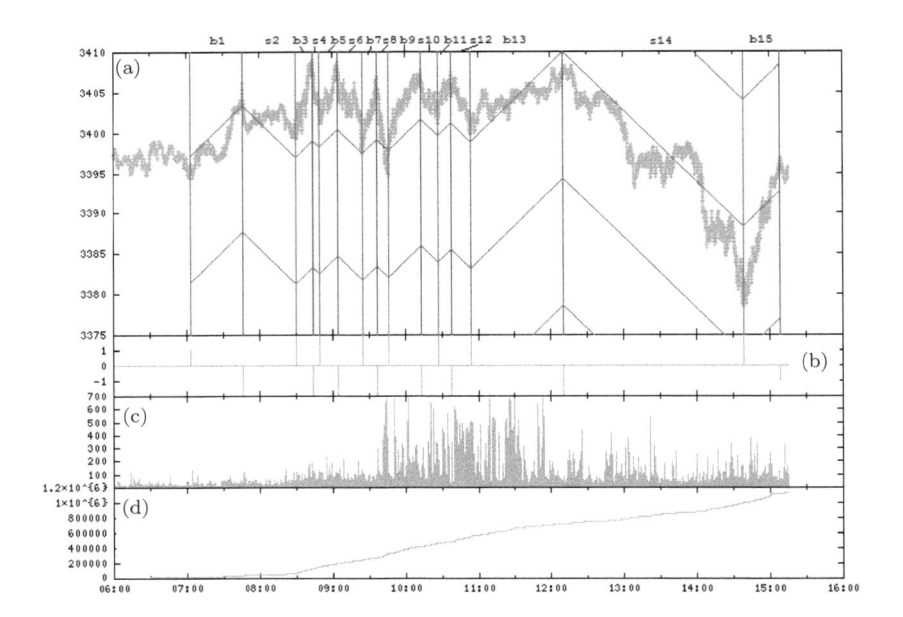

Figure 10. The CME Group Time & Sales trade ticks: prices (a), MPS0 with $FC = \$149.99$ (b), volume (c), and cumulative volume (d) of the E-mini S&P 500 December, Z, 2020 futures, ESZ20, traded on Tuesday September 15, 2020. The interval [06:00:00, 15:15:00] contains 15 alternating buy/sell OTEs. The vertical and oblique lines are drawn to indicate better locations and types of the buy and sell OTEs. They have no other meaning.

Table 6. ESZ20, Tuesday September 15, 2020, [06:00:00, 15:15:00], Optimal Trading Elements, OTE, extracted with $FC = \$149.99$ and evaluated with $C = \$4.68$: beginning t_b and ending t_e times of the optimal intervals, corresponding optimal prices P_b and P_e, durations Δt is seconds, profits in dollars per contract per trade, OTE types.

#	t_b	P_b	t_e	P_e	Δt	Profit	Type
1	07:03:05	3394.25	07:46:35	3406.25	2610	590.64	buy
2	07:46:35	3406.25	08:30:22	3399.25	2627	340.64	sell
3	08:30:22	3399.25	08:43:43	3409.00	801	478.14	buy
4	08:43:43	3409.00	08:49:43	3401.00	360	390.64	sell
5	08:49:43	3401.00	09:04:44	3408.75	901	378.14	buy
6	09:04:44	3408.75	09:24:36	3397.50	1192	553.14	sell
7	09:24:36	3397.50	09:36:18	3407.00	702	465.64	buy
8	09:36:18	3407.00	09:46:14	3394.50	596	615.64	sell
9	09:46:14	3394.50	10:13:29	3407.75	1635	653.14	buy
10	10:13:29	3407.75	10:26:54	3401.00	805	328.14	sell
11	10:26:54	3401.00	10:37:48	3407.50	654	315.64	buy
12	10:37:48	3407.50	10:54:00	3399.75	972	378.14	sell
13	10:54:00	3399.75	12:10:39	3408.25	4599	415.64	buy
14	12:10:39	3408.25	14:37:55	3378.50	8836	1478.14	sell
15	14:37:55	3378.50	15:08:54	3396.75	1859	903.14	buy

Table 7. ESZ20, Tuesday September 15, 2020, [06:00:00, 15:15:00], Empirical distributions and sample statistics of b-increments corresponding to OTEs in Table 6 and Figure 10.

1	2	3	4	5	6	7
Bin	BOTE1	SOTE2	BOTE3	SOTE4	BOTE5	SOTE6
-5		2			1	
-4	1	2			1	1
-3	2	2	1		3	2
-2	9	18	22	11	3	21
-1	386	559	1034	394	737	986
0	4688	6255	10848	4062	7749	9268
1	438	557	1081	375	784	985
2	6	9	15	1	4	4
3	1	4	3	1		
4	1	1		1		
5	1					
a_1	8.67522E-03	-3.77919E-03	2.99908E-03	-6.60475E-03	3.33980E-03	-3.99396E-03
Size	5533	7409	13004	4845	9282	11267
min	-4	-5	-3	-2	-5	-4
max	5	4	3	4	2	2
S	0.175	0.186	0.177	0.174	0.174	0.187
S^2	0.418	0.431	0.420	0.417	0.417	0.432
SK	0.249	-0.537	0.025	0.025	-0.343	-0.253
EK	11.425	12.843	4.458	5.877	6.801	4.084
$Q4(-2)$	1.504077397	2.197224577	3.091042453		0	2.351375257
$Q4(-1)$	3.758612792	3.435777715	3.850147602	3.578455636	5.503975604	3.849133917
$Q5(-1)$	2.498949057	1.563689825	1.245582246			1.636971345
Q^*_-		0.349248274	2.608643515			0.153020034
S^*_-		9.688057095	19.606065869			10.089758259
D^*_-		0.727797086	1.486735E-10			2.716517308
Err_-		0.000E+00	8.310E-12			7.659E-12
$Q4(1)$	4.290459441	4.125340663	4.277591617	5.926926026	5.278114659	5.50634728
$Q4(2)$	1.791759469	0.810930216	1.609437912	0		
$Q5(1)$	2.394551007	5.08717098	2.657817107			

Bin	BOTE7	SOTE8	BOTE9	SOTE10	BOTE11	SOTE12
-5						
-4		2		1		1
-3	1	2	1	2		2
-2	9	21	14	9	4	13
-1	515	654	953	383	274	496
0	4649	7189	11466	4691	2990	6581
1	551	648	1008	379	304	489
2	10	6	8	1	2	6
3	1		3	1		
4			1			
5						

(*Continued*)

270 V. Salov

Table 7. (*Continued*)

1	2	3	4	5	6	7
a_1	6.62483E-03	−5.86717E-03	3.93935E-03	−4.93872E-03	7.27476E-03	−4.08540E-03
Size	5736	8522	13454	5467	3574	7588
min	−3	−4	−3	0	−2	−4
max	3	2	4	3	2	2
S	0.202	0.171	0.156	0.155	0.168	0.144
S^2	0.450	0.414	0.395	0.393	0.410	0.380
SK	0.040	−0.467	0.121	−0.441	0.004	−0.403
EK	3.531	6.605	5.815	7.863	3.651	7.789
$Q4(-2)$	2.197224577	2.351375257	2.639057330	1.504077397		1.871802177
$Q4(-1)$	4.046942323	3.438584914	4.220557574	3.750810412	4.226833745	3.641626569
$Q5(-1)$	1.841842825	1.462371820	1.599267104	2.493761571		1.945518930
Q^*_-		0.726549189	0.247787412			
S^*_-		11.005377724	11.468046093			
D^*_-		0.031968318	1.114842941			
Err_-		−1.439E-11	4.421E-12			
$Q4(1)$	4.009149716	4.682131227	4.836281907	5.937536205	5.023880521	4.400603020
$Q4(2)$	2.302585093		0.980829253	0		
$Q5(1)$	1.741151599		4.930809203			

Bin	BOTE13	SOTE14	BOTE15	ALL	ALL-BOTE	ALL-SOTE
−5	1	1	1	6	3	3
−4		2	1	12	3	9
−3		7	4	29	12	17
−2	3	44	15	216	79	137
−1	1215	4183	1844	14613	6958	7655
0	18212	45520	17934	162102	78536	83566
1	1239	4095	1915	14848	7320	7528
2	8	33	18	131	71	60
3		4	3	21	11	10
4		2	2	8	4	4
5	1	1		3	2	1
a_1	1.64418E-03	−2.20812E-03	3.35833E-03	5.20863E-05	3.67746E-03	−3.35387E-03
Size	20679	53892	21737	191989	92999	98990
min	−5	−5	−5	−5	−5	−5
max	5	5	4	5	5	5
S	0.123	0.163	0.185	0.407	0.406	0.409
S^2	0.351	0.404	0.430	0.166	0.165	0.167
SK	0.058	−0.056	−0.019	−0.100	0.007	−0.198
EK	9.359	5.822	5.375	6.430	6.294	6.550
$Q4(-2)$		1.838279485	1.32175584	2.007982578	1.884541203	2.086767582
$Q4(-1)$	6.003887067	4.554594338	4.811642203	4.214388415	4.478199503	4.023133382
$Q5(-1)$		2.477639758	3.640341172	2.098817222	2.376281026	1.927925955
$Q4(1)$	5.042618340	4.821014435	4.667101144	4.730423132	4.635685730	4.832040119
$Q4(2)$		2.110213200	1.791759469	1.830674885	1.864784604	1.791759469
$Q5(1)$		2.284610121	2.604758743	2.583977728	2.485909482	2.696812938

An addition to the approximation flexibility

Experimental values of S and Q obtained here and earlier [86, p. 48, Table 6] make interesting a fragment of the Hurwitz Zeta function on Figure 11 showing the surface and projections of its contour lines on the coordinate S, Q-plane. Since $\ln(x)$ is a monotonically increasing function, the areas of one height of the $0 < \zeta(S,Q)$ and $\ln(\zeta(S,Q))$ surfaces must be observed for the same S and Q. It might be useful that contour lines permit keeping $\zeta(S,Q)$ constant and still have a degree of freedom changing S or Q bound now by some function $Q(S)$.

For $y(x) = u(x)^{v(x)}$, $0 < u(x)$, $y'(x) = [u(x)^{v(x)}]' = [e^{v(x)\ln(u(x))}]' = u(x)^{v(x)}\left(v'(x)\ln(u(x)) + \frac{v(x)u'(x)}{u(x)}\right)$. "This formula was first established by Leibniz and Johann Bernoulli" [23, p. 206]. For $u(S) = g+Q(S)$, $v(S) = -S$, $u'(S) = Q'(S)$, $\frac{d[(g+Q(S))^{-S}]}{dS} = -\frac{(g+Q(S))\ln(g+Q(S))+SQ'(S)}{(g+Q(S))^{S+1}}$. For a contour line,

$$\zeta(S,Q) = \zeta(S,Q(S)) = \sum_{g=0}^{g=\infty} \frac{1}{(g+Q(S))^S} = \text{constant},$$

$$\frac{\partial \zeta(S,Q(S))}{\partial S} = \sum_{g=0}^{g=\infty} \frac{(g+Q(S))\ln(g+Q(S))+SQ'(S)}{(g+Q(S))^{S+1}} = 0,$$

Natural Logarithm of the Hurwitz Zeta Function

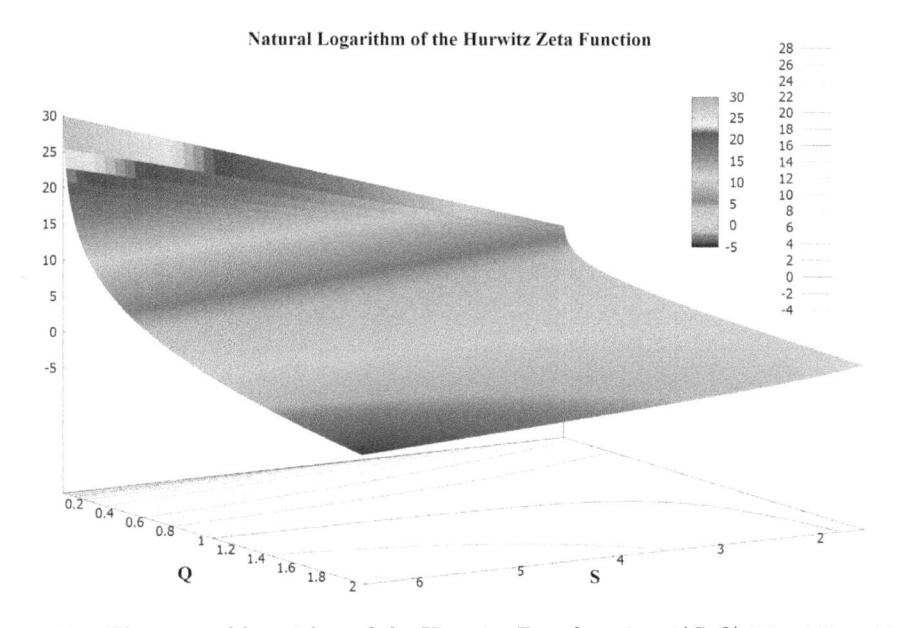

Figure 11. The natural logarithm of the Hurwitz Zeta function $\zeta(S,Q)$ $51 \times 199 = 10149$ values computed on the rectangle grid $(1.5 \leq S \leq 6.5) \times (0.01 \leq Q \leq 1.99)$ with the steps $\Delta S = 0.1$, $\Delta Q = 0.01$. The $\zeta(S,Q)$ is evaluated in C++ using [86, p. 48, Formula with $N = 30$, $M = 15$]. The wgnuplot 5.2 http://www.gnuplot.info is applied for plotting.

where $Q(S)$ is a normal projection of the contour on the S, Q-plane, and $Q'(S)$ is its first derivative. An ordinary differential equation for a summand's numerator with g^* has a solution (check by substitution) zeroing only this summand

$$(g^* + Q(S))\ln(g^* + Q(S)) + SQ'(S) = 0; \quad Q(S) = e^{\frac{C_1}{S}} - g^*.$$

Formulating the task of finding $Q(S)$ zeroing the derivative sum, the author did not find a solution including a literature search and leaves the task unsolved.

Distributions with non-zero H_- and/or H_+

The author came up with (17), fitting many b-increments, and later faced with $\frac{\ln(2)}{\ln(\frac{3}{2})} \leq Q5(-1|1)$, Property (H), where $0 < Q_{-|+}$ cannot be found. Generalization (16) makes (17) a particular case replacing S_-, S_+ with $1 < S_- A_-(g < 0)$, $1 < S_+ A_+(0 < g)$: $PMF(g < 0) = \frac{1}{D(-g+Q_-)^{S_- A_-(g)}}$, $PMF(0 < g) = \frac{1}{D(g+Q_+)^{S_+ A_+(g)}}$, $D = \sum_{g=-\infty}^{g=-1} PMF(g) + PMF(0) + \sum_{g=1}^{g=\infty} PMF(g)$. Then,

$$Q5(g < 0) = \frac{A_-(g-1)\ln(-g+1+Q_-) - A_-(g)\ln(-g+Q_-)}{A_-(g-2)\ln(-g+2+Q_-) - A_-(g-1)\ln(-g+1+Q_-)},$$

$$Q5(0 < g) = \frac{A_+(g+1)\ln(g+1+Q_+) - A_+(g)\ln(g+Q_+)}{A_+(g+2)\ln(g+2+Q_+) - A_+(g+1)\ln(g+1+Q_+)},$$

do not depend on D, $S_{-|+}$ because $S_{-|+} A_{-|+}(g)$ is a product. For $A_-(g < 0) = 1 + H_- e^{K-g} = 1 + e^{K-g+\ln(H_-)}$, $A_+(0 < g) = 1 + H_+ e^{-K+g} = 1 + e^{-K+g+\ln(H_+)}$,

$$Q5(g < 0) = \frac{\ln\left(\frac{-g+1+Q_-}{-g+Q_-}\right) + H_- e^{K-g}(e^{-K} \ln(-g+1+Q_-) - \ln(-g+Q_-))}{\ln\left(\frac{-g+2+Q_-}{-g+1+Q_-}\right) + H_- e^{K-(g-1)}(e^{-K} \ln(-g+2+Q_-) - \ln(-g+1+Q_-))},$$

$$Q5(0 < g) = \frac{\ln\left(\frac{g+1+Q_+}{g+Q_+}\right) + H_+ e^{-K+g}(e^{-K_+} \ln(g+1+Q_+) - \ln(g+Q_+))}{\ln\left(\frac{g+2+Q_+}{g+1+Q_+}\right) + H_+ e^{-K+(g+1)}(e^{-K_+} \ln(g+2+Q_+) - \ln(g+1+Q_+))}.$$

When $\dfrac{\ln(\frac{-g+1+Q_-}{-g+Q_-})}{\ln(\frac{-g+2+Q_-}{-g+1+Q_-})} < Q5(g < 0)$, $\dfrac{\ln(\frac{g+1+Q_+}{g+Q_+})}{\ln(\frac{g+2+Q_+}{g+1+Q_+})} < Q5(0 < g)$? For constant K_-, K_+, denominators may become zero for some g (non-integer). Let

$K_+ = K_+(0 < g) = \ln(\frac{\ln(g+2+Q_+)}{\ln(g+1+Q_+)})$, then $e^{-K_+} = \frac{\ln(g+1+Q_+)}{\ln(g+2+Q_+)}$ and

$$Q5(0 < g) = \frac{\ln\left(\frac{g+1+Q_+}{g+Q_+}\right)}{\ln\left(\frac{g+2+Q_+}{g+1+Q_+}\right)} + \frac{\left[\frac{\ln(g+1+Q_+)}{\ln(g+2+Q_+)}\right]^g \left(\frac{\ln^2(g+1+Q_+)}{\ln(g+2+Q_+)} - \ln(g+Q_+)\right)}{\ln\left(\frac{g+2+Q_+}{g+1+Q_+}\right)} H_+.$$

Proposition 2. *If* $1 < x$, $0 < \Delta x$, *then* $\ln^2(x+\Delta x) - \ln(x)\ln(x+2\Delta x) > 0$.

Proof. For $0 < x \in \mathbb{R}$, $\ln(x)$ is a monotonic, concave, increasing with x function. $[\ln(x)]' = \frac{1}{x}$. From geometry of its chart, $\ln(x+\Delta x) + \frac{\Delta x}{x+\Delta x} > \ln(x+2\Delta x) > 0$, $\ln(x+\Delta x) - \frac{\Delta x}{x+\Delta x} > \ln(x) > 0$. Thus, $\ln^2(x+\Delta x) - \frac{\ln(x+\Delta x)\Delta x}{x+\Delta x} + \frac{\ln(x+\Delta x)\Delta x}{x+\Delta x} - [\frac{\Delta x}{(x+\Delta x)}]^2 = \ln^2(x+\Delta x) - [\frac{\Delta x}{(x+\Delta x)}]^2 > \ln(x)\ln(x+2\Delta x) > 0$. Even more so $\ln^2(x+\Delta x) > \ln(x)\ln(x+2\Delta x) > 0$ after "dropping" $0 < [\frac{\Delta x}{(x+\Delta x)}]^2$. ■

Under our conditions with $1 < x = g+Q_+$ and $\Delta x = 1$, $\ln^2(g+1+Q_+) - \ln(g+Q_+)\ln(g+2+Q_+)) > 0$. Other factors in front of H_+ are positive too. And $0 < H_+$ is the answer.

Similarly, let $K_- = K_-(g < 0) = \ln(\frac{\ln(-g+2+Q_-)}{\ln(-g+1+Q_-)})$, then $e^{-K_-} = \frac{\ln(-g+1+Q_-)}{\ln(-g+2+Q_-)}$. Now, $0 < H_-$ is the answer for

$$Q5(g < 0) = \frac{\ln\left(\frac{-g+1+Q_-}{-g+Q_-}\right)}{\ln\left(\frac{-g+2+Q_-}{-g+1+Q_-}\right)} + \frac{\left[\frac{\ln(-g+1+Q_-)}{\ln(-g+2+Q_-)}\right]^{-g} \left(\frac{\ln^2(-g+1+Q_-)}{\ln(-g+2+Q_-)} - \ln(-g+Q_-)\right)}{\ln\left(\frac{-g+2+Q_-}{-g+1+Q_-}\right)} H_-.$$

In particular, for $g = \pm 1$,

$$Q5(-1) = \frac{\ln\left(\frac{2+Q_-}{1+Q_-}\right)}{\ln(\frac{3+Q_-}{2+Q_-})} + \frac{\left[\frac{\ln(2+Q_-)}{\ln(3+Q_-)}\right] \left(\frac{\ln^2(2+Q_-)}{\ln(3+Q_-)} - \ln(1+Q_-)\right)}{\ln\left(\frac{3+Q_-}{2+Q_-}\right)} H_-,$$

$$Q5(1) = \frac{\ln\left(\frac{2+Q_+}{1+Q_+}\right)}{\ln\left(\frac{3+Q_+}{2+Q_+}\right)} + \frac{\left[\frac{\ln(2+Q_+)}{\ln(3+Q_+)}\right] \left(\frac{\ln^2(2+Q_+)}{\ln(3+Q_+)} - \ln(1+Q_+)\right)}{\ln\left(\frac{3+Q_+}{2+Q_+}\right)} H_+. \quad (30)$$

Even, if $Q_{-|+}$ are fixed, both quantities can be increased by increasing $H_{-|+}$.

Limits are proved using *L'Hopital's (Johann Bernoulli's) Rule* and other techniques for evaluation of *indeterminate forms* [22, pp. 314–320], [104]. $0 < Q_\pm$ and differentiation by $x = |g|$ allow us to combine expressions for Q_-, Q_+.

Proposition 3. $\lim_{|g|\to\infty}[\frac{\ln(|g|+1+Q_\pm)}{\ln(|g|+2+Q_\pm)}]^{|g|} = 1.$

Proof.

$$\lim_{|g|\to\infty} y(|g|) = \lim_{|g|\to\infty}\left[\frac{\ln(|g|+1+Q_\pm)}{\ln(|g|+2+Q_\pm)}\right]^{|g|} = 1^\infty \Rightarrow \lim_{|g|\to\infty}\ln(y(|g|)) = \infty \times 0$$

$$= \lim_{|g|\to\infty}\frac{\ln\left(\frac{\ln(|g|+1+Q_\pm)}{\ln(|g|+2+Q_\pm)}\right)}{\frac{1}{|g|}} = \frac{0}{0} = \lim_{|g|\to\infty}\frac{\frac{\ln(|g|+2+Q_\pm)}{\ln(|g|+1+Q_\pm)} \times \frac{\frac{\ln(|g|+2+Q_\pm)}{|g|+1+Q_\pm} - \frac{\ln(|g|+1+Q_\pm)}{|g|+2+Q_\pm}}{\ln^2(|g|+2+Q_\pm)}}{\frac{-1}{|g|^2}}$$

$$= \lim_{|g|\to\infty} |g|^2 \frac{(|g|+1+Q_\pm)\ln(|g|+1+Q_\pm) - (|g|+2+Q_\pm)\ln(|g|+2+Q_\pm)}{(|g|+1+Q_\pm)(|g|+2+Q_\pm)\ln(|g|+1+Q_\pm)\ln(|g|+2+Q_\pm)}$$

$$= \lim_{|g|\to\infty}\frac{(|g|+1+Q_\pm) - (|g|+2+Q_\pm)}{\ln(|g|)} = \lim_{|g|\to\infty}\frac{-1}{\ln(|g|)} = 0; \Rightarrow \lim_{g\to\infty} y(|g|) = e^0 = 1.$$

∎

Proposition 4. $\lim_{Q_\pm\to\infty}[\frac{\ln(|g|+1+Q_\pm)}{\ln(|g|+2+Q_\pm)}]^{|g|} = 1.$

Proof. $\lim_{Q_\pm\to\infty}[\frac{\ln(|g|+1+Q_\pm)}{\ln(|g|+2+Q_\pm)}]^{|g|} = \lim_{Q_\pm\to\infty}[\frac{\ln(Q_\pm)}{\ln(Q_\pm)}]^{|g|} = 1.$ ∎

Proposition 5. $\lim_{|g|\to\infty}\frac{\ln(\frac{|g|+1+Q_\pm}{|g|+Q_\pm})}{\ln(\frac{|g|+2+Q_\pm}{|g|+1+Q_\pm})} = 1.$

Proof. $\lim_{|g|\to\infty}\frac{\ln(\frac{|g|+1+Q_\pm}{|g|+Q_\pm})}{\ln(\frac{|g|+2+Q_\pm}{|g|+1+Q_\pm})} = \frac{0}{0} = \lim_{|g|\to\infty}\frac{\frac{|g|+Q_\pm}{|g|+1+Q_\pm} \times \frac{-1}{(|g|+Q_\pm)^2}}{\frac{|g|+1+Q_\pm}{|g|+2+Q_\pm} \times \frac{-1}{(|g|+1+Q_\pm)^2}} = 1.$ ∎

Proposition 6. $\lim_{Q_\pm\to\infty}\frac{\ln(\frac{|g|+1+Q_\pm}{|g|+Q_\pm})}{\ln(\frac{|g|+2+Q_\pm}{|g|+1+Q_\pm})} = 1.$

Proof. $\lim_{Q_\pm\to\infty}\frac{\ln(\frac{|g|+1+Q_\pm}{|g|+Q_\pm})}{\ln(\frac{|g|+2+Q_\pm}{|g|+1+Q_\pm})} = \frac{0}{0} = \lim_{Q_\pm\to\infty}\frac{\frac{|g|+Q_\pm}{|g|+1+Q_\pm} \times \frac{-1}{(|g|+Q_\pm)^2}}{\frac{|g|+1+Q_\pm}{|g|+2+Q_\pm} \times \frac{-1}{(|g|+1+Q_\pm)^2}} = 1.$ ∎

Proposition 7. $\lim_{|g|\to\infty}\frac{\frac{\ln^2(|g|+1+Q_\pm)}{\ln(|g|+2+Q_\pm)} - \ln(|g|+Q_\pm)}{\ln(\frac{|g|+2+Q_\pm}{|g|+1+Q_\pm})} = 0.$

Proof. It is similar to the proof of Proposition 8. ∎

Proposition 8. $\lim_{Q_\pm\to\infty}\frac{\frac{\ln^2(|g|+1+Q_\pm)}{\ln(|g|+2+Q_\pm)} - \ln(|g|+Q_\pm)}{\ln(\frac{|g|+2+Q_\pm}{|g|+1+Q_\pm})} = 0.$

Proof.

$$\lim_{Q_\pm \to \infty} \frac{\frac{\ln^2(|g|+1+Q_\pm)}{\ln(|g|+2+Q_\pm)} - \ln(|g|+Q_\pm)}{\ln\left(\frac{|g|+2+Q_\pm}{|g|+1+Q_\pm}\right)} = \frac{0}{0} = \lim_{Q_\pm \to \infty}\left[\frac{(|g|+1+Q_\pm)\ln^2(|g|+1+Q_\pm)}{\ln^2(|g|+2+Q_\pm)}\right.$$

$$\left. - \frac{2(|g|+2+Q_\pm)\ln(|g|+1+Q_\pm)\ln(|g|+2+Q_\pm)}{\ln(|g|+2+Q_\pm)} + \frac{(|g|+1+Q_\pm)(|g|+2+Q_\pm)}{|g|+Q_\pm}\right]$$

$$= \lim_{Q_\pm \to \infty}\left[\frac{Q_\pm \ln^2(Q_\pm) - 2Q_\pm \ln^2(Q_\pm)}{\ln^2(Q_\pm)} + Q_\pm\right] = -Q_\pm + Q_\pm = 0. \quad \blacksquare$$

Applying Proposition 3, it is easy to prove that $\lim_{|g| \to \infty} Q4(|g|) = 0$ and $\lim_{|g| \to \infty} Q8(|g|) = 1$ and support Properties (D), (E).

Power $S_\pm(1 + H_\pm[\frac{\ln(|g|+1+Q_\pm)}{\ln(|g|+2+Q_\pm)}]^{|g|})$ eliminates $K_\pm$, supports $\frac{\ln(2)}{\ln(\frac{3}{2})} \leq Q5(\pm 1)$ with equation (30), contains an important particular constant power case $H_\pm = 0$. For $g = 0$, the power is $S_\pm(1 + H_\pm)$. From Proposition 3, for $|g| \to \infty$, the limit is again $S_\pm(1 + H_\pm)$. Convergence to the limit can be slow. For $0 \leq x$,

$$\frac{d\left(S_\pm\left(1 + H_\pm\left[\frac{\ln(x+1+Q_\pm)}{\ln(x+2+Q_\pm)}\right]^x\right)\right)}{dx} = S_\pm H_\pm \left(\frac{\ln(x+1+Q_\pm)}{\ln(x+2+Q_\pm)}\right)^x$$

$$\times \left(\frac{x}{(x+1+Q_\pm)\ln(x+1+Q_\pm)} - \frac{x}{(x+2+Q_\pm)\ln(x+2+Q_\pm)}\right.$$

$$\left. + \ln\left(\frac{x+1+Q_\pm}{x+2+Q_\pm}\right)\right).$$

The first derivative can be equal to zero, if the last brackets factor is zero. These conditions imply a minimum, Figure 12. Choosing $A_\pm(g)$ for the product $S_\pm A_\pm(g)$ may require more data.

Mysterious Power Laws

Mysterious power laws are fascinating [57], [2, pp. 36–41/37–43], [65, 67]. Arnold recollected that Kolmogorov, studying turbulence, was applying logarithmic plots [2, pp. 29/30–31]: *"Do not ... look for any mathematical meaning in my achievements in hydrodynamics. There is none. I derive nothing from basic axioms or definitions ... My results are not proved, they are true, and this is much more important! ... Kolmogorov obtained his laws of similarity by covering the floors at the dacha in Komarovka with sheets of paper containing thousands of experimental data (mainly received from Prandtl) ... rather than by dimensional considerations ..."*. Arnold presents *seven* other known examples including the *Olof* Arrhenius law and one new from own research on the *little Fermat theorem*.

The author was not lazy to enter the *data* from the Arrhenius's paper of 1921 into the *Microsoft Excel* in order to "see" the law connecting the number of species on an island with its area [86, p. 50, Figure 22] and continued Arnold's list [86, p. 49]: *the author adds the first name to distinguish the son and his father — Svante Arrhenius (Nobel Prize in Chemistry 1903 for "... electrolytic theory of dissociation"), who's equation for the temperature dependence of the constant of the chemical reaction rate is also known under the name "Arrhenius Law". It can be added to Arnold's list due to good straight lines on plots of logarithm of the constant vs. the reciprocal temperature in Kelvin degrees.*

Semi-empirical *Svante* Arrhenius equation for the constant of a chemical reaction rate [95, p. 9] $k = Ae^{-\frac{E_A}{RT}}$, where T is the absolute temperature, E_A is the activation energy constant within a narrow T interval, A is a coefficient, $R \approx 8.3145 \frac{\text{Joule}}{\text{mole} \times \text{Kelvin}}$ is the gas constant, yields $\ln(k) = \ln(A) - \frac{E_A}{RT}$, a straight line of $\frac{1}{T}$. The power law quantities are $\ln(\frac{E_A}{R\ln(\frac{A}{k})}) = \ln(T)$. In other theories, E_A is replaced with $\Delta G = \Delta H - T\Delta S$, where G, H, S are the Gibbs energy, enthalpy, entropy of activation. In $\ln(k) = \left[\ln(A) + \frac{\Delta S}{R}\right] - \frac{\Delta H}{RT}$, the entropy change is "captured" by the pre-exponential factor and the enthalpy change determines an exponential growth/decline with temperature. The

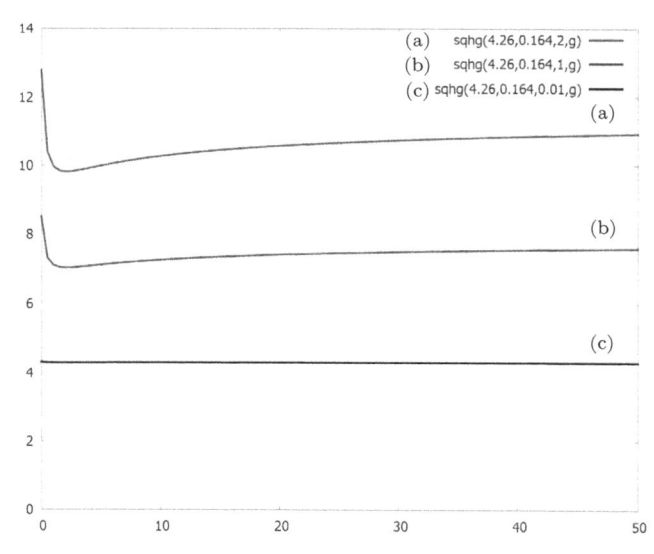

Figure 12. Power $S_+(1 + H_+[\frac{\ln(g+1+Q_+)}{\ln(g+2+Q_+)}]^g)$. $S_+ = 4.26$, $Q_+ = 0.164$. $H_+ = 2$, $H_+ = 1$, $H_+ = 0.01$. The wgnuplot 5.2. http://www.gnuplot.info is applied for plotting.

known representations $A = T^n$ overcome limitations of the Arrhenius approach. The word "entropy" is a link with a logarithm of the number of states of a system and probabilistic interpretations. *Chemical kinetics is applied to the Corn futures* [92].

Zipf's law in linguistics [59,60], and stock price fluctuations [25], so different at the first glance, are other examples. *Such a universalism requires a scalable explanation and the attempt [65] deserves full attention.*

The word Ubuntu

is pronounced several times in a conversation between two people installing an *Ubuntu Server* and included in the instructions for repeating the steps after a next electrical outage corrupting boot sectors of a hard disk. But the author did not find Ubuntu/Убунту in Leo Tolstoy's "War and Peace"/"Война и мир", [60, p. 277, Figure 2. Distribution of words by length]. Ubuntu frequencies in the instructions and masterpiece differ. In the combined text, short instructions, *a fly in the Tolstoy's ointment*, will hardly affect the novel's Zipf's line.

The mental experiment with the word Ubuntu also shows that selection from a general population is a delicate matter and can create a bias. The number of b-increments for any futures contract is finite due to the listing, expiration, and discreteness. If we consider this multiset as a general population, then increasing and combining samples from it will yield statistics closer and closer approaching the statistics of the general population. But this "closeness" would hide that the market *chooses from the multiset differently on different time intervals*. Alternatively, the market could pick up a- and b-increments *uniformly* but then we should recognize that it selects from *different multisets*. In any case, we do not want to exhaust a finite multiset of b-increments and return the chosen objects back to the pool after each selection because we are interested in the future behavior but not a concrete fixed multiset of b-increments.

An MPS is "opposite" to combining samples. It divides a chain of prices on intervals maximizing the profit and does not care how the latter is made and whether the sum of the b-increments has a deterministic contribution, named by some "Trend", or this is a random summation, named by others "Volatility", obeying the limits following from the finite variances of the summands [26]. *An MPS creates a bias in BOTE and SOTE sample means. But it is Market works so and MPS only accurately notices and measures its offers.*

Explanations

Reference [57]. Let g and q are integers, $1 \leq g \leq q$, and S_g and C_g are symbols and their costs. A chain of symbols transfers information and [57, p. 494] *"... one wanted ... to know which values should take the probabilities p_g of the symbols S_g of cost C_g if the information per average cost is to be the largest possible. To solve this, one minimizes $\frac{C}{H} = \sum p_g C_g [-\sum p_g \log p_g]^{-1}$ which gives $p_g = Ae^{-BC_g}$ ....(3). B must be equal to H/C, which leads to the characteristic equation $A^{-1} = \sum_1^q e^{BC_g} = 1$ or $\sum_1^q M_{-C_g} = 1$, with $M = e^{-B}$. Then the average cost per unit of information becomes $C/H = (\log M)^{-1}$. For this to be a minimum, M must be the largest root of the characteristic equation, which is the name number as in the beginning of this section"*. How not to compare S. Arrhenius's $k = Ae^{-\frac{E_A}{RT}}$ with Mandelbrot's $p_g = Ae^{-BC_g}$.

Reference [60]. In revisiting Mandelbrot (above) with slightly different denominations, the paper applies the method of Lagrange multipliers to find the minimum of $C^* = C/H$. This favorably distinguishes [60] from [57] and helps to understand better two Ansatzs [60, p. 280] *"$C_k = C_0 \log_2 k$ (9)"* requiring the Riemann Zeta function probability normalizer and [60, p. 282] *"$C_k = C_0 \log_2(k + k_0)$ (14)"* leading to the Hurwitz Zeta function normalizer. Indeed, years later, [60, p. 275]: *"To account the low-rank flattening, Mandelbrot (1966) proposed ... a modified formula, known as the Zipf–Mandelbrot law: $f_k \propto (k_0 + k)^{-B}$ (2), where k_0 is a parameter ..."*. Clearly that $\lim_{k\to\infty}(k_0+k)^{-B} = k^{-B}$. Mandelbrot focuses on the truncated (Hurwitz) Zeta series, see equation (12), where *independent* k_0 and B promise a better fitting. From the two analyzed Ansatz, the former leads to a degenerate case, where [60, p. 281]: *"... the minimum cost per unit information is achieved when there is only one word in use"*. But the latter allowed to conclude [60, p. 284]: *"... the Zipf–Mandelbrot law can be obtained from a model optimizing the information/cost ratio with no assumptions about word lengths. ... However, the two parameters of the resulting distribution, B and k_0, are not independent"*.

A Comment to [60]. Applying the Hurwitz Zeta function and χ^2-tests to empirical frequencies of b-increments [86], the author did not know about the quantitative linguistics [59,60]. Some zero b-increments deviated from the main dependencies. *If a zero rank point with the largest weight systematically falls out of a line well describing other ranks, then selecting the line as a common model is choosing a step-sister instead of Cinderella when a crystal* slipper cannot be put on her leg. Equations (16), (17), (18) treat $PMF(0)$ separately.

If a dependence between k_0 and B "unexpectedly" follows from a model assumption and is observed experimentally, then this is a discovery allowing not to worry about the loss of flexibility. If experiments do not support the dependence, then the model and loss should cause a concern.

Reference [67]. *"The fact that ... apparently unrelated systems display the same statistical pattern points towards some fundamental, unifying principal"* [67, p. 6]. The authors begin from consideration of stochastic objects observed in time, where the observations are independent and from the same probability distribution. They characterize the *string* of the observations $\boldsymbol{x} = x_1, \ldots, x_m$ by Kolmogorov's complexity $K(\boldsymbol{x})$ defined as the length of a shortest program reproducing $\boldsymbol{x}$, where $\lim_{m\to\infty} \frac{K(\boldsymbol{x})}{m} = \mu \in (0,1].(3)$ [67, p. 2] and replace the complexity with the normalized entropy of the distribution of $X(n)$, namely, $h(n) = \frac{H(X(n))}{\log(n)}, (7) \cdots \lim_{n\to\infty} h(n) = \mu. (9)$ [67, pp. 2, 3]. Their conclusion is *"...this equation ... encodes the concept of growing and ... stabilization of complexity properties in an intermediate point between order and disorder, a feature observed in many systems displaying Zipf-like statistics. From this equation, we derive Zipf's law as the natural outcome of systems belonging to this class of stochastic systems. ... However, ... a system satisfying equation (9) does not necessarily exhibit Zipf's law* [67, p. 6].

A Comments to [67]. Equation (3) [67, p. 2] implies a linearly compressible binary representation of a stochastic object. The consideration assumes non-compressible further objects, which are "almost random". This is noticed in [65, p. 53]. Sometimes, the Market exhibits during *trading at the limit* a rather long chain of trades conducted at the same price yielding zero b-increments characterized by a large defect of randomness [86, p. 81]. Formally, such a chain of zeros is compressed by providing its length and perceived to be "simple" or "non-random". However, the length of the chain is unknown in advance making Strategy II [92, p. 26] risky.

Reference [65]. It convinces that phenomenologically Zipf's law follows from two conditions [65, pp. 52/116–117]: *"(A) Rank ordering coincides with the ordering by growing (exponential) Kolmogorov complexity $K(w)$ up to a factor $\exp(O(1))$"*; and *"(B) The probability distribution producing Zipf's law (with exponent -1) is (an approximation to) the Levin maximal distribution computable from below"*. At first, it defines *complexity of a natural number* choosing a Kolmogorov *optimal encoding* and after that defines *complexity $K(w)$ of a constructive element w as the complexity of its number in a fixed structural numbering*. In contrast with [67], it applies the theory and notation of the *partial functions* and does not limit a consideration by the "almost random" objects whose size is comparable with Kolmogorov complexity and which cannot be compressed. But all this is only an intermediate step to formalization of "software libraries and their reuse", where *"the standard Kolmogorov complexity of partial recursive functions is defined relative to the admissible set of functions ..."* [65, p. 64/125]. The *admissible sets* are defined earlier in the paper and to its end are added by the *oracles for uncomputable functions*. The thought of Harold Edwards cited in Introduction is applicable to this explanation serving mainly to attract attention.

Comments to [65]. In contrast to Zipf's power -1, the experimental values of S found in this research are greater than four. This supports convergence not only for the suggested PMFs but a few beginning and central moments including variances. The numbers of words used for plotting Zipf's lines (straight segments slightly curved at both ends) are $\approx 10^7$. The b-increments do not exceed a few dozens of sizes. We may deal with an initial fragment of the main dependence.

Reusing the libraries of classes and functions such as the Standard C++ [32,101,102] or Boost C++ is a recognized way of accelerating software development [93,108]. The author thinks that *writing a paper* referencing previous results is a similar way of reusing knowledge. The *language of directed graphs* and *basic operators* that make up the *formalization* in [65] has a broader relevance to reuse and sharing of knowledge. Thus, [61] is a logical continuation. Nature, as it were, organizes systems in a hierarchy, supplying each level with its own laws. Otherwise, [28, p. 87]: *"In order to model a bulldozer, we would need to be careful to model its constituent quarks!"* That is why, while the "majority" of the levels is unknown, human beings have "some successes".

How do we know, if an algorithm for a task is impossible or not yet found, that it is impossible to compute *all* E-mini ticks for the next 10 minutes? Is the last part of the question scary, or naive? Rephrasing the title of [63]: "Computable *OR* non-computable" (... "that is the question"). A trader would be happy to know just the maximum and minimum liquid futures prices for any next 10 minutes. Under such conditions making money is boring. One can get a feel using a trading simulator with 10 minutes delayed quotes together with a hint from a real prices stream. The modern cryptographic keys and *secure hash algorithms*, SHA, such as SHA-512 is a basis for cybersecurity. Peter Shor's algorithm breaks them and its realization is "only" a matter of amplifying the baby skills of the already existing *quantum computers* [64,91]. But who thought about *quantum computations* prior [62, pp. 87–108], [63, p. 15]? Is it possible that a solution exists in a branch of science and technology unknown yet? Or maybe the human brain *hypothetically* acting as a quantum computer [74, pp. 400–402, Is there a role for quantum mechanics in brain activity. Quantum computers], [75, pp. 348–392 Quantum theory and the brain], [64, pp. 1–3, (ii)], [91, pp. 11–12, Figure 2] under room temperatures and which is always with us is enough and already has some advantage over classical computers?

A comment within the comment: The author expressed an idea about quantum coherent states of the brain in [91] independently on [64, 74, 75] and as a pure speculation. After that, from an email exchange, he knew about [64] from its author. We agreed about independence and I "recognized his 20 years priority". Only after that, I read [74, 75] and found that the former, mentioning the 1985 work of Deutsche, and latter, referencing the 1982, 1985, 1986 papers of Feynman, are silent with regard to, perhaps, less known in that time 1979 [62], 1980 [63] *considering quantum computations*. In a "plain English" about when quantum parallelism is effective [74, p. 402]: *"... the real gain for the quantum computer might come when a very large number of parallel computations are required - perhaps an indefinitely large number - whose individual answers would not interest us, but where some suitable combination of all the results would"*. The [64] is a short mathematics of this. These words accurately describe the main idea of Steps 1–5 of the quantum algorithm in [91], where unitary transformations keeping a coherent state experience on the last step a measurement in a partial basis.

Back to the main comment: The study of trading strategies with position limits [88, 89, 91] among other tasks counts their number. The paths of these strategies are more sophisticated than Dyck's and Motzkin's paths [89, p. 26, Figure 3]. Any trading rule or their combination essentially yield one from a set with the cardinality $(2W + 1)^{N-1}$, where W is a trading position limit expressed by the number of contracts, and N is the number of ticks. Keeping on long or short positions no more than one contract, with the 11893137 b-increments, $N - 1$ ticks, used to draw Figure 8, we arrive to $3^{11893137} \approx 10^{5673468} \approx 2^{18850176}$. But this is only 20 sessions. A quantum register with 18850176 *qubits* would be able to "hold" all strategies as a superposition. While 18 Mb of memory today is "nothing", the modern *general* quantum computers have 65 qubits and specialized *annealing* systems 2000 qubits. *"IBM promises 1000-qubit quantum computer - a milestone - by 2023"*, https://www.sciencemag.org/news/2020/09/ibm-promises-1000-qubit-quantum-computer-milestone-2023.

The class of chaotic sequences C [43, pp. 430–433/393–395], [107, pp. 126–133/146–154] coincides with the class of typical sequences T [43, pp. 429–430/392–393], [107, pp. 118–126/136–146], see the Levin–Schnorr theorem [43, p. 434/396], [107, pp. 133–137/154–160]. The class of random sequences R is identified with the $T = C$ [43, p. 430/396]. Kolmogorov complexity plays the central role in the definition of C. Hence, [65, 67] attempting to explain Zipf's law using Kolmogorov complexity relate to the topic of randomness defined using typicalness or chaoticness.

A music complement

The author suggests that the *perfect music tuning* [31, p. 28, Figure 12], where *"every fifth corresponds to a $\sqrt[12]{2^7} = 2^{\frac{7}{12}}$ ratio"*, also complements the Arnold's list: C/до 1, C#/до# $\sqrt[12]{2}$, D/ре $\sqrt[12]{2^2}$, D#/ре# $\sqrt[12]{2^3}$, E/ми $\sqrt[12]{2^4}$, F/фа $\sqrt[12]{2^5}$, F#/фа# $\sqrt[12]{2^6}$, G/соль $\sqrt[12]{2^7}$, G#/соль# $\sqrt[12]{2^8}$, A/ля $\sqrt[12]{2^9}$, A#/ля# $\sqrt[12]{2^{10}}$, B/си $\sqrt[12]{2^{11}}$, C/до 2. Recollect that a Steinway piano has 88 keys, A/ля of the *first octave* is the key 49, which supposed to have the frequency $\nu(49) = 440$ Hz. Let us generalize for the nth $= 1,\ldots, 88$ key: $\nu(n) = \sqrt[12]{2^{n-49}} \times 440 = 2^{\frac{n-13}{12}} \times 55$ Hz. For instance, $\nu(1) = 27.50$ Hz corresponds to A/ля of the *subcontroctave*; $\nu(88) = 4186.01$, Hz corresponds to C/до of the *fifth octave*. Then,

$$\log_2(\nu(n)) = \left(\log_2(55) - \frac{13}{12}\right) + \frac{n}{12} = 4.69802638019132627 + \frac{n}{12} \quad (31)$$

is the straight line of the piano key rank n. The power law quantities are

$$\ln([\log_2(\nu(n)) - 4.69802638019132627] \times 12) = \ln(n). \quad (32)$$

While the law is artificially constructed, the *pleasant feelings* caused by this tuning is its *true basis*. Why *"Western harmony is based on the major triad, which is chord composed of the following intervals: the unison, a major third, and a perfect fifth"* is discussed in [100, p. 106, Harmony and Its Evolution].

The Birth of a New Futures Tick

attracts more attention than the birth of a new man. It is announced by an Exchange and retranslated by information services of Bloomberg,

Thomson Reuters, myriads of brokers' data streams and Internet sites. Traders including robots react on it by submitting new or canceling existing orders, or waiting. Once born, it becomes a part of the history and nobody can trade at that time-price-size. Some programs *backtesting* performance of *trading rules* ignore the last circumstance. A "less severe" backtesting "executes" an order on ticks *after* getting the tick triggering an event of interest. Some trading platforms arrange orders to be sent and/or programs of clients in *data centers* of Exchanges in a proximity of *matching engines*. This reduces latency and can submit an order based not on the last tick but current best bid or ask prices. Then, if an order is executed, these prices become the last tick prices. This is when a backtesting program could "assume" that execution at the arrived "event tick" occurs. Simulating latency is a hard task. Analysis shows that the futures market gives opportunities for all types of traders whether they rely on Level I and/or II data, are intraday, or are long-term traders or scalpers. The purpose of gathering empirical distributions of the a-b-increments and volumes changing in time is to get a clue *when* and *what* may happen next. *A modern futures tick is born electronically after interaction of a trading order with the Futures Limit Order Book, FLOB.*

Types of the futures trading orders

A trader without a direct access to the market selects from the broker's order types. Brokers and/or trading platforms use either one available at an Exchange or create own types and combinations, which can be implemented with the Exchange types. For instance, currently, the CME Group has five futures order types: limit, market-limit, market order with protection, stop-limit, and stop-order with protection. They associate with a required duration qualifier such as Good Till Cancel, GTC, or Good Till Date, GTD, and optional minimum execution quantity and display quantity, https://www.cmegroup.com/confluence/display/EPICSANDBOX/Order+Management#OrderManagement-OrderType. The site also describes the terms Fill and Kill, FAK, Fill or Kill, FOK. A market limit order bid with No Market for it can be rejected by *iLink*. Hence, the widely explained Market Order must be implemented by means of the market with protection or market limit orders. This is a reliability challenge for a broker and additional risk of slippage or not filling for a trader. Such combinations as One-Cancels-the-Other, OCO, One-Triggers-Two, OTT, submitting a Stop Loss order conditionally on the filling of an entry Stop-Limit order help. But one has to remember that some of them could be filled partially or not filled, monitor the results of filling and current positions, and cancel orders appropriately. Trading without studying chosen platforms and markets is similar to driving a car without understanding what are the break and gas pedals for. A high liquidity of a nearby E-mini S&P futures makes one forgetting these details under "normal" trading conditions and recollecting them on a "fast" market. *Trading is risky. Why?* Elder [21, p. 49]: *"Trading means trying to rob other people while they are trying to rob you. It is a hard business"*.

FLOB

The limit, market-limit, market orders with protection are either executed at the best available in the book price or become limit orders in FLOB for the part of the original size, which did not find a match in the book. The stop-limit and stop-order with protection require triggering the stop price after which they become either limit or market orders with protection. Finally, all reside as limit orders in the FLOB forming its immediate state. In the benefits of traders, liquidity can be improved by aggregating FLOBs of the related futures constituting, for instance, calendar spreads. This complicates an analyst's task and computer simulations. When new orders arrive, the *matching algorithms* determine what is executed, in which order, what is placed to the book or books, or rejected. Browse https://www.cmegroup.com/confluence/display/EPICSANDBOX/Supported+Matching+Algorithms. See [90], [92, pp. 38–39, 6.9 Influence of the FLOB]. The second compares the work of an FLOB with an interaction of an *incoming signal function* with a *device instrument function*, where the *output function* is a *convolution* of the two. In this system, the instrument function determined by the FLOB's state changes in time.

Predictable or unpredictable ...

Similarly to the already cited [43], authors in [66, p. 267] emphasize: *"Our opinion about the randomness of a sequence depends on accepted probabilistic model, which must be chosen in advance"*. While "symmetrical and asymmetrical coin arguments" for the thought in both papers are convincing, a discouraging remainder is left: We do not explain "randomness" or its origins, leaving alone Borel's and Renyi's philosophy or a quantum mechanical measurement collapsing coherent states in agreement with the squares of the absolute amplitudes constituting a superposition. Instead, we select what we know or assume to be "random" and compare a sequence, as and individual object, with it. But is not it the same way how we apply statistical tests? We chose a probability distribution and decide, using Pearson's or

Kolmogorov–Smirnov test and statistics, if our sample obeys the distribution. *We have no an ensemble?*

Well: While a futures chain is a unique individual object, it is still viewed as a chain of b-increments. Evaluating their empirical distribution, we decompose the chain on parts. What we get statistically describes our unique chain. If we assume that probabilities associate with the found empirical distribution, then our chain is "random" by definition. There are plenty of other distributions, which the chain does not obey, and then it is recognized non-random with respect to them. One may say that empirical distribution is not a model but the data itself. If we apply it to a next chain, "making from data a model", then we can say, if the new chain is the "same" or not with respect to this empirical distribution. Irrespectively on these manipulations with distributions, conceptually, Kolmogorov complexity seems "continue working" for the chain.

The authors of [66, p. 267] gently extend the original typicalness-chaoticness-stochasticness by *unpredictability*. Personally, [66] is interesting and sharpens a few topics. Its seventh section is already titled "Predictable and unpredictable sequences". It is not only a mathematical refinement of the mentioned triad but a set of theorems considering games based on such sequences and chains.

MPS0 as a measure of Market complexity

For any interval containing $1 < N$ ticks, and constant (for simplicity) transaction and filtering costs, MPS0 returns a vector of transactions — strategy — and profit. The r,l-algorithms [80, Chapter 3] are $O(N)$ efficient (see Acknowledgments) and can be applied to a "moving" interval or an "accumulating" interval growing from the first tick. This was first suggested in [80, p. 153 Moving Versions of Strategies]. Such an approach allows us to compare different time intervals and monitor the changing and/or accumulating properties, such as the maximum profit or *maximum profit divided by N*, in terms well understood by traders and market analysts. It is simpler than evaluation of entropy and directly relates to the goals of trading. The OTEs (BOTE, SOTE) organize information in a natural manner and are opened for extension by definition.

Conclusions

Reference [36, p. 12 point 5]: "*The golden rule in publishing work on time-series is to give the original data. To do so in this case would have* *required about thirty pages of journal"*. Since 1953, the sizes of prices time-series have increased. The session of ESM20, Wednesday April 15, 2020, has 505007 total and 504822 trade ticks. A text file contains 19725556 bytes. Tables 7 and A1–A3 together with Formulas (A1) is a "compromise" between Sir Kendall's "imperative" and *impossible* presentation of gigabytes of stored ticks. This, as well as PMFs Formulas 16, 17, are *in sync* with Fisher's reduction of data, Epigraph.

No matter how complex and unpredictable the economic and social factors that determine the state of free markets are, their quintessence — time chains of price increments and algebraic sums of increments, i.e. the prices themselves, along with the limited capital, lead to the existence of a trading strategy, generating a maximum profit. This strategy divides the chains on the alternating buying and selling optimal trading elements. Distributions of price increments in these elements are biased comparing with their total sum. This bias is an essential market property offering a profit. Both, the strategy and elements, are objective market properties that reveal the frequency and magnitude of market offerings. The high frequency and magnitude observed until today, time and time again, are prerequisites for the existence of speculative markets tomorrow. Are ES-mini S&P futures prices random?

Acknowledgments

I would like to thank Oscar Sheynin for emails on the history of statistics, preprint of [97], and other materials on Karl Pearson helping to understand better mathematics and history of the χ^2-criterion, Yuri Manin for attracting attention to [59, 65, 108] on Zipf's law and possible relationship to the Kolmogorov complexity, for sharing [64] and Author's Proof of [61] with email comments, a favorable email on [91], Andrei Linde for sharing [54], where the estimate (3) on page 3, $\sim 10^{10^{77}}$, for *"the number of universes with different geometrical properties"*, which may be created *"as a result of 60 e-folds of the slow-roll inflation"* is incomparably greater than the number of strategies, $\sim 10^{5673468}$, trading futures with no more than one contract on a long or short position, Bjarne Stroustrup for attracting attention to [102], email on `std::mdspan` with respect to matrix classes and friendly comments-suggestions on [93], my children Victor for advices on Python data mining and machine learning modules, Maria and Dmitri for editing, Dmitri for setting up the Ubuntu server *Linux 5.4.0-48-generic #52-Ubuntu x86_64 GNU/Linux*, and Timur Misirpashaev for an email discussion on *dynamic programming*, where he has proposed a different MPS0 algorithm than two in [80] and shared, as a concept of proof, a Python program (specifically,

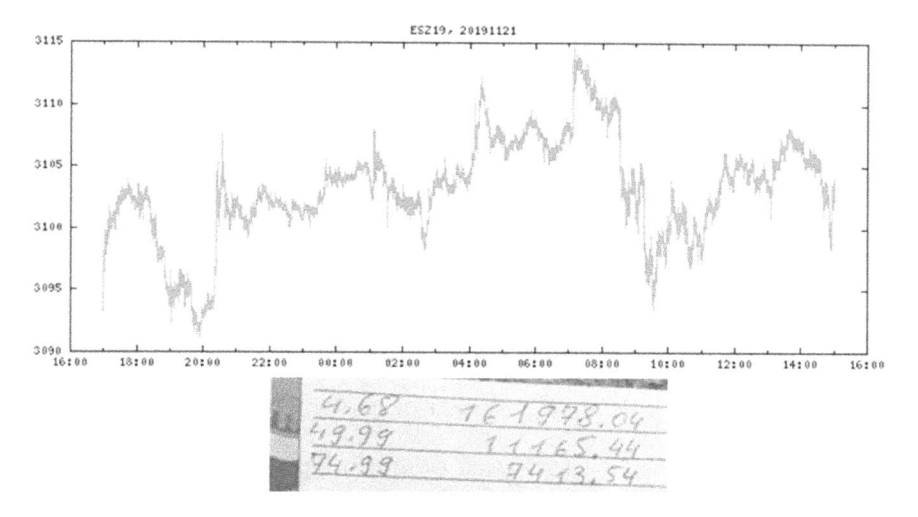

Figure 13. Artifacts: Transaction cost $4.68, three filtering costs, and maximum profits for ESZ19 Thursday November 21, 2019 with 212514 trade ticks.

he proposed to compare optimal profits obtained by his and l,r-algorithms. I applied r-algorithm to ESZ19 traded on Thursday November 21, 2019 with transaction cost $4.68, three filtering costs $4.68, $49.99, $74.99, and recorded the "secrete numbers", Figure 13. After getting the data file, he applied his algorithm and the results matched *exactly*. The 212514 ticks make errors on both sides unlikely. After this, I concluded that the l,r-algorithms [80, pp. 39–54, Chapter 3] can be considered within Bellman's framework [4], even, in 1993–2007, I was not thinking about such a possibility. The l,r-algorithms are based on the introduced s-function, s-interval, and left and right polarities, where "s" stays for "stability", and corresponding proofs. Availability of the $O(N)$ and/or $O(N\log(N))$ algorithms returning MPS0 and maximum profit with transaction costs was one of the keys for the quantum algorithm and explanation of the accelerating effectiveness of quantum parallelism in [91]); Igor Halperin for a LinkedIn exchange of messages on critique of some theories of price processes used for pricing derivatives, [16, pp. 521–528, 2 Market Dynamics, IRL, and Physics], [30], [80, pp. 128–130, Collapse of the Theory?].

Disclaimer

The views, thoughts, and opinions expressed in the text and errors belong to the author and not necessarily to the author's current or past employers, organizations, committees, groups of people, or individuals. This research is carried out at the time free from work duties.

Appendix A: Sample Statistics of b-Increments

Statistics for the samples $x_1, \ldots, x_n$ of the size n in Table A1 are computed using the formulas from [44, p. 611]

$a_s = \dfrac{\sum_{i=1}^{i=n} x_i^s}{n}$, the sample beginning moments of the order s,

$m_s = \dfrac{\sum_{i=1}^{i=n}(x_i - a_1)^s}{n}$, the sample central moments of the order s,

$S^2 = \dfrac{n}{n-1} m_2$, the sample variance — the unbiased and consistent estimate for the central moment of the order 2,

$\mu_3 = \dfrac{n^2}{(n-1)(n-2)} m_3$, the unbiased and consistent estimate for the central moment of the order 3,

$\mu_4 = \dfrac{n(n^2 - 2n + 3)m_4 - 3n(2n-3)m_2^2}{(n-1)(n-2)(n-3)}$, the unbiased and consistent estimate for the central moment of the order 4,

$S = \sqrt{S^2}$, the sample standard deviation,

$SK = \dfrac{\mu_3}{S^3}$, the skewness,

$EK = \dfrac{\mu_4}{S^4} - 3$, the excess kurtosis, \hfill (A1)

where $a_0 = 1$, $m_0 = 1$. For $n = 1$, $m_2 = 0$ and $S = 0$. SK and EK are defined for $2 \leq n$ and $3 \leq n$. Absence of a particular intermediate date in Table A1 indicates a weekend, a holiday, or an illiquid session with $N_d = 0$, or data missing by technical reasons.

Appendix B: Distributions of b-Increments

The S&P 500 index drop starting on Thursday February 20, 2020 and triggered by the *coronavirus events* and continuing until the minimum on Monday March 23, 2020 and the further partial recovery towards Monday April 27, 2020 were accompanied by the corresponding ES-mini futures prices. In the middle of March 2020, the nearby March futures ESH20 passes the baton to the June contract ESM20. This is where the open interest, daily volume, and number of ticks per session for ESH20 decrease and move to ESM20. Despite on strong price correlations between a nearby and next contract, Figure 3, the goal was to avoid mixing sessions for nearby and next

Table A1. Sample statistics of the ESH19, ESM19, ESU19, ESZ19 b-increments in $\delta = 0.25$ or \$12.50. $d = \text{Date}(t^{d,r}_{N_{d,r}})$, $r = 1 = [\text{Date - 1 17:00:00, Date 15:15:00}]$, $r = 2 = [\text{Date 15:30:00, Date 16:00:00}]$, $\Delta t = t^{d,r}_{N_{d,r}} - t^{d,r}_1$ in seconds, $N_{d,r} - 1$ is the sample size, min and max are the smallest and largest b-increment in a sample and # indicates its repetitions. The lines with "*" in the r-column show statistics for all combined ranges corresponding to a single contract. The last line with T, total, in in the r-column combines b-increments for all contracts in one sample.

r	$t^{d,r}_1$	$t^{d,r}_{N_{d,r}}$	Δt	$N_{d,r} - 1$	a_1	min	#	max	#	S	SK	EK
					ESH19							
1	20171214 06:41:23	20171214 14:53:04	29501	4	−14.5	−24	1	−6	1	8.8506	−0.13	−3.825
1	20180123 04:53:05	20180123 07:06:46	8021	1	−28	−28	1	−28	1	0	n/a	n/a
1	20180124 12:36:22	20180124 12:36:22	0	1	3	3	1	3	1	0	n/a	n/a
1	20180129 13:40:55	20180129 14:39:17	3502	1	−8	−8	1	−8	1	0	n/a	n/a
1	20180129 21:54:09	20180130 15:09:23	62114	3	−19	−57	1	0	2	32.909	−1.73	n/a
1	20180201 06:17:29	20180201 07:19:35	3726	1	−10	−10	1	−10	1	0	n/a	n/a
1	20180202 11:40:16	20180202 14:31:34	10278	4	−30	−45	1	5	1	23.58	1.88	0.7583
1	20180205 12:05:13	20180205 12:05:14	1	1	0	0	1	0	1	0	n/a	n/a
1	20180205 17:49:56	20180206 09:35:26	56730	4	20	−101	1	100	1	85.405	−1.33	0.1854
1	20180302 08:23:45	20180302 08:36:33	768	4	12	0	2	40	1	19.044	1.78	0.5242
1	20180319 08:47:50	20180319 09:08:34	1244	3	−0.66667	−4	1	2	1	3.0551	−0.935	n/a
1	20180323 13:57:38	20180323 13:57:38	0	4	0	0	4	0	4	0	n/a	n/a
1	20180402 09:02:59	20180402 09:02:59	0	3	0	0	3	0	3	0	n/a	n/a
1	20180403 09:22:34	20180403 09:29:49	435	1	−4	−4	1	−4	1	0	n/a	n/a
1	20180410 04:45:00	20180410 04:45:05	5	9	0	0	9	0	9	0	n/a	n/a
1	20180411 04:48:18	20180411 05:58:28	4210	2	−4.5	−9	1	0	1	6.364	n/a	n/a
1	20180427 02:55:41	20180427 02:55:41	0	1	−12	−12	1	−12	1	0	n/a	n/a
1	20180430 14:29:00	20180430 14:29:00	0	1	0	0	1	0	1	0	n/a	n/a
1	20180503 09:11:01	20180503 09:26:01	900	2	−21.5	−32	1	−11	1	14.849	n/a	n/a
1	20180509 09:54:13	20180509 09:54:13	0	1	0	0	1	0	1	0	n/a	n/a
1	20180510 07:30:01	20180510 13:48:13	22692	6	12.333	−7	1	49	1	20.5	1.41	0.1879

(Continued)

Table A1. (Continued)

r	$t_1^{d,r}$	$t_{N_{d,r}}^{d,r}$	Δt	$N_{d,r}-1$	a_1	min	#	max	#	S	SK	EK
					ESH19							
2	20180510 15:56:45	20180510 15:58:04	79	1	−13	−13	1	−13	1	0	n/a	n/a
1	20180521 09:43:34	20180521 13:51:52	14898	1	−34	−34	1	−34	1	0	n/a	n/a
1	20180523 15:09:37	20180523 15:14:37	300	3	−1.3333	−4	1	0	2	2.3094	−1.73	n/a
1	20180524 09:12:03	20180524 11:54:08	9725	8	1.25	−20	1	39	1	21.933	0.799	−1.226
1	20180525 08:38:10	20180525 12:27:29	13759	5	−1	−11	1	13	1	8.9722	0.935	−0.1241
1	20180529 03:30:53	20180529 14:55:57	41104	26	−1.3462	−73	1	39	1	18.48	−1.92	8.519
1	20180530 10:53:01	20180530 14:31:52	13131	4	4.25	−7	1	26	1	14.796	1.76	0.6195
1	20180531 08:29:33	20180531 13:48:40	19147	12	−4.9167	−35	1	14	1	12.347	−1.31	1.661
1	20180601 01:51:15	20180601 13:23:55	41560	5	9	−18	1	68	1	34.095	1.88	1.284
1	20180604 08:46:28	20180604 12:40:35	14047	10	−0.9	−6	1	1	1	2.2336	−1.88	1.384
1	20180606 06:55:26	20180606 13:35:12	23986	1	23	23	1	23	1	0	n/a	n/a
1	20180607 08:06:50	20180607 08:10:40	230	5	−1.6	−6	1	4	1	3.8471	0.59	−1.013
2	20180612 15:54:48	20180612 15:54:48	0	1	0	0	1	0	1	0	n/a	n/a
1	20180613 02:30:23	20180613 13:45:17	40494	11	−1.0909	−32	1	14	1	13.194	−1.38	1.233
1	20180614 12:05:37	20180614 12:09:17	220	2	5	2	1	8	1	4.2426	n/a	n/a
1	20180615 09:31:05	20180615 13:20:05	13740	6	4.5	−15	1	16	1	11.113	−1.23	0.0138
1	20180617 18:09:59	20180618 11:27:21	62242	8	−5.5	−38	1	23	1	17.436	−0.423	0.5654
1	20180619 01:00:27	20180619 14:51:58	49891	15	4.0667	−23	1	55	1	17.182	2.03	4.426
1	20180620 08:50:32	20180620 14:16:04	19532	14	0.071429	−19	1	24	1	8.5975	0.945	4.87
1	20180620 20:52:00	20180621 14:46:58	64498	11	−10.455	−54	1	34	1	25.343	0.0968	−0.5203
1	20180625 08:45:51	20180625 14:59:52	22441	13	−5.0769	−48	1	78	1	29.613	1.74	3.883
1	20180625 17:10:04	20180626 09:19:47	58183	35	0.28571	−12	1	24	1	4.6054	3.66	21.12
1	20180627 01:02:12	20180627 14:57:07	50095	15	−4.5333	−144	1	114	1	49.881	−0.705	4.955
1	20180627 22:41:47	20180628 14:27:55	56768	9	5.6667	−41	1	107	1	40.512	2.25	4.093
1	20180628 21:36:51	20180629 14:56:00	62349	12	0.66667	−28	1	43	1	19.476	0.673	0.4866
1	20180701 20:53:29	20180702 14:50:00	64591	6	−2.3333	−71	1	38	1	41.215	−1.08	−0.7456
1	20180703 01:33:57	20180703 11:31:22	35845	35	−0.82857	−45	1	16	1	8.7026	−3.85	18.85

1	20180705 01:43:47	20180705 14:50:41	47214	1	86	86	1	86	0	n/a	n/a
1	20180705 19:14:54	20180706 14:26:18	69084	4	28.75	0	1	59	24.116	0.186	−0.3832
1	20180709 08:34:54	20180709 14:59:39	23085	9	4.7778	−9	1	19	9.7183	0.363	−1.492
1	20180709 19:48:41	20180710 09:07:11	47910	2	0	−10	1	10	14.142	n/a	n/a
2	20180710 15:51:32	20180710 15:56:01	269	2	−30.5	−61	1	0	43.134	n/a	n/a
1	20180710 17:18:05	20180711 15:13:30	78925	16	−1.0625	−52	1	28	18.201	−1.2	2.567
1	20180712 04:02:29	20180712 14:05:40	36191	2	27	20	1	34	9.8995	n/a	n/a
1	20180717 12:46:28	20180717 14:57:51	7883	2	−6	−8	1	−4	2.8284	n/a	n/a
1	20180718 03:18:47	20180718 11:22:27	29020	6	4.3333	−6	1	34	15.095	2.09	2.007
1	20180719 09:24:12	20180719 14:23:57	17985	2	1	0	1	2	1.4142	n/a	n/a
1	20180719 20:33:25	20180720 14:07:56	63271	6	1.3333	−33	1	24	19.18	−1.16	0.6851
1	20180723 20:49:57	20180724 07:20:51	37854	2	16.5	9	1	24	10.607	n/a	n/a
1	20180725 12:22:45	20180725 14:49:59	8834	2	42	0	1	84	59.397	n/a	n/a
1	20180725 17:47:41	20180726 14:57:09	76168	6	1.3333	−33	1	40	23.83	0.355	0.2073
1	20180727 12:09:11	20180727 14:30:48	8497	4	2	−10	1	18	11.662	0.989	−0.0705
1	20180730 01:59:59	20180730 11:52:31	35552	2	−21	−47	1	5	36.77	n/a	n/a
1	20180731 08:25:27	20180731 11:04:40	9553	1	12	12	1	12	0	n/a	n/a
1	20180802 05:24:42	20180802 11:23:50	21548	3	39	−1	1	104	56.789	1.6	0.392
1	20180802 19:44:21	20180803 14:12:04	66463	3	12.667	−11	1	49	31.943	1.5	n/a
1	20180806 04:49:02	20180806 13:03:38	29676	4	9	−6	1	46	24.739	1.96	0.9399
1	20180806 21:33:55	20180807 14:50:00	62165	12	3.25	−10	1	27	9.3529	1.3	2.312
1	20180808 06:37:53	20180808 09:35:01	10628	1	−28	−28	1	−28	0	n/a	n/a
1	20180809 14:33:06	20180809 14:43:38	632	1	0	0	1	0	0	n/a	n/a
1	20180809 19:34:27	20180810 14:25:07	67840	12	−4.5833	−36	1	20	15.629	−0.967	0.392
1	20180812 23:12:19	20180813 10:32:28	40809	4	9.75	0	2	31	14.66	1.64	0.1863
1	20180814 09:04:25	20180814 13:18:10	15225	1	48	48	1	48	0	n/a	n/a
1	20180815 06:32:46	20180815 12:24:42	21116	16	−2.8125	−21	1	29	11.485	1.15	2.336
1	20180816 10:01:01	20180816 14:45:29	17068	4	0.75	−23	1	43	29.33	1.55	0.175
1	20180816 20:37:50	20180817 14:43:02	65112	3	10.333	−8	1	36	22.898	1.29	n/a
1	20180820 01:33:45	20180820 14:18:44	45899	6	1.6667	−4	2	6	4.9666	−0.305	−2.562
1	20180821 08:48:18	20180821 13:11:11	15773	6	2.1667	−4	1	15	7.3598	1.25	−0.2414

(*Continued*)

Table A1. (Continued)

r	$t_1^{d,r}$	$t_{N_{d,r}}^{d,r}$	Δt	$N_{d,r}-1$	a_1	min	# min	max	# max	S	SK	EK
				ESH19								
1	20180821 18:02:33	20180822 14:00:18	71865	17	3.2941	−16	1	18	1	10.641	−0.327	−1.059
1	20180823 08:39:33	20180823 11:40:28	10855	6	−0.66667	−37	1	12	1	18.403	−2.11	2.148
1	20180823 18:43:03	20180824 15:00:53	73070	5	13.2	−7	1	31	1	16.498	0.023	−2.357
1	20180826 17:02:57	20180827 14:51:25	78508	13	5.3077	−7	1	26	1	9.8352	0.867	−0.3461
1	20180827 19:01:30	20180828 15:00:07	71917	10	−0.4	−19	1	15	1	12.195	−0.362	−1.559
1	20180829 09:31:19	20180829 14:57:26	19567	16	2.25	−10	1	20	1	8.5907	0.417	−0.87
1	20180829 21:21:41	20180830 14:59:12	63451	52	−0.76923	−24	1	18	1	6.9723	−0.753	3.19
1	20180830 18:28:27	20180831 14:52:50	73463	56	0.16071	−18	1	16	1	6.6573	−0.229	0.5734
1	20180903 19:41:29	20180904 14:59:19	69470	59	−0.61017	−26	1	15	1	7.4394	−0.85	1.726
1	20180904 20:20:33	20180905 15:09:00	67707	56	−0.75	−38	1	24	1	9.1139	−0.992	4.385
1	20180905 19:42:25	20180906 15:06:43	69858	38	−0.78947	−23	1	26	1	10.249	0.463	0.3717
1	20180906 17:30:41	20180907 13:29:03	71902	78	−0.10256	−43	1	23	1	9.0734	−1.81	8.475
1	20180909 17:00:00	20180910 14:29:24	77364	42	0.28571	−23	1	22	1	8.8463	0.187	0.9537
1	20180911 02:39:02	20180911 14:23:08	42246	39	0.76923	−37	1	31	1	10.027	−0.8	5.576
1	20180912 19:06:29	20180913 15:06:21	71992	61	1.1803	−14	1	22	1	7.6081	0.467	0.4733
1	20180913 17:00:00	20180914 14:57:20	79040	66	0.16667	−19	1	17	1	5.9526	−0.316	1.748
1	20180916 17:15:31	20180917 15:14:58	79167	160	−0.3	−13	1	14	1	3.7025	0.325	2.293
1	20180917 17:36:19	20180918 15:13:40	77841	228	0.38158	−19	1	21	1	4.4956	0.393	6.102
1	20180918 17:24:43	20180919 14:49:49	77106	124	0.072581	−16	1	10	3	4.1185	−0.988	4.284
1	20180919 17:20:04	20180920 15:14:44	78880	140	0.67143	−14	1	9	1	3.4567	−0.418	1.895
1	20180920 17:34:04	20180921 15:12:59	77935	220	−0.10909	−13	1	10	2	3.3169	−0.0981	2.304
2	20180921 15:30:00	20180921 15:59:24	1764	13	−0.61538	−5	2	2	1	2.0223	−1.77	1.628
1	20180923 17:00:00	20180924 15:10:00	79800	283	−0.074205	−15	1	13	1	2.9616	−0.213	4.146
1	20180924 17:01:19	20180925 15:05:23	79444	190	−0.10526	−23	1	14	1	3.3459	−1.03	13.18
2	20180925 15:47:25	20180925 15:58:59	694	1	3	3	1	3	1	0	n/a	n/a
1	20180925 18:18:51	20180926 15:14:05	75314	278	−0.18345	−10	2	14	1	3.1472	0.112	2.142
2	20180926 15:30:00	20180926 15:54:39	1479	10	−0.2	−8	1	4	1	3.6454	−1.06	0.2544

Are E-mini S&P 500 Futures Prices Random? 293

1	20180926 17:21:39	20180927 15:12:18	78639	233	0.15021	−9	1	11	1	3.2108	0.474	1.493
2	20180927 15:34:13	20180927 15:59:44	1531	3	1.3333	−1	1	4	1	2.5166	0.586	n/a
1	20180927 17:23:19	20180928 15:12:25	78546	224	−0.026786	−12	1	13	1	3.6665	0.00834	1.191
1	20180930 17:00:00	20181001 15:10:58	79858	333	0.072072	−12	1	12	1	2.7857	0.204	3.137
2	20181001 15:30:17	20181001 15:57:29	1632	4	−1.25	−5	1	1	1	2.63	−1.44	0.0292
1	20181001 17:32:13	20181002 15:11:39	77966	216	−0.013889	−12	1	13	1	3.6197	0.0517	1.5
1	20181002 17:00:04	20181003 15:14:39	80075	284	0.03169	−13	1	15	1	3.326	−0.2	3.577
2	20181003 15:51:19	20181003 15:58:20	421	4	−2.5	−7	1	0	2	3.3166	−1.1	−1.227
1	20181003 17:00:11	20181004 15:14:48	80077	773	−0.093144	−12	2	15	1	3.0282	0.0566	3.268
2	20181004 15:50:07	20181004 15:56:25	378	3	1	0	2	3	1	1.7321	1.73	n/a
1	20181004 17:24:02	20181005 15:13:36	78574	812	−0.077586	−17	1	29	1	3.6003	0.6	7.335
2	20181005 15:30:00	20181005 15:51:31	1291	17	−0.058824	−6	1	3	3	2.2492	−0.814	1.266
1	20181007 17:00:43	20181008 15:07:51	79628	936	−0.021368	−22	1	14	1	2.9265	−0.537	6.111
2	20181008 15:30:00	20181008 15:53:04	1384	5	−1.4	−5	1	1	2	2.881	−0.59	−2.71
1	20181008 17:01:44	20181009 15:10:19	79715	552	−0.043478	−15	1	13	1	3.5431	−0.242	2.05
2	20181009 17:20:50	20181010 15:14:59	78849	1916	−0.22495	−23	1	18	1	3.2613	−0.0242	4.172
2	20181010 15:30:12	20181010 15:59:40	1768	60	−0.71667	−17	1	5	2	3.5753	−2.06	6.43
1	20181010 17:00:00	20181010 15:14:57	80097	3481	−0.033611	−23	1	26	2	4.2517	0.238	4.549
2	20181011 15:30:41	20181011 15:54:01	1400	21	−0.095238	−9	1	9	1	4.1582	0.167	0.081
1	20181011 17:00:00	20181012 15:14:50	80090	1270	0.062205	−22	1	22	2	4.9275	0.218	2.897
2	20181012 15:30:00	20181012 15:55:47	1547	3	5	−2	1	12	1	7	0	n/a
1	20181014 17:00:02	20181015 15:14:54	80092	841	−0.10107	−26	1	25	1	4.5844	−0.393	4.876
2	20181015 15:30:42	20181015 15:58:11	1649	12	−0.58333	−6	1	5	1	3.3155	−0.296	−0.7856
1	20181015 17:03:37	20181016 15:14:57	79880	763	0.37615	−12	2	14	1	3.2323	−0.0297	2.087
2	20181016 15:30:08	20181016 15:41:06	658	6	−0.5	−5	1	3	1	2.9496	−0.666	−1.252
1	20181016 17:00:00	20181017 15:04:20	79460	922	−0.016269	−17	1	21	1	3.7985	0.28	3.732
2	20181017 15:34:39	20181017 15:56:16	1297	5	0.4	−1	1	3	1	1.5166	1.75	1.234
1	20181017 17:00:16	20181018 15:14:28	80052	1259	−0.13264	−17	1	19	1	4.011	0.00229	2.612
1	20181018 17:00:01	20181019 15:14:40	80079	984	−0.033537	−15	1	24	1	3.8873	0.529	4.715
2	20181019 15:30:22	20181019 15:47:15	1013	8	−0.125	−4	2	3	1	2.6959	−0.641	−1.428
1	20181021 17:00:00	20181022 15:14:05	80045	1516	−0.0032982	−13	2	17	1	3.1125	0.226	4.008

(Continued)

Table A1. (Continued)

r	$t_1^{d,r}$	$t_{N_{d,r}}^{d,r}$	Δt	$N_{d,r}-1$	a_1	min	#	max	#	S	SK	EK
					ESH19							
2	20181022 15:30:01	20181022 15:53:53	1432	2	0.5	0	1	1	1	0.70711	n/a	n/a
1	20181022 17:10:36	20181023 08:46:15	56139	998	−0.16232	−15	1	15	2	3.244	0.139	3.356
1	20181023 17:11:34	20181024 15:14:46	79392	1813	−0.17926	−25	1	21	2	4.0802	−0.109	4.026
2	20181024 15:30:00	20181024 15:58:06	1686	23	1.087	−6	1	12	1	4.4915	0.9	0.6637
1	20181024 17:00:00	20181025 15:14:58	80098	1535	0.020847	−22	1	28	1	4.2906	0.27	4.64
2	20181025 15:30:00	20181025 15:57:20	1640	57	−0.14035	−11	1	12	1	4.3893	0.685	1.451
1	20181025 17:00:05	20181026 15:14:53	80088	2116	−0.015595	−34	1	23	1	5.0551	−0.271	4.031
2	20181026 15:30:00	20181026 15:59:26	1766	51	−0.66667	−12	1	6	2	3.7452	−0.878	0.9133
1	20181028 17:04:46	20181029 15:14:50	79804	1550	−0.066452	−27	1	35	1	4.8869	−0.0787	5.498
2	20181029 15:30:00	20181029 15:59:55	1795	8	−1	−7	1	4	1	3.0706	−0.592	1.124
1	20181029 17:01:11	20181030 15:14:57	80026	1789	0.10006	−37	1	25	1	4.6166	−0.378	5.372
2	20181030 15:31:00	20181030 15:59:19	1699	18	0.72222	−6	1	6	1	3.2685	−0.188	−0.6145
1	20181030 17:00:00	20181031 15:14:49	80089	1246	0.05297	−18	1	14	1	3.7368	−0.505	3.036
2	20181031 15:30:59	20181031 15:55:09	1450	37	−0.13514	−13	1	11	1	3.9381	−0.26	3.126
1	20181031 17:00:14	20181101 15:11:52	79898	861	0.15099	−21	1	18	1	3.9741	−0.227	2.6
2	20181101 15:30:02	20181101 15:59:57	1795	23	−0.86957	−11	1	10	1	5.5046	0.217	−0.4873
1	20181101 17:00:01	20181102 15:14:52	80091	1648	0.0018204	−20	1	23	1	3.9641	0.0641	3.495
2	20181102 15:30:00	20181102 15:59:42	1782	6	−1.3333	−5	2	3	2	3.8297	0.254	−2.531
1	20181104 17:00:00	20181105 15:14:43	80083	794	0.086902	−18	1	19	1	3.2861	−0.116	3.971
2	20181105 15:30:03	20181105 15:59:01	1738	7	0.42857	−5	1	8	1	4.504	0.587	−0.7926
1	20181105 17:03:06	20181106 15:14:36	79890	663	0.1267	−17	1	12	2	3.3224	−0.423	2.926
2	20181106 15:48:35	20181106 15:58:59	624	4	−0.5	−4	1	6	1	4.5092	1.57	0.1294
1	20181106 17:01:25	20181107 15:14:57	80012	1501	0.16855	−18	1	20	2	3.3622	0.386	6.05
2	20181107 15:30:00	20181107 15:58:57	1737	11	0.36364	−3	1	5	1	2.2033	0.737	0.3287
1	20181107 17:00:00	20181108 15:09:31	79771	683	−0.04246	−22	1	21	1	3.7172	−0.216	4.369
2	20181108 15:30:15	20181108 15:51:32	1277	19	−0.63158	−3	2	2	2	1.6059	0.221	−1.312
1	20181108 17:00:00	20181109 15:14:28	80068	1406	−0.073257	−14	3	19	1	2.959	−0.139	3.944

2	20181109 15:30:13	20181109 15:57:56	1663	8	1.125	0	4	5	1	1.7269	1.97	2.211
1	20181111 17:04:51	20181112 15:14:55	79804	1146	−0.17016	−18	2	19	1	3.6859	0.095	4.19
2	20181112 15:30:02	20181112 15:59:55	1793	21	0.28571	−10	1	9	1	3.3933	−0.645	4.395
1	20181113 17:01:19	20181114 15:14:55	80016	2236	−0.055009	−27	1	18	1	3.3416	−0.366	5.462
2	20181114 15:46:20	20181114 15:56:34	614	5	3.8	−1	1	9	1	4.0866	0.312	−2.068
1	20181114 17:00:00	20181115 15:14:59	80099	2442	0.056511	−19	1	20	1	3.6893	0.183	4.011
2	20181115 15:30:01	20181115 15:59:17	1756	43	0	−11	1	8	1	3.4365	−0.255	1.719
1	20181115 17:02:22	20181116 15:14:52	79950	1565	0.031949	−18	1	18	1	3.3299	0.21	3.473
2	20181116 15:30:00	20181116 15:58:59	1739	3	−2.3333	−6	1	0	1	3.2146	−1.55	n/a
1	20181119 17:01:24	20181120 15:14:54	80010	2473	−0.090578	−19	1	29	1	3.5727	0.0773	5.419
2	20181120 15:30:15	20181120 15:59:53	1778	19	−0.68421	−8	1	2	4	2.7091	−1.25	0.9934
1	20181120 17:00:00	20181121 15:14:35	80075	1117	0.033124	−13	4	17	2	3.469	0.0316	3.502
2	20181121 15:30:00	20181121 15:52:42	1362	13	0.15385	−8	1	4	1	3.1845	−1.54	1.888
1	20181122 17:00:59	20181123 12:14:59	69240	854	−0.037471	−15	1	13	1	2.9499	−0.063	3.33
2	20181125 17:00:00	20181126 15:14:57	80097	1386	0.10029	−12	1	15	1	2.4848	−0.0237	3.422
1	20181126 15:30:00	20181126 15:58:59	1739	35	−0.74286	−10	1	9	1	2.9936	0.392	4.462
2	20181126 17:01:12	20181127 15:14:41	80009	1532	0.063316	−22	1	18	1	3.1725	−0.238	6.218
2	20181127 15:31:07	20181127 15:57:46	1599	35	0.22857	−4	2	7	1	2.1974	0.467	1.454
1	20181127 17:00:37	20181128 15:14:55	80058	1735	0.12622	−12	1	12	2	2.6321	0.0687	2.183
2	20181128 15:30:43	20181128 15:57:09	1586	31	0.064516	−4	1	6	1	1.8786	1.13	3.5
1	20181128 17:00:00	20181129 15:14:20	80060	1454	0.011692	−15	1	12	1	2.6353	0.087	3.024
2	20181129 15:32:17	20181129 15:59:15	1618	17	0	−5	1	5	1	2.3452	−0.132	0.2181
1	20181129 17:00:00	20181130 15:14:33	80073	1393	0.061019	−16	2	17	1	2.524	0.142	8.94
2	20181130 15:30:00	20181130 15:59:53	1793	19	0	−4	1	5	1	2.357	0.0569	−0.3023
1	20181202 17:00:00	20181203 15:14:47	80087	3083	0.004541	−11	1	13	2	2.0893	0.158	3.669
2	20181203 15:30:00	20181203 15:59:28	1768	40	−0.05	−5	1	4	2	1.894	0.0754	0.4093
1	20181203 17:00:00	20181204 15:14:58	80098	5579	−0.063811	−25	1	14	2	2.1264	−0.056	6.745
2	20181204 15:30:00	20181204 15:59:43	1783	79	−0.050633	−17	1	11	1	3.3889	−1.05	7.979
1	20181204 17:00:00	20181205 08:29:27	55767	1575	0.048889	−13	1	8	2	1.7187	−0.21	5.961
2	20181205 17:00:00	20181206 15:14:58	80098	17657	−0.0053237	−29	1	30	1	1.7896	0.296	20.48
1	20181206 15:30:03	20181206 15:59:55	1792	49	0.061224	−5	2	9	1	2.3577	0.64	3.532

(*Continued*)

Table A1. (Continued)

r	$t_1^{d,r}$	$t_{N_{d,r}}^{d,r}$	Δt	$N_{d,r}-1$	a_1	min	#	max	#	S	SK	EK
\multicolumn{13}{c}{ESH19}											
1	20181206 17:00:00	20181207 15:14:55	80095	12034	-0.019611	-19	1	18	1	1.7112	-0.0678	9.256
2	20181207 15:30:00	20181207 15:59:53	1793	32	0.375	-4	1	7	1	2.254	0.58	0.9499
1	20181209 17:00:00	20181210 15:14:55	80095	17936	0.0040143	-16	1	12	1	1.4377	-0.044	6.968
2	20181210 15:30:00	20181210 15:59:54	1794	35	-0.14286	-8	1	5	1	2.3657	-0.724	2.54
1	20181210 17:00:00	20181211 15:14:58	80098	25260	-0.00047506	-11	1	8	2	1.0139	-0.0157	5.643
2	20181211 15:30:05	20181211 15:59:56	1791	221	-0.16742	-4	2	6	1	1.2771	0.714	2.824
1	20181211 17:00:00	20181212 15:14:54	80094	33580	0.0022633	-9	1	14	1	0.81258	0.234	8.379
2	20181212 15:30:00	20181212 15:59:41	1781	300	-0.01	-3	1	4	1	0.87891	0.139	1.826
1	20181212 17:00:00	20181213 15:14:59	80099	83755	-0.00046564	-7	4	9	1	0.61235	0.0418	6.308
2	20181213 15:30:00	20181213 15:59:47	1787	875	0.0057143	-3	2	5	1	0.63909	0.311	6.211
1	20181213 17:00:00	20181214 15:14:59	80099	312307	-0.00057315	-6	1	6	1	0.37628	-0.0138	7.626
2	20181214 15:30:00	20181214 15:59:59	1799	1744	-0.0091743	-2	9	2	7	0.54191	-0.0502	1.683
1	20181216 17:00:00	20181217 11:09:36	65376	215885	0.0001297	-6	2	6	1	0.39074	0.00491	8.577
1	20181217 17:00:00	20181218 02:11:11	33071	43926	0.00025042	-6	1	4	2	0.41357	-0.0659	6.741
1	20181218 17:00:00	20181219 15:14:59	80099	600484	-0.00018152	-33	1	37	1	0.54884	0.148	199.2
2	20181219 15:30:00	20181219 15:59:59	1799	4535	-0.0022051	-6	1	8	1	0.75784	0.122	6.748
1	20181219 17:00:00	20181220 15:14:59	80099	819645	-8.7843e-05	-19	1	19	1	0.42777	-0.0521	46.31
2	20181220 15:30:00	20181220 15:59:59	1799	3596	0.0038932	-5	1	4	3	0.67992	0.0271	4.133
1	20181220 17:00:00	20181221 15:14:59	80099	688932	-0.00041949	-20	1	19	1	0.44997	-0.0393	27.62
2	20181221 15:30:00	20181221 15:59:59	1799	3596	0.0083426	-4	1	4	1	0.60818	0.0256	2.388
1	20181223 17:00:00	20181224 12:14:59	69299	375832	-0.00069712	-22	1	21	1	0.53131	0.146	38.76
1	20181225 17:00:00	20181226 15:14:59	80099	557778	0.00086414	-15	3	15	4	0.51902	-0.0427	35.21
2	20181226 15:30:00	20181226 15:59:59	1799	3961	-0.00025246	-7	1	7	1	0.77215	0.0828	7.507
1	20181226 17:00:00	20181227 03:27:40	37660	63909	-0.0023314	-13	2	13	1	0.53137	-0.199	42.88
1	20181227 17:00:00	20181228 15:14:59	80099	545913	-3.2972e-05	-19	1	20	1	0.46345	0.0337	31.26
2	20181228 15:30:00	20181228 15:59:59	1799	1376	0.005814	-3	1	2	18	0.68753	0.0329	0.4845
1	20190101 17:00:00	20190102 15:14:59	80099	497511	2.412e-05	-11	1	10	1	0.42768	-0.00803	11.71

2	20190102	15:30:00	20190102	15:59:59	1799	10353	−0.0077272	−12	1	14	1	0.83602	0.223	17.31
1	20190102	17:00:00	20190103	15:14:59	80099	542368	−0.000025075	−16	1	14	2	0.44965	−0.0164	23.89
2	20190103	15:30:00	20190103	15:59:59	1799	1810	0.0055249	−6	1	5	2	0.75455	−0.00911	5.561
1	20190103	17:00:00	20190104	15:14:59	80099	520281	0.00060736	−19	1	15	1	0.445	−0.00793	34.64
2	20190104	15:30:00	20190104	15:59:59	1799	1963	−0.0020377	−3	1	2	18	0.62238	0.0522	0.9712
1	20190104	17:00:00	20190107	15:14:59	80099	358277	0.00014514	−13	1	15	1	0.41061	0.0683	15.76
2	20190107	15:30:00	20190107	15:59:59	1799	1634	0.0079559	−2	7	3	1	0.59643	0.206	1.43
1	20190107	17:00:00	20190108	15:14:59	80099	349498	0.00021459	−8	2	9	2	0.40765	0.00716	9.846
2	20190108	15:30:00	20190108	15:59:59	1799	1483	−0.0067431	−3	1	3	1	0.57497	−0.0856	1.804
1	20190108	17:00:00	20190109	15:14:59	80099	343164	0.00014862	−8	1	7	1	0.40578	−0.0222	8.015
2	20190109	15:30:00	20190109	15:59:59	1799	2519	−0.00079397	−2	8	2	4	0.58577	−0.0473	0.4015
1	20190109	17:00:00	20190110	15:14:59	80099	306309	0.00017956	−6	2	6	2	0.40697	0.00221	5.322
2	20190110	15:30:00	20190110	15:59:59	1799	1240	0.0024194	−2	2	2	2	0.51997	0.00327	1.234
1	20190110	17:00:00	20190111	15:14:59	80099	233277	1.7147e−05	−6	1	2	2	0.40207	−0.00343	4.707
2	20190111	15:30:00	20190111	15:59:58	1798	916	0.0021834	−2	4	5	1	0.55515	0.0394	1.218
1	20190113	17:00:00	20190114	15:14:59	80099	222739	−0.00013918	−4	1	2	4	0.41833	−0.00421	3.712
2	20190114	15:30:00	20190114	15:59:59	1799	1018	0.00098232	−2	2	2	2	0.56877	0.000163	0.3206
1	20190114	17:00:00	20190115	15:14:59	80099	289153	0.00031817	−11	1	10	1	0.4148	0.0147	10.29
2	20190115	15:30:00	20190115	15:59:59	1799	1117	0.0026858	−3	1	2	3	0.54788	−0.0966	1.653
1	20190115	17:00:00	20190116	15:14:59	80099	226911	0.00013662	−9	1	9	1	0.41156	−0.0146	7.285
2	20190116	15:30:00	20190116	15:59:59	1799	1221	0.004095	−2	4	2	4	0.55735	0.00163	1.039
1	20190116	17:00:00	20190117	15:14:59	80099	263726	0.0003223	−6	3	8	1	0.42539	0.0229	6.666
2	20190117	15:30:00	20190117	15:59:59	1799	1780	0.0095506	−2	4	3	1	0.57434	0.143	1.141
1	20190117	17:00:00	20190118	15:14:59	80099	286089	0.00048586	−13	1	12	1	0.42576	0.0156	14.61
2	20190118	15:30:00	20190118	15:59:59	1799	2302	0.0095569	−3	2	4	1	0.61199	0.00624	2.536
1	20190123	17:00:00	20190124	13:18:09	73089	205880	−3.4e−05	−6	4	6	1	0.42849	−0.0343	5.614
1	20190124	17:00:00	20190125	15:14:59	80099	242174	0.00044596	−7	1	7	1	0.40931	0.03	6.479
2	20190125	15:30:00	20190125	15:59:59	1799	994	0.0080483	−2	2	2	3	0.55687	0.0384	0.8558
1	20190127	17:00:00	20190128	15:14:59	80099	249984	−0.00028802	−6	1	6	1	0.41443	0.0206	5.792
2	20190128	15:30:00	20190128	15:59:59	1799	1023	−0.0058651	−4	1	5	1	0.60167	0.164	6.78
1	20190128	17:00:00	20190129	15:14:59	80099	230703	1.3004e−05	−6	1	6	2	0.39999	0.0176	5.402

(*Continued*)

Table A1. (Continued)

r	$t_1^{d,r}$	$t_{N_{d,r}}^{d,r}$	Δt	$N_{d,r}-1$	a_1	min	#	max	#	S	SK	EK
					ESH19							
2	20190129 15:30:00	20190129 15:59:59	1799	4845	0.004128	−5	1	5	1	0.75482	−0.0212	4.52
1	20190129 17:00:00	20190130 15:14:59	80099	274069	0.00049623	−22	1	22	1	0.46077	0.414	147.2
2	20190130 15:30:00	20190130 15:59:59	1799	1715	0.00058309	−3	1	3	1	0.53194	0.0239	1.851
1	20190130 17:00:00	20190131 15:14:59	80099	287779	0.00032316	−6	4	6	3	0.3991	0.0233	6.63
2	20190131 15:30:00	20190131 15:59:59	1799	3624	0.0066225	−4	3	5	1	0.59759	0.0754	5.048
1	20190131 17:00:00	20190201 15:14:59	80099	261488	−4.5891e-05	−16	1	17	1	0.41539	0.0999	33.97
2	20190201 15:30:00	20190201 15:59:59	1799	1033	0	−2	9	2	6	0.61945	−0.0735	0.7959
1	20190203 17:00:00	20190204 15:14:59	80099	170877	0.00050914	−6	2	6	1	0.38105	−0.0818	7.748
2	20190204 15:30:00	20190204 15:59:59	1799	1435	−0.00069686	−2	3	2	1	0.57554	−0.044	0.3266
1	20190204 17:00:00	20190205 15:14:59	80099	191586	0.00024532	−4	1	3	8	0.3949	0.0067	4.018
2	20190205 15:30:00	20190205 15:59:59	1799	922	−0.0010846	−1	141	2	2	0.55628	0.0754	0.5092
1	20190205 17:00:00	20190206 15:14:59	80099	184337	−2.1699e-05	−6	1	5	2	0.38265	−0.0118	5.245
2	20190206 15:30:00	20190206 15:59:59	1799	985	−0.0081218	−3	1	2	2	0.56303	−0.173	1.486
1	20190206 17:00:00	20190207 15:14:59	80099	292935	−0.00030382	−7	4	7	1	0.4073	−0.0668	7.024
2	20190207 15:30:00	20190207 15:59:59	1799	1257	−0.0039777	−3	1	4	1	0.58032	0.147	2.34
1	20190207 17:00:00	20190208 15:14:59	80099	244943	6.5321e-05	−5	2	5	2	0.39936	0.00438	5.177
2	20190208 15:30:00	20190208 15:59:59	1799	1527	0.005239	−2	6	2	5	0.58931	−0.0203	0.6003
1	20190208 17:00:00	20190211 15:14:59	80099	181170	−4.9677e-05	−6	1	6	1	0.39619	0.00437	4.799
2	20190211 15:30:00	20190211 15:59:59	1799	792	−0.0050505	−2	1	2	1	0.58422	0.000577	0.1946
1	20190211 17:00:00	20190212 15:14:59	80099	195361	0.00078317	−7	1	8	1	0.39104	0.0492	6.434
2	20190212 15:30:00	20190212 15:59:59	1799	1063	−0.0018815	−2	2	2	2	0.57898	4.57e-05	0.3893
1	20190212 17:00:00	20190213 15:14:59	80099	184857	0.00013524	−23	1	21	1	0.40464	−0.25	102.8
2	20190213 15:30:00	20190213 15:59:59	1799	1143	−0.0026247	−2	5	2	3	0.58213	−0.0532	0.6864
1	20190213 17:00:00	20190214 15:14:59	80099	259321	−8.4837e-05	−7	1	6	1	0.4087	−0.0128	5.922
2	20190214 15:30:00	20190214 15:59:59	1799	1410	−0.0099291	−2	3	2	2	0.56253	−0.0268	0.5861
1	20190214 17:00:00	20190215 15:14:59	80099	213990	0.00066826	−5	2	6	1	0.40523	0.0384	5.076
2	20190215 15:30:00	20190215 15:59:59	1799	1416	0.0021186	−2	3	2	4	0.57694	0.0221	0.543

Are E-mini S&P 500 Futures Prices Random? 299

#	DateTime1	DateTime2											
1	20190218 17:00:00	20190219 15:14:59	80099	151896	2.6334e-05	-3	4	4	1		0.38581	0.00301	4.139
2	20190219 15:30:00	20190219 15:59:59	1799	933	-0.0021436	-2	2	2	3		0.54418	0.0385	1.118
1	20190219 17:00:00	20190220 15:14:59	80099	196002	0.00018877	-4	2	4	2		0.38273	0.0068	4.874
2	20190220 15:30:00	20190220 15:59:59	1799	1039	-0.0019249	-3	3	3	1		0.54825	-0.0363	2.255
1	20190220 17:00:00	20190221 15:14:59	80099	222233	-0.00022949	-5	2	6	1		0.38388	-0.0113	5.982
2	20190221 15:30:00	20190221 15:59:59	1799	1000	-0.003	-2	6	2	3		0.58935	-0.0876	0.7786
1	20190221 17:00:00	20190222 15:14:59	80099	174469	0.00041268	-4	1	5	1		0.3911	0.0158	4.452
2	20190222 15:30:00	20190222 15:59:59	1799	1448	0.010359	-1	198	2	5		0.5393	0.141	0.9208
1	20190222 17:00:00	20190225 15:14:59	80099	174854	-5.7191e-05	-4	6	5	1		0.40565	-0.00455	4.375
2	20190225 15:30:00	20190225 15:59:59	1799	1173	-0.0051151	-2	3	2	2		0.53061	-0.0397	1.202
1	20190225 17:00:00	20190226 15:14:59	80099	175592	-9.6815e-05	-8	1	7	1		0.40874	-0.00371	5.336
2	20190226 15:30:00	20190226 15:59:59	1799	1333	-0.0075019	-2	9	2	4		0.55068	-0.139	1.569
1	20190226 17:00:00	20190227 15:14:59	80099	164726	0.00016998	-9	1	8	1		0.42566	0.0326	6.858
2	20190227 15:30:00	20190227 15:59:59	1799	1318	0.0045524	-3	1	3	1		0.54971	0.0575	2.31
1	20190227 17:00:00	20190228 15:14:59	80099	170540	-0.00014659	-7	1	7	1		0.42006	-0.00805	4.505
2	20190228 15:30:00	20190228 15:59:59	1799	1308	-0.0053517	-2	2	2	2		0.51076	-0.0433	1.241
1	20190228 17:00:00	20190301 15:14:59	80099	169619	0.00047165	-8	1	9	1		0.4168	-0.00323	6.599
2	20190301 15:30:00	20190301 15:59:59	1799	1361	0.0051433	-2	10	3	2		0.58722	0.0646	1.908
1	20190301 17:00:00	20190304 15:14:59	80099	252799	-0.00035601	-13	1	14	1	125	0.42614	0.00404	12.22
2	20190304 15:30:00	20190304 15:59:59	1799	846	-0.0023641	-1	127	1	4	1	0.54609	-0.00155	0.358
1	20190304 17:00:00	20190305 15:14:59	80099	172778	2.3151e-05	-3	7	4	1		0.39861	0.00622	3.891
2	20190305 15:30:00	20190305 15:59:59	1799	964	-0.010373	-2	1	1	1	146	0.56176	-0.0384	0.2942
1	20190305 17:00:00	20190306 15:14:59	80099	195641	-0.0003118	-5	1	4	2		0.40642	-0.00554	3.749
2	20190306 15:30:00	20190306 15:59:59	1799	1431	0.00069881	-3	1	2	4		0.549	-0.0758	1.524
1	20190306 17:00:00	20190307 15:14:59	80099	248492	-0.00032999	-9	1	8	1		0.41999	-0.0552	8.277
2	20190307 15:30:00	20190307 15:59:59	1799	960	-0.0010417	-2	1	1	1	132	0.52765	-0.0439	0.7584
1	20190307 17:00:00	20190308 15:14:59	80099	166771	-4.1974e-05	-18	1	18	1		0.49591	0.631	68.68
2	20190308 15:30:00	20190308 15:59:59	1799	992	0.0060484	-3	1	2	5		0.58225	-0.062	1.434
1	20190308 17:00:00	20190311 15:14:59	80099	62657	0.00049476	-7	1	9	1		0.45306	0.11	9.14
2	20190311 17:00:00	20190312 15:14:59	1799	579	0.0086356	-2	5	2	4		0.53422	-0.0601	2.821
1	20190312 15:30:00	20190312 15:59:59	80099	53016	0.0016787	-3	4	5	1		0.43934	0.051	3.602

(Continued)

Table A1. (Continued)

r	$t_1^{d,r}$	$t_{N_{d,r}}^{d,r}$	Δt	$N_{d,r}-1$	a_1	min	#	max	#	S	SK	EK
					ESH19							
2	20190313 15:30:00	20190313 15:59:59	1799	389	0.010283	−2	1	2	1	0.50757	0.0181	1.833
1	20190313 17:00:00	20190314 15:14:59	80099	41350	−0.0007497	−4	1	3	1	0.4258	−0.0289	4.096
2	20190314 15:30:00	20190314 15:59:59	1799	247	−0.016194	−2	6	2	5	0.63737	−0.0814	2.748
1	20190314 17:00:00	20190315 08:29:55	55795	7087	0.0029632	−4	1	6	1	0.57157	0.209	5.772
		ESH19, total 315 = 211 (1) + 104 (2) ranges, $\Delta t = 39404912$										
*	20171214 06:41:23	20190315 08:29:55	15687303	1.5044e-05	−144	1	114	1	0.51853	0.838	1396
					ESM19							
1	20171214 06:41:23	20171214 14:53:04	29501	4	−14.5	−24	1	−6	1	8.8506	−0.13	−3.825
1	20180801 08:41:28	20180801 08:59:51	1103	1	2	2	1	2	1	0	n/a	n/a
1	20180806 09:53:06	20180806 11:34:04	6058	1	20	20	1	20	1	0	n/a	n/a
1	20180906 10:03:59	20180906 14:08:39	14680	2	15	−14	1	44	1	41.012	n/a	n/a
1	20180907 05:49:01	20180907 10:04:37	15336	1	39	39	1	39	1	0	n/a	n/a
1	20180918 08:54:54	20180918 12:30:00	12906	5	4.4	−3	1	25	1	11.589	2.16	1.86
1	20180918 22:19:01	20180919 10:10:00	42659	1	20	20	1	20	1	0	n/a	n/a
1	20180920 08:31:35	20180920 14:01:52	19817	2	26	16	1	36	1	14.142	n/a	n/a
1	20180921 01:19:30	20180921 11:28:43	36553	2	6.5	0	1	13	1	9.1924	n/a	n/a
1	20180924 09:30:18	20180924 10:33:38	3800	4	2.25	−4	1	10	1	5.909	0.68	−0.8667
1	20180925 08:49:18	20180925 14:44:01	21283	6	−1	−12	1	12	1	8.1731	0.514	−0.4388
1	20180926 13:03:38	20180926 15:07:20	7422	4	−19	−44	1	4	1	19.9	−0.284	−0.7783
1	20180926 21:50:19	20180927 09:01:55	40296	4	6	−27	1	52	1	33.116	1.12	0.0138
1	20181001 09:06:43	20181001 13:08:56	14533	3	−9.6667	−29	1	0	2	16.743	−1.73	n/a
1	20181002 08:54:59	20181002 14:37:46	20567	11	−0.090909	−16	2	16	1	9.4176	−0.426	−0.0571
1	20181002 21:00:46	20181002 21:00:46	0	2	0	0	2	0	2	0	n/a	n/a
2	20181003 15:57:43	20181003 15:57:43	0	2	0	0	2	0	2	0	n/a	n/a
1	20181004 04:32:24	20181004 12:35:29	28985	10	−10	−36	1	12	1	15.535	−0.401	−0.908
1	20181004 20:38:32	20181005 13:09:40	59468	8	−16.75	−88	1	17	1	35.78	−1.19	0.1583
1	20181007 19:24:02	20181008 13:31:50	65268	13	−2.1538	−74	1	62	1	28.965	−0.454	3.303

1	20181009 05:54:35	20181009 09:25:53	12678	6	7.8333	−17	1	34	1	16.29	0.179	0.5522
1	20181009 21:38:12	20181010 15:13:20	63308	37	−11.568	−87	1	29	1	23.936	−0.943	1.56
1	20181010 20:57:16	20181011 15:14:32	65836	46	−0.82609	−160	1	90	1	41.123	−0.983	4.209
1	20181011 19:13:55	20181012 14:26:11	69136	16	0.0625	−50	1	76	1	29.368	0.711	1.646
1	20181016 03:19:31	20181016 14:45:02	41131	11	18.545	−22	1	59	1	25.727	0.105	−1.44
1	20181016 19:50:42	20181017 13:53:30	64968	13	−0.69231	−53	1	33	1	25.992	−0.679	−0.6157
1	20181018 04:30:42	20181018 14:59:59	37757	11	−12.545	−110	1	23	1	37.423	−2.1	3.103
1	20181018 18:12:45	20181019 15:00:08	74843	17	−0.76471	−106	1	57	1	31.523	−2.14	7.038
1	20181021 17:01:06	20181022 11:05:08	65042	7	0	−49	1	74	1	41.649	0.933	−0.4347
1	20181022 19:07:08	20181023 13:42:10	66902	40	−0.425	−50	1	68	1	22.656	0.187	1.296
1	20181024 09:16:49	20181024 15:07:27	21038	121	−2.0992	−38	1	50	1	12.475	0.461	3.258
1	20181024 18:07:38	20181025 15:10:20	75762	60	1.0833	−43	1	80	1	21.213	0.72	2.461
1	20181025 17:06:10	20181026 13:48:39	74549	51	−0.62745	−74	1	54	1	27.098	−0.297	0.6021
1	20181028 17:13:12	20181029 14:56:50	78218	43	−2.5581	−226	1	60	1	40.641	−4.01	20.62
1	20181029 22:09:33	20181030 14:51:27	60114	7	15.429	−32	1	105	1	44.211	1.52	1.487
1	20181030 17:52:36	20181031 13:15:34	69778	7	26.571	−7	1	61	1	23.373	−0.031	−1.305
1	20181101 05:08:17	20181101 15:10:25	36128	15	3.4667	−41	1	35	1	18.689	−0.397	0.7918
1	20181102 00:04:15	20181102 14:10:32	50777	29	−4.931	−64	1	46	1	21.64	−0.867	1.608
1	20181104 20:29:15	20181105 14:25:09	64554	7	9.7143	−10	1	34	1	17.614	0.586	−1.794
1	20181106 09:43:44	20181106 13:22:26	13122	3	0	−34	1	30	1	32.187	−0.551	n/a
1	20181106 19:49:57	20181107 15:05:02	69305	15	13.267	−20	1	52	1	17.194	0.344	0.4585
1	20181107 17:02:42	20181108 14:49:16	78394	11	−4.9091	−39	1	20	1	18.934	−0.292	−0.9635
1	20181109 02:25:06	20181109 11:25:09	32403	10	−8	−22	1	7	1	8.2999	0.157	−0.4326
1	20181111 21:08:51	20181112 14:50:46	63715	6	−41.833	−171	1	27	1	73.412	−1.29	−0.0372
1	20181112 19:05:53	20181113 14:44:54	70741	67	0.13433	−44	1	73	1	15.429	1.04	7.529
1	20181114 19:05:55	20181115 14:54:30	71315	24	4.5833	−87	1	179	1	50.459	1.8	5.034
1	20181116 09:07:40	20181116 12:07:13	10773	9	3	−27	1	37	1	25.588	0.14	−2.008
1	20181119 22:56:30	20181120 14:53:08	57398	53	−3.5283	−73	1	40	1	19.275	−1.09	2.761
1	20181120 17:06:44	20181121 15:00:03	78799	13	4.4615	−52	1	80	1	28.78	1.03	3.352
1	20181122 20:32:18	20181123 10:50:17	51479	7	0.28571	−20	1	11	1	10.92	−1.31	−0.1033
1	20181125 17:00:04	20181126 15:00:49	79245	21	8.3333	−36	1	59	1	20.662	0.33	1.26

(*Continued*)

Table A1. (Continued)

r	$t_1^{d,r}$	$t_{N_{d,r}}^{d,r}$	Δt	$N_{d,r}-1$	a_1	min	#	max	#	S	SK	EK
					ESM19							
2	20181126 15:31:44	20181126 15:57:36	1552	4	-3	-5	1	0	1	2.1602	1.19	-0.375
1	20181126 19:00:34	20181127 15:14:55	72861	39	2.8462	-31	1	53	1	15.51	0.645	2.527
1	20181127 19:58:26	20181128 14:51:57	68011	53	4.5283	-27	1	43	1	12.78	0.299	1.316
1	20181129 00:15:48	20181129 15:00:03	53055	27	0.81481	-25	1	36	1	13.056	0.478	0.4778
1	20181129 17:59:22	20181130 15:12:48	76406	34	2.7647	-21	1	34	1	11.048	0.482	0.9577
2	20181130 15:32:39	20181130 15:58:01	1522	3	0	0	3	0	3	0	n/a	n/a
1	20181202 17:02:12	20181203 15:14:50	79958	106	0.13208	-48	1	40	1	11.975	-0.0397	3.538
1	20181203 22:46:33	20181204 15:14:30	59277	118	-2.4915	-76	1	55	1	14.165	-0.188	7.618
1	20181204 17:33:09	20181205 01:38:57	29148	31	1	-3	1	21	1	4	4.42	20.02
1	20181205 17:00:24	20181206 15:01:47	79283	307	-0.052117	-168	1	89	1	16.462	-2.83	38.62
1	20181206 17:17:29	20181207 15:13:55	78986	162	-1.4877	-50	1	63	1	14.86	0.473	2.799
1	20181209 17:01:34	20181210 15:14:29	79975	167	0.61677	-69	1	50	1	15.025	-0.332	3.42
1	20181210 17:03:41	20181211 15:07:08	79407	173	0.057803	-52	1	26	1	10.272	-1.45	6.879
2	20181211 15:37:22	20181211 15:51:45	863	3	-9	-28	1	2	1	16.523	-1.67	n/a
1	20181211 17:09:29	20181212 15:00:55	78686	218	0.21101	-64	1	75	1	10.344	0.96	19.47
1	20181212 17:17:16	20181213 15:05:08	78472	183	-0.15847	-35	1	31	1	10.148	-0.0765	0.9918
2	20181213 15:42:10	20181213 15:56:11	841	1	7	7	1	7	1	0	n/a	n/a
1	20181213 17:22:07	20181214 15:13:03	78656	246	-0.72764	-23	1	27	1	7.5384	0.0846	2.64
1	20181216 17:00:00	20181217 15:10:04	79804	593	-0.26307	-31	1	32	1	5.8591	-0.31	4.966
2	20181217 15:48:04	20181217 15:59:34	690	2	-3.5	-4	1	-3	1	0.70711	n/a	n/a
1	20181218 17:05:06	20181219 15:14:39	79773	804	-0.16045	-66	1	47	1	7.8985	-0.247	8.719
2	20181219 15:30:00	20181219 15:59:45	1785	16	-0.5625	-9	1	25	1	8.0164	2.25	5.457
1	20181219 17:00:00	20181220 15:14:38	80078	1317	-0.05391	-33	1	29	1	6.2486	0.0867	2.905
2	20181220 15:30:00	20181220 15:59:50	1790	2	5	-1	1	11	1	8.4853	n/a	n/a
1	20181220 17:01:48	20181221 15:14:59	79991	2159	-0.13988	-21	1	27	1	4.551	0.0987	2.813
2	20181221 15:30:24	20181221 15:57:05	1601	18	1.2222	-5	1	9	1	2.7559	0.764	3.135
1	20181223 17:00:05	20181224 12:14:17	69252	2066	-0.12488	-27	1	20	1	4.3799	-0.219	4.209

1	20181225 17:00:00	20181226 15:14:59	80099	2647	0.19116	−21	39	1	4.4853	0.264	5.546
2	20181226 15:30:01	20181226 15:59:00	1739	22	−0.22727	−10	16	1	5.7729	1.24	1.932
1	20181226 17:00:24	20181227 15:14:48	80064	2185	0.046682	−31	22	1	4.7134	−0.0556	3.165
2	20181227 15:30:00	20181227 15:58:59	1739	32	−0.375	−17	10	1	4.6956	−0.871	4.021
1	20181227 17:00:00	20181228 15:14:44	80084	1860	−0.0053763	−27	27	1	4.6854	0.00302	3.753
2	20181228 15:30:21	20181228 15:58:59	1718	14	−0.14286	−5	4	2	2.6849	0.122	−0.9015
1	20181228 17:00:00	20181231 15:14:18	80058	1109	0.051398	−17	25	1	4.1937	0.638	4.049
2	20181231 15:30:01	20181231 15:51:34	1293	30	−0.16667	−3	5	2	1.3667	1.36	6.445
1	20181231 17:00:04	20190102 15:12:52	79968	1742	0.012055	−19	25	1	3.8863	−0.248	3.905
2	20190102 15:30:05	20190102 15:59:47	1782	103	−0.71845	−16	19	1	5.4493	−0.0137	1.59
1	20190102 17:00:00	20190103 15:14:53	80093	1637	−0.087355	−23	30	2	4.5999	0.0114	4.319
2	20190103 15:30:01	20190103 15:50:00	1199	17	0.52941	−7	7	1	3.891	−0.217	−0.7743
1	20190103 17:00:00	20190104 15:14:55	80095	1394	0.22812	−21	30	1	4.4753	0.447	4.795
2	20190104 15:32:40	20190104 15:58:47	1567	2	−2	−4	0	1	2.8284	n/a	n/a
1	20190106 17:00:00	20190107 15:14:42	80082	777	0.081081	−18	18	1	4.543	−0.0168	2.11
2	20190107 15:30:10	20190107 15:58:13	1683	4	3.25	0	6	1	2.7538	−0.323	−2.868
1	20190107 17:00:00	20190108 15:10:00	79800	838	0.095465	−15	19	1	4.2492	0.0123	1.957
2	20190108 15:31:09	20190108 15:59:18	1689	7	−0.57143	−4	6	3	3.3094	1.48	0.9925
1	20190108 17:05:04	20190109 15:14:50	79786	1203	0.038238	−22	16	1	3.2775	−0.0771	4.368
2	20190109 15:32:54	20190109 15:59:30	1596	6	0.33333	−6	6	1	5.0067	−0.109	−2.335
1	20190109 17:10:08	20190110 15:14:31	79463	667	0.092954	−17	19	1	4.2891	−0.0984	2.403
2	20190110 15:31:24	20190110 15:58:59	1655	1	0	0	0	1	0	n/a	n/a
1	20190110 17:00:00	20190111 15:09:21	79761	442	0.0181	−19	15	1	4.3654	−0.413	2.013
2	20190111 15:30:00	20190111 15:58:41	1721	8	0.25	−2	2	1	1.5811	0.0361	−1.858
1	20190113 17:00:00	20190114 15:13:00	79980	491	−0.081466	−14	21	1	3.7123	0.213	3.74
2	20190114 15:31:12	20190114 15:59:59	1727	4	−0.75	−7	5	1	4.9244	−0.299	−0.459
1	20190114 17:00:09	20190115 15:14:48	80079	796	0.12563	−28	17	1	3.5511	−0.296	6.498
2	20190115 15:35:14	20190115 15:58:59	1425	4	2	−1	7	1	3.4641	1.54	0.3889
1	20190115 17:02:41	20190116 15:14:44	79923	521	0.044146	−19	15	1	3.5696	−0.279	3.794
2	20190116 15:31:23	20190116 15:49:49	1106	2	1	−6	8	1	9.8995	n/a	n/a
1	20190116 17:01:27	20190117 15:10:42	79755	819	0.10867	−22	15	1	3.4809	−0.395	3.944

(*Continued*)

Table A1. (Continued)

r	$t_1^{d,r}$	$t_{N_{d,r}}^{d,r}$	Δt	$N_{d,r}-1$	a_1	min	#	max	#	S	SK	EK
					ESM19							
2	20190117 15:31:21	20190117 15:53:33	1332	10	0.9	−3	1	9	1	3.7253	1.32	0.4511
1	20190117 17:01:39	20190118 15:14:33	79974	735	0.19048	−23	1	18	1	3.6096	−0.218	4.216
2	20190118 15:33:01	20190118 15:58:59	1558	12	1.3333	−3	1	6	1	2.5702	0.443	−0.5169
1	20190123 17:00:00	20190124 15:14:54	80094	589	−0.0050934	−18	1	22	1	4.2796	−0.202	3.045
2	20190124 15:30:00	20190124 15:57:50	1670	13	0.69231	−2	2	4	2	1.9742	0.353	−0.9051
1	20190124 17:00:09	20190125 15:14:20	80051	568	0.1919	−13	1	17	2	3.4824	0.323	3.608
2	20190125 15:40:13	20190125 15:59:36	1163	2	−1	−3	1	1	1	2.8284	n/a	n/a
1	20190127 17:00:27	20190128 15:13:16	79969	646	−0.13777	−25	1	13	1	3.8282	−0.618	4.558
2	20190128 15:36:01	20190128 15:58:59	1378	6	0.5	−3	1	4	1	2.3452	−0.0698	−0.3322
1	20190128 17:03:36	20190129 15:14:58	79882	440	0.020455	−13	2	16	1	3.8274	0.596	2.716
2	20190129 15:30:05	20190129 15:59:26	1761	19	1.6842	−6	1	9	2	4.4976	0.0475	−1.037
1	20190129 17:00:00	20190130 15:14:57	80097	716	0.18017	−15	1	30	1	3.9305	0.514	6.65
2	20190130 15:45:38	20190130 15:58:59	801	3	3	2	1	4	1	1	0	n/a
1	20190130 17:00:48	20190131 15:14:58	80050	868	0.11521	−14	1	14	1	3.1249	−0.0148	2.088
2	20190131 15:30:39	20190131 15:57:36	1617	26	0.65385	−7	1	8	1	3.2489	−0.21	0.8901
1	20190131 17:00:57	20190201 15:13:55	79978	645	−0.023256	−16	1	15	1	3.5953	−0.0934	2.391
2	20190201 15:32:34	20190201 15:59:55	1641	8	−0.125	−6	1	3	1	2.6424	−1.74	2.299
1	20190203 17:00:00	20190204 15:13:32	80012	317	0.28076	−14	1	11	1	3.2365	0.112	2.047
2	20190204 15:31:04	20190204 15:52:59	1315	3	0	−5	1	3	1	4.3589	−1.63	n/a
1	20190204 17:00:11	20190205 15:10:05	79794	399	0.13033	−11	1	11	2	3.0643	−0.0785	1.315
2	20190205 17:02:00	20190206 15:13:58	79918	436	−0.016055	−16	1	15	1	2.9598	−0.11	4.702
1	20190206 15:30:00	20190206 15:58:59	1739	9	−0.11111	−4	1	3	3	2.6667	−0.000209	−1.65
2	20190206 18:34:25	20190207 15:05:18	73853	879	−0.10353	−16	1	18	1	3.41	−0.113	3.152
1	20190207 15:53:55	20190207 15:58:06	251	3	0.33333	−2	1	3	1	2.5166	0.586	n/a
2	20190207 17:00:00	20190208 15:14:17	80057	706	0.015581	−13	2	14	1	3.1503	−0.063	2.929
1	20190208 15:30:49	20190208 15:58:41	1672	12	0.25	−4	1	3	1	1.9129	−0.894	0.36
2	20190210 17:00:00	20190211 15:13:21	80001	520	−0.017308	−14	1	17	1	3.3765	0.106	3.462

2	20190211 15:33:47	20190211 15:58:59	1512	4		0	2	4	1	1.893	1.66	0.2385
1	20190211 17:00:10	20190212 15:14:52	80082	738	1.25	−10	1	13	1	2.4791	0.0941	2.999
2	20190212 15:52:36	20190212 15:58:04	328	2	0.20325	−5	1	2	1	4.9497	n/a	n/a
1	20190212 17:01:40	20190213 15:13:54	79934	647	−1.5	−9	4	10	1	2.6187	0.00704	1.988
2	20190213 15:35:23	20190213 15:58:59	1416	3	0.03864	−5	1	5	1	5.0332	0.586	n/a
1	20190213 17:02:19	20190214 15:14:59	79960	799	−0.33333	−14	1	17	1	3.421	−0.00604	2.144
2	20190214 15:30:00	20190214 15:59:58	1798	7	−0.027534	−8	1	2	1	3.1472	−0.71	0.2157
1	20190214 17:00:06	20190215 15:14:10	80044	793	−2.2857	−18	1	13	2	3.1753	−0.296	3.426
2	20190215 15:52:34	20190215 15:59:50	436	8	0.18411	−3	1	2	1	1.685	0.168	−1.313
1	20190218 17:00:05	20190219 15:13:11	79986	490	−0.625	−15	1	12	1	2.6484	−0.31	4.559
2	20190219 15:31:13	20190219 15:59:58	1725	10	0.016327	−1	5	2	2	1.2472	0.859	−1.233
1	20190219 17:01:47	20190220 15:09:59	79692	605	0.071074	−14	1	12	2	2.9465	0.0876	3.087
2	20190220 15:34:19	20190220 15:51:46	1047	3	−0.66667	−1	2	0	1	0.57735	1.73	n/a
1	20190220 17:10:11	20190221 15:13:03	79372	776	−0.056701	−13	1	16	1	2.9075	0.392	3.4
2	20190221 15:30:00	20190221 15:55:26	1526	6	0	−4	1	3	1	2.4495	−0.612	−0.45
1	20190221 17:03:27	20190221 15:14:47	79880	801	0.094881	−12	1	14	1	2.5746	−0.112	4.638
2	20190222 15:30:00	20190222 15:59:39	1779	16	1	−1	1	4	1	1.2111	0.772	0.6492
1	20190224 17:00:00	20190225 15:14:42	80082	1453	−0.0055058	−8	3	6	8	1.7663	−0.123	2.132
2	20190225 15:30:00	20190225 15:57:26	1646	13	−0.30769	−4	1	4	1	1.9742	0.429	0.6341
1	20190225 17:00:00	20190226 15:14:58	80098	1245	−0.011245	−10	2	13	2	2.0819	0.402	5.292
2	20190226 15:30:26	20190226 15:59:25	1739	24	−0.45833	−5	1	2	3	1.6676	−0.724	0.6854
1	20190226 17:00:28	20190227 15:14:06	80018	1275	0.022745	−11	1	11	1	2.0842	−0.276	4.004
2	20190227 15:30:00	20190227 15:47:38	1058	15	0.6	−1	3	7	1	1.9567	2.76	7.044
1	20190227 17:01:54	20190228 15:14:34	79960	1858	−0.010764	−13	1	8	2	1.5535	−0.798	8.026
2	20190228 15:30:16	20190228 15:59:38	1762	17	−0.29412	−4	1	2	3	1.7235	−0.479	−0.4282
1	20190228 17:00:00	20190301 15:14:52	80092	2357	0.033517	−8	1	10	1	1.3371	0.135	4.442
2	20190301 15:30:00	20190301 15:59:35	1775	44	0.045455	−5	2	3	2	1.5545	−1.32	3.413
1	20190303 17:00:00	20190304 15:14:50	80090	5498	−0.016551	−12	1	12	1	1.2835	0.0773	6.152
2	20190304 15:30:27	20190304 15:59:58	1771	23	0	−2	1	2	1	0.8528	0	0.6839
1	20190304 17:00:00	20190305 15:14:57	80097	11488	0.00087047	−6	2	5	3	0.70751	−0.0293	5.149
2	20190305 15:30:00	20190305 15:59:58	1798	44	−0.20455	−3	2	2	1	0.97836	−0.972	1.821

(*Continued*)

Table A1. (Continued)

r	$t_1^{d,r}$	$t_{N_{d,r}}^{d,r}$	Δt	$N_{d,r}-1$	a_1	min	#	max	#	S	SK	EK
					ESM19							
1	20190305 17:00:00	20190306 15:14:58	80098	17655	−0.0035684	−4	5	5	1	0.6192	0.0438	3.11
2	20190306 15:30:00	20190306 15:59:58	1798	121	−0.0082645	−2	4	2	3	0.74717	−0.108	1.085
1	20190306 17:00:00	20190307 15:14:59	80099	67159	−0.0012359	−12	1	11	1	0.49862	0.0249	17.74
2	20190307 15:30:00	20190307 15:59:59	1799	582	−0.0051546	−2	7	2	1	0.57633	−0.325	1.509
1	20190307 17:00:00	20190308 15:14:59	80099	180072	−5.553e-05	−12	1	8	1	0.42046	−0.437	24.79
2	20190308 15:30:00	20190308 15:59:58	1798	1361	0.005878	−2	1	2	2	0.46491	0.0646	2.197
1	20190311 17:00:00	20190312 15:14:59	80099	168327	0.00017822	−5	3	6	2	0.37662	0.0186	6.059
2	20190312 15:30:00	20190312 15:59:59	1799	1109	0.0045086	−2	4	2	2	0.57394	−0.0571	0.6407
1	20190312 17:00:00	20190313 15:14:59	80099	176082	0.00049977	−6	1	6	1	0.38124	0.0137	5.539
2	20190313 15:30:00	20190313 15:59:59	1799	1503	0.001996	−2	1	2	2	0.51798	0.0316	1.064
1	20190313 17:00:00	20190314 15:14:59	80099	160844	−0.00020517	−3	3	4	1	0.38062	0.00263	4.369
2	20190314 15:30:00	20190314 15:59:59	1799	1006	−0.0019881	−2	2	2	2	0.52971	−0.00213	1.177
1	20190314 17:00:00	20190315 15:14:59	80099	183384	0.00036535	−8	1	8	1	0.40439	0.00826	6.612
2	20190315 15:30:00	20190315 15:59:59	1799	1886	0	−1	278	2	3	0.54602	0.0587	0.5714
1	20190317 17:00:00	20190318 15:14:59	80099	147945	0.000365	−3	5	3	5	0.39794	0.0069	3.671
2	20190318 15:30:00	20190318 15:59:59	1799	1335	0.0029963	−2	3	2	2	0.60976	−0.0214	0.0176
1	20190318 17:00:00	20190319 15:14:59	80099	203088	−0.00010833	−8	1	8	1	0.40854	0.00744	4.877
2	20190319 15:30:00	20190319 15:59:59	1799	1191	0.006717	−2	1	2	1	0.51527	0.01	1.057
1	20190319 17:00:00	20190320 15:14:59	80099	237672	−0.00017671	−23	1	23	1	0.44827	−0.0515	68.31
2	20190320 15:30:00	20190320 15:59:59	1799	2182	−0.0064161	−2	2	1	259	0.4957	−0.0591	1.252
1	20190320 17:00:00	20190321 15:14:59	80099	226453	0.0006933	−5	2	5	1	0.41584	0.0098	3.758
2	20190321 15:30:00	20190321 15:59:59	1799	1379	−0.0072516	−1	235	2	1	0.57898	0.0227	0.0636
1	20190321 17:00:00	20190322 15:14:59	80099	308920	−0.00063123	−5	2	4	2	0.43718	−0.0128	2.897
2	20190322 15:30:00	20190322 15:59:59	1799	2104	−0.0052281	−5	1	4	1	0.66175	−0.0732	3.513
1	20190324 17:00:00	20190325 15:14:59	80099	292134	−7.8731e-05	−9	1	10	1	0.4308	0.0299	8.437
2	20190325 15:30:00	20190325 15:59:59	1799	1259	0.0023828	−2	7	3	1	0.56317	−0.0261	1.582
1	20190325 17:00:00	20190326 15:14:59	80099	224731	0.00025809	−7	1	5	1	0.41868	−0.022	4.236

Are E-mini S&P 500 Futures Prices Random?

2	20190326 15:30:00	20190326 15:59:59	1799	1975	0.0020253	−2	3	2	1	0.50928	−0.0427	1.221
1	20190326 17:00:00	20190327 15:14:59	80099	268240	−0.00019758	−6	3	6	3	0.42863	−0.00624	4.757
2	20190327 15:30:00	20190327 15:59:59	1799	1057	−0.012299	−3	1	2	2	0.58131	−0.115	0.9489
1	20190327 17:00:00	20190328 15:14:59	80099	213714	0.00026671	−6	1	5	1	0.41241	−0.0366	5.248
2	20190328 15:30:00	20190328 15:59:59	1799	993	0.0040282	−2	2	2	1	0.5626	−0.0329	0.5271
1	20190328 17:00:00	20190329 15:14:59	80099	206293	0.00037326	−9	1	10	1	0.41692	0.0599	8.623
2	20190329 15:30:00	20190329 15:59:59	1799	1735	0	−2	6	2	8	0.58232	0.0351	0.7947
1	20190331 17:00:00	20190401 15:14:59	80099	185279	0.00057211	−13	1	12	1	0.41625	−0.0595	19.66
2	20190401 15:30:00	20190401 15:59:59	1799	1309	−0.0015279	−1	185	2	1	0.53186	0.029	0.6539
1	20190401 17:00:00	20190402 15:14:59	80099	135649	−3.686e-05	−3	3	2	23	0.39785	−0.00382	3.519
2	20190402 15:30:00	20190402 15:59:59	1799	889	−0.0022497	−2	1	2	1	0.57735	−1.33e-05	0.2475
1	20190402 17:00:00	20190403 15:14:59	80099	184679	0.00030323	−8	1	7	1	0.40911	−0.013	5.679
2	20190403 15:30:00	20190403 15:59:59	1799	955	−0.0010471	−2	1	1	139	0.54272	−0.0402	0.5446
1	20190403 17:00:00	20190404 15:14:59	80099	158731	5.67e-05	−5	1	6	1	0.39796	0.00527	4.206
2	20190404 15:30:00	20190404 15:59:59	1799	1033	−0.00096805	−2	1	2	1	0.55771	−0.000379	0.4596
1	20190404 17:00:00	20190405 15:14:59	80099	162835	0.00034391	−8	2	8	3	0.39551	−2.78e-05	13.42
2	20190405 15:30:00	20190405 15:59:59	1799	1177	0.0042481	−2	3	2	2	0.56618	−0.0272	0.6208
1	20190405 17:00:00	20190408 15:14:59	80099	145826	6.1717e-05	−7	2	9	2	0.38013	0.153	13.64
2	20190408 15:30:00	20190408 15:59:59	1799	938	0.0031983	−1	120	2	19	0.51133	0.0532	1.017
1	20190408 17:00:00	20190409 15:14:59	80099	173044	−0.00036407	−3	2	2	2	0.39202	−0.0107	3.669
2	20190409 15:30:00	20190409 15:59:59	1799	762	−0.0026247	−1	110	9	1	0.54011	0.0982	0.8076
1	20190409 17:00:00	20190410 15:14:59	80099	138745	0.00036758	−7	1	2	3	0.40043	0.173	9.258
2	20190410 15:30:00	20190410 15:59:59	1799	1089	0.0018365	−3	1	6	1	0.55572	−0.0314	1.286
1	20190410 17:00:00	20190411 15:14:59	80099	138478	−9.3878e-05	−8	1	3	1	0.39899	−0.0813	6.817
2	20190411 15:30:00	20190411 15:59:59	1799	739	0.0040595	−4	1	10	1	0.63863	−0.254	2.797
1	20190411 17:00:00	20190412 15:14:59	80099	156595	0.00053003	−9	1	2	3	0.4139	0.133	13.96
2	20190412 15:30:00	20190412 15:59:59	1799	1057	−0.0028382	−2	7	2	1	0.60223	−0.103	0.6238
1	20190414 17:00:00	20190415 15:14:59	80099	132930	−6.7705e-05	−4	1	3	1	0.38507	−0.018	4.163
2	20190415 15:30:00	20190415 15:59:59	1799	1015	0.0029557	−2	3	2	1	0.59001	−0.0584	0.2681
1	20190415 17:00:00	20190416 15:14:59	80099	156269	3.1996e-05	−4	1	4	1	0.36908	−0.0111	5.024
2	20190416 15:30:00	20190416 15:59:59	1799	893	−0.0044793	−2	1	3	1	0.5357	0.127	1.641

(Continued)

Table A1. (Continued)

r	$t_1^{d,r}$	$t_{N_{d,r}}^{d,r}$	Δt	$N_{d,r}-1$	a_1	min	# min	max	# max	S	SK	EK
					ESM19							
1	20190416 17:00:00	20190417 15:14:59	80099	192109	−0.00020301	−9	1	10	3	0.38607	0.195	14.96
2	20190417 15:30:00	20190417 15:59:59	1799	1343	0	−2	5	3	1	0.5726	0.0239	1.136
1	20190417 17:00:00	20190418 15:14:59	80099	184989	0.00020542	−14	1	15	1	0.38797	0.328	43.09
2	20190418 15:30:00	20190418 15:59:59	1799	1116	0.00089606	−2	1	2	2	0.51611	0.0405	1.215
1	20190418 17:00:00	20190422 15:14:59	80099	103987	0.00015387	−5	1	5	3	0.39737	0.0353	5.087
2	20190422 15:30:00	20190422 15:59:59	1799	1286	0.0007776	−2	1	2	2	0.54737	0.029	0.6535
1	20190422 17:00:00	20190423 15:14:59	80099	159689	0.00067005	−4	2	5	1	0.37853	0.0147	4.759
2	20190423 15:30:00	20190423 15:59:59	1799	1647	−0.0024287	−1	233	1	229	0.52979	−0.00259	0.5656
1	20190423 17:00:00	20190424 08:35:36	56136	39974	−0.00027518	−3	2	2	5	0.37252	−0.0263	4.55
1	20190424 17:00:00	20190425 15:14:59	80099	184853	−8.6555e-05	−3	7	3	5	0.37508	−0.00464	4.475
2	20190425 15:30:00	20190425 15:59:59	1799	1837	−0.0070768	−5	1	4	1	0.54778	−0.243	5.848
1	20190425 17:00:00	20190426 15:14:59	80099	172467	0.00042327	−11	1	12	1	0.3895	0.0363	13.25
2	20190426 15:30:00	20190426 15:59:59	1799	1211	−0.0066061	−1	173	2	1	0.53005	0.0263	0.6902
1	20190426 17:00:00	20190429 15:14:59	80099	124219	0.00010465	−4	1	3	6	0.39494	0.00404	3.848
2	20190429 15:30:00	20190429 15:59:59	1799	1475	0.0033898	−2	3	1	203	0.52546	−0.0803	0.9481
1	20190429 17:00:00	20190430 15:14:59	80099	183576	0.00017431	−7	1	6	1	0.38727	−3.73e-05	6.28
2	20190430 15:30:00	20190430 15:59:59	1799	4357	0.0041313	−12	1	13	1	0.71996	−0.165	48.97
1	20190430 17:00:00	20190501 15:14:59	80099	198505	−0.00064482	−8	1	8	1	0.41228	−0.0274	7.158
2	20190501 15:30:00	20190501 15:59:59	1799	3124	−0.0073624	−2	8	2	6	0.54064	−0.0301	1.051
1	20190501 17:00:00	20190502 15:14:59	80099	263304	2.2787e-05	−5	1	5	3	0.39355	−0.000548	4.946
2	20190502 15:30:00	20190502 15:59:59	1799	1079	0.0046339	−3	1	2	2	0.55744	−0.0947	1.265
1	20190502 17:00:00	20190503 15:14:59	80099	177422	0.00065381	−13	1	15	1	0.40861	0.0593	24.02
2	20190503 15:30:00	20190503 15:59:59	1799	1343	0.0044676	−1	162	2	2	0.49886	0.0813	1.309
1	20190503 17:00:00	20190506 15:14:59	80099	317780	0.00019196	−21	1	16	1	0.46104	−0.221	50.9
2	20190506 15:30:00	20190506 15:59:59	1799	3308	−0.028416	−7	1	6	4	0.79409	−0.102	10.79
1	20190506 17:00:00	20190507 15:14:59	80099	405523	−0.00020714	−12	1	13	1	0.41678	0.00824	8.707
2	20190507 15:30:00	20190507 15:59:59	1799	2014	0.0019861	−4	1	5	1	0.54867	0.2	5.789

1	20190507 17:00:00	20190508 15:14:59	80099	328909	−3.9525e−05	−8	1	9	1	0.42838	−0.013	9.235
2	20190508 15:30:00	20190508 15:59:59	1799	1844	−0.0043384	−2	1	2	1	0.47963	−0.0122	1.597
1	20190508 17:00:00	20190509 15:14:59	80099	417700	−0.00010773	−8	1	8	1	0.40342	−0.0039	5.91
2	20190509 15:30:00	20190509 15:59:59	1799	1400	−0.00071429	−2	4	3	1	0.56652	0.142	1.454
1	20190509 17:00:00	20190510 15:14:59	80099	487260	0.00013956	−10	1	9	3	0.44738	0.0031	11.89
2	20190510 15:30:00	20190510 15:59:59	1799	1963	0.0010188	−2	7	2	4	0.5493	−0.0548	1.057
1	20190510 17:00:00	20190513 15:14:59	80099	444748	−0.00046543	−11	1	11	1	0.4301	−0.0168	10.89
2	20190512 17:00:00	20190513 15:59:59	1799	2053	0.0024355	−3	1	3	1	0.58029	0.0748	1.674
1	20190513 15:30:00	20190514 15:14:59	80099	274970	0.00048733	−16	1	16	1	0.42031	0.00808	21.04
2	20190513 17:00:00	20190514 15:59:59	1799	1466	0.0027285	−2	3	2	3	0.56881	0.000446	0.5634
1	20190514 15:30:00	20190515 15:14:59	80099	283493	0.00020459	−19	1	18	1	0.42605	−0.00208	60.14
2	20190514 17:00:00	20190515 15:59:58	1798	1384	−0.0021676	−2	1	1	206	0.5491	−0.0275	0.4146
1	20190515 15:30:00	20190516 15:14:59	80099	241691	0.00041375	−8	1	9	1	0.42015	0.0217	7.459
2	20190515 17:00:00	20190516 15:59:59	1799	1260	0.0031746	−2	7	1	193	0.55226	−0.0267	0.385
1	20190516 15:30:00	20190517 15:14:59	80099	301680	−0.00022872	−7	1	6	1	0.41078	−0.000482	4.935
2	20190516 17:00:00	20190517 15:59:59	1799	1524	0.00065617	−2	4	2	5	0.51822	0.0293	1.712
1	20190517 15:30:00	20190520 15:14:59	80099	286619	−0.00024423	−8	1	7	1	0.41055	−0.0187	5.182
2	20190517 17:00:00	20190520 15:59:59	1799	1511	0.0092654	−2	1	2	1	0.51972	0.0126	0.9222
1	20190519 17:00:00	20190520 15:14:59	80099	199368	0.00036114	−5	1	5	1	0.39546	0.00505	4.63
2	20190520 15:30:00	20190521 15:59:59	1799	1123	−0.0062333	−2	1	1	147	0.51963	−0.0467	0.8526
1	20190520 17:00:00	20190521 15:14:59	80099	197332	−0.00016216	−7	2	7	2	0.41611	0.0217	7.773
2	20190521 15:30:00	20190522 15:59:59	1799	1350	0	−4	1	4	1	0.57241	0	4.207
1	20190521 17:00:00	20190522 15:14:59	80099	293942	−0.00053072	−6	2	5	6	0.42149	−0.0148	4.528
2	20190522 15:30:00	20190523 15:59:58	1798	1808	−0.0011062	−4	1	4	1	0.54157	−0.0427	4.432
1	20190522 17:00:00	20190523 15:14:59	80099	190383	0.0002101	−7	1	8	1	0.41822	0.0302	6.576
2	20190523 15:30:00	20190524 15:59:59	1799	1150	−0.0034783	−5	1	5	1	0.61458	−0.0882	7.574
1	20190523 17:00:00	20190524 15:14:59	80099	214143	−0.00047165	−9	1	10	1	0.41331	0.0825	8.439
2	20190524 15:30:00	20190528 15:59:59	1799	2002	0.001998	−4	1	3	1	0.59399	−0.115	1.574
1	20190527 17:00:00	20190528 15:14:59	80099	308260	−0.00031467	−10	1	9	1	0.41948	−0.0439	8.751
2	20190528 15:30:00	20190529 15:59:59	1799	1187	0	−1	172	1	172	0.53856	0	0.4514
1	20190528 17:00:00	20190529 15:14:59	80099	203715	0.0002209	−5	2	4	7	0.424	−0.011	4.269

(*Continued*)

Table A1. (Continued)

r	$t_1^{d,r}$	$t_{N_{d,r}}^{d,r}$	Δt	$N_{d,r}-1$	a_1	min	#	max	#	S	SK	EK
					ESM19							
2	20190530 15:30:00	20190530 15:59:58	1798	914	−0.0043764	−3	1	2	2	0.54781	−0.203	1.941
1	20190530 17:00:00	20190531 15:14:59	80099	300088	−0.00048986	−9	1	7	1	0.42448	−0.11	6.94
2	20190531 15:30:00	20190531 15:59:59	1799	1982	0.0010091	−6	1	6	1	0.62909	−0.0373	9.25
1	20190602 17:00:00	20190603 06:40:54	49254	69664	0.00028709	−8	1	7	1	0.40978	−0.0155	10.37
1	20190603 17:00:00	20190604 15:14:59	80099	265170	0.00077309	−14	1	15	1	0.39866	0.0639	18.13
2	20190604 15:30:00	20190604 15:59:59	1799	1928	0	−2	3	3	1	0.52937	0.042	1.365
1	20190604 17:00:00	20190605 15:14:59	80099	270112	0.00031098	−18	1	18	1	0.4082	−0.017	33.09
2	20190605 15:30:00	20190605 15:59:59	1799	1525	0.0013115	−3	1	3	2	0.56355	0.0444	1.715
1	20190605 17:00:00	20190606 15:14:59	80099	239441	0.00041764	−5	3	6	1	0.39719	0.00591	5.449
2	20190606 15:30:00	20190606 15:59:59	1799	3339	−0.0059898	−3	2	2	7	0.54291	−0.128	1.589
1	20190606 17:00:00	20190607 15:14:59	80099	236073	0.00059727	−13	1	9	1	0.41351	−0.118	11.98
2	20190607 15:30:00	20190607 15:59:59	1799	1417	0.0084686	−2	8	2	8	0.60008	−0.00313	0.8261
1	20190609 17:00:00	20190610 15:14:59	80099	201496	−6.948e-05	−11	1	12	1	0.39564	0.0234	12.75
2	20190610 15:30:00	20190610 15:59:59	1799	1427	0.0014015	−7	1	5	2	0.63001	−0.692	18.85
1	20190610 17:00:00	20190611 15:14:59	80099	209440	−6.6845e-05	−3	5	4	1	0.38887	0.00181	3.981
2	20190611 15:30:00	20190611 15:59:59	1799	1203	0.003325	−3	1	1	178	0.55029	−0.208	1.293
1	20190611 17:00:00	20190612 15:14:59	80099	180030	−0.00012776	−7	1	8	1	0.37999	0.0297	6.627
2	20190612 15:30:00	20190612 15:59:59	1799	1007	0	−2	3	2	2	0.54609	−0.0367	1.03
1	20190612 17:00:00	20190613 15:14:59	80099	193762	0.00025805	−6	3	7	1	0.38182	0.00428	7.528
2	20190613 15:30:00	20190613 15:59:59	1799	1276	−0.0007837	−1	151	2	1	0.48749	0.0387	1.38
1	20190613 17:00:00	20190614 15:14:59	80099	131218	−9.9072e-05	−4	1	4	1	0.37947	0.00566	4.95
2	20190614 15:30:00	20190614 15:59:59	1799	851	0.0058754	−2	2	2	1	0.50055	−0.0447	1.677
1	20190616 17:00:00	20190617 15:14:59	80099	64264	9.3365e-05	−10	1	5	1	0.39035	−0.211	12.26
2	20190617 15:30:00	20190617 15:59:59	1799	335	0.014925	−1	44	1	49	0.52747	0.0172	0.6049
1	20190617 17:00:00	20190618 15:14:59	80099	88398	0.0012896	−9	1	10	1	0.44064	−0.0437	18.73
2	20190618 15:30:00	20190618 15:59:59	1799	281	−0.014235	−1	40	2	1	0.52761	0.131	1.181
1	20190618 17:00:00	20190619 15:14:59	80099	62699	0.00051037	−16	1	18	1	0.55937	−0.0244	117.6

Are E-mini S&P 500 Futures Prices Random?   311

2	20190619 15:30:00	20190619 15:59:59	1799	272	−0.018382	−1	47	2	2	0.58552	0.225	0.7144
1	20190619 17:00:00	20190620 15:14:59	80099	59598	0.0019632	−3	4	3	4	0.40643	0.0177	4.028
2	20190620 15:30:00	20190620 15:59:55	1795	310	−0.012903	−4	1	2	7	0.82428	−0.395	1.669
1	20190620 17:00:00	20190621 08:29:59	55799	8077	−0.0011143	−6	1	5	1	0.65879	0.118	4.916

ESM19, total 314 = 189 (1) + 125 (2) ranges, $\Delta t = 47872116$

*	20171214 06:41:23	20190621 08:29:59		15283536	7.8778e-05	−226	1	179	1	0.52805	−8.83	6859

ESU19

1	20180907 01:43:17	20180907 02:29:56	2799	1	−20	−20	1	−20	1	0	n/a	n/a
1	20180921 05:51:39	20180921 08:27:32	9353	1	18	18	1	18	1	0	n/a	n/a
1	20181011 08:48:33	20181011 13:35:03	17190	1	−126	−126	1	−126	1	0	n/a	n/a
1	20181107 02:22:01	20181107 07:44:45	19364	2	14	0	1	28	1	19.799	n/a	n/a
1	20181116 02:00:36	20181116 09:02:34	25318	11	0	−31	1	31	1	13.864	0	3.364
1	20181127 08:59:14	20181127 12:15:02	11748	1	20	20	1	20	1	0	n/a	n/a
1	20181202 17:33:35	20181203 14:15:23	74508	2	−37	−74	1	0	1	52.326	n/a	n/a
1	20181206 05:59:39	20181206 12:19:11	22772	3	0.66667	−60	1	55	1	57.761	−0.487	n/a
1	20181206 20:00:14	20181207 08:41:21	45667	3	15.667	−25	1	76	1	53.295	1.43	n/a
1	20181209 18:27:35	20181210 07:11:19	45824	2	58	8	1	108	1	70.711	n/a	n/a
1	20181219 14:11:00	20181219 14:36:30	1530	3	2.3333	−1	1	8	1	4.9329	1.65	n/a
1	20181220 08:53:53	20181220 13:45:33	17500	1	−4	−4	1	−4	1	0	n/a	n/a
1	20181220 20:00:11	20181221 14:55:34	68123	2	−110.5	−125	1	−96	1	20.506	n/a	n/a
1	20181224 08:49:05	20181224 12:12:06	12181	7	−22.857	−90	1	49	1	46.117	0.0744	−0.9579
1	20181226 09:37:21	20181226 15:09:29	19928	10	50.7	−43	1	156	1	54.63	0.333	0.2067
1	20181227 05:49:38	20181227 15:12:35	33777	14	17.429	−107	1	166	1	67.491	0.76	0.6238
2	20181227 15:33:29	20181227 15:49:48	979	7	−0.57143	−10	1	6	1	4.7208	−1.23	1.814
1	20181227 17:47:34	20181228 13:53:54	72380	17	9.7647	−88	1	77	1	37.733	−0.678	1.313
2	20181228 15:55:33	20181228 15:55:33	0	3	0	−1	1	1	1	1	0	n/a
1	20181230 19:15:21	20181231 15:00:00	71079	7	6.7143	−2	1	22	1	9.4994	0.74	−1.616
1	20190102 02:22:21	20190102 14:28:57	43596	10	17.4	−34	1	129	1	47.563	1.59	1.547
1	20190102 17:04:09	20190103 10:38:52	63283	2	−1	−2	1	0	1	1.4142	n/a	n/a
1	20190103 17:08:30	20190104 14:48:40	78010	55	5.9273	−42	1	160	1	29.147	2.98	13.35
1	20190106 17:28:16	20190107 15:11:21	78185	34	1.8824	−37	1	40	1	17.486	−0.166	0.5185

(Continued)

Table A1. (Continued)

r	$t_1^{d,r}$	$t_{N_{d,r}}^{d,r}$	Δt	$N_{d,r}-1$	a_1	min	#	max	#	S	SK	EK
					ESU19							
1	20190107 19:50:00	20190108 15:00:36	69036	133	0.58647	−27	1	44	1	9.0018	1.75	9.064
1	20190108 18:57:24	20190109 13:30:56	66812	42	0.5	−39	1	61	1	15.941	0.734	4.595
1	20190109 18:40:09	20190110 15:02:48	73359	21	4	−73	1	42	1	27.393	−1.04	1.476
1	20190111 03:01:49	20190111 11:36:11	30862	16	−0.375	−36	1	25	1	17.2	−0.521	−0.306
1	20190113 17:03:16	20190114 14:42:23	77947	7	−3.4286	−36	1	36	1	23.888	0.513	−0.7202
1	20190114 17:14:57	20190115 15:08:52	78835	35	2.8857	−51	1	45	1	17.31	−0.193	2.16
1	20190115 18:06:13	20190116 14:26:45	73232	45	1.5778	−37	1	19	1	8.4625	−1.67	8.857
1	20190117 01:24:51	20190117 14:02:44	45473	12	9.9167	−41	1	90	1	30.458	1.45	3.315
1	20190118 00:36:48	20190118 15:00:14	51806	26	4.5769	−21	1	22	1	9.8962	−0.503	0.5173
1	20190124 03:33:25	20190124 10:49:45	26180	23	0.69565	−8	1	11	1	4.4562	0.882	0.598
1	20190124 21:40:35	20190125 15:01:21	62446	29	2.2069	−26	1	39	1	11.428	0.786	3.587
1	20190127 20:49:41	20190128 10:47:41	50280	14	−8.6429	−99	1	34	1	30.396	−2.11	4.761
1	20190128 17:56:44	20190129 11:10:56	62052	6	2.8333	−5	1	22	1	9.6003	2.19	2.487
1	20190129 20:36:00	20190130 15:14:53	67133	44	3.7955	−23	1	46	1	14.019	1.16	1.854
1	20190130 19:01:51	20190131 15:01:20	71969	39	2.0513	−21	1	22	1	8.1497	−0.364	1.137
1	20190131 17:03:31	20190201 14:55:00	78689	36	−0.69444	−24	1	39	1	11.081	1.01	3.43
1	20190203 17:38:33	20190204 15:03:02	77069	27	3.037	−12	1	19	1	6.9198	0.536	0.3824
2	20190204 15:30:20	20190204 15:50:42	1222	1	−5	−5	1	−5	1	0	n/a	n/a
1	20190204 17:06:40	20190205 15:05:21	79121	28	1.5714	−20	1	31	1	9.6242	0.898	2.618
1	20190205 19:44:09	20190206 12:44:34	61225	12	0.41667	−13	1	17	1	7.9138	0.473	0.1116
1	20190206 20:00:28	20190207 14:42:23	67315	20	−4.9	−55	1	26	1	20.347	−0.658	0.2687
1	20190208 04:42:13	20190208 15:00:00	37067	26	2.0769	−48	1	34	1	14.257	−1.1	5.358
2	20190208 15:52:29	20190208 15:52:29	0	2	1	1	2	1	2	0	n/a	n/a
1	20190211 02:28:13	20190211 15:13:37	45924	13	−0.61538	−23	1	14	1	10.34	−0.891	0.2586
1	20190211 18:25:56	20190212 14:48:00	73324	20	6.05	−16	1	45	1	14.2	1.17	1.445
2	20190212 15:39:27	20190212 15:52:46	799	5	2.4	1	3	7	1	2.6077	2.09	1.65
1	20190212 17:32:59	20190213 15:08:57	77758	39	0.38462	−30	1	34	1	10.622	−0.173	3.373

2	20190213 15:51:51	20190213 15:51:51	0	2	−0.5	−2	1	1	1	2.1213	n/a	n/a
1	20190213 17:17:58	20190214 14:48:33	77435	12	0.33333	−60	52	1	1	29.218	−0.152	0.4047
1	20190214 20:18:11	20190215 13:48:35	63024	25	5.52	−22	72	2	1	20.362	2.16	4.957
1	20190219 10:05:55	20190219 14:32:57	16022	32	1.1875	−9	20	1	1	5.3608	1.62	3.893
1	20190219 17:31:04	20190220 14:02:58	73914	37	0.81081	−18	22	1	1	8.6147	0.293	0.1996
1	20190220 17:31:16	20190221 14:00:16	73740	23	−3.3913	−43	28	1	1	16.053	−0.532	0.3708
1	20190221 17:06:23	20190222 15:06:22	79199	88	0.89773	−22	35	1	1	6.3518	2.1	11.44
2	20190222 15:42:02	20190222 15:57:04	902	1	5	5	5	1	1	0	n/a	n/a
1	20190224 17:18:03	20190225 14:53:33	77730	82	−0.14634	−23	15	1	1	6.2383	−0.921	2.656
1	20190225 17:10:28	20190226 14:56:08	78340	42	0.33333	−23	18	1	1	9.3487	−0.466	0.3679
1	20190226 17:16:46	20190227 15:07:11	78625	34	0.52941	−24	27	1	1	9.4844	0.0406	1.686
1	20190227 17:00:01	20190228 14:37:56	77875	24	−0.58333	−15	20	1	1	7.9887	0.481	0.6761
1	20190228 17:35:49	20190301 14:57:08	76879	67	1.0896	−19	31	1	1	7.5292	1.36	4.247
1	20190303 17:00:00	20190304 15:10:27	79827	116	−0.68103	−32	28	1	1	8.4951	−0.0562	2.499
1	20190304 15:35:32	20190304 15:52:46	1034	1	−1	−1	−1	1	1	0	n/a	n/a
2	20190304 18:12:54	20190305 15:02:12	74958	31	0.12903	−21	29	1	1	9.8209	0.657	1.638
1	20190305 17:02:41	20190306 14:51:36	78535	76	−0.51316	−24	19	1	1	7.0833	−0.485	2.531
1	20190306 15:53:25	20190306 15:53:25	0	4	−0.25	−4	1	1	3	2.5	−2	1
2	20190306 17:00:58	20190307 14:59:27	79109	140	−0.60714	−35	35	1	1	8.5087	−0.155	5.179
2	20190307 15:45:49	20190307 15:59:03	794	2	0.5	0	1	1	1	0.70711	n/a	n/a
2	20190307 17:28:43	20190308 15:09:38	78055	209	−0.057416	−34	28	1	1	7.3659	−0.396	4.433
1	20190308 15:30:00	20190308 15:54:18	1458	5	1	−2	4	1	2	2.8284	0.331	−2.753
2	20190311 17:07:32	20190312 15:11:21	79429	263	0.11407	−16	17	1	1	4.3758	−0.129	3.123
1	20190312 15:30:43	20190312 15:59:11	1708	7	0.85714	−2	6	1	1	2.5448	1.58	1.4
2	20190312 17:17:55	20190313 15:06:34	78519	174	0.5	−16	18	1	1	5.0291	0.128	1.585
1	20190313 15:33:47	20190313 15:58:21	1474	4	0	−3	4	1	1	2.9439	0.941	−0.375
1	20190313 17:00:10	20190314 15:14:56	80086	227	−0.12775	−20	12	1	2	4.2693	−0.733	3.274
1	20190314 17:02:16	20190315 15:14:57	79961	305	0.21639	−22	13	1	1	3.62	−0.981	5.861
2	20190315 15:30:01	20190315 15:58:59	1738	11	0.27273	−3	3	1	1	1.6787	−0.524	−0.0365
1	20190317 17:00:00	20190318 15:14:57	80097	290	0.19655	−10	24	1	1	3.3358	1.51	9.455
2	20190318 15:33:06	20190318 15:58:59	1553	9	0.33333	−2	4	1	1	1.8708	0.851	−0.2964

(*Continued*)

Table A1. (Continued)

r	$t_1^{d,r}$	$t_{N_{d,r}}^{d,r}$	Δt	$N_{d,r}-1$	a_1	min	#	max	#	S	SK	EK
					ESU19							
1	20190318 17:21:34	20190319 15:14:58	78804	564	-0.047872	-12	1	13	1	2.9404	-0.0766	2.954
2	20190319 15:30:03	20190319 15:53:02	1379	4	0.75	0	2	2	1	0.95743	0.855	-1.909
1	20190319 17:08:53	20190320 15:13:36	79483	686	-0.058309	-16	1	12	2	3.4896	-0.351	3.338
2	20190320 15:30:00	20190320 15:53:32	1412	12	-1	-5	2	5	1	2.7634	0.372	0.4235
1	20190320 17:00:11	20190321 15:12:08	79917	765	0.19739	-13	1	15	1	2.7636	-0.121	2.694
2	20190321 15:34:55	20190321 15:59:36	1481	7	-1.4286	-10	1	3	1	4.0356	-1.91	2.5
1	20190321 17:07:46	20190322 15:14:51	79625	1008	-0.19742	-16	3	18	1	3.693	0.0149	2.497
2	20190322 15:30:09	20190322 15:59:48	1779	29	-0.44828	-5	1	4	2	2.1143	0.0323	0.1449
1	20190324 17:00:00	20190325 15:14:54	80094	839	-0.030989	-21	1	20	1	4.3571	-0.0915	2.962
2	20190325 15:30:01	20190325 15:55:46	1545	8	0.5	-2	1	3	1	1.6036	0	-0.887
1	20190325 17:02:19	20190326 15:12:54	79835	643	0.082426	-15	1	16	2	3.475	0.0463	3.364
2	20190326 15:30:00	20190326 15:57:54	1674	11	0.27273	-3	1	4	1	2.0538	0.399	-0.5247
1	20190326 17:02:28	20190327 15:13:56	79888	656	-0.065549	-12	1	23	1	3.9428	0.408	2.676
2	20190327 15:30:00	20190327 15:56:02	1562	5	-1.6	-6	1	1	1	2.881	-1.08	-1.037
1	20190327 17:10:19	20190328 15:14:10	79431	488	0.1168	-21	1	23	1	3.9914	-0.156	5.223
2	20190328 15:30:41	20190328 15:41:55	674	3	1	-1	1	3	1	2	0	n/a
1	20190328 17:13:51	20190329 15:14:42	79251	491	0.15886	-13	1	23	1	3.4135	0.509	5.22
2	20190329 15:31:39	20190329 15:54:34	1375	7	0.28571	-3	1	4	1	2.1381	0.342	0.2854
1	20190331 17:00:00	20190401 15:14:46	80086	512	0.20898	-10	2	16	1	2.7092	0.561	4.505
2	20190401 15:31:16	20190401 15:59:39	1703	10	-0.1	-2	4	2	2	1.7288	-0.132	-2.061
1	20190401 17:00:00	20190402 15:14:34	80074	300	-0.02	-9	2	10	1	2.9294	0.444	1.544
2	20190402 17:19:03	20190403 15:14:48	78945	456	0.10307	-12	2	18	1	3.4393	0.305	3.559
1	20190403 15:32:05	20190403 15:48:32	987	4	0.5	-1	1	2	1	1.291	0	-1.86
2	20190403 17:01:28	20190404 15:13:46	79938	279	0.032258	-17	1	12	1	3.7898	-0.514	2.65
1	20190404 15:41:48	20190404 15:59:15	1047	2	0	-3	1	3	1	4.2426	n/a	n/a
2	20190404 17:19:58	20190405 15:14:49	78891	350	0.16286	-14	1	13	1	3.0051	0.066	3.031
1	20190405 15:51:02	20190405 15:57:59	417	1	-3	-3	1	-3	1	0	n/a	n/a

	Date	Time									
1	20190407 17:00:33	20190408 15:08:01	79648	232	0.025862	−9	3	12	1	2.9687	0.0638	2.192
2	20190408 15:33:17	20190408 15:59:04	1547	2	1.5	1	1	2	1	0.70711	n/a	n/a
1	20190408 17:07:52	20190409 15:05:35	79063	314	−0.20701	−14	1	11	1	3.0666	0.109	2.373
1	20190409 18:05:39	20190410 15:03:16	75457	190	0.23684	−19	1	12	1	4.0567	−0.572	2.908
2	20190410 15:30:27	20190410 15:58:59	1712	8	0.5	−1	1	2	2	1.069	0.468	−1.255
1	20190410 17:00:00	20190411 15:14:01	80041	228	−0.057018	−14	1	11	1	3.4921	−0.162	1.603
2	20190411 15:30:02	20190411 15:59:44	1782	4	0.25	−3	1	5	1	3.4034	1.2	−0.1113
1	20190411 17:11:18	20190412 15:13:25	79327	300	0.26667	−10	1	9	1	2.8943	−0.217	0.9462
2	20190412 15:47:52	20190412 15:51:17	205	1	2	2	1	2	1	0	n/a	n/a
1	20190414 17:00:20	20190415 15:12:55	79955	215	−0.07907	−7	2	10	1	2.6295	0.231	1.116
2	20190415 15:35:23	20190415 15:48:24	781	8	0.125	0	7	1	1	0.35355	2.83	5
1	20190415 19:15:04	20190416 15:14:52	71988	249	−0.0040161	−9	1	15	1	3.0181	0.524	2.688
2	20190416 15:30:36	20190416 15:46:11	935	2	−3	−9	1	3	1	8.4853	n/a	n/a
1	20190416 17:27:52	20190417 15:14:26	78394	306	−0.11438	−14	1	16	1	3.5681	0.638	4.445
2	20190417 15:48:34	20190417 15:52:08	214	3	−1	−2	1	0	1	1	0	n/a
1	20190417 17:14:14	20190418 15:13:13	79139	274	0.12774	−16	1	14	2	3.7983	0.00856	3.115
2	20190418 15:30:00	20190418 15:59:00	1740	6	0.16667	−1	2	3	1	1.472	1.84	1.658
1	20190421 17:00:01	20190422 15:14:32	80071	176	0.079545	−17	1	9	3	3.2767	−0.78	4.723
2	20190422 15:31:15	20190422 15:58:59	1664	14	0.071429	−5	1	5	1	2.0178	−0.113	4.013
1	20190422 17:03:52	20190423 15:13:45	79793	416	0.2476	−9	2	11	1	2.4036	0.16	2.931
2	20190423 15:32:14	20190423 15:51:10	1136	2	−2	−4	1	0	1	2.8284	n/a	n/a
1	20190423 17:00:00	20190424 15:13:16	79996	245	−0.069388	−11	2	10	1	2.9083	−0.504	2.204
2	20190424 15:30:10	20190424 15:58:59	1729	7	−1.4286	−7	1	1	2	3.2071	−1.28	−0.7976
1	20190424 17:00:00	20190425 15:14:43	80083	383	−0.041775	−14	1	12	1	3.0959	−0.35	3.42
2	20190425 15:30:00	20190425 15:58:59	1739	14	−0.85714	−8	1	5	1	2.8516	−0.609	2.264
1	20190425 17:00:13	20190426 15:14:16	80043	300	0.24333	−17	1	13	1	4.1273	−0.436	2.136
2	20190426 15:30:00	20190426 15:58:36	1716	13	−0.15385	−4	1	4	1	2.2303	0.0142	−0.6624
1	20190428 17:00:00	20190429 15:12:22	79942	245	0.040816	−11	1	9	2	3.0633	−0.163	2.143
2	20190429 15:31:37	20190429 15:59:45	1688	13	0.23077	−3	1	5	1	2.0064	1.02	1.284
1	20190429 17:03:56	20190430 15:14:52	79856	412	0.07767	−11	1	10	2	2.9113	−0.0939	1.641
2	20190430 15:30:00	20190430 15:58:59	1739	37	0.35135	−6	1	8	1	2.627	0.294	1.12

(*Continued*)

Table A1. (Continued)

r	$t_1^{d,r}$	$t_{N_{d,r}}^{d,r}$	Δt	$N_{d,r}-1$	a_1	min	#	max	#	S	SK	EK
					ESU19							
1	20190430 17:00:55	20190501 15:13:40	79965	526	−0.23954	−14	1	15	1	3.152	0.297	2.995
2	20190501 15:30:00	20190501 15:59:14	1754	14	−1.6429	−13	1	1	4	3.6502	−2.57	5.952
1	20190501 17:00:00	20190502 15:13:27	80007	644	−0.0031056	−17	1	16	1	3.4726	−0.321	3.963
2	20190502 15:30:52	20190502 15:56:46	1554	10	0.4	−2	3	6	1	2.5906	1.13	0.2403
1	20190502 17:07:47	20190503 15:14:52	79625	383	0.28198	−14	1	11	3	3.3502	−0.149	2.493
2	20190503 15:30:02	20190503 15:57:40	1658	8	0.25	−3	1	1	5	1.3887	−2.29	3.251
1	20190505 17:00:00	20190506 15:14:45	80085	1311	0.033562	−56	1	21	1	3.9204	−2.21	34.41
2	20190506 15:30:00	20190506 15:59:58	1798	24	−3.625	−19	1	3	1	5.5389	−1.31	0.9588
1	20190506 17:00:06	20190507 15:14:57	80091	1111	−0.081008	−23	1	18	2	3.9745	0.01	3.732
2	20190507 15:30:09	20190507 15:59:59	1790	7	0.42857	−10	1	5	2	5.0615	−1.69	1.717
1	20190507 17:00:23	20190508 15:14:50	80067	1225	−0.010612	−17	1	26	1	3.5476	0.0942	5.007
2	20190508 15:30:16	20190508 15:55:34	1518	7	−1	−5	1	2	1	2.1602	−0.833	0.5296
1	20190508 17:00:42	20190509 15:14:02	80000	1323	−0.028723	−15	2	20	2	3.7772	0.248	3.54
2	20190509 15:30:00	20190509 15:56:21	1581	6	0	−3	1	6	1	3.0984	1.88	1.863
1	20190509 17:00:00	20190510 15:14:46	80086	1919	0.03283	−25	1	33	1	4.183	0.251	6.894
2	20190510 15:31:21	20190510 15:59:50	1709	15	−0.13333	−8	1	5	1	2.9488	−1.03	2.269
1	20190512 17:00:00	20190513 15:14:50	80090	1492	−0.14276	−20	1	24	1	3.8157	0.00978	4.155
2	20190513 15:30:00	20190513 15:59:21	1761	10	0	−5	1	5	1	3.0912	0.169	−0.8942
1	20190513 17:00:43	20190514 15:14:58	80055	858	0.16084	−18	1	19	1	3.2977	−0.181	3.873
2	20190514 15:30:19	20190514 15:58:46	1707	12	−0.33333	−5	1	3	1	2.3868	−0.729	−0.392
1	20190514 17:06:45	20190515 15:14:12	79647	1272	0.047956	−33	1	12	4	3.0816	−0.719	11.91
2	20190515 15:30:00	20190515 15:55:40	1540	7	−0.28571	−10	1	5	1	5.0238	−1.3	0.5984
1	20190515 17:02:34	20190516 15:14:58	79944	1385	0.072202	−13	1	18	1	2.4256	0.334	5.795
2	20190516 15:30:11	20190516 15:38:40	509	2	3	2	1	4	1	1.4142	n/a	n/a
1	20190516 17:01:00	20190517 15:13:59	79979	1240	−0.054839	−19	1	14	1	3.2004	−0.198	3.234
2	20190517 15:30:06	20190517 15:54:24	1458	23	0.043478	−3	1	4	2	1.7704	0.79	0.2018
1	20190519 17:00:50	20190520 15:12:42	79912	1286	−0.070762	−15	1	13	2	3.0844	−0.0811	3.021

2	20190520 15:30:00	20190520 15:55:54	1554	18	0.88889	-2	1	4	1	1.5297	-0.0132	-0.4098
1	20190520 17:00:40	20190521 15:14:06	80006	1043	0.067114	-21	1	14	2	2.4936	-0.266	8.904
2	20190521 15:30:00	20190521 15:59:59	1799	48	-0.22917	-5	1	2	1	1.0567	-2.23	8.276
1	20190521 17:00:04	20190522 15:14:47	80083	1206	-0.024046	-10	2	16	1	2.4247	0.212	4.103
2	20190522 15:30:04	20190522 15:58:59	1735	5	1.2	0	2	3	1	1.3038	0.541	-1.893
1	20190522 17:01:26	20190523 15:14:55	80009	1739	-0.089707	-17	1	14	2	2.616	-0.113	4.179
2	20190523 15:30:00	20190523 15:57:41	1661	26	-0.11538	-2	2	5	1	1.4234	1.93	4.959
1	20190523 17:01:00	20190524 15:14:58	80038	1177	0.033985	-16	1	12	1	2.2925	-0.369	4.927
2	20190524 15:30:50	20190524 15:59:40	1730	10	-0.2	-2	1	3	1	1.3984	1.35	1.308
1	20190527 17:00:55	20190528 15:14:59	80044	1336	-0.077844	-13	1	19	1	2.5666	0.0322	4.839
2	20190528 15:30:00	20190528 15:58:06	1686	36	-0.027778	-6	1	5	2	2.1711	0.0909	1.124
1	20190528 17:00:29	20190529 15:14:59	80070	2388	-0.038526	-14	1	11	2	2.1848	-0.0176	3.548
2	20190529 15:30:20	20190529 15:58:59	1719	22	0.045455	-3	1	2	1	0.99892	-1.04	2.732
1	20190529 17:00:00	20190530 15:14:57	80097	1637	0.025657	-17	1	15	1	2.2766	-0.101	6.992
2	20190530 15:30:00	20190530 15:59:50	1790	8	-0.125	-6	1	3	2	3.0443	-1.1	-0.0276
1	20190530 17:01:35	20190531 15:14:55	80000	2683	-0.056653	-17	1	12	1	2.2041	-0.0653	5.033
2	20190531 15:30:33	20190531 15:59:40	1747	37	0.081081	-4	2	4	1	1.7854	-0.562	0.1231
1	20190602 17:00:00	20190603 15:14:00	80040	4116	0.0063168	-13	1	13	1	2.0454	0.126	4.312
2	20190603 15:30:00	20190603 15:59:54	1794	52	0.34615	-5	1	4	2	1.7364	-0.233	0.7551
1	20190603 17:01:06	20190604 15:14:58	80032	5436	0.03716	-8	1	9	1	1.3395	-0.0899	3.335
2	20190604 15:30:05	20190604 15:58:51	1726	63	0.015873	-5	1	3	1	1.2635	-0.922	3.35
1	20190604 17:00:00	20190605 15:14:59	80099	4991	0.01683	-13	1	10	1	1.4223	-0.0552	5.291
2	20190605 15:30:00	20190605 15:59:58	1798	46	0.086957	-4	1	3	2	1.5323	-0.191	-0.2205
1	20190605 17:00:00	20190606 15:14:53	80093	8944	0.011404	-8	1	14	1	1.0059	0.477	8.769
2	20190606 15:30:00	20190606 15:59:56	1796	137	-0.15328	-4	1	4	5	1.4496	0.522	1.218
1	20190606 17:00:00	20190607 15:14:59	80099	9205	0.014992	-11	1	14	1	0.9628	0.0935	10.69
2	20190607 15:30:00	20190607 15:59:54	1794	86	0.12791	-2	3	3	1	0.93048	0.278	0.6088
1	20190609 17:00:00	20190610 15:14:59	80099	9925	-0.0012091	-18	1	23	1	0.96157	1.03	67.82
2	20190610 15:30:00	20190610 15:59:56	1796	69	-0.014493	-3	2	3	1	1.0913	-0.32	0.7635
1	20190610 17:00:00	20190611 15:14:59	80099	12626	-0.00095042	-5	1	5	1	0.69544	-0.0425	3.477
2	20190611 15:30:00	20190611 15:59:58	1798	110	0.036364	-2	1	2	3	0.68973	0.294	1.033

(*Continued*)

Table A1. (Continued)

r	$t_1^{d,r}$	$t_{N_{d,r}}^{d,r}$	Δt	$N_{d,r}-1$	a_1	min	#	max	#	S	SK	EK
					ESU19							
1	20190611 17:00:00	20190612 15:14:59	80099	16787	−0.0014297	−7	1	6	2	0.62885	0.00395	5.491
2	20190612 15:30:00	20190612 15:59:48	1788	107	0.0093458	−2	1	2	2	0.67984	0.172	0.7952
1	20190612 17:00:00	20190613 15:14:59	80099	50712	0.0010057	−4	1	3	4	0.43967	−0.00617	3.212
2	20190613 15:30:00	20190613 15:59:59	1799	411	0.0024331	−2	1	1	52	0.50605	−0.109	1.371
1	20190613 17:00:00	20190614 15:14:59	80099	129888	−8.4688e-05	−5	1	6	1	0.33718	0.00336	8.655
2	20190614 15:30:00	20190614 15:59:58	1798	727	0.0082531	−2	6	3	1	0.59607	0.0757	1.917
1	20190614 17:00:00	20190617 15:14:59	80099	124290	5.632e-05	−4	1	4	1	0.33637	−0.0206	6.947
2	20190617 15:30:00	20190617 15:59:59	1799	740	0.0054054	−1	121	2	1	0.57927	0.0417	0.1271
1	20190617 17:00:00	20190618 15:14:59	80099	227000	0.00050661	−5	1	6	1	0.36516	0.0138	6.254
2	20190618 15:30:00	20190618 15:59:59	1799	1023	−0.0039101	−2	4	2	2	0.53267	−0.0817	1.404
1	20190618 17:00:00	20190619 15:14:59	80099	214268	0.00015868	−12	1	11	1	0.38443	−0.061	20.85
2	20190619 15:30:00	20190619 15:59:59	1799	996	−0.0040161	−2	2	1	149	0.55455	−0.0727	0.5092
1	20190619 17:00:00	20190620 15:14:59	80099	235244	0.00051011	−6	1	7	1	0.37581	0.0214	6.201
2	20190620 15:30:00	20190620 15:59:59	1799	1649	−0.0024257	−2	5	2	4	0.54304	−0.0245	1.148
1	20190620 17:00:00	20190621 15:14:59	80099	210761	−0.00012811	−6	2	7	1	0.37874	0.0332	6.644
2	20190621 15:30:00	20190621 15:59:59	1799	2145	−0.0051282	−3	1	2	2	0.54761	−0.157	1.265
1	20190621 17:00:00	20190624 15:14:59	80099	144750	0.00011744	−5	1	5	1	0.37321	0.0141	5.872
2	20190624 15:30:00	20190624 15:59:59	1799	690	−0.0043478	−2	3	2	2	0.58151	−0.0441	0.7249
1	20190624 17:00:00	20190625 11:38:10	67090	125953	−0.00053988	−8	1	7	1	0.38337	−0.0282	9.251
1	20190625 17:00:00	20190626 15:14:59	80099	181242	−9.9315e-05	−4	3	3	8	0.38787	−0.00831	4.253
2	20190626 15:30:00	20190626 15:59:58	1798	1476	−0.0067751	−1	209	2	2	0.52847	0.0477	0.795
1	20190626 17:00:00	20190627 15:14:59	80099	158666	0.00040967	−7	2	7	1	0.39961	−0.0298	7.61
2	20190627 15:30:00	20190627 15:59:59	1799	2017	0.0049579	−2	5	2	6	0.54279	0.0222	1.151
1	20190627 17:00:00	20190628 15:14:59	80099	162710	0.00054084	−7	2	8	1	0.39993	−0.01	7.418
2	20190628 15:30:00	20190628 15:59:59	1799	2521	0.0019833	−3	1	4	1	0.58925	0.0462	1.541
1	20190630 17:00:00	20190701 15:14:59	80099	195528	−3.58e-05	−7	2	8	2	0.42239	0.00507	7.716
2	20190701 15:30:00	20190701 15:59:59	1799	1088	−0.0018382	−4	1	3	1	0.61863	0.118	2.413

1	20190701 17:00:00	20190702 15:14:59	80099	145037	0.00032406	−7	1	6	1	0.40428	−0.0382	5.891
2	20190702 15:30:00	20190702 15:59:59	1799	1245	−0.00080321	−2	1	1	172	0.52815	−0.0337	0.7129
1	20190702 17:00:00	20190703 12:14:59	69299	110306	0.00079778	−3	6	4	3	0.41557	0.0364	3.72
1	20190704 17:00:00	20190705 15:14:59	80099	168248	−0.00029124	−8	1	7	1	0.40359	−0.118	8.118
2	20190705 15:30:00	20190705 15:59:59	1799	1194	0.0041876	−3	1	2	7	0.63231	−0.103	0.8298
1	20190707 17:00:00	20190708 15:14:59	80099	141559	−0.00035321	−4	2	7	1	0.39374	0.0198	4.794
2	20190708 15:30:00	20190708 15:59:59	1799	1055	0	−2	2	2	2	0.53528	0	1.051
1	20190708 17:00:00	20190709 15:14:59	80099	138327	0.00013013	−17	1	16	1	0.40556	−0.772	111.2
2	20190709 15:30:00	20190709 15:59:59	1799	833	0.0012005	−1	100	2	1	0.49395	0.0627	1.349
1	20190709 17:00:00	20190710 15:14:59	80099	193637	0.00031502	−4	1	4	5	0.39345	0.0557	4.613
2	20190710 15:30:00	20190710 15:59:59	1799	830	−0.0060241	−3	1	2	1	0.63185	−0.11	0.2312
1	20190710 17:00:00	20190711 15:14:59	80099	166673	0.00019799	−9	1	4	1	0.38727	−0.12	6.594
2	20190711 15:30:00	20190711 15:59:59	1799	905	0	−2	2	2	2	0.57028	0	0.5821
1	20190711 17:00:00	20190712 15:14:59	80099	128969	0.00036443	−13	3	14	1	0.42321	0.356	122.3
2	20190712 15:30:00	20190712 15:59:59	1799	1188	−0.0025253	−2	4	2	5	0.54841	0.0292	1.337
1	20190712 17:00:00	20190715 15:14:59	80099	113773	7.9105e-05	−5	1	5	1	0.39335	−0.00884	4.807
2	20190715 15:30:00	20190715 15:59:58	1798	1018	−0.00098232	−2	2	2	1	0.51621	−0.0444	1.258
1	20190715 17:00:00	20190716 15:14:59	80099	124911	−0.00033624	−4	1	4	1	0.39638	0.0133	3.804
2	20190716 15:30:00	20190716 15:59:59	1799	1098	−0.0045537	−3	1	2	1	0.56885	−0.12	0.9274
1	20190717 17:00:00	20190718 15:14:59	80099	176023	0.00038631	−5	2	6	1	0.39506	0.0149	4.563
2	20190718 15:30:00	20190718 15:59:59	1799	2614	0.0080337	−2	5	3	2	0.5201	0.158	2.134
1	20190718 17:00:00	20190719 15:14:59	80099	197831	−0.00053581	−8	1	7	1	0.40143	0.00229	5.71
2	20190719 15:30:00	20190719 15:59:59	1799	2943	−0.0091743	−9	1	8	1	0.64469	−0.258	22.44
1	20190721 17:00:00	20190722 15:14:59	80099	156193	0.00037774	−6	1	6	1	0.38159	−0.00309	5.752
2	20190722 15:30:00	20190722 15:59:59	1799	848	−0.004717	−1	138	1	134	0.56667	−0.000983	0.1179
1	20190722 17:00:00	20190723 15:14:59	80099	148428	0.00051877	−6	2	5	1	0.38244	−0.0237	5.304
2	20190723 15:30:00	20190723 15:59:59	1799	3445	−0.0063861	−2	6	2	4	0.50314	−0.0395	1.494
1	20190723 17:00:00	20190724 15:14:59	80099	149275	0.00049573	−3	2	3	3	0.37753	0.0195	4.463
2	20190724 15:30:00	20190724 15:59:59	1799	1761	−0.0011357	−2	8	3	1	0.58968	0.0169	0.9507
1	20190724 17:00:00	20190725 15:14:59	80099	195530	−0.00025572	−7	1	6	1	0.39882	−0.0839	6.603
2	20190725 15:30:00	20190725 15:59:59	1799	1561	−0.0025625	−3	2	3	1	0.62017	0.0985	1.687

(*Continued*)

Table A1. (Continued)

r	$t_1^{d,r}$	$t_{N_{d,r}}^{d,r}$	Δt	$N_{d,r}-1$	a_1	min	#	max	#	S	SK	EK
					ESU19							
1	20190725 17:00:00	20190726 15:14:59	80099	127858	0.00054748	−8	1	9	1	0.38848	0.0755	12.07
2	20190726 15:30:00	20190726 15:59:59	1799	1117	−0.003581	−5	1	2	3	0.59114	−0.651	5.41
1	20190728 17:00:00	20190729 15:14:59	80099	114937	0	−5	1	5	1	0.37369	0.008	5.393
2	20190729 15:30:00	20190729 15:59:59	1799	763	0.0052425	−2	2	2	2	0.58635	−0.000786	0.447
1	20190730 17:00:00	20190731 15:14:59	80099	293878	−0.00051042	−14	1	10	1	0.45378	−0.192	20.19
2	20190731 15:30:00	20190731 15:59:59	1799	3140	−0.0050955	−4	3	3	1	0.6042	−0.466	3.342
1	20190731 17:00:00	20190801 15:14:59	80099	416613	−0.00018482	−11	1	9	1	0.43478	−0.114	12.55
2	20190801 15:30:00	20190801 15:59:59	1799	2790	−0.0050179	−6	2	5	3	0.72053	−0.315	10.01
1	20190801 17:00:00	20190802 15:14:59	80099	408011	−0.00018137	−12	1	13	1	0.43765	−0.0502	19.64
2	20190802 15:30:00	20190802 15:59:59	1799	1386	0.007215	−4	1	2	5	0.60439	−0.259	2.029
1	20190805 17:00:00	20190806 15:14:59	80099	488087	0.00048967	−9	2	9	1	0.45317	−0.0124	9.76
2	20190806 15:30:00	20190806 15:59:59	1799	2057	−0.0029169	−2	5	2	7	0.55966	0.0323	0.9099
1	20190807 17:00:00	20190808 15:14:59	80099	275392	0.00089327	−6	1	7	1	0.42004	0.0268	5.103
2	20190808 15:30:00	20190808 15:59:59	1799	1875	−0.0021333	−2	4	3	1	0.53779	0.0806	1.535
1	20190808 17:00:00	20190809 15:14:59	80099	323484	−0.00011129	−17	1	17	1	0.4348	0.0538	31.73
2	20190809 15:30:00	20190809 15:59:59	1799	1767	0.008489	−6	1	5	1	0.63445	−0.233	6.68
1	20190811 17:00:00	20190812 15:14:59	80099	281522	−0.00055058	−7	4	7	2	0.41487	−0.0381	8.008
2	20190812 15:30:00	20190812 15:59:59	1799	1749	0.004574	−1	220	1	228	0.50623	0.00818	0.9049
1	20190812 17:00:00	20190813 15:14:59	80099	400376	0.00049953	−10	1	10	2	0.42609	0.0324	10.46
2	20190813 15:30:00	20190813 15:59:59	1799	1468	−0.0020436	−2	8	2	4	0.55569	−0.0964	1.271
1	20190813 17:00:00	20190814 15:14:59	80099	449042	−0.00081061	−6	5	7	1	0.42817	0.00301	5.496
2	20190814 15:30:00	20190814 15:59:59	1799	2658	−0.006772	−3	1	3	3	0.57546	0.0828	1.715
1	20190814 17:00:00	20190815 15:14:59	80099	511803	9.1832e-05	−9	1	10	1	0.43253	0.0196	8.72
2	20190815 15:30:00	20190815 15:59:59	1799	1704	0.0058685	−3	1	2	3	0.53087	−0.0646	1.531
1	20190815 17:00:00	20190816 15:14:59	80099	264216	0.00061692	−9	1	9	1	0.4114	0.0197	7.812
2	20190816 15:30:00	20190816 15:59:59	1799	1603	−0.0018715	−2	2	2	2	0.53527	−0.00172	0.8587
1	20190818 17:00:00	20190819 15:14:59	80099	202159	0.00060349	−10	1	10	1	0.41798	−0.00785	9.226

2	20190819 15:30:00	20190819 15:59:59	1799	1734	−0.0040369	−2	7	2	8	0.53335	0.0189	1.804
1	20190819 17:00:00	20190820 15:14:59	80099	229739	−0.0003961	−6	1	6	1	0.41392	0.0168	3.917
2	20190820 15:30:00	20190820 15:59:59	1799	1963	−0.0061131	−4	1	4	1	0.54554	−0.00406	3.543
1	20190820 17:00:00	20190821 15:14:59	80099	185149	0.00079396	−6	1	6	2	0.41437	0.0204	4.963
2	20190821 15:30:00	20190821 15:59:59	1799	2192	0.004562	−2	3	2	3	0.52589	0.00536	1.048
1	20190821 17:00:00	20190822 15:14:59	80099	243901	−0.0001558	−6	1	7	1	0.40981	0.00674	4.294
2	20190822 15:30:00	20190822 15:59:59	1799	1297	−0.0038551	−3	1	2	6	0.57264	−0.0744	1.517
1	20190822 17:00:00	20190823 15:14:59	80099	487329	−0.00052942	−10	1	7	3	0.45695	−0.0303	8.519
2	20190823 15:30:00	20190823 15:59:59	1799	4456	−0.0029174	−4	2	4	1	0.60765	−0.0346	2.871
1	20190825 17:00:00	20190826 15:14:59	80099	371208	0.00056303	−13	1	13	2	0.47672	0.00747	16.22
2	20190826 15:30:00	20190826 15:59:59	1799	1417	−0.0028229	−1	157	1	153	0.46789	−0.00949	1.573
1	20190826 17:00:00	20190827 15:14:59	80099	278219	−0.00024082	−8	1	8	1	0.43085	0.0132	6.295
2	20190827 15:30:00	20190827 15:59:59	1799	2100	0.00095238	−2	1	2	4	0.53465	0.0571	0.8504
1	20190827 17:00:00	20190828 15:14:59	80099	252835	0.00037574	−9	1	7	1	0.41508	−0.015	7.497
2	20190828 15:30:00	20190828 15:59:59	1799	3149	−0.0082566	−3	2	3	1	0.54638	0.0181	1.55
1	20190828 17:00:00	20190829 15:14:59	80099	240944	0.00072216	−7	1	9	4	0.42979	0.175	10.53
2	20190829 15:30:00	20190829 15:59:59	1799	1521	−0.0046022	−2	7	2	7	0.50846	−0.00789	2.528
1	20190829 17:00:00	20190830 15:14:59	80099	242148	−3.3038e−05	−5	3	5	1	0.40924	−0.0154	4.363
2	20190830 15:30:00	20190830 15:59:59	1799	1854	0.0070119	−2	7	3	1	0.59086	0.077	0.9793
1	20190830 17:00:00	20190903 15:14:59	80099	284304	9.1451e−05	−7	1	7	1	0.42218	−0.0157	6.493
2	20190903 15:30:00	20190903 15:59:59	1799	1945	−0.0030848	−2	9	2	10	0.57466	0.0161	1.107
1	20190903 17:00:00	20190904 15:14:59	80099	202813	0.00071001	−7	1	7	3	0.39068	0.0476	7.008
2	20190904 15:30:00	20190904 15:59:59	1799	1148	−0.0043554	−1	179	1	174	0.55474	−0.00198	0.2525
1	20190904 17:00:00	20190905 15:14:59	80099	261051	0.00052097	−9	1	8	1	0.40261	0.0104	9.986
2	20190905 15:30:00	20190905 15:59:59	1799	1613	−0.0061996	−2	6	2	4	0.56134	−0.0441	0.9247
1	20190905 17:00:00	20190906 15:14:59	80099	196645	0.00020341	−17	1	18	1	0.41004	0.187	53.85
2	20190906 15:30:00	20190906 15:59:59	1799	1350	0.0066667	−2	3	2	3	0.60449	−0.00289	0.1386
1	20190908 17:00:00	20190909 15:14:59	80099	185333	−3.777e−05	−7	2	6	1	0.38811	−0.0629	7.334
2	20190909 15:30:00	20190909 15:59:59	1799	920	0.0043478	−3	1	2	2	0.52363	−0.0857	2.049
1	20190909 17:00:00	20190910 15:14:59	80099	225347	−4.4376e−06	−6	1	4	4	0.39102	0.0129	5.01
2	20190910 15:30:00	20190910 15:59:59	1799	1621	0	−2	3	2	6	0.54772	0.0677	1.078

(*Continued*)

Table A1. (Continued)

r	$t_1^{d,r}$	$t_{N_{d,r}}^{d,r}$	Δt	$N_{d,r}-1$	a_1	min	#	max	#	S	SK	EK
					ESU19							
1	20190910 17:00:00	20190911 15:14:59	80099	197544	0.00047078	−7	2	6	1	0.37212	−0.0147	8.821
2	20190911 15:30:00	20190911 15:59:59	1799	1919	0	−3	1	7	2	0.59455	1.79	22
1	20190911 17:00:00	20190912 15:14:59	80099	262444	0.00012574	−8	2	13	1	0.41388	0.261	17.9
2	20190912 15:30:00	20190912 15:59:59	1799	1174	−0.0076661	−4	1	4	1	0.53912	−0.104	5.658
1	20190912 17:00:00	20190913 15:14:59	80099	118972	−0.00010927	−6	1	7	1	0.38607	0.0927	7.591
2	20190913 15:30:00	20190913 15:59:59	1799	1578	0.0019011	−2	2	2	5	0.45675	0.127	3.023
1	20190915 17:00:00	20190916 15:14:59	80099	92286	0.00026006	−13	1	12	2	0.45005	0.0696	27.23
2	20190916 15:30:00	20190916 15:59:59	1799	697	0.017217	−2	2	1	83	0.47615	−0.109	2.109
1	20190916 17:00:00	20190917 15:14:59	80099	55565	0.00028795	−5	1	5	1	0.4074	0.00374	5.432
2	20190917 15:30:00	20190917 15:59:59	1799	453	0.0022075	−1	60	1	61	0.51739	0.00319	0.7472
1	20190917 17:00:00	20190918 15:14:58	80098	59536	3.3593e-05	−15	1	10	1	0.49987	−0.508	46.07
2	20190918 15:30:00	20190918 15:59:59	1799	565	0.0035398	−3	2	3	1	0.68156	−0.308	2.431
1	20190918 17:00:00	20190919 15:14:59	80099	45898	−0.0001743	−4	2	4	2	0.41519	−0.00118	4.332
2	20190919 15:30:00	20190919 15:59:58	1798	233	−0.017167	−2	2	2	1	0.57958	−0.134	1.365
1	20190919 17:00:00	20190920 08:29:59	55799	4840	0.0068182	−9	2	16	1	0.76943	1.59	57.76
	ESU19, total 329 = 193 (1) + 136 (2) ranges, $\Delta t = 32683602$											
*	20180907 01:43:17	20190920 08:29:59		14114418	0.00029799	−126	1	166	1	0.49594	8.5	5168
					ESZ19							
1	20181127 01:31:49	20181127 10:42:16	33027	1	0	0	1	0	1	0	n/a	n/a
1	20181204 10:22:09	20181204 12:41:41	8372	1	−242	−242	1	−242	1	0	n/a	n/a
1	20181227 09:07:26	20181227 15:04:08	21402	8	33.25	−41	2	280	1	102.65	2.52	4.069
1	20181227 20:14:39	20181228 10:00:54	49575	2	−9	−18	1	0	1	12.728	n/a	n/a
1	20190103 00:14:39	20190103 14:59:41	53102	3	−28.333	−85	1	0	2	49.075	−1.73	n/a
1	20190106 18:32:22	20190107 12:12:04	63582	7	11.571	−29	1	79	1	43.343	0.778	−1.499
1	20190110 19:30:55	20190111 11:50:12	58757	2	5.5	−12	1	23	1	24.749	n/a	n/a
1	20190117 13:39:11	20190117 13:41:19	128	1	30	30	1	30	1	0	n/a	n/a
1	20190118 09:53:20	20190118 10:59:59	3999	1	40	40	1	40	1	0	n/a	n/a

Are E-mini S&P 500 Futures Prices Random?   323

	Start	End										
1	20190124 02:20:35	20190124 12:10:35	35400	3	0	−33	1	44	1	39.661	1.15	n/a
1	20190125 01:02:03	20190125 09:08:16	29173	1	99	99	1	99	1	0	n/a	n/a
1	20190128 09:14:28	20190128 09:14:29	1	1	0	0	1	0	1	0	n/a	n/a
1	20190206 14:49:03	20190206 14:49:55	52	10	0	0	10	0	10	0	n/a	n/a
1	20190212 04:54:23	20190212 04:54:30	7	1	0	0	1	0	1	0	n/a	n/a
1	20190219 13:31:21	20190219 14:27:15	3354	3	−0.33333	−8	1	4	1	6.6583	−1.69	n/a
1	20190225 08:45:32	20190225 12:02:48	11836	2	−18	−36	1	0	1	25.456	n/a	n/a
1	20190226 05:22:18	20190226 09:18:57	14199	6	10.167	0	2	37	1	14.02	1.85	1.431
1	20190301 08:49:27	20190301 09:48:33	3546	1	−44	−44	1	−44	1	0	n/a	n/a
1	20190303 17:04:08	20190304 11:07:41	65013	2	−60	−120	1	0	1	84.853	n/a	n/a
1	20190307 18:42:08	20190308 12:05:00	62572	10	−9.1	−34	2	29	1	19.433	0.409	−0.3414
2	20190308 15:31:12	20190308 15:31:12	0	1	−12	−12	1	−12	1	0	n/a	n/a
1	20190312 05:43:29	20190312 12:52:36	25747	2	16	5	1	27	1	15.556	n/a	n/a
1	20190312 21:40:24	20190313 12:55:44	54920	10	14.1	−2	1	59	1	21.147	1.4	0.04917
1	20190314 09:46:24	20190314 11:37:37	6673	3	7.6667	−16	1	29	1	22.591	−0.46	n/a
1	20190314 17:31:33	20190315 15:11:31	77998	13	4.9231	−22	1	35	1	15.185	0.237	−0.258
1	20190317 18:44:00	20190318 10:41:33	57453	4	4.75	−25	1	29	1	25.617	−0.32	−3.233
1	20190318 18:08:33	20190319 14:25:11	72998	8	−2.625	−63	1	30	1	29.78	−1.19	0.5387
1	20190320 08:58:28	20190320 14:33:49	20121	15	0.66667	−24	1	31	1	13.589	0.637	0.5345
1	20190320 18:40:16	20190321 14:34:01	71625	18	7.1111	−57	1	68	1	27.14	0.555	2.246
1	20190321 17:31:07	20190322 14:50:57	76790	44	−4.7045	−54	1	32	1	14.245	−1.25	3.557
1	20190324 17:00:02	20190325 15:00:08	79206	26	−1.9231	−35	3	51	1	21.433	0.667	0.7922
1	20190326 05:19:06	20190326 14:58:46	34780	7	2.8571	−31	1	60	1	35.69	0.744	−1.387
1	20190326 22:26:47	20190327 14:50:58	59051	13	−3.3077	−115	1	54	1	40.948	−1.73	3.14
1	20190327 18:41:23	20190328 11:55:17	62034	7	5.1429	−39	1	31	1	25.9	−0.862	−1.124
1	20190328 18:16:28	20190329 15:01:10	74682	15	4.1333	−28	1	27	1	14.643	−0.426	−0.2446
1	20190331 17:02:01	20190401 15:00:00	79079	28	2.6786	−15	1	28	1	9.6073	0.902	0.6835
1	20190402 02:30:10	20190402 15:10:22	45612	9	1.1111	−26	1	19	1	17.142	−0.415	−1.802
1	20190402 20:29:54	20190403 14:30:00	64806	16	−0.125	−50	1	27	1	16.268	−1.77	4.647
1	20190403 18:05:13	20190404 14:55:00	74987	12	1.6667	−22	1	19	1	13.76	−0.377	−1.356
1	20190404 19:05:00	20190405 14:45:02	70802	19	1.5789	−14	1	17	1	7.8054	0.273	−0.4197

(Continued)

Table A1. (Continued)

r	$t_1^{d,r}$	$t_{N_{d,r}}^{d,r}$	Δt	$N_{d,r}-1$	a_1	min	#	max	#	S	SK	EK
					ESZ19							
1	20190407 17:17:25	20190408 14:48:59	77494	11	0.36364	−32	1	35	1	18.408	−0.361	0.4813
1	20190409 08:34:25	20190409 13:43:06	18521	13	−1.2308	−40	1	14	1	13.368	−2.1	4.595
1	20190410 05:53:12	20190410 15:00:07	32815	22	0.40909	−14	1	17	1	5.6117	0.462	3.695
1	20190410 18:41:42	20190411 10:45:54	57852	22	−1.5455	−26	1	26	1	10.211	−0.115	2.627
1	20190411 17:11:18	20190412 15:00:00	78522	30	2.6333	−29	1	29	1	11.019	0.0424	2.137
1	20190414 18:00:05	20190415 11:08:05	61680	10	−3.1	−31	1	16	1	11.893	−1.18	2.302
1	20190416 03:03:44	20190416 15:04:21	43237	21	−1.5714	−18	1	16	1	8.818	−0.00871	−0.7878
1	20190416 19:47:56	20190417 14:58:10	69014	25	−1.24	−16	1	13	1	7.4122	−0.0605	−0.4265
1	20190417 19:53:35	20190418 15:14:58	69683	14	2.4286	−44	1	57	1	23.595	0.256	1.487
1	20190421 17:35:08	20190422 14:30:00	75292	15	−1	−32	1	24	1	11.898	−0.583	2.678
1	20190422 19:29:24	20190423 14:27:18	68274	13	7.6923	−12	1	36	1	14.665	0.43	−1.011
1	20190424 01:10:06	20190424 14:01:55	46309	9	1.1111	−28	1	12	2	12.634	−1.8	1.915
1	20190424 18:07:07	20190425 15:08:21	75674	30	−0.63333	−39	1	19	1	11.199	−1.23	3.132
2	20190425 15:34:23	20190425 15:48:22	839	1	−17	−17	1	−17	1	0	n/a	n/a
1	20190425 17:10:57	20190426 15:10:47	79190	30	2.1	−12	2	26	1	8.5191	0.892	1.169
1	20190428 17:38:00	20190429 14:45:48	76068	31	0.83871	−12	1	18	1	6.8707	0.714	0.6268
1	20190429 17:12:18	20190430 14:56:10	78232	32	0.625	−21	1	42	1	10.453	1.67	6.602
2	20190430 15:30:00	20190430 15:38:03	483	1	3	3	1	3	1	0	n/a	n/a
1	20190430 17:51:00	20190501 15:06:01	76501	29	−4.2759	−37	1	25	1	13.627	−0.81	0.806
1	20190501 17:04:19	20190502 13:50:35	74776	21	−1.1429	−85	1	39	1	24.81	−1.82	5.046
1	20190502 17:17:41	20190503 13:09:49	71528	48	2.2083	−19	1	14	2	6.2158	−0.62	1.835
1	20190505 17:22:04	20190506 13:54:18	73934	48	2.7917	−55	1	72	1	18.725	0.856	4.615
2	20190506 15:32:33	20190506 15:59:23	1610	2	−28	−61	1	5	1	46.669	n/a	n/a
1	20190507 01:20:16	20190507 15:00:19	49203	129	−1.2093	−77	1	33	1	11.856	−2.25	14.47
1	20190507 19:07:40	20190508 15:12:06	72266	75	−0.04	−23	1	32	1	9.38	0.565	1.771
1	20190508 19:34:01	20190509 15:09:36	70535	90	−0.27778	−40	1	44	1	13.831	0.15	3.061
1	20190509 18:51:40	20190510 15:13:46	73326	128	0.42188	−61	1	37	1	13.63	−0.765	3.866

1	20190512 17:02:36	20190513 15:13:03	79827	95	−2.5474	−63	1	39	1	15.178	−1.15	3.599
1	20190513 17:01:15	20190514 14:56:57	78942	65	1.9077	−35	1	40	1	12.466	0.513	1.929
1	20190514 17:00:13	20190515 15:00:49	79236	171	0.2807	−38	1	38	1	8.2078	0.403	7.467
1	20190515 17:35:10	20190516 15:01:56	77206	30	3.5333	−36	1	45	1	15.978	0.181	1.553
1	20190516 17:01:00	20190517 13:07:41	72401	31	−0.64516	−48	1	32	1	14.91	−0.461	2.605
1	20190519 17:02:32	20190520 14:29:00	77188	33	−3.2727	−53	1	20	1	16.121	−1.32	1.951
1	20190520 18:56:11	20190521 14:21:01	69890	37	1.7568	−26	1	22	1	9.1117	−0.189	1.76
1	20190521 17:29:43	20190522 14:11:35	74512	33	−0.66667	−38	1	26	1	11.191	−0.852	3.289
1	20190522 20:03:29	20190523 14:59:01	68132	58	−1.8448	−40	1	50	1	14.799	0.525	2.413
1	20190523 18:57:28	20190524 15:00:00	72152	32	0.71875	−46	1	29	1	14.787	−0.76	2.273
1	20190528 01:47:37	20190528 14:59:40	47523	33	−3.6667	−54	1	14	1	13.261	−2.42	6.027
1	20190528 18:51:59	20190529 14:40:51	71332	86	−0.83721	−43	1	26	1	10.88	−0.83	3.512
1	20190529 17:01:46	20190530 14:58:51	79025	55	0.6	−46	1	30	1	11.546	−1.13	4.44
1	20190530 18:07:43	20190531 15:11:11	75808	92	−1.7065	−48	1	24	1	10.275	−1.11	3.989
2	20190531 15:30:02	20190531 15:31:58	116	1	2	2	3	2	1	0	n/a	n/a
1	20190602 17:10:41	20190603 15:13:26	79365	88	0.375	−54	1	39	1	15.776	−0.518	1.913
1	20190603 17:21:14	20190604 15:08:54	78460	129	1.5349	−29	1	43	1	8.1068	1.05	7.542
1	20190604 17:01:06	20190605 15:01:56	79250	123	0.4878	−32	1	37	1	8.502	0.851	5.003
2	20190605 15:32:13	20190605 15:46:46	873	4	1	0	3	4	1	2	2	1
1	20190605 17:31:13	20190606 15:08:17	77824	104	1.1442	−30	1	34	1	7.9139	0.0681	5.227
1	20190606 17:01:45	20190607 15:07:36	79551	111	1.3153	−17	2	31	1	8.1441	0.945	2.739
2	20190607 15:31:09	20190607 15:45:53	884	1	2	2	1	2	2	0	n/a	n/a
1	20190609 17:00:00	20190610 14:56:59	79019	122	−0.098361	−18	2	18	1	5.6208	−0.522	2.248
1	20190610 18:24:57	20190611 14:54:44	73787	90	−0.066667	−28	1	25	1	8.1844	−0.338	2.161
1	20190611 18:29:48	20190612 14:55:00	73512	51	−0.56863	−24	1	22	1	8.5633	−0.671	1.378
2	20190612 15:34:13	20190612 15:51:39	1046	3	−0.33333	−4	1	2	1	3.2146	−1.55	n/a
1	20190612 17:04:50	20190613 15:02:32	79062	129	0.31008	−44	1	36	1	8.3972	−0.808	8.068
1	20190613 17:13:17	20190614 15:02:36	78559	74	−0.054054	−43	1	20	2	8.5414	−1.34	7.977
2	20190614 15:30:05	20190614 15:41:12	667	1	−1	−1	1	−1	1	0	n/a	n/a
1	20190616 19:03:16	20190617 15:14:00	72644	138	−0.10145	−20	1	17	1	3.9639	−0.338	7.726
1	20190617 17:35:37	20190618 15:03:03	77246	224	0.48214	−26	1	38	1	5.7113	0.755	11.01

(*Continued*)

Table A1. (Continued)

r	$t_1^{d,r}$	$t_{N_{d,r}}^{d,r}$	Δt	$N_{d,r}-1$	a_1	min	#	max	#	S	SK	EK
					ESZ19							
2	20190618 15:31:22	20190618 15:40:36	554	1	2	2	1	2	1	0	n/a	n/a
1	20190618 19:21:27	20190619 15:04:01	70954	176	0.073864	−19	1	28	1	5.5591	0.305	4.48
2	20190619 15:30:00	20190619 15:43:23	803	3	1	−1	1	4	1	2.6458	1.46	n/a
1	20190619 17:03:46	20190620 15:07:18	79412	250	0.448	−17	1	20	1	4.71	0.133	3.251
2	20190620 15:32:25	20190620 15:45:32	787	2	1	−1	2	1	2	0	n/a	n/a
1	20190620 17:01:02	20190621 15:14:31	80009	477	−0.041929	−17	1	26	1	3.4273	0.612	11.76
2	20190621 15:34:23	20190621 15:52:53	1110	11	0	−4	1	3	1	2.1448	−0.595	−1.002
1	20190623 17:00:01	20190624 15:13:42	80021	167	0.083832	−12	1	16	1	3.7002	0.311	2.885
2	20190624 15:30:17	20190624 15:54:06	1429	3	0	−1	1	1	1	1	0	n/a
1	20190624 17:01:37	20190625 15:14:05	79948	405	−0.2963	−19	1	14	1	3.9522	−0.779	3.715
2	20190625 15:34:23	20190625 15:42:11	468	4	−0.5	−2	1	0	3	1	−2	1
1	20190625 17:05:01	20190626 15:06:42	79301	380	−0.047368	−13	1	12	2	3.4737	0.00354	1.884
2	20190626 15:34:57	20190626 15:55:01	1204	7	−1.2857	−7	1	3	1	3.1997	−0.706	−0.0563
1	20190626 17:10:36	20190627 15:14:00	79404	268	0.20149	−14	1	19	1	4.1782	0.126	2.364
2	20190627 15:30:00	20190627 15:49:50	1190	22	0.5	−3	3	5	1	2.2625	0.204	−0.6304
1	20190627 17:00:31	20190628 15:14:58	80067	307	0.25081	−13	2	16	1	3.4011	0.248	3.989
2	20190628 15:30:19	20190628 15:57:48	1649	8	0.125	−3	1	6	1	2.8504	1.36	0.7363
1	20190630 17:00:00	20190701 15:08:44	79724	349	0.011461	−16	1	14	1	3.734	0.21	1.65
2	20190701 15:46:48	20190701 15:59:57	789	2	−0.5	−1	1	0	1	0.70711	n/a	n/a
1	20190701 17:03:19	20190702 15:13:26	79807	256	0.17969	−16	1	13	1	3.9832	−0.391	2.484
2	20190702 15:30:48	20190702 15:45:21	873	3	1	0	1	2	1	1	0	n/a
1	20190702 17:02:37	20190703 12:14:59	69142	213	0.41784	−11	1	11	1	2.9666	−0.0706	2.352
2	20190704 17:00:00	20190705 15:13:07	79987	382	−0.10733	−20	1	14	1	3.6474	−0.302	3.792
1	20190705 15:30:01	20190705 15:58:59	1738	12	0.5	−4	1	5	1	2.1532	−0.131	1.356
2	20190707 17:00:00	20190708 15:08:44	79724	246	−0.15854	−16	1	15	2	4.037	0.0496	2.448
1	20190708 15:32:36	20190708 15:54:17	1301	8	0.5	−1	2	4	1	1.6036	1.66	1.757
2	20190708 17:00:57	20190709 15:05:23	79466	257	0.070039	−10	2	11	1	3.1442	0.077	1.941

Are E-mini S&P 500 Futures Prices Random?

2	20190709 15:32:55	20190709 15:51:15	1100		2	2	2	2	0	n/a	n/a
1	20190709 17:02:47	20190710 15:01:34	79127	0.18553	−15	2	18	1	3.9929	0.108	3.049
2	20190710 15:35:52	20190710 15:57:57	1325		0	1	0	1	0	n/a	n/a
1	20190710 17:00:19	20190711 15:13:52	80013	0.11715	−18	1	13	2	3.7071	−0.119	4.064
2	20190711 15:47:08	20190711 15:49:17	129		2	1	2	1	0	n/a	n/a
1	20190711 17:01:33	20190712 15:13:00	79887	0.22705	−11	1	12	1	3.0958	0.217	2.311
2	20190712 15:43:20	20190712 15:58:59	939	−0.4	−5	1	2	2	3.0496	−1.04	−1.252
1	20190714 17:00:00	20190715 15:11:56	79916	0.019231	−10	2	10	1	2.9846	0.132	1.445
1	20190715 17:00:52	20190716 15:14:43	80031	−0.18483	−14	1	13	1	3.7466	0.151	1.837
2	20190716 15:34:52	20190716 15:58:50	1438	−0.28571	−1	2	0	5	0.48795	−1.23	−1.32
1	20190717 17:00:09	20190718 15:12:43	79954	0.15851	−14	1	13	1	3.3437	0.0613	3.498
2	20190718 15:30:59	20190718 15:42:16	677	1.3636	−4	1	4	1	2.0627	−1.85	3.226
1	20190718 17:04:51	20190719 15:14:55	79804	−0.29442	−29	1	17	1	4.1895	−0.904	8.167
2	20190719 15:30:08	20190719 15:59:55	1787	−0.76923	−5	1	5	1	2.1332	0.557	0.5765
1	20190721 17:00:00	20190722 15:14:08	80048	0.27727	−20	1	22	1	4.926	0.532	4.434
2	20190722 15:32:08	20190722 15:45:07	779		0	1	0	1	0	n/a	n/a
1	20190722 17:01:35	20190723 15:13:55	79940	0.28731	−12	1	12	4	3.4342	0.222	2.391
2	20190723 15:30:01	20190723 15:55:01	1500	−0.95	−8	1	8	1	3.7763	0.1	0.2906
1	20190723 17:13:43	20190724 15:14:39	79256	0.22727	−14	1	12	1	2.9314	−0.646	4.628
2	20190724 15:30:00	20190724 15:59:56	1796	−0.15789	−3	1	4	1	1.6077	0.734	0.8105
1	20190724 17:00:00	20190725 15:14:50	80090	−0.10156	−16	2	13	1	3.4148	−0.244	3.184
2	20190725 15:30:10	20190725 15:55:19	1509	−0.15385	−5	1	3	2	2.6092	−0.866	−0.7193
1	20190725 17:03:53	20190726 15:12:32	79719	0.29289	−9	2	12	1	2.691	0.307	2.769
2	20190726 15:30:03	20190726 15:59:21	1758	−0.13333	−4	1	6	1	2.3865	0.982	1.451
1	20190728 17:02:11	20190729 15:09:54	79663	−0.076923	−8	2	12	1	2.9074	0.896	3.193
2	20190729 15:30:08	20190729 15:57:26	1638	0.57143	−2	1	7	1	3.0472	1.96	2.086
1	20190730 17:15:17	20190731 15:12:30	79033	−0.16368	−45	1	26	1	4.3248	−1.1	15.98
2	20190731 15:30:00	20190731 15:55:28	1528	−1.0625	−8	1	4	1	3.0653	−1.01	0.6378
1	20190731 17:00:00	20190801 15:14:59	80099	−0.060496	−23	1	28	1	4.1846	0.0121	4.268
2	20190801 15:30:10	20190801 15:59:33	1763	−0.82609	−13	1	7	1	4.1522	−1.19	2.945
1	20190801 17:00:00	20190802 15:14:55	80095	−0.053287	−19	1	22	1	3.8523	0.0143	3.89

(Continued)

Table A1. (Continued)

r	$t_1^{d,r}$	$t_{N_{d,r}}^{d,r}$	Δt	$N_{d,r}-1$	a_1	min	#	max	#	S	SK	EK
					ESZ19							
2	20190802 15:31:26	20190802 15:46:58	932	9	0.88889	0	5	4	1	1.3642	1.77	1.624
1	20190805 17:00:00	20190806 15:13:53	80033	1966	0.12055	−28	1	23	1	4.4547	0.094	3.86
2	20190806 15:36:25	20190806 15:59:26	1381	7	−1.4286	−8	1	4	1	4.5408	−0.597	−1.529
1	20190807 17:00:40	20190808 15:14:35	80035	1158	0.21762	−21	1	25	1	3.5239	0.303	7.242
2	20190808 15:30:00	20190808 15:59:40	1780	24	−0.70833	−13	1	5	1	3.3425	−2.12	6.574
1	20190808 17:00:00	20190809 15:14:13	80053	1612	−0.021712	−18	1	20	1	3.1442	0.0678	4.172
2	20190809 15:30:06	20190809 15:59:31	1765	14	0.78571	−4	2	6	1	3.1422	−0.0994	−1.07
1	20190811 17:00:00	20190812 15:14:39	80079	1014	−0.15187	−18	1	28	1	3.8025	0.295	5.577
2	20190812 15:30:27	20190812 15:56:31	1564	14	0.71429	−2	2	4	1	1.9779	0.254	−1.392
1	20190812 17:00:09	20190813 15:10:50	79841	1989	0.097034	−23	1	18	1	3.3937	−0.361	4.657
2	20190813 15:31:07	20190813 15:57:55	1608	11	−0.27273	−6	1	3	1	2.4936	−1.09	0.9142
1	20190813 17:00:34	20190814 15:14:51	80057	2046	−0.17595	−16	2	28	1	3.7023	0.228	4.311
2	20190814 15:30:01	20190814 15:56:39	1598	21	−0.85714	−4	2	2	2	1.8244	−0.397	−1.036
1	20190814 17:00:00	20190815 15:14:41	80081	2778	0.017639	−21	1	27	1	3.822	0.347	4.956
2	20190815 15:30:29	20190815 15:57:48	1639	15	0.86667	−11	1	15	1	5.2081	0.7	3.925
1	20190815 17:00:00	20190816 15:14:29	80069	1501	0.10593	−13	1	12	4	2.6907	−0.0676	3.773
2	20190816 15:30:54	20190816 15:59:00	1686	13	−0.69231	−9	1	1	2	2.5621	−3.29	8.696
1	20190818 17:00:00	20190819 15:14:11	80051	1139	0.10623	−10	1	13	1	2.4438	0.198	3.181
2	20190819 15:30:00	20190819 15:55:07	1507	8	−1.25	−6	1	3	1	2.6592	−0.311	0.0036
1	20190819 17:00:04	20190820 15:14:46	80082	1166	−0.07976	−15	1	12	1	2.5386	−0.1	3.709
2	20190820 15:30:00	20190820 15:57:19	1639	26	−0.38462	−8	1	4	1	2.3337	−1.44	3.518
1	20190820 17:00:24	20190821 15:14:40	80056	895	0.15531	−16	1	13	1	2.6052	−0.106	5.579
2	20190821 15:30:43	20190821 15:59:06	1703	30	0.3	−3	1	3	1	1.2905	−0.297	0.2386
1	20190821 17:00:31	20190822 15:14:55	80064	1383	−0.023861	−13	1	13	1	2.5223	−0.12	4.282
2	20190822 15:30:28	20190822 15:59:36	1748	22	−0.18182	−3	1	4	1	1.7358	0.788	−0.003
1	20190822 17:00:01	20190823 15:14:55	80094	3560	−0.072753	−20	1	19	1	2.9725	−0.0202	4.523
2	20190823 15:30:00	20190823 15:59:55	1795	72	−0.23611	−6	1	5	1	1.9319	0.357	0.7649

1	20190825 17:00:00	20190826 15:14:44	80084	3750	0.056267	−26	1	13	3	2.9393	−0.48	5.076
2	20190826 15:31:19	20190826 15:58:55	1656	16	−0.1875	−2	3	4	1	1.5586	1.32	1.716
1	20190826 17:00:27	20190827 15:14:50	80063	1950	−0.033333	−16	1	21	1	2.7101	0.228	5.48
2	20190827 15:33:37	20190827 15:53:10	1173	31	0.48387	−3	1	10	1	2.3219	2.39	7.898
1	20190827 17:00:00	20190828 15:14:50	80090	1706	0.0551	−20	1	15	1	2.7526	−0.193	6.89
2	20190828 15:30:00	20190828 15:59:02	1742	29	−0.96552	−7	3	5	1	2.9819	−0.583	0.1491
1	20190828 17:00:00	20190829 15:14:55	80095	2088	0.083812	−14	1	12	2	2.344	−0.11	3.591
2	20190829 15:30:02	20190829 15:59:24	1762	11	−0.63636	−4	1	3	1	1.8586	−0.0802	0.299
1	20190829 17:00:00	20190830 15:14:52	80092	2326	−0.0068788	−12	1	11	1	1.9181	0.186	4.571
2	20190830 15:30:00	20190830 15:58:59	1739	16	0.5625	−3	1	6	1	2.1593	1.06	1.177
1	20190902 17:00:00	20190903 15:14:58	80098	6221	0.0038579	−9	3	9	1	1.6399	−0.00881	3.43
2	20190903 15:30:00	20190903 15:59:33	1773	39	−0.051282	−3	2	4	1	1.3945	0.403	1.163
1	20190903 17:00:00	20190904 15:14:55	80095	5738	0.024747	−6	3	6	1	1.1008	−0.186	3.349
2	20190904 15:30:00	20190904 15:59:28	1768	31	−0.096774	−2	5	2	2	1.1062	−0.272	−0.3685
1	20190904 17:00:20	20190905 15:14:59	80079	13030	0.010361	−8	1	10	1	0.85415	0.097	6.901
2	20190905 15:30:20	20190905 15:59:38	1758	57	−0.15789	−3	2	3	2	1.207	0.251	0.7469
1	20190905 17:00:06	20190906 15:14:59	80093	11680	0.0034247	−16	1	14	1	0.82286	−0.196	29.72
2	20190906 15:30:00	20190906 15:59:58	1798	46	0.19565	−2	2	2	5	1.0671	−0.0654	−0.7905
1	20190906 17:00:00	20190909 15:14:59	80099	12863	−0.0005442	−4	3	8	1	0.61666	0.257	6.129
2	20190909 15:30:01	20190909 15:59:58	1797	46	0.1087	−2	1	2	2	0.79522	0.0768	0.5683
1	20190909 17:00:00	20190910 15:14:59	80099	21266	0	−5	2	4	3	0.58245	−0.134	4.107
2	20190910 15:30:00	20190910 15:59:56	1796	131	0.0076336	−2	5	2	2	0.71786	−0.391	1.447
1	20190910 17:00:00	20190911 15:14:59	80099	18398	0.0050549	−5	1	4	7	0.58447	0.0239	5.019
2	20190911 15:30:00	20190911 15:59:54	1794	301	0.0066445	−2	5	3	1	0.69279	0.233	1.881
1	20190911 17:00:00	20190912 15:14:59	80099	67113	0.00053641	−8	1	15	1	0.55535	0.405	21.84
2	20190912 15:30:00	20190912 15:59:58	1798	542	−0.016605	−2	2	2	2	0.50295	−0.0318	2.352
1	20190912 17:00:00	20190913 15:14:59	80099	129635	−9.2568e-05	−4	2	5	1	0.3271	0.026	8.716
2	20190913 15:30:00	20190913 15:59:59	1799	1298	0.0023112	−4	1	4	1	0.47367	−0.0365	9.754
1	20190913 17:00:00	20190916 15:14:59	80099	176416	0.00012471	−8	2	8	1	0.40068	−0.102	17.16
2	20190916 15:30:00	20190916 15:59:59	1799	1189	0.0084104	−2	1	2	7	0.494	0.271	2.444
1	20190916 17:00:00	20190917 15:14:59	80099	147844	0.00010822	−5	1	6	1	0.36327	0.0318	8.428

(*Continued*)

Table A1. (Continued)

r	$t_1^{d,r}$	$t_{N_{d,r}}^{d,r}$	Δt	$N_{d,r}-1$	a_1	min	#	max	#	S	SK	EK
					ESZ19							
2	20190917 15:30:00	20190917 15:59:59	1799	1217	0.00082169	−2	2	3	2	0.55087	0.237	2.016
1	20190917 17:00:00	20190918 15:14:59	80099	201943	4.9519e-06	−10	1	8	1	0.41366	−0.0906	12.09
2	20190918 15:30:00	20190918 15:59:59	1799	2347	0.00085215	−5	1	8	1	0.65484	2.11	22.61
1	20190918 17:00:00	20190919 15:14:59	80099	183879	−3.8069e-05	−7	1	5	3	0.38539	−0.042	6.747
2	20190919 15:30:00	20190919 15:59:59	1799	1373	−0.0029133	−2	2	6	1	0.46604	−0.0533	2.166
1	20190919 17:00:00	20190920 15:14:59	80099	199173	−0.00035647	−8	1	2	8	0.41003	−0.0804	8.043
2	20190920 15:30:00	20190920 15:59:59	1799	2479	−0.00080678	−3	1	6	1	0.52615	−0.0509	1.942
1	20190920 17:00:00	20190923 15:14:59	80099	161204	4.3423e-05	−5	3	4	1	0.40967	0.035	5.141
2	20190923 15:30:00	20190923 15:59:59	1799	1897	0.0079072	−3	1	6	1	0.53707	0.13	2.747
1	20190923 17:00:00	20190924 15:14:59	80099	286627	−0.0004082	−6	1	2	3	0.42327	−0.00277	5.176
2	20190924 15:30:00	20190924 15:59:59	1799	1358	0	−2	1	5	2	0.54157	0.0558	0.8248
1	20190924 17:00:00	20190925 15:14:59	80099	233729	0.00026954	−5	3	2	1	0.40378	0.00287	5.05
2	20190925 15:30:00	20190925 15:59:59	1799	923	−0.0032503	−2	2	2	1	0.55401	−0.0399	0.677
1	20190925 17:00:00	20190926 15:14:59	80099	205569	−0.00010702	−8	1	9	1	0.40832	−0.00507	6.228
2	20190926 15:30:00	20190926 15:59:59	1799	952	−0.0021008	−2	2	2	1	0.54647	−0.0401	0.7781
1	20190926 17:00:00	20190927 15:14:59	80099	252684	−0.00025328	−6	3	7	1	0.41801	−0.0111	5.014
2	20190927 15:30:00	20190927 15:59:59	1799	2235	0.0076063	−3	1	2	5	0.50462	−0.0278	2.092
1	20190927 17:00:00	20190930 15:14:59	80099	162714	0.00022125	−7	1	7	1	0.41875	0.00645	5.85
2	20190930 15:30:00	20190930 15:59:59	1799	1408	−0.0042614	−1	222	2	1	0.5592	0.0229	0.2884
1	20190930 17:00:00	20191001 15:14:59	80099	255152	−0.0007133	−12	1	11	1	0.4226	−0.0284	14.43
2	20191001 15:30:00	20191001 15:59:59	1799	1540	0.01039	−2	4	2	3	0.56301	−0.019	0.7017
1	20191001 17:00:00	20191002 15:14:59	80099	320145	−0.00077465	−9	2	7	1	0.43239	−0.147	7.842
2	20191002 15:30:00	20191002 15:59:59	1799	2367	0.0059147	−3	2	2	11	0.52813	−0.0107	2.35
1	20191002 17:00:00	20191003 15:14:59	80099	336267	0.00033307	−8	2	8	3	0.44166	0.00844	10.65
2	20191003 15:30:00	20191003 15:59:59	1799	1192	−0.0025168	−2	1	2	1	0.52237	−0.00322	0.9401
1	20191003 17:00:00	20191004 15:14:59	80099	232731	0.00066171	−27	1	29	1	0.4478	0.255	149.8
2	20191004 15:30:00	20191004 15:59:59	1799	1374	0.0007278	−2	6	2	3	0.56416	−0.0729	0.9229

1	20191006 17:00:00	20191007 15:14:59	80099	205834	0.00015061	−7	2	6	6	0.42158	0.00288	7.482
2	20191007 15:30:00	20191007 15:59:59	1799	1633	−0.0042866	−2	1	2	2	0.50304	0.0208	1.301
1	20191007 17:00:00	20191008 15:14:59	80099	321917	−0.00052187	−14	1	14	1	0.43076	0.0165	13.8
2	20191008 15:30:00	20191008 15:59:59	1799	1818	0.0027503	−2	6	3	2	0.56546	0.165	1.744
1	20191008 17:00:00	20191009 15:14:59	80099	232934	0.00045077	−29	1	18	1	0.47068	−0.545	112.2
2	20191009 15:30:00	20191009 15:59:59	1799	1852	0.0021598	−2	4	3	1	0.50923	0.00364	1.825
1	20191009 17:00:00	20191010 15:14:59	80099	311214	0.00027634	−19	1	20	1	0.49243	0.0248	37.28
2	20191010 15:30:00	20191010 15:59:59	1799	7472	0.0044165	−7	1	7	1	0.74628	−0.0245	7.463
1	20191010 17:00:00	20191011 15:14:59	80099	296549	0.00028663	−14	1	14	1	0.44508	−0.037	18.48
2	20191011 15:30:00	20191011 15:59:59	1799	3141	0.00031837	−3	1	3	2	0.64517	0.0638	1.191
1	20191011 17:00:00	20191014 15:14:59	80099	158493	−5.0475e-05	−8	1	8	1	0.42384	0.0305	7.616
2	20191014 15:30:00	20191014 15:59:59	1799	1206	0.0024876	−1	123	2	1	0.45639	0.0623	2.037
1	20191014 17:00:00	20191016 15:14:59	80099	189603	−0.00011076	−8	1	7	2	0.40064	−0.124	9.386
2	20191015 17:00:00	20191016 15:59:59	1799	1214	0.0041186	−2	5	2	4	0.5176	−0.0299	1.981
1	20191016 15:30:00	20191017 15:14:59	80099	189478	0.00012139	−6	1	9	1	0.40178	0.0425	8.957
2	20191016 17:00:00	20191017 15:59:59	1799	1112	0.0017986	−2	5	2	7	0.56127	0.0618	1.486
1	20191017 15:30:00	20191018 15:14:59	80099	200385	−0.00018963	−7	1	5	2	0.39261	−0.0259	5.945
2	20191017 17:00:00	20191018 15:59:59	1799	1274	0.0039246	−2	6	2	3	0.5485	−0.0835	1.269
1	20191018 15:30:00	20191022 15:14:59	80099	165417	−0.00030227	−8	1	5	1	0.39524	−0.127	6.891
2	20191018 17:00:00	20191022 15:59:59	1799	2868	−0.0048815	−3	3	3	2	0.54319	−0.095	2.553
1	20191021 15:30:00	20191023 15:14:59	80099	169650	0.0003183	−11	1	12	1	0.39894	0.00708	19.26
2	20191022 15:30:00	20191023 15:59:59	1799	1462	0.002052	−1	200	3	1	0.53039	0.14	1.284
1	20191022 17:00:00	20191024 15:14:59	80099	169890	−4.1203e-05	−5	2	6	1	0.39078	−0.0412	5.654
2	20191023 15:30:00	20191024 15:59:59	1799	1559	−0.0044901	−2	7	3	1	0.54747	0.0207	1.8
1	20191023 17:00:00	20191025 15:14:59	80099	151849	0.00042806	−4	1	4	3	0.39419	0.0224	4.689
2	20191024 15:30:00	20191025 15:59:59	1799	1149	−0.00087032	−3	1	2	3	0.54341	−0.131	1.831
1	20191024 17:00:00	20191028 15:14:59	80099	134126	0.0003877	−6	1	7	1	0.38819	0.012	5.468
2	20191025 15:30:00	20191028 15:59:59	1799	1893	0.002113	−3	3	6	1	0.53916	0.468	10.28
1	20191027 17:00:00	20191029 15:14:59	80099	153325	−7.1743e-05	−6	1	6	1	0.36808	−0.00556	7.008
2	20191028 15:30:00	20191029 15:59:59	1799	1133	0.0044131	−1	128	2	2	0.48382	0.105	1.662
1	20191029 17:00:00	20191030 15:14:59	80099	186378	0.00025218	−11	1	10	1	0.40801	−0.0583	16.54

(*Continued*)

Table A1. (Continued)

r	$t_1^{d,r}$	$t_{N_{d,r}}^{d,r}$	Δt	$N_{d,r}-1$	a_1	min	#	max	#	S	SK	EK
\multicolumn{13}{c}{ESZ19}											
2	20191030 15:30:00	20191030 15:59:59	1799	5415	0.0020314	−18	1	19	1	0.6448	0.668	250.9
1	20191030 17:00:00	20191031 15:14:59	80099	252732	−0.00025323	−7	1	8	1	0.38251	0.0242	10.01
2	20191031 15:30:00	20191031 15:59:59	1799	1422	−0.00070323	−2	1	2	1	0.50958	−0.00118	1.106
1	20191031 17:00:00	20191101 15:14:59	80099	181187	0.00060159	−11	1	13	1	0.39183	0.0885	25.61
2	20191101 15:30:00	20191101 15:59:59	1799	1706	−0.00058617	−2	5	2	3	0.55172	−0.0423	0.896
1	20191101 17:00:00	20191104 15:14:59	80099	158799	0.00027078	−5	1	5	1	0.37144	0.00236	5.705
2	20191104 15:30:00	20191104 15:59:59	1799	1177	0.00084962	−2	3	5	147	0.50423	−0.118	1.414
1	20191104 17:00:00	20191105 15:14:59	80099	160629	−0.00012451	−8	1	8	1	0.39071	0.0427	8.636
2	20191105 15:30:00	20191105 15:59:58	1798	1115	−0.0017937	−1	131	1	129	0.4831	−0.00479	1.291
1	20191105 17:00:00	20191106 15:14:59	80099	150987	0.00011259	−10	1	11	1	0.38744	0.038	12.74
2	20191106 15:30:00	20191106 15:59:59	1799	1274	0.0031397	−3	1	3	1	0.55632	−0.0261	1.714
1	20191106 17:00:00	20191107 15:14:59	80099	184513	0.00019511	−7	2	8	1	0.39289	0.0162	8.842
2	20191107 15:30:00	20191107 15:59:59	1799	1090	−0.00091743	−1	147	1	146	0.5187	−0.00128	0.7215
1	20191107 17:00:00	20191108 15:14:59	80099	149863	0.00010676	−8	1	7	1	0.39593	−0.0133	8.476
2	20191108 15:30:00	20191108 15:59:58	1798	1363	0.0044021	−1	163	1	169	0.4937	0.00987	1.107
1	20191108 17:00:00	20191111 15:14:59	80099	122414	−8.9859e-05	−8	1	7	1	0.36123	−0.135	10.89
2	20191111 15:30:00	20191111 15:59:59	1799	887	−0.0011274	−1	129	1	128	0.53858	−0.000947	0.4524
1	20191111 17:00:00	20191112 15:14:59	80099	154510	0.0001165	−4	2	4	1	0.38186	−0.00928	5.096
2	20191112 15:30:00	20191112 15:59:59	1799	996	−0.0050201	−2	3	1	121	0.50423	−0.151	1.494
1	20191112 17:00:00	20191113 15:14:59	80099	176817	0.0001018	−9	1	6	2	0.39133	−0.16	9.39
2	20191113 15:30:00	20191113 15:59:59	1799	1106	0.0018083	−1	148	3	1	0.52623	0.189	1.607
1	20191113 17:00:00	20191114 15:14:59	80099	156256	2.5599e-05	−11	1	8	1	0.39474	−0.2	14.23
2	20191114 15:30:00	20191114 15:59:59	1799	1009	0.0019822	−1	120	1	122	0.48997	0.00474	1.172
1	20191114 17:00:00	20191115 15:14:59	80099	154350	0.00053774	−9	4	10	1	0.40575	0.173	27.82
2	20191115 15:30:00	20191115 15:59:59	1799	1271	0.0023603	−1	178	2	1	0.53315	0.0335	0.6383
1	20191117 17:00:00	20191118 15:14:59	80099	145134	7.5792e-05	−4	2	5	1	0.39268	0.0177	4.595
2	20191118 15:30:00	20191118 15:59:58	1798	1070	−0.0018692	−1	137	1	135	0.50442	−0.00347	0.9356

1	20191118 17:00:00	20191119 15:14:59	80099	159068	−8.1726e-05	−7	1	6	1	0.38836	−0.0194	5.139
2	20191119 15:30:00	20191119 15:59:59	1799	1146	−0.0026178	−1	154	2	1	0.5178	0.0341	0.8811
1	20191119 17:00:00	20191120 15:14:59	80099	240602	−0.00012884	−10	1	9	1	0.39313	−0.0385	10.98
2	20191120 15:30:00	20191120 15:59:59	1799	1637	0	−2	1	2	2	0.49693	0.0299	1.415
1	20191120 17:00:00	20191121 15:14:59	80099	212513	0.00012235	−11	1	10	1	0.3903	0.00729	16.17
2	20191121 15:30:00	20191121 15:59:59	1799	1155	0	−1	127	1	127	0.46915	0	1.55
1	20191121 17:00:00	20191122 15:14:59	80099	140866	0.00022007	−4	2	4	1	0.39228	−0.0121	4.211
2	20191122 15:30:00	20191122 15:59:59	1799	1028	0.00097276	−1	152	2	1	0.54674	0.0364	0.4801
1	20191122 17:00:00	20191125 15:14:59	80099	115430	0.00046782	−5	2	5	2	0.39632	0.00898	4.69
2	20191125 15:30:00	20191125 15:59:59	1799	959	−0.0052138	−1	115	1	110	0.4846	−0.0136	1.265
1	20191125 17:00:00	20191126 15:14:59	80099	153447	0.00033236	−4	1	4	1	0.35953	0.00774	5.176
2	20191126 15:30:00	20191126 15:59:58	1798	1466	0.00068213	−2	1	1	215	0.5424	−0.0252	0.4969
1	20191126 17:00:00	20191127 15:14:59	80099	126317	0.00031666	−5	2	5	1	0.36918	0.0113	6.035
2	20191127 15:30:00	20191127 15:59:59	1799	919	0.0043526	−1	127	1	131	0.53012	0.00463	0.5632
1	20191127 17:00:00	20191129 12:14:59	69299	105825	−0.00023624	−4	2	5	1	0.38037	0.00169	4.93
1	20191129 17:00:00	20191202 15:14:59	80099	261050	−0.00049416	−14	1	13	1	0.38539	−0.0943	23.36
2	20191202 15:30:00	20191202 15:59:58	1798	1276	−0.0031348	−2	1	1	165	0.51334	−0.0397	0.9338
1	20191202 17:00:00	20191203 11:29:03	66543	230377	−0.00055561	−12	1	11	1	0.39037	−0.0366	15.33
2	20191203 17:00:00	20191204 15:14:59	80099	214060	0.00037373	−8	1	7	3	0.38361	0.0424	8.015
1	20191204 15:30:00	20191204 15:59:59	1799	1063	−0.0028222	−1	132	1	129	0.49574	−0.00612	1.075
2	20191204 17:00:00	20191205 15:14:59	80099	196678	0.00012711	−10	1	9	1	0.36291	−0.0392	13.37
1	20191205 15:30:00	20191205 15:59:59	1799	996	0.003012	−2	1	2	3	0.51994	0.0901	1.364
2	20191205 17:00:00	20191206 15:14:59	80099	175368	0.00065576	−21	1	22	1	0.38007	0.183	128.7
1	20191206 15:30:00	20191206 15:59:58	1798	1240	0.0016129	−2	4	2	5	0.53903	0.0323	1.48
2	20191206 17:00:00	20191209 15:14:59	80099	154868	−0.00030994	−6	1	5	1	0.3293	−0.0308	8.732
1	20191209 15:30:00	20191209 15:59:59	1799	1668	−0.0029976	−3	1	2	1	0.50492	−0.0895	1.7
2	20191209 17:00:00	20191210 15:14:59	80099	220427	4.9903e-05	−5	1	5	1	0.35858	−0.0188	6.17
1	20191210 15:30:00	20191210 15:59:59	1799	985	0.0040609	−1	133	1	137	0.52381	0.00504	0.6494
2	20191210 17:00:00	20191211 15:14:59	80099	165575	0.00018723	−5	1	4	2	0.34779	−0.00233	6.258
1	20191211 15:30:00	20191211 15:59:59	1799	1056	0.0056818	−1	119	1	125	0.48088	0.0157	1.33
2	20191211 17:00:00	20191212 15:14:59	80099	309999	0.00030323	−13	2	13	1	0.37172	−0.044	31.49

(Continued)

Table A1. (Continued)

r	$t_1^{d,r}$	$t_{N_{d,r}}^{d,r}$	Δt	$N_{d,r}-1$	a_1	min	#	max	#	S	SK	EK
				ESZ19								
2	20191212 15:30:00	20191212 15:59:59	1799	2423	0.0024763	−2	4	2	4	0.48681	0.00622	1.929
1	20191212 17:00:00	20191213 15:14:59	80099	219795	−6.8245e-05	−9	1	8	4	0.4426	0.0422	16.57
2	20191213 15:30:00	20191213 15:59:59	1799	703	0.0014225	−1	92	2	1	0.51612	0.0644	1.003
1	20191215 17:00:00	20191216 15:14:59	80099	70900	0.0010296	−9	2	10	1	0.40998	0.227	24.35
2	20191216 15:30:00	20191216 15:59:58	1798	586	−0.0051195	−1	66	2	3	0.48036	0.264	2.519
1	20191216 17:00:00	20191217 15:14:59	80099	123471	−4.0495e-05	−3	9	8	1	0.30829	0.195	14.49
2	20191217 15:30:00	20191217 15:59:59	1799	937	−0.0032017	−2	2	2	4	0.45876	0.121	3.506
1	20191217 17:00:00	20191218 15:14:59	80099	43275	0.00050838	−3	2	3	3	0.34484	0.0824	7.057
2	20191218 15:30:00	20191218 15:59:58	1798	302	0.0033113	−2	1	2	1	0.52511	0.00405	1.702
1	20191218 17:00:00	20191219 15:14:59	80099	37436	0.0012555	−5	1	6	1	0.3479	0.0723	11.17
2	20191219 15:30:00	20191219 15:59:58	1798	231	0.030303	−1	35	2	4	0.60716	0.456	1.178
1	20191219 17:00:00	20191220 08:29:57	55797	4680	0.009188	−4	3	4	2	0.60922	0.166	4.528
	ESZ19, total 342 = 213 (1) + 129 (2) ranges, $\Delta t = 33548288$											
*	20181127 01:31:49	20191220 08:29:57	12858783	0.0001805	−242	1	280	1	0.50726	2.36	13000
	ESH19, ESM19, ESU19, ESZ19, total 1300 = 806 (1) + 494 (2) ranges, $\Delta t = 33548288$											
T	20171214 06:41:23	20191220 08:29:57	57944040	0.00013749	−242	1	280	1	0.51318	0.0694	6226

futures and review one contract, when it has sufficient liquidity measured by the volume and number of trade ticks. The choice fell on the ESM20 traded in the date range [03/12/2020, 04/27/2020]. The sessions, missed due to the technical *data mining* reasons 03/17/2020, 04/02/2020, or with incomplete data 03/12/2020, 03/18/2020, 03/25/2020, 03/26/2020, 04/22/2020, 04/23/2020, are excluded. We skip 03/13/2020, 03/16/2020, 03/23/2020, 03/24/2020, where price limits were set. They adds to b-increments extra zeros at the price limit and significant fluctuations after relaxing the limit, Figure A1. The rest is in Table A2.

If we combine these b-increments sizes in 18 groups-rows $r = 1, \ldots, 18$: $b = [-44, -9], [-8, -7], -6, -5, -4, -3, -2, -1, 0, 1, 2, 3, 4, 5, 6, 7, [8, 9], [10, 51]$ for each of the 20 dates-columns $c = 1, \ldots, 20$: $d = 03/19/2020$, 03/20/2020, 03/27/2020, 03/30/2020, 03/31/2020, 04/01/2020, 04/03/2020, 04/06/2020, 04/07/2020, 04/08/2020, 04/09/2020, 04/13/2020, 04/14/2020, 04/15/2020, 04/16/2020, 04/17/2020, 04/20/2020, 04/21/2020, 04/24/2020, 04/27/2020, then the frequency in each cell of the 18×20 table is ≥ 5. The aggregates $[-44, -9], [-8, -7], [8, 9], [10, 51]$ are in Table A2 together with their breaks down. The sum for each of the 20 columns is in the row "$N_{d,1}-1$". The sum for each of the 18 group rows is in the column "ALL DATES". The total sum in "$N_{d,1} - 1$' and "ALL DATES" is the total number of frequencies in all cells $N^* = 11893137$.

The empirical fraction of a b-increment size to occur in the negative extreme group $[-44, -9]$ is $f([-44, -9]) = \frac{N_{\text{ALL DATES}}([-44,-9])}{N^*} = \frac{852}{11893137} \approx 7.16 \times 10^{-5}$. The empirical fraction to get this on 03/19/2020 is $f(03/19/2020)) = \frac{N_{N_{d,1}-1}(03/19/2020))}{N^*} = \frac{1112600}{11893137} \approx 0.0935$. Would these events be independent, then the fraction to be in the cell should be close to $f([-44, -9] \cap 03/19/2020) = f([-44, -9]) \times f(03/19/2020) \approx 6.70 \times 10^{-6}$. Then, the expected frequency is $E([-44, -9], 03/19/2020) = 6.70 \times 10^{-6} \times 11893137 \approx 79.70$ but the observed is $O([-44, -9], 03/19/2020) = 140$. The famous cell quantity is $\chi^2(1,1) = \chi^2([-44, -9], 03/19/2020)) = \frac{(O([-44,-9],03/19/2020)-E([-44,-9],03/19/2020))^2}{E([-44,-9],03/19/2020)} \approx 45.61$

$$\chi^2(b,d) = \frac{(O(b,d) \times N^* - N_{\text{ALL DATES}}(b) \times N_{N_{d,1}-1}(d))^2}{N_{\text{ALL DATES}}(b) \times N_{N_{d,1}-1}(d) \times N^*}.$$

The $\chi^2(r,c) = \chi^2(b,d)$ was computed for all $18 \times 20 = 360$ cells. Their sum is

$$\chi^2 = \sum_{r=1}^{r=18} \sum_{c=1}^{c=20} \chi^2(r,c) \approx 367819.94.$$

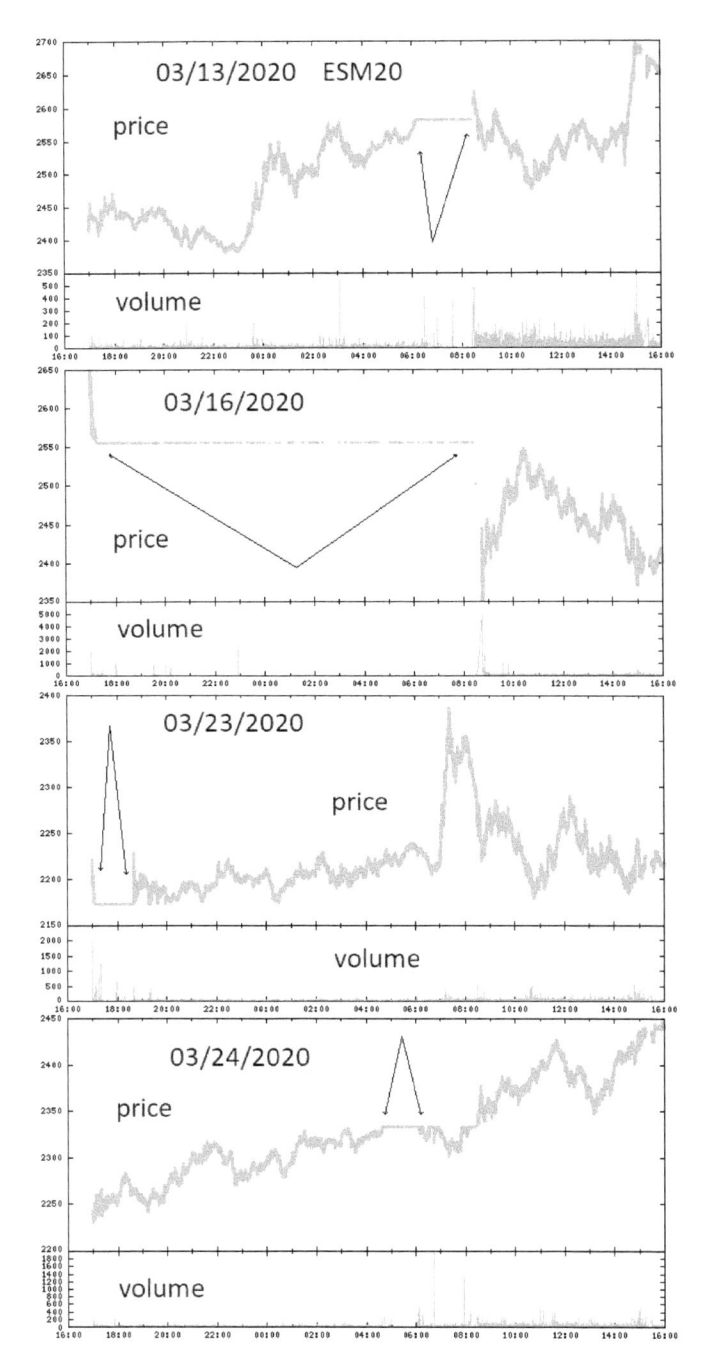

Figure A1. The CME Group Time & Sales trade tick times, prices, and volumes of the E-mini S&P 500 June M, 2020 futures for the sessions with price limits.

Table A2. Distributions of b-increments by size expressed in $\delta = 0.25 = \$12.50$ for the 20 ESM20 sessions for the first time ranges $[t_b^{d,1} = d-1\ 17{:}00{:}00,\ t_e^{d,1} = d\ 15{:}15{:}00]$, CST. The column ALL DATES summarizes 20 data samples in one. For all columns, the mode is zero.

1	2	3	4	5	6	7	8
b	03/19/2020	03/20/2020	03/27/2020	03/30/2020	03/31/2020	04/01/2020	ALL DATES
[−44,−9]	140	83	52	162	13	15	852
−44				2			2
−43				2			2
−36				1			1
−35							1
−34				1			1
−32				1			1
−31				3			3
−30							1
−27							1
−26				1			1
−25							3
−24				1			4
−23							2
−22				1			5
−21				3			6
−20	1			3	1		8
−19				2			6
−18				2			8
−17		1		4	1		19
−16	3	2		5			26
−15	2		2	5			20
−14	6	2		7	2	2	46
−13	8	4	6	12			49

(Continued)

Table A2. (Continued)

1	2	3	4	5	6	7	8
−12	8	4	5	13		2	85
−11	18	14	7	18	1	2	108
−10	40	21	7	30	1	3	171
−9	54	35	25	45	7	6	272
[−8,−7]	246	148	116	138	11	58	1139
−8	90	53	45	51	3	23	402
−7	156	95	71	87	8	35	737
−6	308	169	120	148	22	56	1258
−5	803	339	246	276	45	155	2773
−4	2341	827	523	599	127	282	6772
−3	7462	2790	1884	1874	415	809	21176
−2	28850	19410	11009	9436	2961	4974	111613
−1	197920	188940	112183	90471	30617	79364	1550611
0	636570	674888	517782	439337	131263	465484	8501630
1	197780	189024	111812	89556	31114	79973	1549421
2	28876	19208	10870	9769	2798	4794	111983
3	7372	2867	1889	1919	378	725	20952
4	2395	812	550	658	101	229	6852
5	837	325	225	283	35	96	2745
6	342	154	130	153	13	64	1290
7	162	100	75	101	11	37	751
[8,9]	126	88	63	106	12	25	682
8	83	55	37	65	5	16	430
9	43	33	26	41	7	9	252
[10,51]	70	50	42	136	8	10	637
10	26	24	10	34	1	3	182
11	19	11	9	18	1		100
12	10	5	14	16	3	1	103
13	8	3	3	16	1	2	59

	04/03/2020	04/06/2020	04/07/2020	04/08/2020	04/09/2020	04/13/2020	04/14/2020
14	3	3	3	11	2	3	52
15	2	2	1	5		1	27
16	1			7			33
17	1			3			17
18		2		1			9
19			2	2			6
20				7			13
21				4			9
22				2			6
23				1			2
24				1			3
25							4
26							2
28				1			1
29							1
31				2			2
34							1
44				2			2
45				1			1
46				1			1
51				1			1
a_1	−0.000056600	−0.000228136	−0.000537962	0.000950208	−0.000495139	−0.000715687	0.000015051
$N_{d,1}-1$	1112600	1100222	769571	645122	199944	637150	11893137
min	−20	−17	−15	−44	−20	−14	−44
max	17	17	18	51	14	15	51
S	0.93	0.78	0.73	0.82	0.72	0.62	0.66
S^2	0.86	0.61	0.53	0.67	0.52	0.38	0.44
SK	−0.029	0.002	0.099	0.355	−0.133	−0.042	0.068
EK	10.15	11.45	16.83	208.72	16.99	16.36	52.93
$Q5(-1)$	1.4241	1.1732	1.3150	1.3984	1.1888	1.5251	1.5831
$Q5(1)$	1.4093	1.2021	1.3319	1.3615	1.2033	1.4899	1.5675
b	04/03/2020	04/06/2020	04/07/2020	04/08/2020	04/09/2020	04/13/2020	04/14/2020
[−44,−9]	17	34	51	31	63	68	11

(Continued)

Table A2. (Continued)

1	2	3	4	5	6	7	8
−30						1	
−27		1					
−25				1			
−24						1	
−23		1			1		
−22		1			1		
−21			1		1	1	
−20					2	1	
−19				1	1	1	
−18		2	2		2		
−17		1	3	1	1		
−16		1	2		3	1	1
−15			2	1	6	5	1
−14		5	2	3	2	1	
−13		1	1	1	7	4	
−12	3	4	2	1	2	7	2
−11	3	4	9	3	10	11	
−10	5	3	6	5	4	7	1
−9	6	10	10	14	8	16	3
[−8, −7]	29	38	13	46	10	11	3
−8	10	7	44	15	56	68	19
−7	19	31	15	31	22	20	3
−6	30	36	29	39	34	48	16
−5	93	74	44	93	73	66	13
−4	205	158	105	208	111	134	37
−3	594	446	237	585	251	288	111
−2	3529	2811	622	3695	783	789	303
−1	68638	62369	3855	72278	4055	3554	1674
			91880		87169	56009	50120

0	397163	393659	582477	445815	563286	393698	382542
1	68123	63112	91883	72381	86634	55772	50301
2	3680	2767	3789	3956	4077	3741	1689
3	611	461	626	510	759	858	293
4	204	146	245	196	269	286	105
5	83	68	119	96	129	110	53
6	40	38	45	43	81	43	11
7	13	18	20	25	48	36	14
[8,9]	18	26	32	19	54	28	12
8	13	16	20	14	34	19	7
9	5	10	12	5	20	9	5
[10,51]	15	26	38	26	74	44	12
10	6	4	5	12	20	10	2
11	2	5	7	1	7	7	1
12	3	4	9	1	12	7	1
13	1	3	2	2	2	7	2
14	3	1	2	3	9	4	1
15		2	2	1	2	2	2
16		2	4	2	8		2
17			4		2	4	
18			1	1	2		
19		1			1		
20		1		1	3	1	
21				1	3	1	
22		2			1		
24			1				
25					2		1
26			1	1		1	
34		1					

(Continued)

Table A2. (Continued)

1	2	3	4	5	6	7	8
a_1	−0.000237532	0.001216067	−0.000007730	0.000628289	0.000225944	−0.000397601	0.000681277
$N_{d,1}$-1	543085	526287	776112	600042	747972	515592	487320
min	−12	−27	−21	−25	−25	−30	−15
max	14	34	26	26	25	25	22
S	0.60	0.59	0.58	0.60	0.61	0.63	0.52
S^2	0.36	0.35	0.34	0.36	0.37	0.39	0.27
SK	0.092	0.169	0.192	0.150	0.331	−0.491	0.388
EK	15.86	81.16	49.06	41.71	88.82	75.74	33.78
$Q5(-1)$	1.6656	1.6836	1.7384	1.6133	1.8655	1.8321	1.9887
$Q5(1)$	1.6253	1.7449	1.7708	1.4189	1.8180	1.8349	1.9375
b	04/15/2020	04/16/2020	04/17/2020	04/20/2020	04/21/2020	04/24/2020	04/27/2020
[−44,−9]	18	26	26	6	13	13	10
−35							1
−25				1			
−24							1
−18							1
−17	1	1	1			1	
−16	2	1					
−15	2		1				1
−14	2	2	1				
−13	1	2			1	1	
−12	3	4	2	1	3	1	1
−11	1	5	7	1	4	2	
−10	2	4	3		5	3	2
−9	4	7	11	3		5	3
[−8,−7]	22	29	27	7	10	13	14
−8	7	11	9	4	6	2	6
−7	15	18	18	3	4	11	8

−6	18	24	33	11	19	12	17
−5	34	48	63	23	32	31	31
−4	92	102	165	52	85	64	55
−3	255	355	456	136	230	229	159
−2	1644	2058	2677	1025	1803	1431	1162
−1	54736	56333	60857	48059	70463	39108	33097
0	388442	404461	396576	343317	441072	267802	235996
1	54385	56563	60774	47702	70074	39141	33317
2	1605	2060	2720	1073	1889	1528	1094
3	252	278	438	127	220	211	158
4	95	138	159	49	84	76	55
5	49	51	67	26	31	33	29
6	31	20	28	12	9	11	22
7	18	15	20	6	10	13	9
[8,9]	12	12	17	6	9	8	9
8	10	10	8	3	4	4	7
9	2	2	9	3	5	4	2
[10,51]	13	20	23	6	8	5	11
10	4	6	6	2	2	3	2
11	3	5	2	1	2		
12	3	2	4	1	4	1	2
13		3	2				1
14		1	3				
15	2						
16	1	2	4			1	2
17							
18			2				1
19		1					
23				1			1
24				1			1
25							
29							1

(Continued)

Table A2. (Continued)

1	2	3	4	5	6	7	8
a_1	−0.000554093	0.000147342	0.000045700	−0.000498140	−0.000508479	0.000594746	0.000517617
$N_{d,1} - 1$	501721	522593	525126	441643	586061	349729	305245
min	−17	−17	−17	−25	−13	−17	−35
max	16	19	18	25	12	15	29
S	0.53	0.54	0.57	0.51	0.54	0.54	0.55
S^2	0.28	0.29	0.33	0.26	0.29	0.29	0.30
SK	−0.088	−0.015	0.272	0.303	−0.039	−0.090	−0.130
EK	31.86	34.58	31.87	49.33	13.30	20.53	137.26
$Q5(-1)$	1.8810	1.8832	1.7649	1.9050	1.7802	1.8053	1.6839
$Q5(1)$	1.9028	1.6540	1.7011	1.7781	1.6806	1.6381	1.7655

Since the complete initial data are in Table A2 and the procedure is described, one can repeat the computation and let know to the author, if [s]he disagrees with 367819.94. For the verification, three intermediate $\chi^2(r,c) = \chi^2(b,d)$ are

$$\chi^2(6,9) = \chi^2(-3, 04/07/2020) = \frac{(622 \times 11893137 - 21176 \times 776112)^2}{21176 \times 776112 \times 11893137} \approx 417.85,$$

$$\chi^2(9,6) = \chi^2(0, 04/01/2020) = \frac{(465484 \times 11893137 - 8501630 \times 637150)^2}{8501630 \times 637150 \times 11893137} \approx 220.74,$$

$$\chi^2(11,16) = \chi^2(2, 04/17/2020) = \frac{(2720 \times 11893137 - 111983 \times 525126)^2}{111983 \times 525126 \times 11893137} \approx 1000.76.$$

The author believes that under such circumstances, skipping intermediate quantities for briefness does not compromise the trust to the final sum. It is worth noticing that the least expected $E(18,5) = E([10,51], 03/31/2020) \approx 10.71$ and observed $O(18,19) = O([10,51], 04/24/2020) = 5$ are not less than five. Controversial details of the χ^2 techniques, a few original research papers and textbooks on the topic are discussed and applied for checking non-Gaussian properties of the b-increments in [92, pp. 23–26].

Let us setup two hypothesis. H_0: In Table A2, the groups of b-increments sizes are independent on the observation dates. H_1: H_0 is false with the *level of significance* $\alpha = 0.05, 0.01, 0.001$. H_0 implies homogeneity of the distributions of b-increments sizes in different trading sessions. The number of degrees of freedom, d.f., for our case is $(r-1) \times (c-1) = 17 \times 19 = 323$. The Microsoft Excel's CHIINV(0.05, 323) = 365.91, CHIINV(0.01, 323) = 385.05, CHIINV(0.001, 323) = 407.27 < 367819.94. *We have to reject H_0. For the considered 20 sessions, the observed b-increments sizes distributions differ for different dates.*

Appendix C: Two Hours Distributions of b-Increments

Do distributions of b-increments change within one session or range? Figure A2 breaks down the ESM20 trading range [04/29/2020 17:00:00, 04/30/2020 15:00:00] on 11 two hours intervals.

Combining b-increments sizes in seven groups-rows $r = 1,\ldots,7$: [−17, −3], −2, −1, 0, 1, 2, [3, 14] for each of the 11 two hours intervals-columns $c = 1,\ldots,11$: 17-19 = [04/29/2020 17:00:00, 19:00:00), 19-21 = [19:00:00, 21:00:00), 21-23 = [21:00:00, 23:00:00), 23-01 = [04/29/2020 23:00:00, 04/30/2020 01:00:00), 01-03 = [01:00:00, 03:00:00), 03-05 = [03:00:00, 05:00:00), 05-07 = [05:00:00, 07:00:00), 07-09 = [07:00:00, 09:00:00), 09-11 = [09:00:00, 11:00:00), 11-13 = [11:00:00. 13:00:00), 13-15 = [13:00:00, 04/30/2020 15:00:00) ensures that the least expected frequency

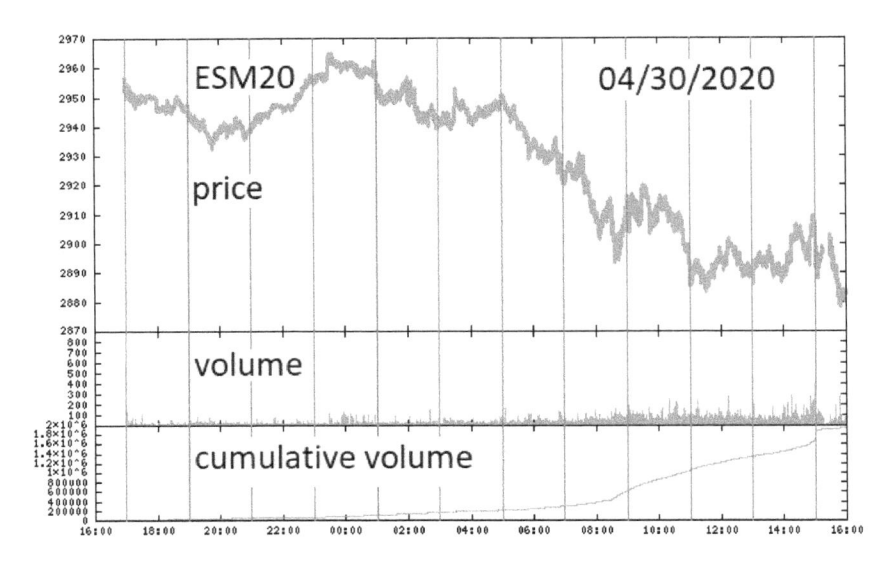

Figure A2. The CME Group Time & Sales trade tick times, prices, and volumes of the E-mini S&P 500 June M, 2020 futures in two hours intervals.

$E(7,1) = E([3,14], 17-19) \approx 5.45$ is >5. The χ^2 formulas are in Appendix B. The number of degrees of freedom is $(7-1) \times (11-1) = 60$ and

$$\chi^2 = \sum_{r=1}^{r=7} \sum_{c=1}^{c=11} \chi^2(r,c) \approx 3774.03.$$

Microsoft Excel's CHIINV($\alpha = 0.05, d.f. = 60$) ≈ 79.08, CHIINV(0.01, 60) ≈ 88.38, CHIINV(0.001, 60) $\approx 99.61 < 3774.03$. H_0: In Table A3, the groups of b-increments are independent on the two hours intervals has to be rejected.

Appendix D: Relationships between Sample Moments of Values and Their Increments

While statistics (A1) are symmetric functions of input, the order of values and increments matters, if both are considered together. Formulas in the following illustrate this. Expressions can be simplified using vectors, matrices, and weighted means.

Statistics with vectors and matrices: The chains $x_1, \ldots, x_i, \ldots, x_n$ and $\Delta x_2, \ldots, \Delta x_i = x_i - x_{i-1}, \ldots, \Delta x_n$ are the n- and $(n-1)$-dimensional vectors $\boldsymbol{x} = (x_1, \ldots, x_n)^T$ and $\boldsymbol{\Delta x} = (\Delta x_2, \ldots, \Delta x_n)^T$. Let $\boldsymbol{\Delta x_1} =$

Table A3. Distributions of b-increments by size expressed in $\delta = 0.25 = \$12.50$ for the 11 two hours CST intervals of the ESM20 Thursday April 30, 2020 session. The column ALL HOURS summarizes 11 data samples in one. For all columns, the mode is zero.

1	2	3	4	5	6	7
b	17–19	19–21	21–23	23–01	01–03	ALL HOURS
[−17, −3]	20	15	21	22	20	320
−17						1
−16						1
−13						1
−12						1
−10						1
−9			2	1		4
−7			1	0		8
−6			0	1		18
−5			2	1	2	18
−4	5	7	6	4	4	59
−3	15	8	10	15	14	208
−2	82	80	45	59	102	1425
−1	1449	1724	1043	1475	2820	48099
0	5659	8621	5357	9517	16445	364381
1	1403	1731	1081	1462	2736	47923
2	91	72	63	55	126	1468
[3, 14]	15	9	22	23	14	290

(Continued)

Table A3. (Continued)

1	2	3	4	5	6	7
3	10	5	13	10	11	194
4	4	3	5	6	3	53
5	1	0	1	2		13
6		1	0	3		11
7			2	1		10
8			0	0		1
9			1	1		4
10						1
11						1
12						1
14						1
a_1	−0.004817066	−0.002285341	0.009433962	−0.000317133	−0.002650137	−0.000478545
N	8719	12252	7632	12613	22263	463906
min	−4	−4	−9	−9	−5	−17
max	5	6	9	9	4	14
S	0.67	0.60	0.67	0.58	0.55	0.50
S^2	0.45	0.36	0.45	0.33	0.31	0.25
SK	0.003	−0.057	−0.180	0.351	−0.061	−0.143
EK	2.95	3.91	21.59	19.75	3.76	18.76
b	03–05	05–07	07–09	09–11	11–13	13–15
[−17, −3]	10	31	80	32	15	54
−17						1
−16						1
−13						1
−12						1
−10						1
−9						0
−7		1	7			4
−6		2	2		1	7
−5		1			1	8
−4	1	3	17		1	6
−3	9	23	54	24	12	24

−2	64	111	347	221	128	186
−1	1740	3205	10245	9954	6437	8007
0	11604	23559	71853	85500	54081	72185
1	1767	3124	10358	9885	6385	7991
2	60	120	319	213	134	215
[3,14]	11	21	56	34	19	66
3	7	17	48	20	11	42
4	2	2	7	8	5	8
5	2	1	0	1	0	5
6		0	1	3	1	2
7		0		2	1	4
8		1			0	0
9					1	1
10						1
11						1
12						1
14						1
a_1	0.001769795	−0.003513307	−0.000503978	−0.000623589	−0.000238099	0.000529852
N	15256	30171	93258	105839	67199	88704
min	−4	−9	−6	−7	−6	−17
max	5	8	6	7	9	14
S	0.53	0.52	0.52	0.46	0.46	0.49
S^2	0.28	0.27	0.27	0.21	0.21	0.24
SK	0.104	−0.307	−0.188	0.074	0.177	−0.688
EK	4.45	12.25	5.61	6.87	7.56	70.81

$(x_1, \Delta x_2, \ldots, \Delta x_n)^T$, $\boldsymbol{x}_{\rightarrow} = (0, x_1, \ldots, x_{n-1})^T$ are the vectors of size n. Let $\mathcal{V}$ is the square $n \times n$ shift (βάρδια [vardia]) matrix: all zeros except the subdiagonal $e_{i,i-1} = 1$, $\forall\, i = 2, \ldots, n$. Then, $\boldsymbol{\Delta x_1} = \boldsymbol{x} - \boldsymbol{x}_{\rightarrow} = \boldsymbol{x} - \mathcal{V}\boldsymbol{x} = (\mathcal{I} - \mathcal{V})\boldsymbol{x}$, where $\mathcal{I}$ is the square $n \times n$ identity matrix: all zeros except the major diagonal $e_{i,i} = 1$, $\forall\, i = 1, \ldots, n$. Let $\mathcal{A}$ is the rectangular $(n-1) \times n$ resize (αλλαγή μεγέθουσ [allagi megethous]) matrix: all zeros except the superdiagonal $e_{i,i+1} = 1$, $\forall\, i = 1, \ldots, n-1$. Then, $\boldsymbol{\Delta x} = \mathcal{A}\boldsymbol{\Delta x_1} = \mathcal{A}(\mathcal{I} - \mathcal{V})\boldsymbol{x} = \mathcal{D}\boldsymbol{x}$, where $\mathcal{D} = \mathcal{A}(\mathcal{I} - \mathcal{V})$ is the $(n-1) \times n$ rectangular matrix: all zeros except $e_{i,i+1} = 1$, $\forall\, i = 1, \ldots, n-1$; $e_{i,i} = -1$, $\forall\, i = 1, \ldots, n-1$. We see that $\mathcal{M} = \mathcal{D}^T\mathcal{D}$ is the square $n \times n$ matrix: all elements are zeros except $e_{1,1} = e_{n,n} = 1$; $e_{i,i+1} = -1$, $\forall\, i = 1, \ldots, n-1$; $e_{i,i-1} = -1$, $\forall\, i = 2, \ldots, n$; $e_{i,i} = 2$, $\forall\, i = 2, \ldots, n-1$. Let the vector $\boldsymbol{1_n} = (1, \ldots, 1_n)^T$ has all unit coordinates, see on *Hadamard identity* in [88, pp. 26–27], and the vector $\boldsymbol{n_2} = (2, 3, \ldots, n)^T$ has $n-1$ coordinates starting from two and increasing by one. Then,

$$a_1^x = \frac{\sum_{i=1}^{i=n} x_i}{n} = \frac{\boldsymbol{1}_n^T \boldsymbol{x}}{n};\ a_1^{\Delta x} = \frac{\sum_{i=2}^{i=n} \Delta x_i}{n-1} = \frac{\boldsymbol{1}_{n-1}^T \boldsymbol{\Delta x}}{n-1} = \frac{x_n - x_1}{n-1},$$

$$\boldsymbol{\Delta x}^T \boldsymbol{\Delta x} = \boldsymbol{x}^T (\mathcal{I}^T - \mathcal{V}^T) \mathcal{A}^T \mathcal{A} (\mathcal{I} - \mathcal{V}) \boldsymbol{x} = \boldsymbol{x}^T \mathcal{D}^T \mathcal{D} \boldsymbol{x} = \boldsymbol{x}^T \mathcal{M} \boldsymbol{x},$$

$$\boldsymbol{n_2}^T \mathcal{D} = (-2, [-1,]_{n-2}, n);\ \frac{\boldsymbol{n_2}^T \mathcal{D}\boldsymbol{x}}{n} = \frac{-2x_1 - x_2 - \cdots - x_{n-1} + nx_n}{n}$$

$$= x_n - a_1^x + \frac{x_n - x_1}{n} = x_n - a_1^x + \frac{n-1}{n} a_1^{\Delta x}.$$

Since $x_i = x_1 + \sum_{j=2}^{j=i} \Delta x_j$,

$x_1 = x_1$,

$x_2 = x_1 + \Delta x_2$,

$\ldots$,

$x_n = x_1 + \Delta x_2 + \cdots + \Delta x_n$,

$$\sum_{i=1}^{i=n} x_i = nx_1 + (n-1)\Delta x_2 + \cdots + \Delta x_n = nx_1 + \sum_{i=2}^{i=n} (n-i+1)\Delta x_i,$$

and

$$a_1^x = x_1 + \frac{n^2 - 1}{n} a_1^{\Delta x} - \frac{\sum_{i=2}^{i=n} i \Delta x_i}{n} = x_1 + \frac{n^2 - 1}{n} a_1^{\Delta x} - \frac{\boldsymbol{n_2}^T \boldsymbol{\Delta x}}{n},$$

$$a_1^{\Delta x} = \frac{n}{n^2 - 1} \left(a_1^x - x_1 + \frac{\boldsymbol{n_2}^T \mathcal{D}\boldsymbol{x}}{n} \right) = \frac{x_n - x_1}{n-1}. \tag{A2}$$

Equation (A2) express the sample mean of values by the sample of increments and their sample mean and vice versa except that x_1 is in both right sides. The scalar products with the $\mathbf{n}_2^T$ indicate that *as a pair* both can be sensitive to the order of values and/or increments. *If x_1, x_n are fixed, then permutations of intermediate x_i do not affect a_1^x, $a_1^{\Delta x}$, and $\frac{\sum_{i=2}^{i=n} i\Delta x_i}{n}$.*

$\sum_{i=1}^{i=n} ix_i$, $\sum_{i=2}^{i=n} i\Delta x_i$, **and some weighted average:** Let us introduce

$$a_s^{*x} = \frac{\sum_{i=1}^{i=n} ix_i^s}{\sum_{i=1}^{i=n} i} = \frac{2\sum_{i=1}^{i=n} ix_i^s}{n(n+1)}; \quad a_s^{*\Delta x} = \frac{\sum_{i=2}^{i=n} i\Delta x_i^s}{\sum_{i=2}^{i=n} i} = \frac{2\sum_{i=2}^{i=n} i\Delta x_i^s}{(n-1)(n+2)}. \tag{A3}$$

The sum in the enumerator for a_1^{*x} coincides with the *static moment of masses about a point*, where i is a distance between the point and location of mass x_i on a *straight line*, Figure A3. For masses $0 < x_i$ on a straight line, the formula for the vector of the *center of inertia* [45, p. 28, equation (8.3)] transforms with our denominations, for $0 \le i$, into the scalar $a_1^{*i} = \frac{\sum_{i=1}^{i=n} ix_i}{\sum_{i=1}^{i=n} x_i}$.

If the weights $\frac{i}{\sum_{i=1}^{i=n} i} = \frac{2i}{n(n+1)}$, $\frac{i}{\sum_{i=2}^{i=n} i} = \frac{2i}{(n-1)(n+2)}$ and x_i^s, Δx_i^s are permuted simultaneously in equation (A3), then the weighted averages a_s^{*x}, $a_s^{*\Delta x}$ remain intact. In [12, p. 63], this property is named *almost symmetry*. It is not applicable, if x_i and Δx_i are considered together. Then,

$$\sum_{i=2}^{i=n} i\Delta x_i = \sum_{i=2}^{i=n} ix_i - \sum_{i=2}^{i=n} ix_{i-1} = \frac{n(n+1)}{2} a_1^{*x} - x_1 - \sum_{i=2}^{i=n} ix_{i-1}$$

$$= (n+1)x_n - x_1 - na_1^x,$$

$$a_1^x = x_1 + \frac{n^2-1}{n} a_1^{\Delta x} - \frac{(n-1)(n+2)}{2n} a_1^{*\Delta x}. \tag{A4}$$

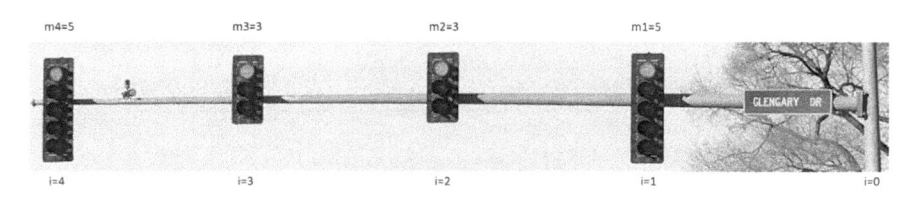

Figure A3. A static moment of masses about the pole, averages: $\sum_{i=1}^{i=4} im_i = 1 \times 5 + 2 \times 3 + 3 \times 3 + 4 \times 5 = 40$, $\sum_{i=1}^{i=4} i = 1 + 2 + 3 + 4 = 10$, $\sum_{i=1}^{i=4} m_i = 5 + 3 + 3 + 5 = 16$, $a_1^{*m} = \frac{40}{10} = 4$, $a_1^{*i} = \frac{40}{16} = 2.5$, $a_1^m = \frac{16}{4} = 4$, $a_1^i = \frac{10}{4} = 2.5$.

Other sample moments: From (A1) and Newton's binomial theorem

$$m_s = \frac{\sum_{i=1}^{i=n}(x_i - a_1)^s}{n} = \frac{\sum_{i=1}^{i=n}\sum_{j=0}^{j=s}\frac{s!}{j!(s-j)!}x_i^{s-j}(-1)^j a_1^j}{n}$$

$$= \sum_{j=0}^{j=s}\frac{s!}{j!(s-j)!}(-1)^j a_1^j \frac{\sum_{i=1}^{i=n}x_i^{s-j}}{n} = \sum_{j=0}^{j=s}\frac{s!}{j!(s-j)!}(-1)^j a_1^j a_{s-j}. \quad (A5)$$

For instance,

$$m_0 = 1,$$
$$m_1 = 0,$$
$$m_2 = a_2 - a_1^2,$$
$$m_3 = a_3 - 3a_1 a_2 + 2a_1^3,$$
$$m_4 = a_4 - 4a_1 a_3 + 6a_1^2 a_2 - 3a_1^4. \quad (A6)$$

Formulas (A6) are similar to those for the theoretical moments α_s, μ_s [26, p. 68, equations (9)]. Both types of formulas allow inversion:

$$a_0 = 1,$$
$$a_1 = 0,$$
$$a_2 = m_2 + a_1^2,$$
$$a_3 = m_3 + 3a_1 m_2 + a_1^3,$$
$$a_4 = m_4 + 4a_1 m_3 + 6a_1^2 m_2 + a_1^4. \quad (A7)$$

Opening the brackets below and using (A3), (A4) yields

$$a_2^x = \frac{\boldsymbol{x}^T \boldsymbol{x}}{n} = \frac{\sum_{i=1}^{i=n} x_i^2}{n} = \frac{\sum_{i=1}^{i=n}(x_1 + \sum_{j=2}^{j=i}\Delta x_j)^2}{n} = \frac{\sum_{i=1}^{i=n} x_1^2}{n}$$

$$+ \frac{2x_1 \sum_{i=1}^{i=n}\sum_{j=2}^{j=i}\Delta x_j}{n} + \frac{\sum_{i=1}^{i=n}(\sum_{j=2}^{j=i}\Delta x_j)^2}{n} = x_1^2 + \frac{2(n^2-1)x_1 a_1^{\Delta x}}{n}$$

$$- \frac{(n-1)(n+2)x_1 a_1^{*\Delta x}}{n} + \frac{(n^2-1)a_2^{\Delta x}}{n} - \frac{(n-1)(n+2)a_2^{*\Delta x}}{2n}$$

$$+ \frac{2}{n}\sum_{i=3}^{i=n}(n-i+1)\Delta x_i \sum_{j=2}^{j=i-1}\Delta x_j. \quad (A8)$$

From (A4), (A6), (A8), and $m_2^{\Delta x} = a_2^{\Delta x} - (a_1^{\Delta x})^2$,

$$m_2^x = a_2^x - (a_1^x)^2 = \frac{n^2-1}{n}m_2^{\Delta x} - \frac{(n^2-1)(n^2-n-1)(a_1^{\Delta x})^2}{n^2}$$

$$- \frac{(n-1)(n+2)a_2^{*\Delta x}}{2n} + \frac{(n^2-1)(n-1)(n+2)a_1^{\Delta x}a_1^{*\Delta x}}{n^2}$$

$$- \left[\frac{(n-1)(n+2)a_1^{*\Delta x}}{2n}\right]^2 + \frac{2}{n}\sum_{i=3}^{i=n}(n-i+1)\Delta x_i \sum_{j=2}^{j=i-1}\Delta x_j. \quad (A9)$$

Geometry of $\sum_{i=1}^{i=n}(\sum_{j=2}^{j=i}\Delta x_j)^2$: This sum arises in equation (A8). The inner is the square of a sum described by a *known formula* $(\sum_{i=1}^{i=n}x_i)^2 = \sum_{i=1}^{i=n}x_i^2 + 2\sum_{i=1}^{i=n-1}\sum_{j=i+1}^{j=n}x_ix_j$, which can be viewed as summation of elements of a square symmetrical matrix with squares on the diagonal. Since Δx_1 is undefined and excluded from summation, the outer sum adds $n-1$ matrices with increasing dimension. Figure A4 shows that the summed

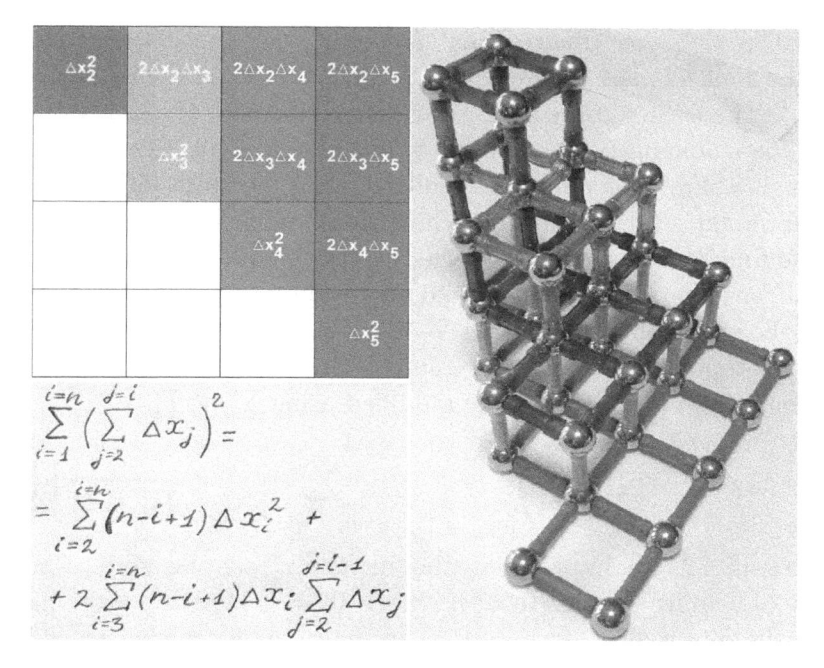

Figure A4. A geometrical illustration to $\sum_{i=1}^{i=n}(\sum_{j=2}^{j=i}\Delta x_j)^2$ for $n=5$.

matrices padded by zeros, white blank squares, can be interpreted as triangular matrices aligned on the left-top corner. Vertical elements of the overlaid matrices coincide and we can simplify with counters regularly changing in rows and columns. Thus,

$$\sum_{i=1}^{i=n}\left(\sum_{j=2}^{j=i}\Delta x_j\right)^2 = \sum_{i=2}^{i=n}(n-i+1)\Delta x_i^2 + 2\sum_{i=3}^{i=n}(n-i+1)\Delta x_i \sum_{j=2}^{j=i-1}\Delta x_j$$

$$= (n^2-1)a_2^{\Delta x} - \frac{(n-1)(n+2)a_2^{*\Delta x}}{2} + 2\sum_{i=3}^{i=n}(n-i+1)\Delta x_i \sum_{j=2}^{j=i-1}\Delta x_j.$$

A comment on the *multinomial formula*: While the known $(\sum_{i=1}^{i=n} x_i)^2 = \sum_{i=1}^{i=n} x_i^2 + 2\sum_{i=1}^{i=n-1}\sum_{j=i+1}^{j=n} x_i x_j$ can be obtained by opening the brackets and grouping summands, we show that it follows from the multinomial formula or *multinomial theorem* [78, p. 3, equation (4), "expansion of $(a+b+c+\cdots\cdots)^n$"], [72, p. 40, equation (1.18)], [103, p. 1058, equation (2)], [5, pp. 33–35, The Multinomial Formula in a Commutative Ring], [10, p. 405, "the multinomial theorem"], [79, p. 292, Exercises, *49 "Prove the multinomial theorem"], [11, pp. 143–146, 5.4 The Multinomial Theorem]. The earlier contributions of Mersenne 1635, Wallis 1685, Leibniz May 16, 1695, Johann Bernoulli June 8, 1695, De Moivre 1698 are mentioned in [19, p. 179, From binomials to multinomials]. See also [56], [39, p. 158]. Leibniz formulates the problem switching from binomials of degrees one, two, three, and four to the trinomial $(x+y+z)^m$ [48, pp. 46–47]. In the letter-reply, Bernoulli considers $(s+x+y+z)^r$ and for the generic product-term $s^a x^b y^c z^e$ presents the multinomial coefficient $\frac{r.r-1.r-2.r-3.r-4.....a+1}{1.2.3...b \times 1.2.3...c \times 1.2.3...e}$ [6, p. 55]. He could not use "!" proposed by Christian Kramp of Strasburg only in 1808 [84, p. 2]. The formula from [5, p. 33] can be modified using $\sum, \forall, \prod$

$$(a_1 + \cdots + a_p)^n = \left(\sum_{i=1}^{i=p} a_i\right)^n = \sum_{\sum_{i=1}^{i=p} n_i = n; \forall\ 1\leq i \leq p:\ 0\leq n_i} \frac{n!}{\prod_{i=1}^{i=p} n_i!} \prod_{i=1}^{i=p} a_i^{n_i}.$$

References [5, 72, 78] imply that the integer n can be zero. References [11, 79, 103] require integer $0 < n$. Since $0! = 1$ and $x^0 = 1$, including $0^0 = 1$, the left is one for $n = 0$. The multinomial coefficient $\frac{0!}{0!...0!} = 1$ and $\prod_{i=1}^{i=p} a_i^0 = 1$ complete the identity. *There is no reason to exclude $n = 0$ from the multinomial formula.*

Neither of the referenced sources comments the case $p > n$. This may create an impression that the formula is only for $p \leq n$. Summation on the right goes for all solutions $(0 \leq n_1, \ldots, 0 \leq n_p)$ of $n_1 + \cdots + n_p = n$. If $p > n$, then in each solution at least one n_i is zero. For $n = 2$, $p = 3$, the six solutions are $2+0+0 = 2$, $0+2+0 = 2$, $0+0+2 = 2$, $1+1+0 = 2$, $1+0+1 = 2$, $0+1+1 = 2$. Then, by the formula $(x_1+x_2+x_3)^2 = \frac{2!}{2!0!0!}x_1^2 x_2^0 x_3^0 + \frac{2!}{0!2!0!}x_1^0 x_2^2 x_3^0 + \frac{2!}{0!0!2!}x_1^0 x_2^0 x_3^2 + \frac{2!}{1!1!0!}x_1^1 x_2^1 x_3^0 + \frac{2!}{1!0!1!}x_1^1 x_2^0 x_3^1 + \frac{2!}{0!1!1!}x_1^0 x_2^1 x_3^1 = x_1^2 + x_2^2 + x_3^2 + 2x_1 x_2 + 2x_1 x_3 + 2x_2 x_3$. But expansion of $(x_1 + x_2 + x_3)^2$ yields the same. $(\sum_{i=1}^{i=n} x_i)^2 = \sum_{i=1}^{i=n} x_i^2 + 2\sum_{i=1}^{i=n-1}\sum_{j=i+1}^{j=n} x_i x_j$ *follows from the multinomial formula as a particular case.*

Example for checking: Let $x = (2, 7, 3, 1, 2, 1)^T$, $\Delta x = (5, -4, -2, 1, -1)^T$. Then, $n = 6$; $a_1^x = \frac{8}{3}$, $a_2^x = \frac{34}{3}$, $a_3^x = \frac{194}{3}$, $a_4^x = \frac{1258}{3}$, $m_1^x = 0$, $m_2^x = \frac{38}{9}$, $m_3^x = \frac{322}{27}$, $m_4^x = \frac{1658}{27}$, $a_1^{*x} = 2.25$, $a_2^{*x} = 7.95$; $a_1^{\Delta x} = -0.2$, $a_2^{\Delta x} = 9.4$, $a_3^{\Delta x} = 10.6$, $a_4^{\Delta x} = 179.8$, $m_1^{\Delta x} = 0$, $m_2^{\Delta x} = 9.36$, $m_3^{\Delta x} = 16.224$, $m_4^{\Delta x} = 190.5312$, $a_1^{*\Delta x} = -0.55$, $a_2^{*\Delta x} = 6.25$, $2\sum_{i=3}^{i=n}(n-i+1)\Delta x_i \sum_{j=2}^{j=i-1} \Delta x_j = -176$.

Useful C++: A Standard C++ Library sequence container, std::vector<T> [101, pp. 902–906, 31.4.1 vector], [32, pp. 840–844, 22.13.11 Class template vector] is a type example, corresponds to an x of size n. The algorithm std::adjacent_difference [101, pp. 1179–1180, 40.6.3 partial_sum() and adjacent_difference()], [32, pp. 1132–1133, 25.10.11 Adjacent difference] is ready to evaluate Δx_1. The std::iota [101, p. 1180, 40.6.4 iota()], [32, p. 1133, 25.10.12 Iota] is good for initializing n_2. Summation and scalar products can be done using std::accumulate [101, pp. 1177–1178, 40.6.1 accumulate()], [32, p. 1125, 10.10.2 Accumulate], and std::inner_product [101, pp. 1178–1179, 40.6.2 inner_product()], [32, p. 1126, 25.10.4 Inner product]. Customization of these algorithms with *lambda expressions* [101, pp. 290–295, 11.4 Lambda Expressions], [32, pp. 98–99, 7.5.5 Lambda expressions] is handy for getting sample moments of the higher order.

Appendix E: Evaluation of $\sum_{g=1}^{g=\infty} \frac{g^s}{(g+Q)^S}$

Let $s \in \mathbb{N}$; $0 \leq x, 1 < S, 0 < Q \in \mathbb{R}$; $f(x) = f^{(0)}(x) = \frac{x^s}{(x+Q)^S} = x^s(x+Q)^{-S} = u(x)v(x)$; $u(x) = u^{(0)}(x) = x^s$, $v(x) = v^{(0)}(x) = (x+Q)^{-S}$. Then, for the $1 \leq k$th derivative, $u^{(k)}(x) = s(s-1)\cdots(s-k+1)x^{s-k}$, [23, p. 233, 1)], $v^{(k)}(x) = -S(-S-1)\cdots(-S-k+1)(x+Q)^{-S-k}$, [23, p. 234, 2)]. We generalize both for $0 \leq k$, setting $\prod_{\text{bottom}}^{\text{top}}(\cdots) = 1$ for top < bottom: $u^{(k)}(x) = x^{s-k} \prod_{j=0}^{j=k-1}(s-j)$, $v^{(k)}(x) = (x+Q)^{-S-k} \prod_{j=0}^{j=k-1}(-S-j)$. Using the Leibniz formula [23, pp. 236–238], $f^{(k)}(x) = \sum_{i=0}^{i=k} \frac{k!}{i!(k-i)!} u^{(k-i)}(x) v^{(i)}(x)$.

Thus,

$$f^{(k)}(x) = \sum_{i=0}^{i=k} \binom{k}{i} \frac{x^{s-k+i}}{(x+Q)^{S+i}} \left(\prod_{j=0}^{j=k-i-1} (s-j) \right) \left(\prod_{j=0}^{j=i-1} (-S-j) \right)$$

$$= \sum_{i=\max(0,k-s)}^{i=k} \binom{k}{i} \frac{x^{s-k+i}}{(x+Q)^{S+i}} \left(\prod_{j=0}^{j=k-i-1} (s-j) \right) \left(\prod_{j=0}^{j=i-1} (-S-j) \right).$$
(A10)

The $i = 0$ is replaced at the bottom of the sum with $i = \max(0, k - s)$ to exclude zero summands. Indeed, $\prod_{j=0}^{j=k-i-1}(s-j) = 0$ for $i \leq k - s - 1$. The simpler $i = k - s$ could create a chain of negative integers and a difficulty of using $\binom{k}{i}$. The difficulty can be resolved: There is one way not to choose objects from an empty set, $\binom{0}{0} = \frac{0!}{0!(0-0)!} = 1$, and no way to select a negative number of objects from any set, $\binom{0 \leq k}{i < 0} = 0$. But in our case, zeroing any factor in the product of a summand vanishes the latter and this suggests using $i = \max(0, k - s)$.

Example A1. From (A10), $f^{(0)}(x) = \frac{x^s}{(x+Q)^S}$, $f^{(1)}(x) = sx^{s-1}(x+Q)^{-S} - Sx^s(x+Q)^{-S-1} = f(x)(\frac{s}{x} - \frac{S}{x+Q})$, $f^{(2)} = f(x)(\frac{s(s-1)}{x^2} - \frac{2sS}{x(x+Q)} + \frac{S(S+1)}{(x+Q)^2})$.

While for great k the $f^{(k)}(x)$ expressions are "unmanageable on paper", formula (A10) is generic, compact, and easily programmable as an outer loop summing on each iteration a product of four factors, where the first is the binomial coefficient $C(k, i) = \binom{k}{i} = \frac{k!}{i!(k-i)!}$, the second is the fraction of powers $\frac{x^{s-k+i}}{(x+Q)^{S+i}}$, and the remaining two are the inner product loops not involving x and Q.

(A10) allows to apply to $f(x)$ the *Euler–Maclaurin formula* [23, pp. 540–542], named in [29, pp. 469–475] *Euler's summation formula*, in the form of [29, p. 474, equation (9.78)] skipping Bernoulli polynomials in the remainder:

$$\sum_{g=L}^{g=U} f(g) = \int_L^U f(x)dx - \frac{1}{2}f(x)|_L^U + \sum_{j=1}^{j=M} \frac{B_{2j}}{(2j)!} f^{(2j-1)}(x)|_L^U + R,$$

$$R = O\left(\frac{1}{(2\pi)^{2M}}\right) \times \int_L^U |f^{(2M)}(x)|dx, \text{ remainder,}$$

$$B_0 = 1, \; B_1 = -\frac{1}{2}, \; B_2 = \frac{1}{6}, \; B_4 = -\frac{1}{30}, \; B_6 = \frac{1}{42} \; B_8 = -\frac{1}{30}, \ldots,$$

$$B_3 = B_5 = B_7 = B_9 = B_{11} = \cdots = 0, \text{ Bernoulli numbers.} \quad \text{(A11)}$$

For a converging series, (A11) is valid with $U \to \infty$. Thus,

$$\sum_{g=1}^{g=\infty} \frac{g^s}{(g+Q)^S} = \sum_{g=1}^{g=L-1} \frac{g^s}{(g+Q)^S} + \frac{L^s}{2(L+Q)^S} + \int_L^\infty \frac{x^s}{(x+Q)^S} dx$$

$$- \sum_{j=1}^{j=M} \frac{B_{2j}}{(2j)!} f^{(2j-1)}(L) - \lim_{x \to \infty} \left(\frac{x^s}{2(x+Q)^S} - \sum_{j=1}^{j=M} \frac{B_{2j}}{(2j)!} f^{(2j-1)}(x) \right) + R. \quad (A12)$$

For $1 \leq s < S$, from (A10), the limit in (A12) is zero. Empirical S are greater than 4, 5. Let us evaluate $\int_L^\infty \frac{x^s}{(x+Q)^S} dx$ for $s = 1, 2, 3, 4$, where $s + 1 < S$.

$s = 1:$ $-\frac{Q + (S-1)x}{(S-2)(S-1)(x+Q)^{S-1}} \Big|_L^\infty = \frac{Q + (S-1)L}{(S-2)(S-1)(L+Q)^{S-1}}$,

$s = 2:$ $\frac{2Q^2 + 2Q(S-1)L + (S^2 - 3S + 2)L^2}{(S-3)(S-2)(S-1)(L+Q)^{S-1}}$,

$s = 3:$ $\frac{6Q^3 + 6Q^2(S-1)L + 3Q(S^2 - 3S + 2)L^2 + (S^3 - 6S^2 + 11S - 6)L^3}{(S-4)(S-3)(S-2)(S-1)(L+Q)^{S-1}}$,

$s = 4:$ $\frac{\frac{(L+Q)^4}{S-5} - \frac{4Q(L+Q)^3}{S-4} + \frac{6Q^2(L+Q)^2}{S-3} - \frac{4Q^3(L+Q)}{S-2} + \frac{Q^4}{S-1}}{(L+Q)^{S-1}}.$ (A13)

The Riemann and Hurwitz Zeta series converge slowly. Summing a few initial terms and applying the Euler–Maclaurin formula to the rest is an acceleration [20, pp. 114–118], [33]. This idea is extended here to moments. We sum the initial terms from lesser to greater. Analysis of R is in [23, pp. 542–544], [29, pp. 472–475]. Together with $f^{(k)}(x)$ in (A10) this is a key to selection of L, M. A few first factors in the R in (A11) are $(M, O((2\pi)^{-2M}))$: $(5, 10^{-8})$, $(10, 10^{-16})$, $(15, 10^{-24})$, $(20, 10^{-32})$.

Since $\binom{2M}{i} \leq \binom{2M}{M}$ and, for $0 < x$, $0 < \frac{x^{s+i-2M}}{(x+Q)^{S+i}} = \frac{x^{s-2M}}{(x+Q)^S}\left(\frac{x}{x+Q}\right)^i \leq \frac{x^{s-2M}}{(x+Q)^S} < \frac{x^{s-2M}}{x^S} = x^{s-2M-S}$, we can write for the second factor

$$\int_L^\infty |f^{(2M)}(x)| dx \leq \sum_{i=\max(0,2M-s)}^{i=2M} \binom{2M}{i} \prod_{j=0}^{j=2M-i-1} |s-j| \prod_{j=0}^{j=i-1} |S+j| \int_L^\infty \frac{x^{s+i-2M}}{(x+Q)^{S+i}} dx$$

$$\leq \binom{2M}{M} \int_L^\infty x^{s-2M-S} dx \sum_{i=\max(0,2M-s)}^{i=2M} \prod_{j=0}^{j=2M-i-1} |s-j| \prod_{j=0}^{j=i-1} |S+j|$$

$$= \binom{2M}{M} \times \frac{L^{s+1-S-2M}}{S+2M-s-1} \times \sum_{i=\max(0,2M-s)}^{i=2M} \prod_{j=0}^{j=2M-i-1} |s-j| \prod_{j=0}^{j=i-1} |S+j|.$$

Recollect that it is assumed that $s + 1 < S$. For $\max(0, 2M - s) \leq i \leq 2M$, $\prod_{j=0}^{j=2M-i-1} |s-j| \leq \prod_{j=0}^{j=s-1} |s-j| = s!$ and $\prod_{j=0}^{j=i-1} |S+j| < \prod_{j=0}^{j=2M-1}$

$(S+j) < \prod_{j=0}^{j=2M-1}(\lceil S \rceil + j) = \frac{(\lceil S \rceil + 2M - 1)!}{(\lceil S \rceil - 1)!}$. Also, $\sum_{i=\max(0,2M-s)}^{i=2M} 1 \leq s$. We get

$$R \leq O\left(\frac{1}{(2\pi)^{2M}}\right) \times \binom{2M}{M} \times \frac{(s+1)! L^{s+1-S-2M}}{S + 2M - s - 1} \times \frac{(\lceil S \rceil + 2M - 1)!}{(\lceil S \rceil - 1)!}. \tag{A14}$$

Example A2. $s = 4$, $S = 5.7$, $L = 100$, $M = 10$, $R < O(10^{-16}) \times \frac{20!}{(10!)^2}$
$\times \frac{5! 100^{-20.7}}{20.7} \times \frac{25!}{5!} = 10^{-16} \times \frac{20! 25!}{10! 10!} \times \frac{100^{-20.7}}{20.7} = 6 \times 10^{-29}$. Therefore, the sum value 6×10^{-13} would be estimated with the accuracy of 16 decimal digits.

Appendix F: Evaluation of $\sum_{g=1}^{g=\infty} \frac{\log_2(g+Q)}{(g+Q)^S}$

Appendix E shows that generic representation of the kth derivative of the function under the sum is important for the application of the Euler–Maclaurin formula. For $\sum_{g=1}^{g=\infty} \frac{\log_2(g+Q)}{(g+Q)^S} = \frac{1}{\ln(2)} \sum_{g=1}^{g=\infty} \frac{\ln(g+Q)}{(g+Q)^S}$, $f(x) = f^{(0)}(x) = u(x)v(x) = u^{(0)}(x)v^{(0)}(x)$, where $u(x) = u^{(0)}(x) = (x+Q)^{-S}$, $v(x) = v^{(0)}(x) = \ln(x+Q)$

$$u^{(k)}(x) = (-1)^k (x+Q)^{-S-k} \prod_{i=0}^{i=k-1}(S+i), \quad 0 \leq k,$$

$$v^{(0)}(x) = \ln(x+Q), \quad v^{(k)}(x) = (-1)^{k-1}(k-1)!(x+Q)^{-k}, \quad 1 \leq k.$$

Again, using the Leibniz formula and treating the 0th summand separately

$$f^{(k)}(x) = u^{(k)}(x)v^{(0)}(x) + \sum_{i=1}^{i=k} \frac{k!}{i!(k-i)!} u^{(k-i)}(x) v^{(i)}(x)$$

$$= \ln(x+Q)(-1)^k (x+Q)^{-S-k} \prod_{i=0}^{i=k-1}(S+i)$$

$$+ \sum_{i=1}^{i=k} \frac{k!}{i!(k-i)!} (-1)^{k-i}(x+Q)^{-S-k+i}$$

$$\times \prod_{j=0}^{j=k-i-1}(S+j)((-1)^{i-1}(i-1)!(x+Q)^{-i})$$

$$= (-1)^{k-1}(x+Q)^{-S-k}$$

$$\times \left(\sum_{i=1}^{i=k} \frac{k!}{i(k-i)!} \prod_{j=0}^{j=k-i-1}(S+j) - \ln(x+Q) \prod_{i=0}^{i=k-1}(S+i) \right).$$

$$\tag{A15}$$

Notice that $\frac{k!}{i(k-i)!}$ is a result of reduction $\frac{k!(i-1)!}{i!(k-i)!}$ and is supported by $1 \le i$.

Example A3. From (A15), check: $f^{(0)}(x) = (x+Q)^{-S}\ln(x+Q)$; $f^{(1)}(x) = (x+Q)^{-S-1}(1 - S\ln(x+Q))$; $f^{(2)}(x) = (x+Q)^{-S-2}(-2S - 1 + (S+1)S\ln(x+Q))$; $f^{(3)}(x) = (x+Q)^{-S-3}(3S^2 + 6S + 2 - (S+2)(S+1)S\ln(x+Q))$; $f^{(4)}(x) = (x+Q)^{-S-4}(-4S^3 - 18S^2 - 22S - 6 + (S+3)(S+2)(S+1)S\ln(x+Q))$.

Equation (A15) is generic, easily programmed, and does not require the illustrative expansions. Integration by parts [23, pp. 31–36, Chapter 8] yields $\int \frac{\ln(x+Q)}{(x+Q)^S} dx = \frac{(-S+1)\ln(x+Q) - 1}{(-S+1)^2 (x+Q)^{S-1}} + C$. See also a remarkable *recurrent formula* for $\int x^k \ln^m(x) dx$, $k \ne -1$, $m = 1, 2, \ldots$, [23, p. 34]. Then, from (A11), (A15), $(-1)^{2j-2} = 1$:

$$\sum_{g=1}^{g=\infty} \frac{\ln(g+Q)}{(g+Q)^S} = \sum_{g=1}^{g=L-1} \frac{\ln(g+Q)}{(g+Q)^S} + \frac{(S-1)\ln(L+Q) + 1}{(S-1)^2(L+Q)^{S-1}} + \frac{\ln(L+Q)}{2(L+Q)^S}$$

$$+ \sum_{j=1}^{j=M} \frac{B_{2j}[\ln(L+Q)\prod_{i=0}^{i=2j-2}(S+i) - \sum_{i=1}^{i=2j-1} \frac{(2j-1)!}{i(2j-1-i)!} \prod_{m=0}^{m=2j-i-2}(S+m)]}{(2j)!(L+Q)^{S+2j-1}}$$

$$+ R. \qquad (A16)$$

The expression for R was not obtained.

Appendix G: HurwitzZetaStart

HurwitzZetaStart(S, Q, start) returns the value of $\zeta_{\text{start}}(S, Q)$ using

$$\zeta_{\text{start}}(S, Q) = \sum_{n=\text{start}}^{n=\infty} \frac{1}{(n+Q)^S} = \sum_{n=\text{start}}^{n=\text{start}+N-1} \frac{1}{(n+Q)^S}$$

$$+ \frac{(Q + \text{start} + N)^{S-1}}{S-1} + \frac{1}{2(Q + \text{start} + N)^S}$$

$$+ \sum_{i=1}^{i=M} \frac{B_{2i}(N + \text{start} + Q)^{1-S-2i}}{(2i)!} \prod_{j=0}^{j=2i-2} (S+j) + R \quad (A17)$$

It follows from the Euler–Maclaurin formula (A11) and [86, p. 48]. $N = 30$, $M = 15$ support sufficient accuracy for R, when the C++ type double is applied.

References

[1] T. M. Apostol. (1976). *Introduction to Analytic Number Theory*. Springer, New York.

[2] V. Arnold, I. A. N. Kolmogorov, and Natural Science. (2004). *Uspehi Matematicheskih Nauk* 59(1), 25–44 (January – February 2004) (Russian); *Russian Mathematical Surveys* 59, 27–46.

[3] H. Bateman, and A. Erdélyi. (1973). *Higher Transcendental Functions*. Vol. 1, 2nd edn. Translated to Russian by N. Y. Vilenkin. Nauka, Moscow (Russian).

[4] R. Bellman. (1954). The theory of dynamic programming. P-550. The RAND Corporation, Santa Monica, California, pp. 1–23 (30 July 1954).

[5] C. Berge. (1971). *Principles of Combinatorics*. Academic Press, New York and London.

[6] J. Bernoulli. (1745). Epistola XI. $\frac{8}{18}$ Junii 1695. In *Virorum celeberr. Got. Gul. Leibnitii et Johan. Bernoullii Commercium. Philosophicum et Mathematicum*. Lausannae & Genevae, pp. 52–64 (Latin). www.e-rara.ch ETH-Bibliothek Zürich, Shelf Mark: Rar 4960, Persistent Link: http://dx.doi.org/10.3931/e-rara-3671.

[7] J. Bollinger. (2001). *Bollinger on Bollinger Bands*. McGraw-Hill, New York.

[8] É. Borel. (1923). *Chance*. Translated from French by U. I. Kosticina. State Publishing, Moscow, Petrograd (Russian).

[9] É. Borel. (1956). *Probabilité et certitude*. Presses Universitaires de France, Paris. Cited by the Russian translation: *Probability and Certainty*, 3rd edn. Translated from the 2nd edn. by I. B. Pogrebyisskii. B. V. Gnedenko (ed.). Nauka, Moscow (1969).

[10] C. B. Boyer. (1991). *A History of Mathematics*, 2nd edn. (Revised by U. C. Merzbach). John Wiley & Sons Inc., New York.

[11] R. A. Brualdi. (2010). *Introductory Combinatorics*, 5th edn. Pearson Education Inc., China Machine Press, Beijing, China.

[12] P. S. Bullen. (2003). *Handbook of Means and Their Inequalities*. Mathematics and Its Applications, Vol. 560. In M. Hazewinkel. Springer-Science+Business Media, B.V., Dordrecht, Netherlands.

[13] P. S. Bullen, D. S. Mitrinović, and P. M. Vasić. (1988). *Means and Their Inequalities*. Mathematics and Its Applications. East European Series. In M. Hazewinkel. Springer Science+Business Media Dordrecht, Amsterdam.

[14] C. S. Calude. (2002). *Information and Randomness. An Algorithmic Perspective*, 2nd edn. (Revised and Extended). Springer, New York.

[15] J.-C. Chen, Y. Zhou, and X. Wang. (2017). Profitability of simple stationary technical trading rules with high-frequency data of Chinese index futures. *ArXiv, Statistical Finance*, 1–24 (October 20). https://arxiv.org/abs/1710.07470

[16] M. F. Dixon, I. Halperin, and P. Bilokon. (2020). *Machine Learning in Finance. From Theory to Practice*. Gewerbestrasse 11, 6330. Springer Nature Switzerland AG, Cham, Switzerland.

[17] V. A. Dobrovolsky. (1981). *Vasily Petrovich Ermakov*. Nauka, Moscow (Russian).

[18] R. G. Downey, and D. R. Hirschfeldt (2010). *Algorithmic Randomness and Complexity*. Springer Science+Business Media LLC, New York.

[19] Edwards, A. W. F. (2013). Chapter 7: The arithmetical triangle, pp. 167–180. In R. Wilson, and J. J. Watkins (eds.) *Combinatorics: Ancient and Modern*. Oxford University Press, Oxford.

[20] Edwards, H. M. (1974, 2001). *Riemann's Zeta Function*. Dover Publications Inc, New York (An unabridged republication of the edition originally published by Academic Press Inc.).

[21] A. Elder. (1993). *Trading for a Living. Psychology. Trading Tactics. Money Management.* John Wiley & Sons Inc., New York.

[22] G. M. Fichtenholz (Fikhtengolts). (1962). *Course of Differential and Integral Calculus,* Vol. I, 5th edn. (Stereotype). Physical-Mathematical Literature State Publisher, Moscow (Russian).

[23] G. M. Fichtenholz (Fikhtengolts). (1970). *Course of Differential and Integral Calculus,* Vol. II, 7th edn. (Stereotype). Nauka, Moscow (Russian).

[24] R. A. Fisher. (1922). On the mathematical foundations of theoretical statistics. *Philosophical Transactions of the Royal Society of London. Series A. Containing Papers of Mathematical or Physical Character* 222, 309–368.

[25] X. Gabaix, P. Gopikrishnan, V. Plerou, and H. E. Stanley. (2003). A theory of power-law distributions in financial market fluctuations. *Nature* 423, 267–270 (May 15). See also: X. Gabaix, P. Gopikrishnan, V. Plerou, and H. E. Stanley. A Theory of Large Fluctuations in Stock Market Activity. Massachusetts Institute of Technology Department of Economics Working Paper Series. Working Paper 03-30, 1–46 (2003, August 16). At the time of this writing, it has been revised on October 2, 2005, and May 12, 2010 with the latest title Institutional Investors and Stock Market Volatility, pp. 1–50. Social Science Research Network (SSRN-Elsevier). http://papers.ssrn.com/abstract_id=442940.

[26] B. V. Gnedenko, and A. N. Kolmogorov. (1949). *Limit Distributions for Sums of Independent Random Variables.* Technico-Theoretical Literature Governmental Press, Moscow, Leningrad (Russian).

[27] B. V. Gnedenko. (1988). *The Probability Theory (Kurs Teorii Veroyatnostei).* Nauka, Moscow (Russian).

[28] N. Goldenfeld, and L. P. Kadanoff (1999). Simple lessons from complexity. *Science. New Series* 284(5411), 87–89 (April 2).

[29] R. L. Graham, D. E. Knuth, and O. Patashnik. (1998). *Concrete Mathematics,* 2nd edn. Addison-Wesley Publishing Company Inc., New York.

[30] I. Halperin. (2020). Non-equilibrium skewness, market crises, and option pricing: Non-linear Langevin model of markets with supersymmetry. *Social Science Research Network* (SSRN-Elsevier) (2020, November 2). https://papers.ssrn.com/sol3/papers.cfm?abstract_id=3724000.

[31] L. Harkleroad. (2008). *The Math behind the Music.* Cambridge University Press, New York (First Published 2006, Reprinted 2007 (twice), 2008).

[32] *ISO/IEC JTC1 SC22 WG21 Working Draft, Standard for Programming Language C++,* N4861 (April 1, 2020).

[33] F. Johansson. (2015). Rigorous high-precision computation of the Hurwitz zeta function and its derivatives. *Arxiv, Computer Science, Symbolic Computation* 1–15 (September 11, 2013). https://arxiv.org/abs/1309.2877. The author could not get the paper under the same title in *Numerical Algorithms* 69(2), 253–270 (June 2015).

[34] A. F. Karr. (1993). *Probability.* Springer-Verlag New York Inc., New York.

[35] P. J. Kaufman. (2005). *New Trading Systems and Methods,* 4th edn. John Wiley & Sons Inc., Hoboken, New Jersey.

[36] M. G. Kendall, and A. B. Hill. (1953). The analysis of economic time-series — Part I: Prices. *Journal of the Royal Statistical Society. Series A (General)* 116(1), 11–34.

[37] M. G. Kendall, and S. Alan. (1973). *The Advanced Theory of Statistics. Volume 2. Inference and Relationship,* 2nd edn. Translated to Russian by L. I. Galchuk, A. T. Terekhin, and A. N. Kolmogorov (ed.). Nauka, Moscow (Russian).

[38] A. Y. Khinchin. (1953). The notion of entropy in probability theory. *Uspehi Matematicheskih Nauk* 8(3), 3–20 (1953, May–June) (Russian).

[39] E. Knobloch. (2013). Chapter 6: The origins of modern combinatorics, pp. 147–166. In R. Wilson and J. J. Watkins (eds.) *Combinatorics: Ancient and Modern*. Oxford University Press, Oxford.

[40] Kolmogorov, A. N. (1930). Sur la notion de la moyenne. *Atti della Reale Accademia Nazionale dei Lincei. Classe di Scienze Fisiche, Matematiche e Naturale. Rendiconti. Serie VI* 12(9), 388–391 (French). Translated to Russian by V. M. Tikhomirov. On definition of average, pp. 136–138. In A. N. Kolmogorov. (1985). *Selected Works. Mathematics and Mechanics*. Nauka, Moscow. Both Russian/French pages are given using the slash sign.

[41] A. N. Kolmogorov. (1974). *The Basic Concepts of Probability Theory*, 2nd edn. Nauka, Moscow (Russian).

[42] A. N. Kolmogorov, I. G. Zhurbenko, and A. V. Prokhorov. (1982). *Introduction to Probability Theory*. Nauka, Moscow (Russian).

[43] A. N. Kolmogorov, and V. A. Uspenskii. (1987). Algorithms and randomness. *Teoriya Veroyatnostei i ee Primeneniya* 32(3), 425–455 (Russian). Translated from Russian Journal to English by Bernard Seckler. (1987). *Theory of Probability and Its Applications* 32(3), 389–412. Both Russian/English pages are given using the slash sign.

[44] G. A. Korn, and T. M. Korn. (1968). *Mathematical Handbook for Scientists and Engineers. Definitions, Theorems, and Formulas for Reference and Review* (2nd Enlarged and Revised Edition). McGraw-Hill Book Company, New York (Pages are cited by the Russian translation, 5th edn., Aramanovich, I. G. (ed.) (1984)). Nauka, Moscow.

[45] L. D. Landau, and E. M. Lifshitz. (1965). *Mechanics. Theoretical Physics*, Vol. I, 2nd edn. Nauka, Moscow (Russian).

[46] Z. Landsman, U. Makov, and T. Shushi. (2016). A new class of distributions based on Hurwitz zeta function with applications for risk management. *The Open Mathematics, Statistics and Probability Journal* 7, 53–62 (August 31).

[47] M. Leeds. (2013). Bollinger bands thirty years later. *ArXiv, Statistical Finance* 1–58 (December 20, 2013). https:/arxiv.org/abs/1212.4890.

[48] G. W. Leibniz. (1745). Epistola X. $\frac{6}{16}$ May 1695. In *Virorum celeberr. Got. Gul. Leibnitii et Johan. Bernoullii Commercium. Philosophicum et Mathematicum* Lausannae & Genevae, pp. 46–51 (Latin). www.e-rara.ch ETH-Bibliothek Zürich, Shelf Mark: Rar 4960, Persistent Link: http://dx.doi.org/10.3931/e-rara-3671.

[49] B. Y. Lemeshko, and S. N. Postovalov. (1998). On dependencies of limiting distributions statistics of Pearson's chi-square and relationships of the likelihood on grouping data. *Industrial Laboratory (Zavodskaya Laboratoriya)* 64(5), 56–63 (Russian).

[50] B. Y. Lemeshko, S. B. Lemeshko, and Postovalov, S. N. (2008). Comparative analysis of the power of goodness-of-fit tests for near competing hypotheses. I. The verification of simple hypotheses. *Siberian Journal of Industrial Mathematics* 11(2), 96–111 (Russian). Translated to English in *Journal of Applied and Industrial Mathematics* 3(4), 462–475 (2009). English translation was not reviewed and is given for convenience.

[51] B. Y. Lemeshko, S. B. Lemeshko, and S. N. Postovalov. (2008). Comparative analysis of the power of goodness-of-fit tests for near competing hypotheses. II. The verification of complex hypotheses. *Siberian Journal of Industrial Mathematics* 11(4), 78–93 (Russian). Translated to English in *Journal of Applied and Industrial Mathematics* 4(1), 79–93 (2010). English translation was not reviewed and is given for convenience.

[52] Leontiev, A. F. (1976). *Series of Exponentials*. Nauka, Moscow (Russian).
[53] M. Li and P. Vitánuy. (2019). *An Introduction to Kolmogorov Complexity and Its Applications*, 4th edn. Springer, New York.
[54] A. Linde, and V. Vanchurin. (2010). How many universes are in the multiverse? *Physical Review D* 81(083525), 083525-1–083525-11. See also *Arxiv, High Energy Physics - Theory*, 1–12 (2009, October 9). https://arxiv.org/abs/0910.1589.
[55] E. Lukacs. (1970). *Characteristic Functions*, 2nd edn. Charles Griffin & Company Limited, London.
[56] L. E. Maistrov, B. A. Rozenfeld, and O. B. Sheynin. (1970). Chapter 5: Combinatorics and theory of probabilities, pp. 81–97. In A.-A. P. Yushkevich (ed.) *The History of Mathematics from the Ancient Times to the Beginning of the XIX Century*. The Academy of Sciences of the USSR. Mathematics of the XVII Century, Vol. 2. Nauka, Moscow (Russian).
[57] Mandelbrot, B. (1953). An information theory of the statistical structure of languages. In W. Jackson (ed.)*Communication Theory: Papers Read at a Symposium on "Applications of Communication Theory" Held at the Institution of Electrical Engineers, London, September 22nd–26th 1952*. Butterworths Scientific Publications, London, pp. 486–502.
[58] S. Mandelbrot. (1973). *Dirichlet Series. Principles and Methods*. Mir, Moscow (Russian).
[59] D. Y. Manin. (2008). Zipf's law and avoidance of excessive synonymy. *Cognitive Science* 32, 1075–1098; See also *Arxiv, Computer Science, Computation and Language* 1–47 (September 30, 2007). https://arxiv.org/abs/0710.0105.
[60] D. Y. Manin. (2009). Mandelbrot's model for Zipf's law: Can Mandelbrot's model explain Zipf's law for language? *Journal of Quantitative Linguistics* 16(3), 274–285 (August 6).
[61] D. Y. Manin, and Y. I. Manin. (2017). Cognitive networks: Brains, internet, and civilizations. In B. Sriraman (ed.) *Humanizing Mathematics and Its Philosophy. Essays Celebrating the 90th Birthday of Reuben Hersh*. Birkhaüser; Springer International Publishing, Cham, Switzerland, pp. 85–96.
[62] Y. I. Manin. (1979). *Provable and Non-provable* (Kibernetika). Sovetskoe Radio, Moscow (Russian).
[63] Y. I. Manin. (1980). *Computable and Non-computable* (Kibernetika). Sovetskoe Radio, Moscow (Russian).
[64] Y. Manin. (1999). Classical computing, quantum computing, and Shor's factoring algorithm. *Arxiv, Quantum Physics* 1–27 (March 2). https://arxiv.org/abs/quant-ph/9903008.
[65] Y. Manin. (2014). Zipf's law and Levin's probability distributions. *Functional Analysis and Its Applications* 48(2), 51–66 (Russian); English translation: Zipf's Law and L. Levin probability distributions. *Functional Analysis and Its Applications* 48(2), 116–127 (2014); Both Russian/English pages are given using the slash sign. See also Zipf's law and L. Levin's probability distributions. *Arxiv, Computer Science, Information Theory* 1–19 (2013, January 3). https://arxiv.org/abs/1301.0427.
[66] A. A. Muchnik, A. L. Semenov, and V. A. Uspensky. (1998). Mathematical metaphysics of randomness. *Theoretical Computer Science* 207, 263–317.
[67] B. Corominas Murtra, and R. Solé. (2010). Universality of Zipf's Law. *Arxiv, Condensed Matter, Statistical Mechanics* 1–10 (January 15). https://arxiv.org/abs/1001.2733.

[68] V. M. Nagumo. (1930). Über eine Klasse der Mittelwerte. *Japanese Journal of Mathematics* 7, 71–79 (German).

[69] J. Neyman, and E. S. Pearson. (1933). On the problem of the most efficient tests of statistical hypotheses. *Philosophical Transactions of the Royal Society of London. Series A, Containing Papers of a Mathematical or Physical Character* 231, 289–337.

[70] J. Neyman. (1937). Outline of a theory of statistical estimation based on the classical theory of probability. *Philosophical Transactions of the Royal Society of London. Series A, Containing Papers of a Mathematical and Physical Sciences* 236(767), 333–380 (August 30).

[71] A. Nies. (2012). *Computability and Randomness*. Oxford Logic Guides 51. University Press, Oxford (First Published 2009, First Published in Paperback 2012, Cited by the Edition of Clarendon Press, Oxford 2008).

[72] E. Parzen. (1960). *Modern Probability Theory and Its Applications*. John Wiley & Sons Inc., New York.

[73] K. Pearson. (1900). On the criterion that a given system of deviations from the probable in the case of a correlated system of variables is such that it can be reasonably supposed to have arisen from random sampling. *Philosophical Magazine* 1, 157–175.

[74] R. Penrose. (1989, 1991). *Emperor's New Mind. Concerning Computers, Minds, and The Laws of Physics*. Penguin Books, Harmondsworth, Middlesex, England (First Published by Oxford University Press).

[75] R. Penrose. (1994). *Shadows of the Mind. A Search for the Missing Science of Consciousness*. Oxford University Press, New York.

[76] S. M. Potirakis, P. I. Zitis, G. Balasis, and K. Eftaxis. (2016). On the use of financial analysis tools for the study of Dst time series in the frame of complex systems. *ArXiv, Data Analysis, Statistics and Probability* 1–36 (January 27). https:/arxiv.org/abs/1601.07334.

[77] A. Rényi. (1980). Trilogy on Mathematics. Dialogues on Mathematics. Letters on Probability. Diary - Notes of a Student on the Theory of Information. Russian translation from Hungarian. Publishing "Mir", Moscow (Russian).

[78] J. Riordan. (1978). *An Introduction to Combinatorial Analysis*. Princeton University Press, Princeton, New Jersey (Originally Published 1958 by John Wiley & Sons, Inc.).

[79] K. H. Rosen. (1995). *Discrete Mathematics and Its Applications*, 3rd edn. McGraw-Hill Inc., New York.

[80] V. V. Salov. (2007). *Modeling Maximum Trading Profits with C++: New Trading and Money Management Concepts*. John Wiley & Sons Inc., Hoboken, New Jersey.

[81] V. V. Salov. (2011). Market profile and the distribution of price. *Futures Magazine* XL(6), 34–36 (June).

[82] V. V. Salov. (2011). Trading system analysis: Learning from perfection. *Futures Magazine* XL(11), 38–39, 43 (November).

[83] V. V. Salov. (2012). High-frequency trading in live cattle futures. *Futures Magazine* XL(6), 26–27, 31 (May).

[84] V. V. Salov. (2012). Notation for iteration of functions, iteral. *Arxiv, Mathematics, Dynamical Systems* 1–23 (June 30). https://arxiv.org/abs/1207.0152.

[85] V. V. Salov. (2012). Inevitable Dottie number. Iterals of cosine and sine. *Arxiv, Quantitative Finance, General Finance* 1–17 (December 1). https://arxiv.org/abs/1212.1027.

[86] V. V. Salov. (2013). Optimal trading strategies as measures of market disequilibrium. *Arxiv, Quantitative Finance, General Finance* 1–222 (December 6). http://arxiv.org/abs/1312.2004.

[87] V. Salov. (2017). The wandering of corn. *Arxiv, Quantitative Finance, General Finance* 1–65 (April 3). https://arxiv.org/abs/1704.01179?context=q-fin.

[88] V. Salov. (2017). Trading strategies with position limits. *Arxiv, Quantitative Finance, General Finance* 1–64 (December 19). https://arxiv.org/abs/1712.07649.

[89] V. Salov. (2018). Profit and loss distributions on a market of single futures contract. *Social Science Research Network* (SSRN-Elsevier) (December 23). http://ssrn.com/abstract=3305964.

[90] V. Salov. (2019). A linear smoothing property of a futures limit order book. *Social Science Research Network* (SSRN-Elsevier) (February 18). http://ssrn.com/abstract=3337345.

[91] V. Salov. (2019). A quantum algorithm for trading strategies with position limits. *Social Science Research Network* (SSRN-Elsevier) (May 27). https://papers.ssrn.com/sol3/papers.cfm?abstract_id=3394838.

[92] V. Salov. (2019). "Explosions" of corn futures prices. *Social Science Research Network* (SSRN-Elsevier) (November 2). http://ssrn.com/abstract=3479685.

[93] V. Salov (2021). Nice sets and a C++ mapping template. *Social Science Research Network* (SSRN-Elsevier) (January 12). https://papers.ssrn.com/sol3/papers.cfm?abstract_id=3764242.

[94] P. A. Samuelson (1965). Proof that properly anticipated prices fluctuate randomly. *Industrial Management Review* 6(2), 41–49.

[95] I. A. Semiokhin, B. V. Strakhov, and A. I. Osipov. (1995). *Kinetics of Chemical Reactions*. Moscow State University, Moscow (Russian).

[96] C. E. Shannon (1948). A mathematical theory of communication. *The Bell System Technical Journal* XXVII(3), 379–423 (July); XXVII(4), 623–656 (October).

[97] O. B. Sheynin. (2010). Karl Pearson a century and half after his birth. *The Mathematical Scientist* 35(1), 1–9.

[98] O. B. Sheynin. (2018). Theory of probability. An elementary treatise against a historical background. *Arxiv, Mathematics, History and Overview* 1–77 (February 10). https://arxiv.org/abs/1802.03485.

[99] O. B. Sheynin. (2018). Theory of probability. A historical essay (Revised and enlarged edition). *Arxiv, Mathematics, History and Overview* 1–316 (February 27). https://arxiv.org/abs/1802.09966.

[100] W. J. Strong, and G. R. Plitnik. (1983). *Music, Speech, High-Fidelity*. 2nd edn. Provo, UT 84604: SOUNDPRINT.

[101] B. Stroustrup. (2013). *The C++ Programming Language*, 4th edn. Addison-Wesley, New York.

[102] B. Stroustrup. (2020). Thriving in a crowded and changing world: C++ 2006–2020. In *Proceedings of the ACM on Programming Languages*, Vol. 4. History of Programming Languages, HOPL, Article 70, pp. 70:1–70:168. Publication date: June 2020.

[103] S. Tauber. (1963). On multinomial coefficients. *The American Mathematical Monthly* 70(10), 1058–1063 (December).

[104] A. E. Taylor. (1952). L'Hospital's rule. *The American Mathematical Monthly* 59(1), 20–24 (January).

[105] E. C. Titchmarsh. (1986). *The Theory of the Riemann Zeta-Function*, 2nd edn. revised by D. R. Heath-Brown. Oxford University Press. Reprinted in 1988. Clarendon Press, Oxford.

[106] V. N. Tutubalin. (1972). *Probability Theory in Natural Science*. Series Mathematics, Cybernetics. Znanie, Moscow (Knowledge) (Russian).

[107] V. A. Uspenskii, A. L. Semenov, and A. K. Shen. (1990). Can an individual sequence of zeros and ones be random? *Uspekhi Matematicheskikh Nauk*, 45(1) (January-February), 105–162 (Russian). Translated from Russian Journal to English by Shen, A. (1990). *Russian Mathematical Surveys* 45(1), 121–189. Both Russian/English pages are given using the slash sign.

[108] T. L. Veldhuizen. (2005). Software libraries and their reuse: Entropy, Kolmogorov complexity, and Zipf's law. *Arxiv, Software Engineering, Information Theory, Programming Languages*, pp. 1–13 (August 3). https://arxiv.org/abs/cs/0508023.

[109] N. K. Vereshchagin, V. A. Uspenskii, and A. K. Shen. (2013). *Kolmogorov Complexity and Algorithmic Randomness*. Moscow Center of Continuous Mathematical Education, MCNMO, Moscow (Russian). A. K. Shen, V. A. Uspensky, and N. K. Vershchagin. (2017). *Kolmogorov Complexity and Algorithmic Randomness*. Mathematical Surveys and Monographs, Vol. 220. American Mathematical Society, Providence, Rhode Island (English). Both Russian/English pages are given using the slash sign.

[110] V. V. V'yugin. (2019). Kolmogorov complexity in the USSR (1975–1982): Isolation and its end. *Arxiv, Computer Science, General Literature, Information Theory* (July 11), pp. 1–26, 1–13 (English), Russian (pp. 14–26). https://arxiv.org/abs/1907.05056.

Index

A
Arnold, Vladimir, 275
Arrhenius equation, 276
Arrhenius, Svante, 276, 278
Asian options, 6, 68, 72

B
backtesting, 283
Bernoulli numbers, 356
Bernoulli polynomials, 356
Bernoulli, Johann, 271, 273, 354
binary barrier options, 3, 6, 56
bollinger bands, 239–240
Borel, Emile, 284

C
C++, Boost Library, 280
C++, Standard Library, 280
call options, 6, 58, 72
cancellation errors, 260
chain, 230
chain, ticks, 230–231, 235
chaoticness, 231
Cinderella, 278
commissions and fees, 262, 266
context matrix, 157
convolution, 284
cross-entropy, 151, 168

D
degree centrality, 213–214, 216, 218–220, 222–223, 226
de Moivre, Abraham, 354
Deutsche, David, 281
direct network, 217, 219
Dirichlet series, 250
distribution, asymmetric, 252
    discrete, 231
    empirical, 242
    entropy, 254
    joint, 230
    Levin maximal, 279
    non-stationary, 240
    probability, 231
    symmetric, 250, 252, 256

E
E-mini S&P 500, 231, 233, 267–268, 284, 336, 346
Edwards, Harold, 231, 280
eigenvector centrality, 216–220, 222–223, 226
embeddings, 151, 182
ESG ratings, 189–191, 194, 199, 201
    strength, 189, 192, 195, 199
    tilted Sharpe ratio, 190, 194, 199

estimator, biased, 240
    consistent, 240
    unbiased, 240
Euler–Maclaurin formula, 356–359

F

Feynman, Richard, 281
filtering cost, 266
first exit times, 54
Fisher, Ronald, 229, 286
function, characteristic, 253
    distribution, 252
    Hurwitz Zeta, 244, 249–251, 255, 271, 278, 357
    probability mass, 243, 252
    Riemann Zeta, 244, 249–251, 357
    symmetric, 239
    zeta(S,Q,H,K), 250–251
    Zipf-Mandelbrot probability mass, 244, 251
futures, contract systematic evolution, 235
    copper, 262
    corn, 242
    crude oil, 243
    daily volume, 235
    E-mini S&P 500, 231
    limit order book, 283–284
    liquid contract, 234
    listing, 235
    maximum prices, 242
    negative prices, 242
    price limits, 242
    trading orders, 283

G

Gnedenko, Boris, 240

H

Hadamard identity, 350
hydrodynamics, 275

I

implied volatility, 6, 62–64, 68, 72
i.i.d., 242

K

Kendall, Maurice, 286
king effect, 244
Kleinberg centrality, 217, 220
Kolmogorov, Andrey, 229–230, 240, 275
    complexity, 279, 282, 285–286
    optimal encoding, 279
Kolmogorov–Smirnov test, 285
Kramp, Christian, 354

L

L'Hopital, Guillaume, 273
Lagrange multipliers, 278
Langevin, 97, 100–101, 107–108, 115, 119
large deviation principles, 2–5, 17, 28, 31, 42
large language models, 152
latency, 283
Leibniz, Gottfried, 271, 354
    formula, 355, 358
little Fermat theorem, 276

M

Mandelbrot, Benoit, 278
Manin, Yuri, 281
market complexity, 285
    profile, 239
Markov generative model for texts, 162
matching algorithms, 284
mathematical expectation, 239
maximum profit strategy, 263, 277, 286
    without reinvesting, MPS0, 266
McKean–Vlasov, 97, 101–102, 115–118, 120, 140–141
mean field, 147–148
mean–variance portfolio (MVP), 190–192
Mersenne, Marin, 354
Microsoft excel's solver, 249
moment, absolute, 239
    beginning, 253
    central, 254
    central absolute, 239

general population, 239
  masses, 351
multinomial theorem, 354
music complement, 282

N

natural language processing, 151
Newton's binomial theorem, 352
non-equilibrium, 100, 102, 111, 144–147

O

open trading elements, 285
optimal ESG portfolio (OESGP), 189–190, 192, 194–195, 199, 201
optimal trading elements, 263
ordinary differential equation, 272

P

partial correlations, 190, 193
paths, Dyck's, 281
  Motzkin's, 281
peacocks' tails, 265
Pearson, Karl, 286
  correlations, 190, 193
  test, 284
portfolio ESG value (PESGV), 189–190, 192–193
power laws, 275–276
price, big daily range, 262
  daily closing, 240
  daily closing, contradiction, 242
  increments, 231
  linear correlations, 235
  non-Gaussian increments, 237
  settlement, 231, 249
  synchrounous moves, 235
process, realizations, 230, 235
Prundtl, Ludwig, 275

Q

quantum computations, 281
qubits, 281

R

randomness, 230
  delta, 231
recurrent formula, 359
Renui, Alfred, 284
robots trade, 262

S

sample, discrete, 242
  distribution, 237, 242
  excess kurtosis, 237
  increments, 239
  moments, 242, 351
  relationship between moments of values and increments, 240–241, 346
  skewness, 237
  statistics, 288
  statistics of values and increments, 238
  values, 239
Samuelson, Paul, 229
scientific credibility, 230
sentiment, 151
  polarity, 177
sequence, 230
SHA-512, 281
Shannon, Claude, 254
Shor, Peter, 281
skip-gram loss function, 158, 169
statistical learning, 151
statistical mechanics, 98, 100, 146
stochastic volatility models, 2–6, 16, 18, 31–32, 35, 52, 58, 62, 72–74
stochasticness, 231
supply chain, 209–213, 218–219, 226–227
synonyms, 151, 166, 181, 183, 185

T

theory of probability, 230
time & sales, 230–231, 233, 267–268, 336, 346
time ranges, 231, 234
Tolstoy, Leo, 277

trading cost, 266
  rules, 283
  sessions, 231, 234
trend, 262, 277
typicalness, 231

U

Ubuntu, 277, 286
Uspenskii, Vladimir, 229–230

V

volatility, 262, 277

Volterra equations, 5
  type, 3–6, 11, 13–14, 19, 31–32, 36, 43, 53–54, 73–74
  type process, 33

W

Wallis, John, 354
weighted average, 351

Z

Zipf's law, 279, 282
  in linguistics, 277, 279